1995

The ELECTROPHYSIOLOGY of NEUROENDOCRINE CELLS

Edited by
Hans Scherübl, M.D., Ph.D.
Klinikum Benjamin Franklin
Freie Universität Berlin
Berlin, Germany

Jürgen Hescheler, Ph.D.
Institut für Pharmakologie
Freie Universität Berlin
Berlin, Germany

CRC Press
Boca Raton   New York   London   Tokyo

Library of Congress Cataloging-in-Publication Data

The electrophysiology of neuroendocrine cells / edited by Hans
    Scherübl and Jürgen Hescheler.
        p.   cm.
    Includes bibliographical references and index.
    ISBN 0-8493-2477-7
    1. Paraneurons.   2. Electrophysiology.   I. Scherübl, Hans.
II. Hescheler, J. K.-J. (Jürgen Karl-Josef), 1959-
QP356.4.E44   1995
599'.0188—dc20   95-4036
                                                                                                    CIP

No claim to original U.S. Government works
International Standard Book Number 0-8493-2477-7
Library of Congress Card Number 95-4036
Printed in the United States of America 1 2 3 4 5 6 7 8 9 0
Printed on acid-free paper

PREFACE

Neuroendocrine cells share a variety of morphological, biochemical, and functional properties, thereby constituting a family of their own. Their common physiological function is to monitor and respond to changes of the "milieu interieur". The neuroendocrine family resides at the functional frontiers between the thyroid- or steroid-hormone-producing endocrine system and the muscular and nervous systems. The similarity to neuronal cells is most obvious in the case of the neuroectodermal adrenal chromaffin cells. The relationship to muscle cells holds true for the myoepitheloid renin-secreting juxtaglomerular cells and the ANF-secreting atrial heart cells. The common features that distinguish the peptide-hormone-producing neuroendocrine cells from the thyroid- or steroid-hormone-producing endocrine cells are (1) the secretory machinery (large dense core vesicles and small synaptic vesicles), (2) the secretory products, and (3) the electrical excitability. With the exception of renin-secreting juxtaglomerular, and possibly parathyroid hormone-secreting cells and Sertoli cells, neuroendocrine cells are able to fire spontaneous action potentials. Thus, the majority of neuroendocrine cells display strong similarities in their electrical activity and express a similar set of voltage-dependent ion channels.

It is the purpose of this volume to outline the general role of electrical activity in both stimulus-secretion coupling and intercellular communication. The specific involvement of ion channels in the sensory mechanisms of neuroendocrine cells is described in a separate chapter for each individual cell type. In view of their close functional similarity to neuroendocrine cells, two neuronal cell types have been included: (1) the osmosensitive, hypothalamic, supraoptic, and paraventricular neurosecretory cells and (2) the oxygen-sensitive carotid body type I cells.

A number of neuroendocrine cells, notably the large group of hormone-producing cells in the gut, have never been investigated electrophysiologically. Moreover, a number of neuroendocrine cells have only been studied in a preliminary way. We hope this book will encourage electrophysiological studies on the neuroendocrine system as a whole and will draw attention to the hitherto neglected neuroendocrine cells of the lungs, the biliary system, and the gut.

Hans Scherübl
Jürgen Hescheler

EDITORS

Hans Scherübl, M.D., Ph.D., works in the Department of Gastroenterology at the Benjamin Franklin Hospital of the Free University of Berlin, Germany. Dr. Scherübl graduated in 1986 from the Ruprecht-Karls-University of Heidelberg, School of Medicine, in Germany and obtained his Ph.D. degree there in 1988. He did his postdoctoral work at the Free University of Berlin in the Institute of Pharmacology from 1989 to 1991.

Dr. Scherübl is a member of the German Society of Gastroenterology, the German Society of Endocrinology and Metabolism, and the German Society of Osteology. Among other awards, he has received the COPP Prize of the German Society of Osteology and the von Recklinghausen Prize of the German Society of Endocrinology and Metabolism for research in neuroendocrinology.

Dr. Scherübl has been the recipient of research grants from the Deutsche Forschungsgemeinschaft, the Sonnenfeld Stiftung, and private sponsors. He has published more than 50 papers. His current interests are in neuroendocrinology of the gastroenteropancreatic system and in the molecular pathogenesis of neuroendocrine tumor disease.

Jürgen Hescheler, M.D., Ph.D., is Chairman of the Department of Neurophysiology and Professor of Physiology at the University of Cologne, Germany. Dr. Hescheler studied medicine at the University of the Saarland, Homburg, Germany. He received his license to practice medicine in 1984 and his Ph.D. degree with *summa cum laude* in 1985, both from the University of the Saarland. He did post-doctoral work there in the Department of Physiology under the direction of Prof. Dr. W. Trautwein. In 1989 he worked as a research assistant in the Department of Pharmacology at the Free University of Berlin. Dr. Hescheler also carried out collaborative research projects with Prof. Dr. Mark Nelson at the University of Vermont and was awarded an Associate Visiting Professorship of that university. In 1994 Dr. Hescheler took over the Chair of the Department of Neurophysiology at the University of Cologne.

Dr. Hescheler is a member of various societies, including the Deutsche Physiologische Gesellschaft and the Biophysical Society. He has been the recipient of several scientific youth competition prizes, the Claude-Bernard Prize for his doctoral thesis, the Copp-Prize of the German Society for Osteology, the Hildegard Doerenkamp Gerhard Zbinden Foundation International Research Prize, as well as the Award of the Bundesminister für Jugend, Familie, Frauen und Gesundheit.

Dr. Hescheler is the author of approximately 100 full papers and reviews and is a reviewer for various international journals. His current research interest relates to the regulation of ionic channels in neuroendocrine cells and other cells, as well as the development of electrophysiological properties during cell differentiation.

CONTRIBUTORS

Helmut Acker, Prof. Dr.
Max-Planck-Institut f. Mol. Physiologie
Postfach 10 26 64
44026 Dortmund, Germany

Gudrun Ahnert-Hilger, Ph.D.
Innere Medizin, Abt. Gastroenterologie
Klinikum Benjamin Franklin
Freie Universität Berlin
Hindenburgdamm 30
D12200 Berlin, Germany

Norio Akaike, Ph.D.
Professor & Chairman
Department of Physiology
Kyushu University Faculty of Medicine
Fukuoka 812, Japan

Antonio R. Artalejo, M.D.
Max-Planck-Institute for Biophysical
  Chemistry
Department of Membrane Physics
Am Fassberg
37077 Göttingen, Germany

Frances Ashcroft, Ph.D.
University of Oxford
University Laboratory of Physiology
Parks Road
Oxford OX1 3PT, Great Britain

Illani Atwater, Ph.D.
Laboratory of Cell Biology and Genetics
National Institutes of Health
Bethesda, Maryland 20892

Christiane K. Bauer, Ph.D.
Physiologisches Institut
Universitäts-Krankenhaus Eppendorf
Universität Hamburg
Martinistr. 52
20246 Hamburg, Germany

Maria Luisa Brandi, M.D., Ph.D.
Associate Professor of Metabolic Diseases
Department of Clinical Physiopathology/
  Endocrine Unit
University of Florence
Viale Pieraccini 6
50139 Florence, Italy

Geir Christensen, M.D., Ph.D.
Institute for Experimental Medical Research
Research Forum, Ullevål Hospital
University of Oslo
0407 Oslo, Norway

Brian J. Corrette, Ph.D.
Physiologisches Institut
Universitäts-Krankenhaus Eppendorf
Universität Hamburg
Martinistr. 52
20246 Hamburg, Germany

Patricia Grasso, Ph.D.
Department of Biochemistry and
  Molecular Biology
Albany Medical College
Albany, New York 12208

Jürgen Hescheler, Prof. Dr.
Universitätsklinikum Benjamin Franklin
Institut für Pharmakologie
Freie Universität Berlin
Thielallee 69/73
D14195 Berlin, Germany
          and
Institut für Neurophysiologic
Universitat Köln
Robert Koch Str. 39
50931 Köln, Germany

Bin Hu, M.D. Ph. D.
Ottawa Civic Hospital
Civic Parkdale Clinic, Rm 455
1053 Carling Avenue
Ottawa, Ontario K1Y 4E9
Canada

Nicole M. Le Douarin, Prof. Dr.
Institut d'Embryologie cellulaire
  et moléculaire du CNRS
Collége de France
49 bis Avenue de la Belle-Gabrielle
94736 Nogent-sur-Marne Cedex, France

David Mears, Ph.D.
Department of Biomedical Engineering
The Johns Hopkins University School
  of Medicine
Baltimore, Maryland 21205

Patrice Mollard, Ph.D.
INSERM Unité 401
Pharmacologie Moléculaire
CCIPE, Rue de la Cardonille
34094 Montpellier Cedex 5, France

Hartmut Osswald, Prof. Dr.
Pharmakologisches Institut
Universität Tübingen
Wilhelmstr. 56
72074 Tübingen, Germany

Ulrich Quast, Prof. Dr.
Pharmakologisches Institut
Universität Tübingen
Wilhelmstr. 56
72074 Tübingen, Germany

Kurt Racké, Prof. Dr.
Department of Pharmacology
University Hospital
J.W. Goethe-University
Theodor-Stern-Kai 7
D60590 Frankfurt, Germany

Friedhelm Raue, Prof. Dr.
Universitäts Klinikum
Abt. Innere Medizin I/Endokrinologie
Luisenstr. 5
69115 Heidelberg, Germany

Leo E. Reichert, Jr., Ph.D.
Department of Biochemistry and
  Molecular Biology
Albany Medical College
Albany, New York 12208

A. Reimann, Ph.D.
Department of Pharmacology
University Hospital
J.W. Goethe-University
Theodor-Stern-Kai 7
D60590 Frankfurt, Germany

Leo P. Renaud, M.D., Ph.D., F.R.C.P.
Director of Neurology/Neuroscience
Ottawa Civic Hospital
Civic Parkdale Clinic, Room 455
1053 Carling Avenue
Ottawa, Ontario K1Y 4E9
Canada

Ernst-Otto Riecken, Prof. Dr.
Innere Medizin, Abt. Gastroenterologie
Klinikum Benjamin Franklin
Freie Universität Berlin
Hindenburgdamm 30
D12200 Berlin, Germany

Eduardo Rojas, Ph. D.
Laboratory of Cell Biology and Genetics
Building 8, Room 326
National Institutes of Health
8 Center Drive
Bethesda, Maryland 20892

Patrik Rorsman, Ph.D.
University of Gothenburg
Department of Physiology and Pharmacology
Division of Medical Biophysics
Medicinaregatan 11
S41390 Göteborg, Sweden

Rosa M. Santos, Ph.D.
Departmento de Bioquimica
Facultade de Ciencias e Tecnologia da
  Universidade de Coimbra
Apartado 3126
Coimbra, Portugal

Hans Scherübl, Ph.D., M.D.
Innere Medizin, Abt. Gastroenterologie
Klinikum Benjamin Franklin
Freie Universität Berlin
Hindenburgdamm 30
D12200 Berlin, Germany

Werner Schlegel, Prof. Dr.
Universite de Geneve
Faculte de Medecine
Fondation pour Recherches Medicales
64, Avenue de la Roseraie
CH-1211 Geneve, Switzerland

Jürgen R. Schwarz, Prof. Dr.
Physiologisches Institut
Universitäts-Krankenhaus Eppendorf
Universität Hamburg
Martinistr. 52
20246 Hamburg, Germany

Harald Schwörer, M.D.
Department of Medicine
Division of Gastroenterology and
  Endocrinology
University of Göttingen
Robert-Koch-Str. 40
37075 Göttingen, Germany

Annalisa Tanini, Prof. Dr.
Associate Professor of Endocrinology
IIIrd Institute of Internal Medicine
University of Florence
Viale Pieraccini 6
50139 Florence, Italy

B. Wiedenmann, Prof. Dr.
Innere Medizin, Abt. Gastroenterologie
Klinikum Benjamin Franklin
Freie Universität Berlin
Hindenburgdamm 30
D12200 Berlin, Germany

Yoshiro Yamashita, Ph.D.
Department of Physiology
Kumamoto University School of
  Medicine
Kumamoto 860, Japan

CONTENTS

Section I
Development and Classification

Chapter 1

FROM THE APUD TO THE NEUROENDOCRINE SYSTEMS: A DEVELOPMENTAL PERSPECTIVE

Nicole M. Le Douarin

CONTENTS

1. INTRODUCTION

High interest has focused during the 1970s to the polypeptide hormone-secreting cells distributed in various sites of the body, including the gut epithelium. Such cells were first described by Feyrter[1] as forming a widespread "system of clear cells" (Helle Zellen). He considered that, some of them, at least, act on their immediate neighbors and are therefore paracrine rather than endocrine in nature. He thought that the ones in the gut epithelium were derived from the enterocytes by a process that he called "endophytie".

Pearse[2,3] revisited the question of the embryonic origin and cytochemical features of the "system of clear cells" that he studied with more modern experimental approaches. He demonstrated that these cells which, in his view, form a diffuse endocrine organ, possess a common set of cytochemical and ultrastructural characteristics (Table 1). This led him to group them under the designation of APUD cells, an acronym derived from the initial letters of their three more constant and important cytochemical properties: amine content and/or amine precursor uptake and decarboxylation (APUD). Pearse considered these cytochemical characteristics to indicate closely related functional features including a common primary function, the production of polypeptide hormones. On that account he proposed that they share a common embryonic origin from the neural crest.

The list of cells belonging to the APUD series, limited to 14 cell types (including pituitary corticotrophs and melanotrophs, pancreatic islet cells, calcitonin-producing cells, carotid body type I cells, adrenal medulla, and various endocrine cells of the gut epithelium) in 1969, had increased to 40 less than 10 years later. By then Pearse[4,5] had included the parathyroid and a number of cells in the gut, lung and skin that had been shown to produce peptide or neuropeptides.

In all these cells the most characteristic properties can be demonstrated by a simple cytochemical test. The L-isomer of either of the two principal amino acid precursors of the fluorogenic monoamines (3,4-dihydroxyphenylananine [DOPA] for catecholamines or

TABLE 1
Cytochemical Characteristics of Polypeptide Hormone-Secreting Cells of the APUD Series

(A) 1. Fluorogenic amine content (catecholamines, 5-HT,[a] or others) (a) primary; (b) secondary uptake
(P) 2. Amine precursor uptake (5-HTP, DOPA)
(U)
(D) 3. Amino and decarboxylase
     4. High side-chain carboxyl or carboxyamide (masked metachromasia)
     5. High nonspecific esterase or cholinesterase (or both)
     6. High α-GPD (α-glycerophosphatase menadione reductase)
     7. Specific immunofluorescence

[a]   5-HT, 5-hydroxytryptamine; 5-HTP, hydroxytryptophan; DOPA, 3,4-dihydroxyphenylanine; α-GPD, α-glyceophosphatase dehydrogenase.

From Pearse, A. G. E., *J. Histochem. Cytochem.*, 17, 303, 1969.[3] With permission.

5-hydroxytryptophan [5-HTP] for serotonin) administered intravenously is taken up and decarboxylated by APUD cells. Some ultrastructural characteristics are also shared by the APUD cells:[2] (1) low levels of rough (granular) endoplasmic reticulum; (2) high levels of smooth endoplasmic reticulum in the form of vesicles; (3) high content of free ribosomes; (4) electron-dense, fixation-labile mitochondria; (5) membrane-bound secretion vesicles with osmiophilic contents (average diameter 100–200 nm).

Finally, Pearse[6] put forward the idea that cells of the APUD series form a "diffuse neuroendocrine system" (DNES) which he viewed as a third branch of the nervous system, "acting with the second, autonomic division, in the control of the function of all the intestinal organs".

One of the implications of the APUD cell concept was that a common origin for a variety of endocrine cells could account for a number of associated endocrine disorders such as medullary thyroid carcinomas concomitant with the pheochromocytoma syndrome,[7] the Zollinger-Ellison syndrome,[8] and other forms of the so-called multiple endocrine tumors. This explains why, although based only on very circumstantial evidence, the APUD cell concept was widely accepted.

The neuroectodermal origin of the endocrine cells of the gut epithelium and of the endocrine glands (pancreatic islet cells) associated with the digestive tract, is, however, a very controversial subject. On the one hand, an impressive number of molecular markers have been shown to be shared by neurons and a variety of these polypeptide hormone-producing cells, thus pleading for the existence of a common embryological ancestor, from which, in a terminal differentiation state, several cell types would have emerged. On the other hand, precise cell-tracing techniques applied on the avian embryo, an experimental model particularly favorable for this type of analysis, have failed to confirm the implication of ectodermal cells in the histogenesis of gut-related endocrine organs. Moreover, new molecular and genetic data obtained on certain forms of associated neoplasia have revealed that deregulation of certain oncogenes such as *c-ret* can affect cell lineages that may or may not be closely related embryologically.

2. COMMON MOLECULAR MARKERS SHARED BY NEUROENDOCRINE CELLS

The fact that different cell types share certain specialized biochemical characteristics is sometimes considered as a marker for their kinship. This is not necessarily true particularly because, when carefully examined, most such markers do not show the strict specificity that should be expected in such a case. For example, the L-amino acid decarboxylase responsible for the APUD properties is not restricted to neural and endocrine cells, but is transiently expressed by other tissues of mesodermal origin such as notochord and muscles as well as in pancreatic acinar cells, the endodermal origin of which is universally accepted.[9]

Another proposed marker for APUD cells, the specific acetylcholinesterase (AChE), is found in developing pancreatic endocrine cells, in some unidentified cells in the gut groove,[10] in the neuroepithelium of the basal plate of the neural tube, and in the neural crest.[11] At later stages, however, AChE has a more widespread distribution; thus its value as a specific molecular marker for DNES is doubtful. Another case is the neurone-specific enolase[12] present in enterochromaffin (EC) cells and pancreatic islet cells,[13,14] but also found in mega-karyocytes.[15]

Some molecular features, however, are strikingly common to neurons and endocrine cells. Such is the case for synaptophysin, an integral membrane glycoprotein localized in presynaptic vesicles of the adrenal medulla. It was also reported to be present in pancreatic islet cells and in a variety of epithelial tumors including islet cell adenomas, neuroendocrine carcinomas of the gastrointestinal and branchial tracts, and medullary carcinomas of the thyroid.[16]

Gangliosides acting as receptors for tetanus toxin and for the monoclonal antibody A2b5 are shared by nerve cells, astrocytes, and endocrine cells of the pancreas.[17]

Similarly, Teitelman and Lee[18] reported that tyrosine hydroxylase (TH), the first enzyme in the catecholamine (CA) biosynthetic pathway, is transiently produced by all pancreatic islet cells, but never in acinar exocrine cells.

Moreover, a detailed lineage analysis of the endocrine cells of the pancreas has been performed during ontogeny of transgenic mice carrying hybrid genes in which the 5′ regulatory sequence of the rat insulin gene was linked to the coding sequence of simian virus (SV40) large T antigen (Tag).[19] Alpert and co-workers[19] found that islet cells synthesizing TH plus glucagon, somatostatin, or pancreatic polypeptide coexpressed the transgene when they first arose. Tag was present transiently in the developing nervous system specifically, in some neural crest cells, and in the basal plate of the neural tube, the rhombencephalon, rostral mesencephalon, and diencephalon. It is noteworthy that monoaminergic systems producing serotonin and CA subsequently develop in some of these locations. These observations were interpreted as supporting the long-standing suggestion that the islet cells may be derived from neurectodermal precursors rather than from the endoderm.

However, although the list of molecular markers common to the nervous system and the polypeptide hormone-producing cells of the APUD series is impressive and may be suggestive that these cells share a common ancestor, this cannot be taken as proof since there is no obligatory ancestral relationship between cells expressing the same structural genes.

This is interestingly illustrated by the implication of the *c-ret* proto-oncogene in the onset of multiple endocrine neoplasia.

3. THE *c-ret* ONCOGENE IN MULTIPLE ENDOCRINE NEOPLASIA

As mentioned above, one of the attractive aspects of Pearse's APUD series and of the theory of their common embryonic origin from a neural crest ancestor was the presence of associated tumors affecting simultaneously various cell types of this series.

These multiple endocrine neoplasia have since been further characterized and recently shown to be linked with mutations in the proto-oncogene *c-ret*. Thus high-level expression of *ret* was found in medullary thyroid carcinomas and in pheochromocytomas which affect, respectively, the calcitonin-secreting C cells of the thyroid and the adrenomedullary cells of the suprarenal gland. Some of these tumors are sporadic, others show familial recurrence. Families can be classified according to phenotype into three categories: familial medullary thyroid carcinomas (FMTC) and multiple endocrine neoplasia type 2A (MEN2A) in which C cell tumors and pheochromocytomas are often associated. This type may include parathyroid hyperplasia, which is a nonmalignant overgrowth of the gland. Thirdly, a more severe form, MEN2B, has all the features of MEN2A plus the incidence of ganglioneuromas (neural cell hyperplasia) of the lips, tongue, and colon.

The *MEN2A* gene has been localized to a 480-kDa region in the chromosome 10q11.2.[20,21] This DNA segment includes the proto-oncogene *c-ret*, a receptor tyrosine kinase gene expressed at low levels in the normal C cells and adrenal medulla. The involvment of *c-ret* in these tumors was supported by the identification of mutations in the *ret* gene of FMTC and MEN2A and 2B patients.[22,23] Mutations of *c-ret* were also found in patients with Hirschprung's disease, which results from a deficiency in the innervation of the posterior bowel. The nature of the mutations is different in the cancer and Hirschprung's disease, the latter being characterized by a large deletion in the *ret* gene.

Interestingly, loss-of-function mutants have been generated in the mouse which result in complete agenesis of the nervous system in the gut and also in the kidney. *C-ret* is thus an essential component of a signaling pathway controlling cell survival and proliferation not only in certain neural crest derivatives, but also in the primordium of the kidney. Its structure as a membrane receptor suggests that regulation of its activity depends on a ligand, the nature of which is still to be determined.

This example stresses the fact that the activity of a particular gene is essential for a subclass (but not all) of neural crest derivatives on the one hand and for the kidney, an organ with an embryonic origin unrelated with the neurectoderm.

Pearse's hypothesis had the advantage of stimulating research in the field of embryology. In the early 1970s, I was personally challenged by Pearse's ideas and tested them by using the newly designed cell-marking technique based on the construction of quail chick chimeras.[24,25]

4. EMBRYOLOGICAL ANALYSIS OF THE ORIGIN OF THE APUD CELLS

Before the quail chick marker system was applied to the problem of the embryonic origin of the APUD cells, Andrew[26] brought about convincing evidence that the differentiation of EC cells of the gut epithelium occurred independently of the immigration of extrinsically derived cells. By explanting chick blastoderms deprived of the neural primordium (including the neural crest) on the chorioallantoic membrane, she showed that the gut which differentiated in the explants was devoid of intrinsic innervation, but not of EC cells. The same type of experiment was performed later on rat embryos and yielded similar results regarding several cell types of the APUD series, namely the pancreatic islet cells.[27]

The quail chick chimera system is based on the particular structure of the interphase nucleus in the quail species where the constitutive heterochromatin is condensed in one (or a few) mass(es) associated with the nucleolus. When quail and chick cells have been experimentally associated *in ovo* or in organotypic culture, they can be identified anytime after the association, thanks to the stable marker provided by the quail nuclei. Staining the chimeric tissues by any procedure revealing DNA allows recognition of quail and chick cells, whatever the phenotype they have acquired during the differentiation process. This method was extensively applied to investigate the development of the neural crest (see Reference 28 for a review). The principle of the experiments was to substitute definite fragments of the chick neural tube by their exact counterpart coming from stage-matched quail embryos (or vice versa). The development of the chimeras was normal; they were shown to be able to hatch and displayed a sensorimotor behavior compatible with long-term survival.[29] In this type of isotopic and isochronic grafts the behavior of the grafted neural crest cells was shown similar in chimeras and control quail and chick embryos.

This method was systematically applied to all levels of the neural axis in order to follow the migration and fate of neural crest-derived cells.[30] The gut was found to receive crest cells from two well-defined levels of the neuraxis: vagal and, to a lesser extent, lumbosacral. These cells build up the intrinsic enteric innervation, but do not reach the gut epithelium where the endocrine and paracrine cells of the APUD series differentiate from host progenitors.

Regarding the pancreatic islet cells, Andrew,[31] using a similar experimental approach, confronted the results of these xenografts of neural epithelium with those of grafts involving neural tubes of the same species (quail or chick, according to the host) labeled by tritiated thymidine. She reached similar conclusions: whatever the technique used, no cells from the neural crest contributed to the pancreatic islets. Some neural crest cells, however, colonize the pancreas where they aggregate in small groups distinct from the endocrine islets. Fontaine et al.[32] studied the type of differentiation expressed by these cells and observed that they did not exhibit the cytochemical properties of pancreatic endocrine islets, but differentiated into parasympathetic ganglia. At the same time, A, B, and D cells revealed by antibodies against glucagon, insulin, and somatostatin never carried the quail marker in these chimeric pancreas.[33]

The possibility that these cells might be derived from the neuroectoderm at a stage preceding the onset of the neural crest structure was thereafter tested experimentally.[34] The endomesoderm of chick embryos was associated with the endodermal germ layer of quail blastoderms at various stages, including the formation of the primitive streak, the head process, and the neural plate. The recombined embryos were either cultured *in vitro* or on the chorioallantoic membrane, and the intestinal structure which developed in the explants was analyzed for chimerism using various cytochemical techniques: formaldehyde induced fluorescence (FIF) technique after L-DOPA injection, lead hematoxylin, silver staining to indicate argentaffinity, and argyrophily, all combined with the Feulgen-Rossenbeck reaction. In all cases enteric ganglia originated from the quail ectoderm, but the EC as well as other APUD cells, which developed normally in the epithelium, were always of chick type. Therefore, no migration of cells from the ectoderm into the endoderm appeared to occur before formation of the neural crest. Thus a neurectodermal origin for the endocrine or paracrine cells of the gastrointestinal tract epithelium had to be excluded. An additional reason to attribute an endodermal origin to the enteroendocrine cells is the demonstration that they produce α-fetoprotein, a reliable marker of endodermal derivatives,[35] even in the adult.

Results were different concerning certain other cell types of the APUD series. The neural crest origin of the adrenal medulla and the melanocytes as well as the sympathetic ganglia and plexuses was fully confirmed by the quail chick chimera system.[28] Interestingly, the calcitonin-producing cells of the ultimobranchial bodies and the type I cells of the carotid bodies were also shown, by the same technique, to be derived from the vagal neural crest.[36,37] In mammals, a different experimental approach applied to the mouse embryo by Fontaine[38] led to the conclusion that the calcitonin producing cells of the thyroid gland did not originate from the endoderm of the ultimobranchial bodies which, during ontogeny, fuse with the thyroid anlage, but rather from the branchial arch mesenchyme whose neural crest origin had been demonstrated in the avian model[39] (see Reference 40 for a review).

An interesting case in point concerns the origin of the pituitary gland, the endocrine cells of which (corticotrope, melanotrope, somatotrope cells) have been claimed by Pearse to belong to the DNES and therefore to share a neuroectodermal origin with many other cells of this series. By applying the quail chick chimera system to the development of the neural primordium at the neurula stage, we could show that the presumptive territory of the Rathke's pouch is located in the anterior neural fold and that therefore its embryonic origin is located very close to the neural crest. In fact, the most rostral extremity of the neural crest corresponds to the presumptive level of the epiphysis.[41-43]

5. CONCLUSIONS

Since it has been shown that all cells of the peripheral ganglia and nerves originate from the neural crest, the origin of the entire nervous system of vertebrates is a fully accepted dogma (see Reference 28 for a review). It has also long been known that the ectodermal germ layer yields the endocrine cells of the adenohypophysis via the Rathke's pouch. Moreover, chromaffin endocrine cells, sympathetic neurons, and the small intensely fluorescent cells of the sympathetic ganglia share common characters to such an extent that they have been proposed to belong to the same lineage.[44] It is therefore evident that neurons, either displaying or not displaying neuroendocrine activity, and certain polypeptide hormone-producing cells share developmental and functional features.

Although endocrinology began with the discovery of a hormone produced by the digestive epithelium, the gastrointestinal tract was recognized as an endocrine organ only considerably later. Progress in this field started in the 1960s when efficient purification procedures were used to isolate and sequence several polypeptide hormones from the digestive tract. The first were mammalian gastrins,[45] then secretin, isolated in 1979 by Mutt and colleagues.[46] A number of others have followed which turned out to be especially concentrated (besides the gut) in various locations of the central and peripheral nervous system.

Pearse's APUD series was conceived during this period of expanding knowledge, and the idea of a link between many cells producing polypeptide hormones and neurons through a common embryonic ancestor originating in the neuroectoderm was very appealing. This idea had the merit of stimulating research on this field and led to the demonstration that some cells of the APUD series, like the C cells of the thyroid as well as the cells of the carotid body, are actually of neural crest origin, whereas others, such as the pancreatic islet cells and those of the digestive epithelium, are not.

The recent discovery that within the peripheral nervous system the development of certain subsets of cells are selectively controlled either by transcription factors or by growth factors is very informative in this respect. The case of *c-ret* is particularly relevant, since it shows that the C cells and the enteric ganglia are particularly affected in *null* mutants of this gene, whereas other neural crest derivatives develop normally. In contrast, in *c-ret* mutants, the kidney, which is a mesodermal derivative, does not develop. The *Mash-1* gene, a transcription factor of the basic-Helix-loop-Helix family with significant levels of homology with the *Achaete-scute* proneural gene of *Drosophila*, has been isolated in the rat by Anderson.[47] Knockout of this gene in the mouse results in the absence of sympathetic neurons, whereas adrenomedullary cells that were supposed to be lineally related to them are not affected.[48] This illustrates the complexity of the developmental regulations controlling the differentiation of apparently related cell types and stresses the point that only appropriate cell lineage tracers applied on embryos can reveal the embryonic origin of a given cell type.

Despite their diverse embryological origins, neuroendocrine cells have, in common, sufficient properties to constitute a fascinating cellular system situated at the functional frontier between the endocrine cells and the nervous system.

REFERENCES

 1. Feyrter, F., *Uber diffuse endokrine epitheliale Organe,* Barth J. A., Leipzig, 1938.
 2. Pearse, A. G. E., Common cytochemical and ultrastructural characteristics of cells producing polypeptide hormones (the APUD series) and their relevance to thyroid and ultimobranchial C cells and calcitonin, *Proc. R. Soc. London, Ser. B,* 170, 71, 1968.

3. Pearse, A. G. E., The cytochemical and ultrastructure of polypeptide hormone-producing cells of the APUD series and the embryologic, physiologic and pathologic implications of the concept, *J. Histochem. Cytochem.,* 17, 303, 1969.

4. Pearse, A. G. E., Peptides in brain and intestine, *Nature,* 262, 92, 1976.

5. Pearse, A. G. E., The diffuse endocrine system and the "common peptides", in *Molecular Endocrinology,* Mac Intyre, I. and Szelke, M., Eds., Elsevier/North-Holland, Amsterdam, 1977, 309

6. Pearse, A. G., APUD: concept, tumors, molecular markers and amyloid, *Mikroskopie,* 36, 257, 1980.

7. Ljungberg, O., Cederquist, E., and Studnitz, W., Medullary thyroid carcinoma and phaeochrocytoma: a familial chromaffinomatosis, *Br. Med. J.,* 279, 1967.

8. Zollinger, R. M. and Ellison, E. H., Primary peptic ulceration of the jejunum associated with islet cell tumors of the pancreas, *Ann. Surg.,* 142, 709, 1955.

9. Teitelman, G., Joh, T. H., and Reis, D. J., Linkage of the brain-skin-gut axis: islet cells originate from dopaminergic precursors, *Peptides,* 2 (Suppl. 2), 157, 1981.

10. Drews, U., Kussather, E., and Usadel, K. H., Histochemischer Nachweis der Cholinesterase in der Frühentwicklung der Hühnerkeimscheibe, *Histochemie,* 8, 65, 1967.

11. Cochard, P. and Coltey, P., Cholinergic traits in the neural crest: acetylcholinesterase in crest cells of the chick embryo, *Dev. Biol.,* 98, 221, 1983.

12. Marangos, P. J., Zis, A. P., Clark, R. L., and Goodwin, F. K., Neuronal, non-neuronal and hybrid forms of enolase in brain: structural, immunological and functional comparisons, *Brain Res.,* 150, 117, 1978.

13. Schmechel, D., Marangos, P. J., and Brightman, M., Neurone-specific enolase is a molecular marker for peripheral and central neuroendocrine cells, *Nature,* 276, 834, 1978.

14. Falkmer, S., Hakanson, R., and Sundler, F., *Evolution and Tumor Pathology of the Neuroendocrine System,* Elsevier, Amsterdam, 1984, 433.

15. Marangos, P. J., Campbell, J. C., Schmechel, D. E., Murphy, D. L., and Goodwin, E. K., Blood platelets contain a neuron-specific enolase subunit, *J. Neurochem.,* 34, 1254, 1980.

16. Wiedenmann, B., Franke, W. W., Kuhn, C., Moll, R., and Gould, V. E., Synaptophysin: a marker protein for neuroendocrine cells and neoplasms, *Proc. Natl. Acad. Sci., U.S.A.,* 83, 3500, 1986.

17. Eisenbarth, G. S., Shimizu, K., Bowring, M. A., and Wells, S., Expression of receptors for tetanus toxin and monoclonal antibody A2B5 by pancreatic islet cells, *Proc. Natl. Acad. Sci. U.S.A.,* 79, 5066, 1982.

18. Teitelman, G. and Lee, J. K., Cell lineage analysis of pancreatic islet cell development: glucagon and insulin cells arise from catecholaminergic precursors present in the pancreatic duct, *Dev. Biol.,* 121, 454, 1987.

19. Alpert, S., Hanahan, D., and Teitelman, G., Hybrid insulin genes reveal a developmental lineage for pancreatic endocrine cells and imply a relationship with neurons, *Cell,* 53, 295, 1988.

20. Gardner, E., Papi, L., Easton, D. F., Cummings, T., Jackson, C. E., Kaplan, M., Love, D. R., Mole, S. E., Moore, J. K., Mulligan, L. M., and Et, A. L., Genetic linkage studies map the multiple endocrine neoplasia type 2 loci to a small interval on chromosome 10q11.2, *Hum. Mol. Genet.,* 2, 241, 1993.

21. Mole, S. E., Mulligan, L. M., Healey, C. S., Ponder, B. A., and Tunnacliffe, A., Localisation of the gene for multiple endocrine neoplasia type 2A to a 480 kb region in chromosome band 10q11.2, *Hum. Mol. Genet.,* 2, 247, 1993.

22. Mulligan, L. M., Kwok, J. B. J., Healey, C. S., Elsdon, M. J., Eng, C., Gardner, E., Love, D. R., Mole, S. E., Moore, J. K., Papi, L., Ponder, M. A., Telenius, H., Tunacliffe, A., and Ponder, B. A. J., Germ-line mutations of the RET proto-oncogene in multiple endocrine neoplasia type 2A, *Nature,* 363, 458, 1993.

23. Hofstra, R. M. W., Landsvater, R. M., Ceccherini, I., Stulp, R. P., Stelwagen, T., Luo, Y., Pasini, B., Höppener, J. W. M., Ploos Van Amstel, H. K., Romeo, G., Lips, C. J. M., and Buys, C. H. C. M., A mutation in the RET proto-oncogene associated with multiple endocrine neoplasia type 2B and sporadic medullary thyroid carcinoma, *Nature,* 367, 375, 1994.

24. Le Douarin, N., Particularités du noyau interphasique chez la Caille japonaise (*Coturnix coturnix japonica*). Utilisation de ces particularités comme "marquage biologique" dans les recherches sur les interactions tissulaires et les migrations cellulaires au cours de l'ontogenèse, *Bull. Biol. Fr. Belg.,* 103, 435, 1969.

25. Le Douarin, N., A biological cell labeling technique and its use in experimental embryology, *Dev. Biol.,* 30, 217, 1973.

26. Andrew, A., A study of the developmental relationship between enterochromaffin cells and the neural crest, *J. Embryol. Exp. Morphol.,* 11, 307, 1963.

27. Pictet, R. L., Rall, L. B., Phelps, P., and Rutter, W. J., The neural crest and the origin of the insulin-producing and other gastrointestinal hormone-producing cells, *Science,* 191, 191, 1976.

28. Le Douarin, N., *The Neural Crest,* Cambridge University Press, Cambridge, 1982.

29. Kinutani, M. and Le Douarin, N. M., Avian spinal cord chimeras, *Dev. Biol.,* 11, 243, 1985.

30. Le Douarin, N. M. and Teillet, M. A., The migration of neural crest cells to the wall of the digestive tract in avian embryo, *J. Embryol. Exp. Morphol.,* 30, 31, 1973.

31. Andrew, A., An experimental investigation into the possible neural crest origin of pancreatic APUD (islet) cells, *J. Embryol. Exp. Morphol.,* 35, 577, 1976.

32. Fontaine, J., Le Lièvre, C., and Le Douarin, N. M., What is the developmental fate of the neural crest cells which migrate into the pancreas in the avian embryo?, Gen. Comp. Endocrinol., 33, 394, 1977.
33. Fontaine-Pérus, J., Le Lièvre, C., and Dubois, M. P., Do neural crest cells in the pancreas differentiate into somatostatin-containing cells?, *Cell Tissue Res.,* 213, 293, 1980.
34. Fontaine, J. and Le Douarin, N. M., Analysis of endoderm formation in the avian blastoderm by the use of quail-chick chimaeras. The problem of the neurectodermal origin of the cells of the APUD series, *J. Embryol. Exp. Morphol.,* 41, 209, 1977.
35. Tyner, A. L., Godbout, R., Compton, R. S., and Tilghman, S. M., The ontogeny of α-fetoprotein gene expression in the mouse gastrointestinal tract, *J. Cell Biol.,* 110, 915, 1990.
36. Le Douarin, N., Le Lièvre, C., and Fontaine, J., Recherches expérimentales sur l'origine embryologique du corps carotidien chez les Oiseaux, *C.R. Acad. Sci. Ser. III, Paris,* 275, 583, 1972.
37. Le Douarin, N., Fontaine, J., and Le Lièvre, C., New studies on the neural crest origin of the avian ultimobranchial glandular cells. Interspecific combinations and cytochemical characterization of C cells based on the uptake of biogenic amine precursors, *Histochemistry,* 38, 297, 1974.
38. Fontaine, J., Multistep migration of calcitonin cell precursors during ontogeny of the mouse pharynx, *Gen. Comp. Endocrinol.,* 37, 81, 1979.
39. Le Lièvre, C. S. and Le Douarin, N. M., Mesenchymal derivatives of the neural crest: analysis of chimaeric quail and chick embryos, *J. Embryol. Exp. Morphol.,* 34, 125, 1975.
40. Le Douarin, N. M. and Fontaine-Pérus, J., Embryonic origin of polypeptide hormone producing cells, in *Markers for Neural and Endocrine Cells,* Gratzl, M. and Langley, K., Eds., VCH Publishers, Weinheim, West Germany, 1991, 3.
41. Couly, G. F. and Le Douarin, N. M., Mapping of the early neural primordium in quail-chick chimeras. I. Developmental relationships between placodes, facial ectoderm, and prosencephalon, *Dev. Biol.,* 110, 422, 1985.
42. Couly, G. F. and Le Douarin, N. M., Mapping of the early neural primordium in quail-chick chimeras. II. The prosencephalic neural plate and neural folds: implications for the genesis of cephalic human congenital abnormalities, *Dev. Biol.,* 120, 198, 1987.
43. Le Douarin, N. M., Fontaine-Pérus, J., and Couly, G., Cephalic ectodermal placodes and neurogenesis, *Trends Neurosci.,* 9, 175, 1986.
44. Landis, S. C. and Patterson, P. H., Neural crest cell lineages, *Trends Neurosci.,* 4, 172, 1981.
45. Gregory, R. A. and Tracy, H. J., The constitution and properties of two gastrins extracted from hog antral mucosa, *Gut,* 5, 103, 1964.
46. Mutt, V., Carlquist, M., and Tatemoto, K., Secretin-like bioactivity in extracts of porcine brain, *Life Sci.,* 25, 1703, 1979.
47. Anderson, D. J., MASH genes and the logic of neural crest cell lineage diversification, *C.R. Acad. Sci., Ser. III, Paris,* 316, 1090, 1993.
48. Guillemot, F., Lo, I. C., Johnson, J. E., Auerbach, A., Anderson, D. J. and Joyner, A. L., Mammalian achaete-scute homolog 1 is required for the early development of olfaction and autonomic neurons, *Cell,* 15, 1, 1993.

Chapter 2

CLASSIFICATION OF NEUROENDOCRINE CELLS

G. Ahnert-Hilger, H. Scherübl, E.-O. Riecken, and B. Wiedenmann

CONTENTS

1. HISTORICAL BACKGROUND

In his medical thesis in 1869, Langerhans[1] described certain cells in the pancreas that are specifically arranged in insular arrays and form morphologically a separate entity. A year later, Heidenhain described cells with similar distinct morphological properties in the intestinal mucosa. Almost 30 years later another organ, the lung, revealed to contain similar specialized cells, which were termed after its discoverer "Kulchitsky cells". Neoplastic epithelial cells of the gut, distinct from the adenocarcinoma cells, were then described by Oberndorfer[2] and termed "carcinoid cells". Functional aspects of these cells were first described in 1902 by Bayliss and Starling[3] who reported that certain cells released blood-borne chemical messengers exerting an influence either on their neighboring cells in a paracrine mode or at more distant sites in an endocrine fashion.

Further aspects were added by new staining procedures using silver reagents that appeared to react specifically with chromaffin cells of the adrenal medulla.[4] In the late 1920s, Scharrer demonstrated for the first time that hypothalamic neurons possessed neurosecretory properties. Ten years later, Feyrter[5] observed, similar to Heidenhain (see above), rather pale cells in the intestinal tract which he termed "Helle Zellen". Because of their disseminated distribution he suggested for the first time that these cells constituted a "diffuse endocrine epithelial organ". Subsequent studies with increasingly sophisticated methods showed that cytoplasmic granules or "droplets" could be selectively stained not only in neurons, but also in these later to be called "neuroendocrine" cells. With the advent of electron microscopy, these structures could be identified with greater precision, and turned out to be cytoplasmic vesicles bounded by membranes with variable diameter and an electron-dense core. These vesicles or granules showed to contain a specific biologically active content and gained considerable interest for the years to come.

The discovery of Pearse in 1968[6] and 1969[7] that amine precursors are taken up and also decarboxylated by these "endocrine APUD (**a**mine **p**recursor **u**ptake and **d**ecarboxylation) cells" demonstrated their neuronal properties. Based on this finding, Pearse went — as we know by now — one step too far by postulating that all endocrine cells with neuronal properties — now termed neuroendocrine cells — must be embryologically derived from the neural crest. As shown very elegantly by Le Douarin[8,10,11] and Le Douarin and Teillet,[19] most neuroendocrine cells are actually of endodermal origin. Using new molecular and cell biological approaches, a large number of molecules were identified that could be assigned morphologically and functionally to various subcellular structures, such as secretory vesicles, the Golgi complex, the trans-Golgi network, and endosomes. It became evident that neurons and neuroendocrine cells share many properties, such as the expression and synthesis of very similar secretory vesicles and molecules, cell adhesion molecules, neuropeptide and amino acid transmitters, and similar receptors.

2. ENDOCRINE-NEURONAL RELATIONSHIPS

Extensive research in the last decades shows that "endocrine" cells represent a rather heterogenous group of different embryonic origin and functional properties. Classically, endocrine cells are defined as hormone-releasing cells. However, a considerable amount of data has been accumulated showing that most endocrine cells share a variety of properties with neurons. Some authors even hypothesize that part of the endocrine cells are — though morphologically distinct — functionally modified neurons. In contrast to neuronal cells, most endocrine cells are, however, derived from the endoderm and not from the neural crest (see Chapter 1 of this volume).

Despite its false embryological approach, the APUD concept has embraced, as a fundamental tenet, a variety of cellular functions shared by endocrine cells and neurons. This functional aspect of the APUD concept has been experimentally extended and further verified by a large number of groups during recent years (for review see References 12 and 13). Based on the cell biological and functional similarities of neurons and this group of endocrine cells, the latter should be termed "neuroendocrine" rather than "endocrine". Hereby the term "neuroendocrine" defines cells expressing and synthesizing molecules which lead in a concerted action to the regulated release of neuropeptides, neurotransmitters, or hormones and which, in contrast to neuronal cells, do not form synapses. The vesicular release of these in most cases electrically excitable cells is mediated — at least in part — by voltage-dependent ion channels, in particular by voltage-dependent calcium channels. Neurons, which release their secretory products not into the synaptic cleft, but into the blood as do the neurosecretory cells of the hypothalamus, also belong to this group. In contrast, the follicular cells of the thyroid and the various steroid hormone-producing cells lack the outlined neuroendocrine properties and should only be named endocrine (see Figure 1).

Neurons
Neuroendocrine cells
- hypothalamic supraoptic and paraventricular neurosecretory cells
- chromaffin cells
- epithelial cells
  - anterior pituitary cells
  - C-cells of the thyroid
  - parathyroid cells
  - Kulchitsky cells of bronchial tract
  - neuroendocrine cells of the thymus
  - G-cells of the gastric antrum
  - the group of enterochromaffin cells of the stomach and intestines
  - pancreatic islet cells
  - Sertoli cells
  - Merkel cells of the skin
- muscle related cells
  - ANF-producing cells of the heart
  - renin-producing cells of the kidney

Endocrine cells
- follicular cells of the thyroid
- steroid hormone producing cells

FIGURE 1. Neuroendocrine and endocrine cells.

3. FEATURES OF NEUROENDOCRINE CELLS

Morphologically, neuroendocrine cells are characterized by uniform nuclei and an abundant granular or faintly staining (clear) cytoplasm. They can exist as single cells interdispersed in the epithelial lining of, for example, the intestine or the lung. When neuroendocrine cells are associated with each other in a multicellular complex, they are found as solid cell clusters or show a trabecular pattern (e.g., pancreatic islets or islets of Langerhans). Whereas the latter can be well identified at light microscopic level by conventional staining methods, neuroendocrine cells interdispersed in epithelial linings are hardly detectable. Various chemical staining methods more or less specific for neuroendocrine cells have been developed during this century. Among them are silver impregnation methods for carcinoids (argentaffin cell tumors) and nerve hyperplasia of the appendicular mucosa[4] and α_2 cells in human pancreatic islets.[14]

3.1 MARKER MOLECULES OF NEUROENDOCRINE CELLS

Neuroendocrine cells express various characteristic polypeptides which reside in different compartments of the cell.

3.1.1. Plasma Membrane

The plasma membrane of neurons and of at least some neuroendocrine cells is characterized by the cell type-specific expression of neural cell adhesion molecules (NCAM) (e.g., NCAM and L1)[15,16] as well as receptors for neuropeptides (e.g., somatostatin) and amino acid neurotransmitters (e.g., γ-aminobutyric acid [GABA]).[17] It is not yet known if the combined expression of these molecules occurs in normal neuroendocrine cells (e.g., chromaffin cells, enterochromaffin-like cells, or C cells of the thyroid) or develops in the course of malignant transformation.

TABLE 1
Molecular Markers of Neuroendocrine Cells

Secretory Machinery
LDCV
  Membrane
    • Amino acid transporters
    • Cytochrome b561
    • Dopamine-β-hydroxylase
  Matrix
    • Granins (esp. chromogranin A)
    • Neuropeptides (e.g., somatostatin)
    • Hormones (e.g., insulin)
SSV
  Membrane
    • Synaptophysin
    • Synaptobrevin
    • Protein SV2
    • Synaptotagmin
    • Synapsins
    • Various transporters
  Matrix
    • Amino acids (e.g., GABA, glycine, glutamate)
Docking complex
    • SNAP 25
    • Syntaxin

Cytoplasm
Enzymes
    • Acetylcholinesterase
    • Tyrosine hydroxylase
    • Dopamine-β-hydroxylase
    • Neuron-specific enolase
    • Glutamic acid decarboxylase
Others
    • Protein gene product 9.5

Cytoskeleton and Plasma Membrane
Intermediate filaments
    • Peripherin
    • Neurofilament proteins
    • Cytokeratins
Cell adhesive molecules
    • NCAMs
    • L1
Receptors
    • For neuropeptides (e.g., somatostatin)
    • For neurotransmitters (e.g., serotonin, GABA)

3.1.2. Regulated Secretory Vesicles

Neuroendocrine cells communicate with the environment by their secretory products. Besides constitutive secretion, which is a cellular process going on in any cell, neurons and neuroendocrine cells are able to release their products in a regulated fashion. Regulated secretion is mediated by two types of secretory vesicles: (1) large dense-core vesicles (LDCV) of neurons and neuroendocrine cells and (2) small synaptic vesicles (SSV) of neurons and their analogs in neuroendocrine cells. LDCV and SSV differ with respect to their membrane composition and content (see Table 1). At an ultrastructural level, neuroendocrine cells are identified electronmicroscopically by the electron-dense LDCV with a diameter ranging from

80–400 nm. LDCV and chromaffin granules of the adrenal medulla are characterized by integral or tightly bound membrane proteins, among them cytochrome b_{561} and dopamine-β-hydroxylase.[12,13] These two polypeptides can be used as marker molecules for normal and neoplastic neuroendocrine cells.[18,19] The electron-microscopically dense content is due to highly concentrated polypeptides of the granin family. Among the granins, chromogranin A has proven to be the most reliable marker molecule for normal and neoplastic cells (for review see Reference 20).

The membrane composition of SSV from brain and neuroendocrine cells comprises integral membrane proteins such as synaptophysin, protein SV2, the synaptotagmins, the synaptobrevins, and associated proteins, among them the synapsins, the small GTP-binding protein rab3a, and heterotrimeric G proteins. Interestingly, LDCV and SSV differ in their content of heterotrimeric G proteins, i.e., SSV contain different α subunits as compared to LDCV.[18]

The SSV are only clearly detectable by the use of vesicle-specific antibodies[12,13] which have proven to be good tools for the identification and characterization of both neurons and neuroendocrine cells (see also Table 1). SSV analogs of neuroendocrine cells are similar in size to constitutive vesicles (40–80 nm) and contain an electron-microscopically clear content. Vesicular storage of neurotransmitters such as monoamines and amino acid transmitters is accomplished by specific transporter molecules located in the plasma and vesicular membrane (e.g., vesicular monoamine transporter or GABA transporters).

3.1.3. Cytosol and Cytoskeleton Proteins

To synthesize and subsequently store vesicular transmitters of low molecular weight, e.g., dopamine, GABA, and glutamate, neurotransmitter-synthesizing enzymes exist in the cytoplasm of neurons and neuroendocrine cells. Some of these enzymes listed in Table 1 are used as marker molecules for normal and neoplastic neuroendocrine cells (e.g., neuron-specific enolase, glutamate decarboxylase, and tyrosine hydroxylase).

In addition to cytosolic enzymes, intermediate filament proteins have proven as reliable marker molecules for neuroendocrine cells. In general, neurofilament proteins occur in neurons, whereas cytokeratins are characteristic for epithelial cells. In accordance to their embryonic origin (see Le Douarin, Chapter 1, this volume), neuroendocrine cells of the gastroenteropancreatic system express cytokeratins. Similarly, chromaffin cells known to be of neuroectodermal origin express neurofilament proteins. However, coexpression of cytokeratins and neurofilament proteins is observed in certain normal and neoplastic cells. Reasons for this coexpression are unclear.

3.2. PHYSIOLOGICAL PROPERTIES OF NEUROENDOCRINE CELLS

The most important characteristic of neuroendocrine cells is their ability to secrete various polypeptides, low-molecular weight neurotransmitters, or both. The formation of regulated vesicles, LDCV and SSV, the vesicular storage and/or release has to be organized in a continuous (constitutive) or an "on demand" (regulated) fashion.

3.2.1. Vesiculogenesis and Vesicle Maturation

During recent years, the molecular mechanisms of polypeptide sorting, vesicle formation, and vesicular release have been well characterized. Detailed accounts of these studies are given in recent reviews.[20-27]

3.2.1.1. *Formation of LDCV and Neuroendocrine Granules*

The formation and maturation of LDCV and neuroendocrine granules have been studied in considerable detail. Proteins of the granin family such as chromogranin A[28] appear to play a crucial role during the early steps in their formation. These polypeptides aggregate in an environment created by the membranes of nascent granules. Their milieu is characterized by acidity and a high calcium and/or zinc concentration, probably leading to a condensation and polymerization of the vesicular matrix polypeptides.[23] Since similar conditions appear to exist in LDCV also *in vivo*, it is postulated that granins mediate the packaging of hormones and neuropeptides in regulated vesicles.

3.2.1.2. *Vesiculogenesis of SSV and Neuroendocrine SSV Analogs*

So far, only limited data on the formation of SSV and neuroendocrine SSV analogs are available. The mature, regulated vesicles appear to be formed after separate passage of all major SSV-membrane proteins through the plasma membrane to the early endosome.[23,29] In differentiating, hypothalamic neurons, immature and mature SSV appear to differ in their association with small GTP-binding proteins such as rab6p.[30] In certain neuroendocrine cells such as the rat pheochromocytoma cell line PC12, neuroendocrine SSV analogs probably evolve out of constitutive vesicles. This suggests that vesicular targeting and release also varies, depending on the state of neuronal and possibly also neuroendocrine differentiation. In nonneuroendocrine epithelial cells, transfected with a gene coding for synaptophysin, a new vesicle type is formed which differs morphologically and biochemically from intrinsic, constitutive vesicles.[31] Based on its intrinsic properties, it is suggested that synaptophysin plays a major role in the vesiculogenesis of SSV and SSV analogs.[31]

3.2.1.3. *Constitutive Vesicles*

To ensure normal cellular functions, every cell type has to secrete a variety of products such as albumin in hepatocytes or certain growth factors in fibroblasts. Secretion of these products is accomplished by a vesicular route which continuously transports secretory products to the plasma membrane and other membrane compartments. It does not possess means for storage or concentration of vesicular contents. In addition, a variety of membrane proteins (e.g., ATPases, transporters, and certain enzymes) are delivered to their final cellular compartment via this vesicular, constitutive pathway. Similar to SSV, formation of these vesicles appears to take place in the endosome.[29] So far, mechanisms controlling the initial step of separation between membranes destined for constitutive or regulated vesicles have not been identified. It remains to be clarified whether the SSV analogs of neuroendocrine cells represent a homologous population of "regulated" vesicles or if they exhibit only transiently or partially regulated properties.

3.2.2. **Vesicular Storage**

The association and packaging of secretory products takes place either in the trans-Golgi network for polypeptide-storing LDCV or is accomplished by specific transporter molecules of the secretory vesicle membrane.

3.2.2.1. *Synthesis of Hormones, Neuropeptides, and Neurotransmitters*

After transcription, synthesis of hormones and neuropeptides continues by a complex, stepwise processing mechanism involving various proteolytic enzymes. These enzymes process the immature hormones or peptides in various locally defined steps to the mature form. Interestingly, hormones and neuropeptides appear to be only sorted into LDCV. Classical neurotransmitters, on the other hand, are primarily found in SSV of neurons and SSV analogs of neuroendocrine cells.

Similar to neurons, synthesis of neurotransmitters takes place in neuroendocrine cells, at least in part, next to the plasma membrane, i.e., the site of release. Transmitter-synthesizing enzymes are found in close proximity to the site of transmitter storage (glutamate decarboxylase) or they are located inside the secretory vesicles (dopamine-β-hydroxylase).

3.2.2.2. *Uptake of Monoamines and Amino Acids in Neuroendocrine Cells*

Various neurotransmitters, such as acetylcholine, norepinephrine, epinephrine, dopamine, and serotonin, as well as amino acid transmitters, such as GABA, glycine, and glutamate, have been identified in neuroendocrine cells. However, in contrast to neuronal and chromaffin cells, an identification and molecular characterization of specific uptake mechanisms by the plasma membrane and regulated vesicles have not yet been done in neuroendocrine cells. Although GABA has been found in various peripherial tissues besides neurons, only recently GABA uptake into neuroendocrine tumor cell lines has been demonstrated.[33-35]

3.2.3. Secretory Machinery

Secretion from regulated vesicles mainly depends on an elevated cytosolic calcium concentration. An opening of the presynaptic voltage-dependent calcium channels triggers the rapid fusion of synaptic vesicles with the presynaptic plasma membrane.[25] Voltage-dependent calcium channels are also involved in regulated secretion from neuroendocrine cells. The fusion of the vesicle and plasma membranes results in the release of neurotransmitters. As outlined by various authors in this volume, similar events also take place in neuroendocrine cells. Following neurotransmitter release, the components of the synaptic vesicle membrane are selectively recycled by the endocytic pathway within the nerve terminal to regenerate new synaptic vesicles.

Studies in recent years have shown that proteins present in the synaptic vesicle membrane are intimately associated with the process of neurotransmitter release. Furthermore, additional soluble, cytosolic polypeptides as well as polypeptides of the presynaptic plasma membrane are required for a specific interaction and/or docking of the synaptic vesicle with the plasma membrane. Only the combined action of all components of the secretory machinery leads to the final fusion event involving the vesicle and the plasma membrane in response to membrane depolarization.

Recent results have implicated a crucial role of a highly conserved protein complex, consisting out of the *N*-ethylmaleimide-sensitive fusion (NSF) protein (an ATPase) and soluble NSF-attachment proteins (α/γSNAP) as well as two synaptic vesicle membrane proteins (synaptotagmin and synaptobrevin), and two synaptic plasma membrane proteins (syntaxin and SNAP25) in the process of synaptic vesicle docking and/or fusion.[36] Syntaxin, synaptobrevin, and SNAP25 are probably the only molecular targets of clostridial neurotoxins.[24] This implies that each of these proteins is required for exocytosis of small synaptic vesicles. It is beyond the scope of this review to discuss the molecular mechanisms studied so far.

Whereas the molecular mechanisms of vesicle docking and release of SSV have been well elucidated in neurons, studies in neuroendocrine cells are only beginning. Probably, exocytosis from LDCV obeys similar rules as shown for bovine adrenal chromaffin cells.[37-38]

4. CONCLUDING REMARKS

Neuroendocrine cells possess a regulated secretory machinery, and the majority of them are electrically excitable. The combination of common morphological, histochemical, and functional criteria classifies this group of cells as "neuroendocrine". By contrast, hormone-releasing cells without these features should only be termed "endocrine". Using this definition of neuroendocrine cells, the embryological origin appears less important. The classification is also applicable to neuroendocrine pathology, where markers specific for, e.g., the regulated secretory vesicles (synaptophysin, chromogranin A, dopamine-β-hydroxylase, etc.) are increasingly used and have led to an improved diagnosis of neuroendocrine tumor disease.

ACKNOWLEDGMENTS

The authors thank the Deutsche Forschungsgemeinschaft (SFB 366/A5, Forschungs-schwerpunkt Wi 617/5-3, Sche 326/3-1), Bundesministerium für Forschung und Technologie (Partnerschaftsprojekt Neurobiologie), Dr. Mildred Scheel Stiftung/Deutsche Krebshilfe (W 31/91/Wi 1), and Verum-Stiftung for support. Finally, we also thank Mrs. Malgorzata Szott for skillful typing of the manuscript.

REFERENCES

1. Langerhans, P., Beiträge zur mikroskopischen Anatomie der Bauchspeicheldrüse, Inaugural-dissertation zur Erlangung der Doktorwürde in der Medicin und Chirurgie vorgelegt der medicinischen Facultät der Friedrich-Wilhelms-Universität zu Berlin, 24, 1869.
2. Oberndorfer, S., Karzinoide Tumoren des Dünndarms, *Frankf. Z. Pathol.* (Wiesbaden), I, 426, 1907.
3. Bayliss, W. M. and Starling, E. H., On the causation of the so-called "peripheral reflex secretion" of the pancreas, *Proc. R. Soc. London,* 69, 352, 1902.
4. Gosset and Masson, Carcinoids (argentaffin cell tumor) and nerve hyperplasia of the appendicular mucosa, *Am. J. Pathol.,* 4, 181, 1928.
5. Feyrter, F., *Über Diffuse Endokrine Epitheliale Organe,* Johann Ambrosius Barth, Leipzig, 1938.
6. Pearse, A. G. E., Common cytochemical and ultrastructural characteristics of cells producing polypeptide hormones (the APUD series) and their relevance to throid and ultimobranchial C cells and calcitonin, *Proc. R. Soc. London, Ser. B,* 170, 71, 1968.
7. Pearse, A. G. E., The cytochemical und ultrastructure of polypeptide hormone-producing cells of the APUD series and the embryologic, physiologic and pathologic implications of the concept, *J. Histochem. Cytochem.,* 17, 303, 1969.
8. Le Douarin, N., Particulartés du noyau interphasique chesz la Caille japonaise (Coturnix coturnix japonica). Utilisation de ces particularités comme marquage biologique dans les recherches sur les interactions tissulaires et les migrations cellulaires au cours de l´ontogenèse, *Bull. Biol. Fr. Belg.,* 103, 435, 1969.
9. Le Douarin, N. M. and Teillet, M. A., The migration of neural crest cells to the wall of the digestive tract in avian embryo, *J. Embryol. Exp. Morphol.,* 30, 31, 973.
10. Le Douarin, N. M., The ontogeny of the neural crest in avian embryo chimaeras, *Nature,* 286, 663, 1980.
11. Le Douarin, N. M., On the origin of pancratic endocrine cells, *Cell,* 53, 169, 1988.
12. Wiedenmann, B. and Huttner, W., Synaptophysin and the secretogranins/chromogranins — widespread components of distinct types of neuroendocrine vesicles and new tools in tumor diagnosis, *Virchows Arch. B.,* 58, 95, 1989.
13. Südhof, T. C. and Jahn, R., Proteins of synaptic vesicles involved in exocytosis and membrane recycling, *Neuron,* 6, 665, 1991.
14. Grimelius, L., A silver stain for alpha 2 cells in human pancreatic islets, *Acta Soc. Med. Upsa.,* 73, 243, 1968.
15. Langley, K. and Gratzl, M., Neural cell adhesion molecules as markers for neural and endocrine cells, in *Molecular and Cell Biology, Diagnostic Applications,* Gratzl, M. and Langley, K., Eds., Verlag Chemie, Weinheim, 1990, 133.
16. Langley, K., The neuroendocrine concept today, *Ann. N.Y. Acad. Sci.,* 733, 1, 1994.
17. von Blankenfeld, G., Kettenmann, G., Turner, J., Ahnert-Hilger, G., John, M., and Wiedenmann, B., Expression of a functional GABAA receptor in neuroendocrine gastropancreatic cells, *Eur. J. Physiol.,* in press.
18. Ahnert-Hilger, G., Schmitt, L., Grube, K., Mönch, E., Kvols, L., Riecken, E. O., Lee, I., and Wiedenmann, B., Gastroenteropancreatic neuroendocrine tumors contain a common set of synaptic vesicle proteins and aminoacid neurotransmitters, *Eur. J. Cancer,* 29A, 14, 1993.
19. Ahnert-Hilger, G., Schäfer, T., Spicher, K., Schultz, G., and Wiedenmann, B., Detection of G protein heterotrimers on large dense core and small synaptic vesicles of neuroendocrine and neuronal cells, *Eur. J. Cell Biol.,* 65, 26, 1994.
20. Klöppel, G. and Heitz, Ph., Classification of normal and neoplastic neuroendocrine cells, *Ann. N.Y. Acad. Sci.,* 733, 18, 1994.
21. Rodman, J., Feng, Y., Myers, M., Zhang, J., Magner, R., and Forgac, M., Comparison of the coated vesicle and synaptic vesicle vacuolar (H+)-ATPases, *Ann. N.Y. Acad. Sci.,* 733, 203, 1994.

22. Roush, D., Gottardi, C. J., and Caplan, M. J., Sorting of the gastric H, K-ATPase in endocrine and epithelial cells, *Ann. N.Y. Acad. Sci.,* 733, 212, 1994.

23. Bauerfeind, R., Ohashi, M., and Huttner, W. B., Biogenesis of secretory granules and synaptic vesicles — facts and hypotheses, *Ann. N.Y. Acad. Sci.,* 733, 233, 1994.

24. Jahn, R. and Niemann, H., Molecular mechanisms of clostridial neurotoxins, *Ann. N.Y. Acad. Sci.,* 733, 255, 1994.

25. Bennett, M., Molecular mechanisms of neurotransmitter release, *Ann. N.Y. Acad. Sci.,* 733, 256, 1994.

26. Schäfer, T., Hode., A., Heuss, C., and Burger, M., The docking protein of chromaffin granules, *Ann. N.Y. Acad. Sci.,* 733, 279, 1994.

27. Ahnert-Hilger, G. and Wiedenmann, B., Requirements for exocytosis in permeabilized neuroendocrine cells: possible involvement of heterotrimeric G-proteins associated with secretory vesicles, *Ann. N.Y. Acad. Sci.,* 733, 298, 1994.

28. O'Connor, D., Wu, H., Gill, B. M., Rozansky, D. J., Tang, K., Mahata, S. K., Mahata, M., Eskeland, N. L., Videen, J. S., Zhang, X., Takiyuddin, M. A., and Parmer, R. J., Hormone storage vesicle proteins: transcriptional basis of the widespread neuroendocrine expression of chromogranin A, and evidence of its diverse biological actions, intracellular and extracellular, *Ann. N.Y. Acad. Sci.,* 733, 36, 1994.

29. Südhof, T. C., DeCamilli, A., Niemann, H., and Jahn, R., Membrane fusion machinery: insights from synaptic protein, *Cell,* 75, 1, 1993.

30. Tixier-Vidal, A., Barret, A., Picart, R., Mayau, V., Vogt, D., Wiedenmann, B., and Goud, B., The small GTP-binding protein, Rab 6p, is associated with both Golgi and synaptic vesicle membranes during synaptogenesis of hypothalamic neurons in culture, *J. Cell Sci.,* 105, 935, 1993.

31. Leube, R. E., Harth, N., Grund, Ch., Franke, W. W., and Wiedenmann, B., Sorting of synaptophysin into special vesicles in transfected non-neuroendocrine cells, *J. Cell Biol.,* 127, 1589, 1994.

32. Erdö, S. L. amd Wolff, J. R., g Aminobutyric acid outside the mammalian brain, *J. Neurochem.,* 54, 363, 1989.

33. Ahnert-Hilger, G. and Wiedenmann, B., The amphicrine pancreatic cell line AR42J secretes GABA and amylase by separate regulated pathways, *FEBS Lett.,* 314, 41, 1992.

34. Thomas-Reetz, A., Hell, J. W., During, M. J., Walch-Solimena, C., Jahn, R., and De Camilli, P., A g-aminobutyric acid transporter driven by a proton pump is present in synaptic-like microvesicles of pancreatic β cells, *Proc. Natl. Acad. Sci, U.S.A.,* 90, 5317, 1993.

35. Thomas-Reetz, A. and De Camilli, P., A role for synaptic vesicles in non-neuronal cells: clues from pancreatic β cells and from chromaffin cells, *FASEB J.,* 8, 209, 1994.

36. Rothman, J. E. and Orci, L., Molecular direction of the secretory pathway, *Nature,* 355, 401, 1992.

37. Hodel, A., Schäfer, T., Gerosa, D., and Berger, M., In chromaffin cells, the mammalian sec 1p-homoloque is a syntaxin 1 A-building protein associated with chromaffin granules, *J. Biol. Chem.,* 259, 8623, 1994.

38. Roth, A. and Burgoyne, R., SNAP 25 is present in a SNARE complex in adrenal chromaffin cells, *FEBS Lett.,* 351, 207, 1994.

Section II
General Principles

Chapter 3

ELECTRICAL ACTIVITY AND STIMULUS-SECRETION COUPLING IN NEUROENDOCRINE CELLS*

Werner Schlegel and Patrice Mollard

CONTENTS

* The authors dedicate this chapter to Professor W. W. Douglas.

1. THE STIMULUS-SECRETION COUPLING CONCEPT: A HISTORICAL PERSPECTIVE

Strikingly common features of the control of the muscle contraction with those of endocrine secretion — among them a central role for Ca^{2+} ions — led to the concept of "stimulus-secretion coupling".[1] The control of endocrine secretion was thus conceived as a chain of biochemical mechanisms linking an initial event, which was the "stimulus", to the secretion. The term "stimulus-secretion coupling" was coined after the concept of "excitation-contraction coupling".[2] Thus, inherent to the terminology was the proposal to follow the lines of thoughts of the research in muscle physiology when studying the regulation of secretion. This idea, summarized with its first results by Douglas[1] in his Gaddum lecture in 1968, led Douglas and his colleagues to investigate endocrine cells with the methods of electrophysiology, methods that were thus far reserved for the "aristocratic" neurons. As a result, it was reported that stimulation of secretion was accompanied by changes in the membrane potential, and that endocrine cells were capable of producing fast coordinated alterations in ion fluxes resulting in rapid reversible changes in the membrane potential, namely action potentials.

Excitability of endocrine cells was described for various cell types in a series of early studies in parallel, or following, Douglas' proposals. While action-potential firing of cells of the posterior pituitary[1] (which is innervated) was possibly less surprising and was based on Na^+ influx,[3] the discoveries of Ca^{2+}-dependent action potentials in anterior pituitary cells[4-7] and the action potentials in chromaffin cells, which had an equally important Ca^{2+} component,[8] marked the line of investigations to follow. In the subsequent series of studies on these endocrine cells — now shown to be excitable — the ionic nature of the currents involved in maintaining the resting potential, or of those required for the generation of action potentials, were defined.

Excitability appeared at first as an "exotic" behavior for a simple endocrine cell. However, interest rose considerably in the following years, when it was shown that action potential firing was controlled by a whole range of secretion regulators.

A few pioneering studies paved the way. Very important was the demonstration by Henquin[9] that glucose inhibited K^+ efflux, which provided a basis for the earlier observations,[10-13] showing that glucose stimulation of insulin secretion was accompanied by depolarization of the membrane potential and subsequent firing of action potentials. In the pituitary cells, stimulation by the releasing factor TRH, was shown to have a biphasic effect on K^+ permeability and to correspond to a dual action on the electrical activity.[7,14,15] In the pars intermedia of pituitary, inhibitory modulation of action potentials was first demonstrated by a study of Douglas and Taraskevich,[5] an observation confirmed for most inhibitors of endocrine secretion.

By 1986, the stimulus-secretion-coupling concept, helped by the advent of the patch clamp (tight seal) electrophysiological recording technique invented in 1976 and extensively described in 1981,[16] had produced a substantial body of literature giving rise to extended reviews by Ozawa and Sand[17] and by Mason and Ingram.[18] It is important to note that the early electrophysiological studies on endocrine cells were only possible due to the considerable progress of the *in vitro* maintenance of endocrine cells due to the pioneering work of Tixier-Vidal et al.,[19] Yasumura et al.,[20] Tashjian et al.,[21] and others.

The invention of fluorescent intracellular Ca^{2+} probes suitable for single-cell work by Tsien and colleagues[22] made it possible to demonstrate that action potentials generated repetitive coherent transient alterations in the cytosolic free Ca^{2+} concentration, $[Ca^{2+}]_i$.[23] Furthermore, it was demonstrated then that all the modulatory control of electrical activity in endocrine cells, e.g., by the inhibitor somatostatin,[23] is reflected by changes in frequency and amplitude of action potential-induced $[Ca^{2+}]_i$ transients.

During the same period, single-cell $[Ca^{2+}]_i$ recordings obtained with the Ca^{2+}-sensitive photoprotein aequorin by Cobbold and colleagues[24] had shown that repetitive $[Ca^{2+}]_i$ spikes occur in "nonexcitable" cells, i.e., cells devoid of an excitable plasma membrane. The cooperative and feedback mechanisms inherent to the functional elements of endoplasmic reticulum/sarcoplasmic reticulum (ER/SR) Ca^{2+} stores — mainly the second messenger or Ca^{2+}-gated release channels — allow for a steady stimulus to produce an oscillatory pattern of Ca^{2+} fluxes.[25–28] A whole host of receptors activate phospholipase C, producing a steady output of the second messenger, inositol trisphosphate ($Ins[1,4,5]P_3$), which triggers Ca^{2+} oscillations based on the rhythmic release of Ca^{2+} from ER Ca^{2+} stores. Thus, the cytoplasm of most cells is "excitable" since it contains the elements to produce rapid, autopropagating $[Ca^{2+}]_i$ transients.

Excitable neuroendocrine cells also have an excitable cytoplasm. Although its role may frequently be limited to amplify $[Ca^{2+}]_i$ oscillations paced by action potentials, in some cells, such as the pituitary gonadotroph, $[Ca^{2+}]_i$ oscillations are paced by the mechanisms of Ca^{2+} mobilization from internal stores. Curiously, among the first reports suggesting excitability of endocrine cells is a paper by Poulsen and Williams[29] which shows rhythmic hyperpolarizations of membrane potential in nonidentified pituitary cells. These slow oscillations can now be easily recognized as the consequence of Ca^{2+} spiking due to internal stores and the ensuing activation of Ca^{2+}-stimulated K^+ efflux.[30–32]

The prime function of neuroendocrine cells, namely, controlled secretion of hormones and neurotransmitters, was — and still is — difficult to assess on the level of a single cell. Immunological approaches such as the reverse hemolytic plaque assay[33] were desperately slow in comparison to the electrophysiological recordings. The electrical correlate of exocytosis is an increase in membrane capacitance due to increased plasma membrane surface following the fusion with secretory granules. Measurements of single-cell capacitance were initiated by Neher and Marty[34] in chromaffin cells, but were first largely applied in mast cells.[35] These large cells exocytose the contents of large granules and release a substantial fraction of the granules upon stimulation. Neuroendocrine cells release from small granules and only from a small fraction of them upon stimulation. The adaptation of the capacitance technique to neuroendocrine cells provided important and detailed information on the role of Ca^{2+} and other factors on exocytosis, but successful application of the technique has remained for a long time limited to a small number of cell preparations.[36,37] Now, together with the use of electrochemical sensor microelectrodes,[38] the capacitance measurements help to elucidate the physiological role of electrical activity in the control of neuroendocrine exocytosis. (See Section 5.)

The demonstration of action potential-linked $[Ca^{2+}]_i$ transients in neuroendocrine cells confirmed Douglas' prediction that mechanisms producing rapid and coherent changes in $[Ca^{2+}]_i$ leading to exocytosis are shared by endocrine and excitable cells of the first category (neurons, muscle cells). Thus, the theoretical concept of "stimulus-secretion coupling" had met its reality. The more recent description of the "excitability" of the cytoplasm breaks down the distinction between excitable and nonexcitable cells. In parallel, the patch clamp technique has made electrophysiology fit for any cell type. One can thus safely conclude that the basic idea of Douglas to widen the horizon of electrophysiology has taken science as far as it possibly could.

2. THE EXCITABLE PLASMA MEMBRANE

2.1. ACTION POTENTIAL GENERATION

Action potential generation is presented below. Neuroendocrine cells use a specific set of ion channels during the various phases of the action potential. Since the main function of the

TABLE 1
Properties of Major Voltage-Gated Calcium Channels

	T	N	L	P
Activation range	Positive to -70 mV[a]	Positive to -30 mV[a]	Positive to -30 mV[a,b]	Positive to -50 mV[a,c]
Inactivation rate	Rapid[a]	Moderate[a]	Very slow[a]	Very slow[a,c]
Single-channel conductance (≥ 80 mM Ba^{2+})	8–10 pS[a,b]	11–21 pS[d]	20–27 pS[d]	10–12 pS[c] 18–20 pS[c]
Drug/toxin sensitivity	?	ω-Conotoxin[a]	DHP[a]	(s)FTX[c] ω-Aga-IVA[c]

References: [a] Tsien et al., 1991;[76] [b] Armstrong and Eckert, 1987;[87] [c] Llinas et al., 1992;[79] [d] Tsien et al., 1988.[107]

action potential in neuroendocrine cells is to signal via $[Ca^{2+}]_i$-sensitive mechanisms voltage-gated Ca^{2+} channels play a more predominant role than in nerve cells. Neuroendocrine cells cover a large range of thresholds for the triggering of action potentials. The relative contribution of low- vs. high-threshold Ca^{2+} channels during the action potential has marked consequences for the microscopic $[Ca^{2+}]_i$ changes in the vicinity of the channels.[39,40] However, this influx is limited due to the steady-state inactivation of these channels.

2.1.1. Action Potentials Cause Characteristic Cytosolic Calcium Transients

The plasma membrane of neuroendocrine cells contains *voltage-gated Na^+, Ca^{2+}, and K^+ channels* that mediate all-or-none action potentials associated with rises in the cytosolic free $[Ca^{2+}]$, $[Ca^{2+}]_i$. This is a brief description of the characteristics of those channels and how they operate.

For each ionic conductance ($Na+$, Ca^{2+}, K^+) distinct types of channels have been described, distinguished by their biophysical properties (such as, ion selectivity, single channel conductance, voltage dependence, and kinetics of activation and inactivation) by their molecular biology, and by their pharmacology.[72–74] *It is remarkable that in most — if not all — excitable neuroendocrine cells, one or several types of Ca^{2+} channels open during action potentials.* Four major types of Ca^{2+} channels have been identified (see Table 1). The T (tiny, or transient) type is activated and inactivated at low voltage. No specific blocker of T type channels has yet been identified. The L (long-lasting) type, N (neither L- nor T-channel) type, and P (originally found in cerebellar Purkinje cells) type all are activated at high membrane potential and inactivate slowly.[75,76] L type channels are blocked by dihydropyridines (nifedipine, nicardipine, isradipine, DHP) and also by phenylalkylamines (verapamil) and benzothiazepines (diltiazem). N type and P type channels are blocked by ω-conotoxin GVIA and the funnel web spider toxins ([s]FTX, ω-AGA-IVA), respectively.[77–79]

Voltage-gated channels open in a concerted manner to cause action potentials. They respond to changes in membrane potential beyond a threshold level, with the probability of channel opening increasing with depolarization. Na^+ channels and/or T type Ca^{2+} channels open first. Na^+ and/or Ca^{2+} flow into the cell along the large electrochemical gradient for these ions. The net cationic influx further depolarizes the plasma membrane, and so more channels are opened, resulting in a rapid depolarization of the membrane and an activation of high-voltage threshold Ca^{2+} channels. Whereas a built-in inactivation process leads Na^+ and T type Ca^{2+} channels to close quickly, L type, N type, and P type channels remain open long enough to cause **a large transient increase in $[Ca^{2+}]_i$ with characteristic kinetic features: a time to peak of less than 1 s and a slower decay phase ($t_{1/2} = 2$ to 4 s).**[23] The return of the potential to the resting level and below (afterhyperpolarization) is accelerated by the rise in $[Ca^{2+}]_i$, which the activities of K^+ channels, such as the BK (*Big* single-channel conductance) and SK

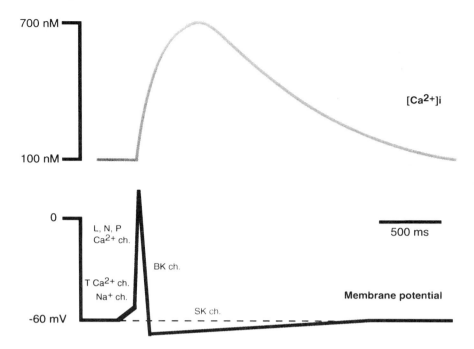

FIGURE 1. One action potential, ionic channels involved, ensuing $[Ca^{2+}]_i$ change.

(Small single-channel conductance) channels; K^+ efflux through the latter mainly causes the repolarization and the lasting afterhyperpolarization, respectively.[80] Voltage- and Ca^{2+}-dependent K^+ channels open more slowly than Na^+ and Ca^{2+} channels. This permits the Na^+/Ca^{2+} entry — causing depolarization — to precede the K^+ efflux, causing repolarization. $[Ca^{2+}]_i$ continues to rise after influx is stopped, because of the slow propagation of Ca^{2+} though the cytosol,[81] which may involve amplification of the signal by Ca^{2+}-induced Ca^{2+} release.[46,60] Activation of Ca^{2+}-exporting ATPases[82] at the plasma membrane by the rise in $[Ca^{2+}]_i$, Ca^{2+} sequestering into intracellular compartments,[46] and possibly Na^+/Ca^{2+} exchange[83] initiates the restoration of prespike $[Ca^{2+}]_i$ levels within seconds (see Figure 1).

2.2. ACTION POTENTIAL MODULATION

Action potential modulation is presented below. Neuroendocrine cells have an extensive arsenal of regulatory mechanisms ready to modulate action potentials. One would thus assume *a priori* that this modulation is of physiological importance. Why does it remain difficult to prove? Neuroendocrine electrical activity is studied most often *in vitro* in dispersed cell preparations. Out of their tissue contact, neuroendocrine cells are no longer under the tonic control by their neighbor cells, which normally may secrete local hormones, produce adenosine, and influence the ionic composition of the fluid environment. Under these circumstances important constraints are relieved, and "spontaneous" electrical activity is observed, which is hardly physiologically meaningful. It is a major challenge for future research on action potential modulation to produce the experimental models that will allow separation of the physiologically relevant from the irrelevant mechanisms. Work on tissue slices and other pluricellular preparations is a promising beginning.

2.2.1. Electrical Activity: A Transduction Mechanism

Neuroendocrine cells use their excitability to respond to physiological stimuli.

Cells exhibit a large variety of action potential patterns in response to secretagogues. The primary consequence of related changes of electrical activity is a rise in $[Ca^{2+}]_i$. The patterns

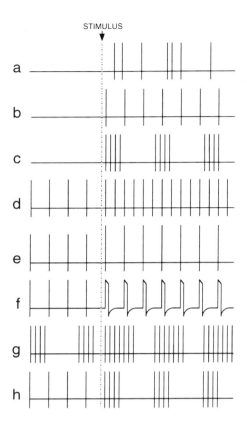

FIGURE 2.   Patterns of electrical responses in neuroendocrine cells.

of electrical responses may be grouped with regard to the degree of excitability of a cell prior to stimulation (see Figure 2). In silent cells, action potentials triggered upon agonist application can appear randomly (random mode, *a*), as single spikes firing at a near constant frequency (pace-making mode, *b*), or as bursts of spikes separated with silent intervals (bursting mode, *c*). In cells which are active prior to stimulation, the stimulatory modulation of electrical activity can be observed as an increase in spike frequency (*d*), an increase in amplitude (*e*) or duration of spikes (*f*), a decrease in silent intervals between bursts or spikes (*g*), and a shift from one mode to another (e.g., pace-making to bursting, *h*). Note that some cell types generate one or two patterns (e.g., B cells of the endocrine pancreas respond to glucose with the patterns *c* and *g*), whereas others (e.g., TRH-stimulated lactotrophs) employ most of them.

Inhibitors of secretion tend to decrease or abolish the voltage-gated Ca^{2+} entry by reducing the amplitude and duration of spikes, or by enhancing the interval between spikes (or bursts).

2.2.2.   Modulation of Ionic Channels Underlying the Changes in Electrical Activity

Neuroendocrine cells engage a large repertory of mechanisms to modulate their excitability. Not only voltage-gated Ca^{2+} channels opened during action potentials, but also channels for K^+, Na^+, Cl^-, and Ca^{2+} that dominate the membrane potential, are targets for regulation, as follows:

1. Ligand-gated channels for Na^+, Ca^{2+}, and Cl^-, which are an integral part of neurotransmitter (e.g., acetylcholine, GABA), or purinergic receptors are opened, causing depolarization (for review see Unwin).[84] In chromaffin cells, activation by acetylcholine of nACh receptors causes a depolarization that brings the membrane potential to the threshold level, triggering action potentials.[8] However, the consequent Ca^{2+} influx is limited because of the acetylcholine-evoked reduction of Ca^{2+} channel activities.[85]

2. G protein-linked receptors cause the modulation of ionic channel activities via different routes. There is evidence for direct G protein gating of ionic channels and G protein α- and $\beta\gamma$-subunits interacting directly with K^+ and Ca^{2+} channels (reviewed by Clapham).[86] G protein-controlling effector enzymes, such as adenyl cyclase and phospholipases, mediate changes in second messengers (cAMP, cGMP, diacylglycerol, eicosanoids), which in turn control ion channel activity via phosphorylation/dephosphorylation cascades.[87–89]

3. Metabolites are potent channel modulators, e.g., in glucose-stimulated β cells of the endocrine pancreas, in which they cause the closure of ATP-sensitive K^+ channels and thereby set the pace of the bursting electrical activity. It is yet elusive whether or not nitric oxide (NO), a free-radical gas formed from arginine by the NO-synthase and involved in neuronal synaptic function,[90] acts on channel activities to control secretion from neuroendocrine cells.[91,92]

2.3. MODULATION OF STEADY-STATE Ca^{2+} INFLUX AT RESTING POTENTIALS

Controlled Ca^{2+} influx in neuroendocrine cells is not limited to action potentials. It can be shown in pituitary cells, e.g., that there is a steady-state Ca^{2+} influx possible at resting potential.[41] This influx is through voltage-gated, dihydropyridine-sensitive Ca^{2+} channels which have yet to be further characterized and has a substantial impact on steady-state $[Ca^{2+}]_i$.[42] Steady-state Ca^{2+} influx is certainly inhibited by inhibitors of pituitary hormone release (somatostatin, dopamine, or adenosine) via their polarizing effects on the resting potential. No evidence has so far been presented for direct modulation of the channels involved.

Steady-state Ca^{2+} influx at resting potential is through low-threshold channels and thus produces locally high amplitude $[Ca^{2+}]$ rises, which may play a specific signaling role.

An important part of regulated Ca^{2+} influx occurring independently of membrane potential is due to a pathway stimulated by the release of Ca^{2+} from internal stores.[43] For corresponding currents the term "I_{crac}" (for Ca^{2+} *release activated current*) is used.[44] The maximal whole-cell current for I_{crac} is usually very small[44] and the molecular equivalent is currently being searched for. Although this current is likely operating in neuroendocrine cells, in view of its anticipated small amplitude I_{crac} is difficult to demonstrate in cells with a normal set of voltage-gated Ca^{2+} channels.

3. THE "EXCITABLE" CYTOPLASM

So far, the term "excitable" has been assigned to cells with the potential to fire action potentials, i.e., rapid and massive, self-promoted, and reversible changes in membrane potential based on voltage-sensitive plasma membrane ion channels with particular characteristics. This excitability thus manifests itself in the electrical properties of a cell. More recently, excitability in the sense of "chemical excitability" has been introduced to cellular biochemistry to highlight the potential of the cytosol to produce a rapid Ca^{2+} spike initiated by Ca^{2+} release from internal stores.[45]

The activation of intracellular processes triggered via $[Ca^{2+}]_i$ elevations frequently requires fairly elevated Ca^{2+} concentrations. To achieve activation of Ca^{2+}-dependent proteins by raising steady-state $[Ca^{2+}]_i$ would be very wasteful, since, in view of the high buffering capacity of the cytoplasm, the Ca^{2+} fluxes required would be very important. Thus it is generally assumed that Ca^{2+} activation of intracellular processes takes place in a limited space and a limited time. To achieve this kind of activation, most cells are equipped to produce $[Ca^{2+}]_i$ spikes.

Ca^{2+} fluxes with rapid onset initiate these spikes. These may be originating at the plasma membrane during the action potential, or at the membrane of the ER. In these two situations, [Ca^{2+}]$_i$ signaling features are quite distinct. In neuroendocrine cells, both possibilities exist, but it appears that cell differentiation leads to an organization that favors either one of the two options. Pituitary gonadotrophs and rat chromaffin cells base their [Ca^{2+}]$_i$ spiking on an excitable cytoplasm; lactotrophs, pancreatic β cells, bovine chromaffin cells, and others pace their spiking via the control of action potentials.

3.1. Ca^{2+} RELEASE FROM SUBCOMPARTMENTS OF THE ENDOPLASMIC RETICULUM

The essential properties of the ER Ca^{2+} stores from which Ca^{2+} is released during cell activation have been recently reviewed.[46] Although functionally and morphologically quite distinct in different cell types, these stores all comprise three common elements: namely, a Ca^{2+} ATPase of the SERCA type; low-affinity, high-capacity Ca^{2+} binding proteins of the calsequestrin or the calreticulin family; and a ligand-gated Ca^{2+} release channel. The latter are of two types: namely, the Ins(1,4,5)P$_3$ receptors[47] and the ryanodine receptors,[48] and share the fundamental molecular architecture.[49] The cooperativity of the release channels and feedback of Ca^{2+} are the basis for generating spikes.[26–28]

3.1.1. Ca^{2+} Release Channels
3.1.1.1. Ins(1,4,5)P$_3$ Receptors

Ins(1,4,5)P$_3$ action to release Ca^{2+} from the ER was first demonstrated in permeabilized pancreatic acinar cells.[50] Ins(1,4,5)P$_3$ receptors that correspond to the release channels were later characterized pharmacologically, and finally purified[51] and cloned[47] from brain and other tissues.[46,49]

The Ins(1,4,5)P$_3$ action on its receptors is cooperative[52,53] and is furthermore likely to be modulated by phosphorylation.[54–56] Incorporated into planar lipid bilayers, Ins(1,4,5)P$_3$ receptor channels are regulated by Ca^{2+} in a bimodal fashion; at lower levels, Ca^{2+} favors opening; at higher levels it is inhibitory.[57,58] This demonstrates that Ca^{2+} released from the store could have both a positive and a negative feedback on the Ins(1,4,5)P$_3$ release channel.[57–59]

3.1.1.2. Ca^{2+}-Induced Ca^{2+} Release (CICR)

A release channel of the ER and in particular of the SR which would be gated by Ca^{2+} itself has been postulated early on.[60] The description of specific inhibitors (ryanodine) as well as modulators of CICR (caffeine) has finally led to the isolation and cloning of an integral membrane protein which likely functions as the corresponding ion channel (reviewed in Pozzan et al.).[46]

CICR is certainly an important feature in excitable endocrine cells. It has been shown, e.g., in rat chromaffin cells that ryanodine- and caffeine-sensitive Ca^{2+} stores are responsible for the generation of Ca^{2+} spikes.[61]

3.2. Ca^{2+} OSCILLATIONS BASED ON INTRACELLULAR Ca^{2+} MOBILIZATION

Oscillatory pattern may be generated by any process that is cooperative and may be controlled by positive or negative feedback. The various models proposed for [Ca^{2+}]$_i$ oscillations produced by the excitable cytoplasm have been reviewed extensively (e.g., References 26–28).

The simpler models explain [Ca^{2+}]$_i$ oscillations at constant — or constantly progressing — Ins(1,4,5)P3 levels. These models imply cooperativity and Ca^{2+} dependency of Ins(1,4,5)P$_3$ action. Oscillations are initiated by a rise in Ins(1,4,5)P$_3$, which beyond a certain threshold causes full activation of the Ins(1,4,5)P$_3$ release channel (cooperativity), which is amplified and spread by the positive feedback from the initial local rise in [Ca^{2+}] on the neighboring Ins(1,4,5)P$_3$ release channels, or by the mechanisms of CICR. Oscillations are terminated — in the presence of still elevated Ins(1,4,5)P$_3$ — due to exhaustion of the Ca^{2+} stores and the negative feedback from further elevated [Ca^{2+}]$_i$ on Ins(1,4,5)P3 receptor channels.

More complicated models evoke the possibility that Ins(1,4,5)P$_3$ may be oscillating as a result of Ca^{2+} interference with phospholipase C, and inositol 3-kinase or inositol phosphate phosphatase, the enzymes that generate and eliminate the second messenger Ins(1,4,5)P$_3$ by phosphorylation or dephosphorylation, respectively. This "cross coupling"[27] would lead to alternative oscillations of [Ca^{2+}]$_i$ and Ins(1,4,5)P$_3$.

The experimental evidence suggests that the simpler models cover many receptor-induced oscillatory [Ca^{2+}]$_i$ changes, since they can be reproduced with constant application of Ins(1,4,5)P$_3$ (e.g., Reference 62). Sophisticated use of fluorescent Ca^{2+} probes, especially in large cells, has furthermore produced striking demonstrations that the spatiotemporal aspects of [Ca^{2+}]$_i$ signals are governed by self-propagating Ca^{2+} waves across the excitable cytoplasm.[45,63]

3.3. SIGNALING VIA [Ca^{2+}]$_i$ CHANGES IN PITUITARY GONADOTROPHS

Gonodotrophs are equipped with voltage-gated Ca^{2+} channels and have the potential to fire action potentials. They are thus bona fide excitable cells, although their excitability may be reduced in certain circumstances.[64]

[Ca^{2+}]$_i$ oscillations in gonadotrophs stimulated with GnRH, endothelin, or PACAP are due to Ins(1,4,5)P$_3$-induced mobilization from internal stores.[31,32,62] The oscillations are accompanied by hyperpolarizations due to Ca^{2+} activation of K$^+$ efflux. Upon the recovery from the hyperpolarization, the membrane potential frequently passes the threshold for action potentials, resulting in a few spikes which have no apparent impact on [Ca^{2+}]$_i$.[30]

The present view of gonadotroph [Ca^{2+}]$_i$ control and its consequence on the membrane potential explain well the early finding by Poulsen and Williams[29] with the description of spontaneous repetitive hyperpolarizations recorded in cells of the anterior pituitary from ovariectomized rats, a model favoring the abundance of gonadotrophs.

More recently, it has been shown that [Ca^{2+}]$_i$ oscillations in gonadotrophs are relevant for hormone release, in that each spike elicits a burst of exocytotic activity.[65] However, in spite of the dominating role of intracellular Ca^{2+} mobilization for the oscillatory [Ca^{2+}]$_i$ signals in gonadotrophs, the participation of voltage-gated Ca^{2+} channels in stimulating secretion has been clearly manifest.[66]

4. MODES OF [CA^{2+}]$_i$ SIGNALING IN NEUROENDOCRINE CELLS

4.1. ACTION POTENTIAL-PACED [Ca^{2+}]$_i$ TRANSIENTS

In this mode [Ca^{2+}]$_i$ transients are initiated by Ca^{2+} influx during the action potential. The excitable cytoplasm propagates and amplifies the signal. Regulatory input is on the membrane potential and on the voltage-gated Ca^{2+} channels. Steady-state influx between action potentials may regenerate stores and keep the cytoplasm "excitable".

4.2. [Ca^{2+}]$_i$ OSCILLATIONS INITIATED BY INTRACELLULAR Ca^{2+} MOBILIZATION

Receptor stimulation of phospholipase C raises Ins(1,4,5)P$_3$ and initiates Ca^{2+} oscillations as described above. Voltage-gated Ca^{2+} channels open in between transients to refill the intracellular stores. Regulatory input is on Ins(1,4,5)P$_3$, on Ca^{2+} and ryanodine receptors (phosphorylation), and on Ca^{2+} influx replenishing intracellular Ca^{2+} stores.

4.3. [Ca²⁺]ᵢ CONTROLLED BY STEADY-STATE CA²⁺ INFLUX

In the absence of, or in between action potentials, $[Ca^{2+}]_i$ can vary considerably due to changes in steady-state Ca^{2+} influx. As this Ca^{2+} influx is voltage gated, slowly maintained potential changes, as they occur in some cells prior to bursts of action potentials, may translate into $[Ca^{2+}]_i$ changes. Note that steady-state Ca^{2+} influx may also serve to refill the intracellular Ca^{2+} stores and its control, thereby amplifying or attenuating subsequent $[Ca^{2+}]_i$ signals generated in the above modes.

5. Ca²⁺-DEPENDENT ENDOCRINE CELL FUNCTIONS

5.1. EXOCYTOSIS

The relation between action potentials and exocytosis is addressed below. Ca^{2+} control of exocytosis illustrates general features of Ca^{2+}-controlled mechanisms, namely the multiplicity of Ca^{2+} control sites and their diversification. It is anticipated — but yet to be established in detail — that the various steps involved in exocytosis are sensitive to distinct aspects of the $[Ca^{2+}]_i$ signal: amplitude, kinetics, multiple spikes, localization.

There is some evidence for the existence of a Na^+-controlled system which would function similarly to, but independently of, the Ca^{2+}-controlled exocytosis.[67] The control of intracellular Na^+ levels and of Na^+ fluxes might be worth more attention than was devoted to this question in the past.

5.1.1. The Action Potential: A Trigger for Exocytosis

Exocytosis corresponds to the formation of fusion pores between membrane-bound vesicles and the plasma membrane and the subsequent release of the vesicular contents into the extracellular space. Secretory products (peptides, catecholamines, ATP, etc.) are primarily packaged in large (diameter >70 nm) dense-core vesicles (LDCV) in neuroendocrine cells, which undergo two pathways of exocytosis — constitutive and regulated.

In constitutive exocytosis, the vesicle membrane fuses with the plasma membrane as soon as the vesicle migrates from the *trans*-Golgi network and approaches the cell surface, so that the molecules stored within these vesicles are readily released into the extracellular milieu. The factors which control the fusion of vesicle and plasma membranes are yet unknown.

In regulated exocytosis, vesicles are waiting beneath the inner face of the plasma membrane until a cytosolic messenger triggers the fusion event (see Figure 3). *In many neuroendocrine cells, where the two exocytotic pathways coexist, the rise in cytosolic Ca^{2+} due to voltage-gated Ca^{2+} influx through the plasma membrane is sufficient to trigger exocytosis.* Recent measurements of secretion at the single-cell level have demonstrated that action potentials (or simulated spikes) effectively induce release.[93,94]

After the action potential occurs, how long does it take before exocytosis begins? Ca^{2+} influx through voltage-gated channels can initially cause close to the channels quite elevated $[Ca^{2+}]$ (several micromoles),[40,95] which later spread throughout the cytosol and are attenuated.[81] Although the calcium dependences of exocytotic processes resemble those of fast synaptic transmission,[37,40,96] the delay between "excitation" and secretion in neuroendocrine cells is much longer (about 200 µs for fast synapses vs. 3–50 ms for endocrine cells).[37,38,97] The difference between these releasing processes most likely relates to the different organization of the exocytotic "machinery", as described below.

How far are secretory vesicles from their site of fusion at the plasma membrane and the relevant Ca^{2+} channels? Neurons undergo exocytosis from small clear vesicles which are associated ("docked" vesicles) with specialized regions of the plasma membrane, the "active zones".[98] Synaptic vesicles are tied up close to Ca^{2+} channels, so that the transmitter release sites overlap with the $[Ca^{2+}]$ bulks generated around opening Ca^{2+} channels.[39,95,99] In neuroendocrine cells, vesicles of the readily releasable ("release-ready") pool (1–20% of all LCDV)[100]

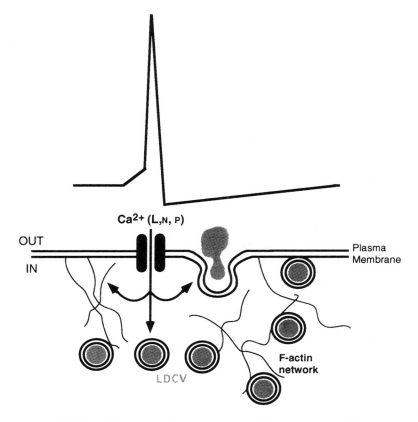

FIGURE 3. Voltage-gated Ca^{2+} entry commands the exocytotic machinery.

are located close to the plasma membrane[5,96,101] (<1 vesicle radius), but are not obviously further organized. Although "release-ready" LCDV in neuroendocrine cells have no proven selective association with Ca^{2+} channels, there is clear evidence that not all channel types are equally effective in causing exocytosis. In chromaffin cells, L-type Ca^{2+} channels are more efficient than P-type or N-type channels to trigger catecholamine release.[102] The majority of LCDV are further away from the plasma membrane, and serve to refill/overfill the "release-ready pool". Disruption of the submembranous cortical actin network which can be triggered by a rise in $[Ca^{2+}]_i$ seems to be a prerequisite for the migration of the "reserve" LCDV towards the release sites.[103]

The search for the elements which confer Ca^{2+} sensitivity to exocytosis has been on for decades and still continues. The complexity of the problem relates to the fact that vesicle release requires occupancy of more than a single site for Ca^{2+},[104] and likely depends on multiple Ca^{2+}-dependent mechanisms.[96,100] Several known Ca^{2+}-binding proteins have been proposed to be involved, among them calmodulin, annexin II, synaptotagmin, and others (for review see References 101, 105, 106). However, the identification of the specific Ca^{2+}-binding molecule(s) which control a given Ca^{2+}-dependent step is still tantalizing.

5.2. OTHER Ca^{2+}-DEPENDENT FUNCTIONS
5.2.1. Gene Transcription

$[Ca^{2+}]_i$ signals have been shown to enhance the expression of immediate early genes such as c-*fos* in neuroendocrine cells. The pathways involved include the Ca^{2+}/calmodulin-activated protein kinase (CAM-kinase), and protein kinase C (PKC).[68] The former enzyme acts on the same target as protein kinase A (PKA), namely, the cAMP response element binding protein.[69] The pathways by which protein kinase(s) C (PKC) activate(s) gene transcription, and in particular whether PKC and CAM-kinase translocated to the nucleus upon activation — as does PKA — is intensively investigated at present. What has already been shown is that the type of Ca^{2+} channel involved in generating the signal may determine the pathway of transcriptional activation.[70]

Programming of gene transcription has its consequences on cell proliferation. It is of interest to note that, e.g., in pituitary cells, electrical activity does not occur in all phases of the cell cycle, notably not in the S-phase, as if the signals generated by action potentials could perturb cell cycle progression.

5.2.2. Protein Synthesis

The site of protein synthesis — rough ER — coincides with a region which may be heavily involved in $[Ca^{2+}]_i$ signaling. Ca^{2+} dependence of protein synthesis has been demonstrated in pituitary tumor and other mammalian cells.[71] However, it has not been demonstrated that the accelerated hormone synthesis which accompanies stimulation of neuroendocrine secretion is the consequence of Ca^{2+}-stimulated translation.

The multiplicity of functions controlled by $[Ca^{2+}]_i$ signals is astonishing. How could a single variable exert precise control over many central neuroendocrine cell functions? $[Ca^{2+}]_i$ signals always occur in a context of other signaling pathways. Thus it is the combined action of Ca^{2+} with various degrees of activation of other signaling cascades that is the key to the diversification, resulting in clear intracellular "messages".

6. CONCLUSIONS AND OUTLOOK

The stimulus-secretion coupling concept has led to the discovery and description of action potentials and their modulation in neuroendocrine and endocrine cells. It was shown that electrical activity produces coherent $[Ca^{2+}]_i$ signals: concerted opening of voltage-gated Ca^{2+} channels during the firing of a single action potential produces a transient elevation of $[Ca^{2+}]_i$ with characteristic kinetics and localization. Control of electrical activity is an important signal-transduction mechanism utilized by cell-surface receptors for hormones and neurotransmitters and engaged by metabolic activators of endocrine cells. However, the regulatory potential in neuroendocrine cells is not limited to the control of "noisy" electrical events, and marked $[Ca^{2+}]_i$ transients may be generated, accompanied by slow and small changes in membrane potential. Even oscillations of $[Ca^{2+}]_i$ that are not dissimilar to those associated with action potentials are seen in pituitary cells without activation of voltage-gated Ca^{2+} channels, reflecting the "excitability" of the cytoplasm. Furthermore, Ca^{2+} influx through voltage-gated channels is not limited to the action potential, but occurs also at resting potentials.

Ca^{2+} influx may serve two functions, namely either to trigger the $[Ca^{2+}]_i$ transients that are the intracellular signals, or to sustain the intracellular stores from which Ca^{2+} is mobilized for the signaling.

Issued from the same ectodermic parent cells, neurons and endocrine cells share the same inventory of molecular entities (receptors, G proteins, voltage- and ligand-gated ion channels, ion pumps) and the same fundamental organization of Ca^{2+} stores. These elements may be used differentially in the generation and reversal of $[Ca^{2+}]_i$ transients. It is thus proposed that excitability is not a permanent label characteristic of a given cell type. It rather appears as a mode in which regulation may occur. With these various modes, neuroendocrine cells are capable of producing a large repertory of $[Ca^{2+}]_i$ signals.

The challenge for further research on "stimulus-secretion coupling" in neuroendocrine cells will be to link a precise $[Ca^{2+}]_i$ signal pattern to a defined element of the secretory pathways. The aim now is to elucidate the physiological role of electrical activity, and relevant progress will come from studies that take on the challenge to study this role in a situation as close to *in vivo* as possible. Given the tissular and humoral context of neuroendocrine cells functioning in our bodies, the task of maintaining this context while reading the fine print of intracellular signals is not a small one. Indeed, to work out the coupling of stimuli to secretion relevant to the *in vivo* situation might keep a further generation of physiologists busy and happy.

REFERENCES

1. **Douglas, W. W.,** Stimulus-secretion coupling: the concept and clues from chromaffin and other cells, *Br. J. Pharmacol.,* 34, 451, 1968.
2. **Sandow, A.,** Excitation-contraction coupling in muscular response, *Yale J. Biol. Med.,* 25, 176, 1952.
3. **Dreifuss, J. J., Kalnins, I., Kelly, J. S., and Ruf, K. B.,** Action potentials and release of neurohypophysial hormones in vitro, *J. Physiol.,* 215, 805, 1971.
4. **Kidokoro, Y.,** Spontaneous Ca^{2+} action potentials in a clonal pituitary cell line and their relationship to prolactin secretion, *Nature,* 258, 741, 1975.
5. **Douglas, W. W. and Taraskevich, P. S.,** Action potentials in gland cells of rat pituitary pars intermedia: inhibition by dopamine, an inhibitor of MSH secretion, *J. Physiol.,* 285, 171, 1978.
6. **Dufy, B., Vincent, J.-D., Fleury, H., Du Pasquier, P., Gourdji, D., and Tixier-Vidal, A.,** Dopamine inhibition of action potentials in a prolactin secreting cell line is modulated by oestrogen, *Nature,* 282, 855, 1979.
7. **Dufy, B., Vincent, J.-D., Fleury, H., Du Pasquier, P., Gourdji, D., and Tixier-Vidal, A.,** Membrane effects of thyrotropin-releasing hormone and estrogen shown by intracellular recording from pituitary cells, *Science,* 204, 509, 1979.
8. **Brandt, B. L., Hagiwara, S., Kidokoro, Y., and Miyazaki, S.,** Action of potentials in the rat chromaffin cell and effects of acetylcholine, *J. Physiol.,* 263, 417, 1976.
9. **Henquin, J. C.,** D-Glucose inhibits potassium efflux from pancreatic islet cells, *Nature,* 271, 271, 1978.
10. **Dean, P. M. and Matthews, E. K.,** Electrical activity in pancreatic islet cells, *Nature,* 389, 390, 1968.
11. **Dean, P. M. and Matthews, E. K.,** Glucose-induced electrical activity in pancreatic islet cells, *J. Physiol.,* 210, 255, 1970.
12. **Meissner, H. P. and Schmelz, H.,** Membrane potential of beta-cells in pancreatic islets, *Pfluegers Arch.,* 351, 195, 1974.
13. **Meissner, H. P.,** Electrical characteristic of the beta-cells in pancreatic islets, *J. Physiol., Paris,* 72, 757, 1976.
14. **Ozawa, S.,** Biphasic effect of thyrotropin-releasing hormone on membrane K$^+$ permeability in rat clonal pituitary cells, *Brain Res.,* 209, 240, 1981.
15. **Dubinsky, J. M. and Oxford, G. S.,** Dual modulation of K$^+$ channels by thyrotropin-releasing hormone in clonal pituitary cells, *Proc. Natl. Acad. Sci. U.S.A.,* 82, 4282, 1985.
16. **Hamill, O. P., Marty, A., Neher, E., Sakmann, B., and Sigworth, F. J.,** Improved patch-clamp techniques for high-resolution current recording from cells and cell-free membrane patches, *Pfleugers. Arch. Eur. J. Physiol.,* 311, 538, 1981.
17. **Ozawa, S. and Sand, O.,** Electrophysiology of excitable endocrine cells, *Physiol. Rev.,* 66, 887, 1986.
18. **Mason, W. T. and Ingram, C. D.,** Techniques for studying the role of electrical activity in control of secretion by normal anterior pituitary cells, *Methods Enzymol.,* 124, 207, 1982.
19. **Tixier-Vidal, A., Gourdji, D., and Tougard, C.,** Cell-culture approach to study of anterior pituitary cells, *Int. Rev. Cytol.,* 41, 173, 1975.
20. **Yasumura, Y., Tashjian, A. H., Jr., and Sato, G. H.,** Establishment of four functional clonal strains of animal cells in culture, *Science,* 154, 1186, 1966.
21. **Tashijian, A. H., Jr., Yasamura, Y., Levine, C., Sato, G. H., and Parker, M. L.,** Establishment of clonal strains of rat pituitary tumour cells that secrete growth hormone, *Endocrinology,* 82, 342, 1968.
22. **Tsien, R. Y., Rink, T. J., and Poenie, M.,** Measurement of cytosolic free Ca^{2+} in individual small cells using fluorescence microscopy with dual excitation wavelengths, *Cell Calcium,* 6, 145, 1985.
23. **Schlegel, W., Winiger, B. P., Mollard, P., Vacher, P., Wuarin, F., Zahnd, G. R., Wollheim, C. B., and Dufy, B.,** Oscillations of cytosolic Ca^{2+} in pituitary cells due to action potentials, *Nature,* 329, 719, 1987.
24. **Woods, N. M., Cuthbertson, K. S. R., and Cobbold, P. H.,** Repetitive transient rises in cytoplasmic free calcium in hormone-stimulated hepatocytes, *Nature,* 319, 600, 1986.
25. **Berridge, M. J.,** Calcium oscillations, *J. Biol. Chem.,* 265, 9583, 1990.
26. **Berridge, M. J.,** Inositol trisphosphate and calcium signalling, *Nature,* 361, 315, 1993.

27. **Meyer, T. and Stryer, L.,** Calcium spiking, *Annu. Rev. Biophys. Chem.,* 20, 153, 1991.
28. **Tsien, R. W. and Tsien, R. Y.,** Calcium channels, stores and oscillations, *Annu. Rev. Cell. Biol.,* 6, 715, 1990.
29. **Poulsen, J. H. and Williams, J. A.,** Spontaneous repetitive hyperpolarisations from cells in the rat adeno-hypophysis, *Nature,* 263, 156–158, 1976.
30. **Tse, A. and Hille, B.,** GnRH-induced Ca^{2+} oscillations and rhythmic hyperpolarizations of pituitary gonadotropes, *Science,* 255, 462, 1992.
31. **Stojilkovic, S. S., Merelli, F., Iida, T., Krsmanovic, L. Z., and Catt, K. J.,** Endothelin stimulation of cytosolic calcium and gonadotrophin secretion in anterior pituitary cells, *Science,* 248, 1663, 1992.
32. **Guérineau, N. C., Bouali-Benazzouz, R., Corcuff, J.-B., Audy, M.-C., Bonnin, M., and Mollard, P.,** Transient but not oscillating component of the calcium mobilizing response to gonadotropin-releasing hormone depends on calcium influx in pituitary gonadotrophs, *Cell Calcium,* 13, 521, 1992.
33. **Frawley, L. S. and Neill, J. D.,** A reverse hemolytic plaque assay for microscopic visualization of growth hormone release from individual cells: evidence for somatotrope heterogeneity, *Neuroendocrinology,* 39, 484, 1985.
34. **Neher, E. and Marty, A.,** Discrete changes of cell membrane capacitance observed under conditions of enhanced secretion in bovine adrenal, *Proc. Natl. Acad. Sci. U.S.A.,* 79, 6712, 1982.
35. **Neher, E.,** The influence of intracellular calcium concentration on degranulation of dialyzed mast cells from rat peritoneum, *J. Physiol.,* 395, 193, 1988.
36. **Zorec, R., Sikdar, S. K., and Mason, W. T.,** Increased cytosolic calcium stimulates exocytosis in bovine lactotrophs, *J. Gen. Physiol.,* 97, 4763, 1991.
37. **Thomas, P., Wong, J. G., and Almers, W.,** Millisecond studies of secretion in single rat pituitary cells stimulated by flash photolysis of caged Ca^{2+}, *EMBO J.,* 12, 303, 1993.
38. **Chow, R. H., von Rüden, L., and Neher, E.,** Delay in vesicle fusion revealed by electrochemical monitoring of single secretory events in adrenal chromaffin cells, *Nature,* 356, 60, 1992.
39. **Simon, S. M. and Llinás, R. R.,** Compartmentalization of the submembrane calcium activity during calcium influx and its significance in transmitter release, *Biophys. J.,* 48, 485, 1985.
40. **Llinás, R., Sugimori, M., and Silver, R. B.,** Microdomains of high calcium concentration in a presynaptic terminal, *Science,* 256, 677, 1992.
41. **Scherübl, H. and Hescheler, J.,** Steady-state currents through voltage-dependent, dihydropyridine-sensitive Ca^{2+} channels in GH3 pituitary cells, *Proc. R. Soc. London Ser. B,* 245, 127, 1991.
42. **Mollard, P., Theler, J.-M., Guérineau, N., Vachner, P., Chiavaroli, C., and Schlegel, W.,** Cytosolic Ca^{2+} of excitable pituitary cells at resting potentials is controlled by steady-state Ca^{2+} currents sensitive to dihydropyridines, *J. Biol. Chem.,* 269, 25158, 1994.
43. **Putney, J. W.,** A model for receptor-regulated calcium entry, *Cell Calcium,* 7, 1, 1989.
44. **Hoth, M. and Penner, R.,** Depletion of intracellular calcium stores activates a calcium current in mast cells, nature, 355, 353, 1992.
45. **Lechleitner, J. D. and Clapham, D. E.,** Molecular mechanisms of intracellular calcium excitability in X. leavis oocytes, *Cell,* 69, 283, 1992.
46. **Pozzan, T., Rizzuto, R., Volpe, P., and Meldolesi, J.,** Molecular and cellular physiology of intracellular Ca^{2+} stores, *Physiol. Rev.,* 74, 595, 1994.
47. **Furuichi, T. S., Yoshikawa, S. A., Miyawaki, K., Wada, N., Maeda, and Mikoshiba, K.,** Primary structure and functional expression of the inositol 1,4,5-trisphosphate-binding protein P400, *Nature,* 342, 32, 1989.
48. **Takeshima, H., Hishimura, S., Matsumoto, T., Ishida, H., Kangawa, K., Minamino, N., Matsuo, H., Ueda, M., Hanaoka, M., Hirose, T., and Numa, S.,** Primary structure and expression from complementary DNA of skeletal muscle ryanodine receptor, *Nature,* 339, 439, 1989.
49. **Mignery, G. A., Südhof, T. C., Takei, K., and De Camilli, P.,** Putative receptor for inositol 1,4,5-trisphosphate similar to ryanodine receptor, *Nature,* 342, 192, 1989.
50. **Streb, H., Irvine, R. F., Berridge, M. J., and Schulz, I.,** Release of Ca^{2+} from a nonmitochondrial intracellular store in pancreatic acinar cells by inositol-1,4,5-trisphosphate, *Nature,* 306, 67, 1983.
51. **Supattapone, S., Worley, P. F., Baraban, J. M., and Snyder, S. H.,** Solubilization, purification, and characterization of an inositol trisphosphate receptor, *J. Biol. Chem.,* 263, 1530, 1988.
52. **Meyer, T., Holowka, D., and Stryer, L.,** Highly cooperative opening of calcium channels by inositol 1,4,5-trisphosphate, *Science,* 240, 653, 1988.
53. **Parker, I.,** A threshold level of inositol trisphophate is required to trigger intracellular calcium release in Xenopus oocytes, *J. Physiol.,* 407, 95P, 1988.
54. **Mauger, J.-P., Claret, M., Piietri, F., and Hilly, M.,** Hormonal regulation of inositol 1,4,5-trisphosphate receptor in rat liver, *J. Biol. Chem.,* 264, 8821, 1989.
55. **Supattapone, S., Danoff, S. K., Theibert, A., Joseph, S. K., Steiner, J., and Snyder, S. H.,** Cyclic AMP-dependent phosphorylation of a brain inositol trisphosphate receptor decreases its release of calcium, *Proc. Natl. Acad. Sci. U.S.A.,* 85, 8747, 1988.

56. **Parker, I. and Ivorra, I.,** Inhibition by Ca^{2+} of inositol trisphosphate-mediated Ca^{2+} liberation: a possible mechanism for oscillatory release of Ca^{2+}, *Proc. Natl. Acad. Sci. U.S.A.,* 87, 260, 1990.

57. **Bezprozvanny, I., Watras, J., and Ehrlich, B. E.,** Bell-shaped calcium-response curves of Ins(1,4,5)P3- and calcium-gated channels from endoplasmic reticulum of cerebellum, *Nature,* 351, 751, 1991.

58. **Finch, E. A., Turner, T. J., and Goldin, S. M.,** Calcium as a coagonist of inositol 1,4,5-trisphosphate-induced calcium release, *Science,* 252, 443, 1991.

59. **Iino, M. and Endo, M.,** Calcium-dependent immediate feedback control of inositol 1,4,5-trisphosphate-induced Ca^{2+} release, *Nature,* 360, 76, 1992.

60. **Endo, M.,** Calcium release from the sarcoplasmic reticulum, *Physiol. Rev.,* 57, 71, 1977.

61. **Malgaroli, A., Fesce, R., and Meldolesi, J.,** Spontaneous $[Ca^{2+}]_i$ fluctuations in rat chromaffin cells do not require inositol 1,4,5-trisphosphate elevations but are generated by a caffeine- and ryanodine-sensitive intracellular Ca^{2+} store, *J. Biol. Chem.,* 265, 3005, 1990.

62. **Rawlings, S. R., Demaurex, N., and Schlegel, W.,** Pituitary adenylate cyclase-activating polypeptide increases $[Ca^{2+}]_i$ in rat gonadotrophs through an inositol trisphosphate-dependent mechanism, *J. Biol. Chem.,* 269, 5680, 1994.

63. **DeLisle, S. and Welsh, M. J.,** Inositol triphosphate is required for the propagation of calcium waves in Xenopus oocytes, *J. Biol. Chem.,* 267, 7963, 1992.

64. **Sidkar, S. K., McIntosh, R. P., and Mason, W. T.,** Differential modulation of Ca^{2+}-activated K^+ channels in ovine pituitary gonadotrophs by GnRH, Ca^{2+} and cyclic AMP, *Brain Res.,* 496, 113, 1989.

65. **Tse, A., Tse, F. W., Almers, W., and Hille, B.,** Rhythmic exocytosis stimulated by GnRH-induced calcium oscillations in rat gonadotropes, *Science,* 260, 82, 1993.

66. **Stojilkovic, S. S., Izumi, S., and Catt, K. J.,** Participation of voltage-sensitive calcium channels in pituitary hormone release, *J. Biol. Chem.,* 263, 13054, 1988.

67. **Nordmann, J. J. and Stuenkel, E. L.,** Ca^{2+} independent regulation of neurosecretion by intracellular Na^+, *FEBS Lett.,* 292, 37, 1991.

68. **Herschman, H. R.,** Primary response genes induced by growth factors and tumor promoters, *Annu. Rev. Biochem.,* 60, 281, 1991.

69. **Meyer, T. E. and Habener, J. F.,** Cyclic Adenosine 3',5'-monophosphate response element binding protein (CREB) and related transcription-activating deoxyribonucleic acid-binding proteins, *Endocrine Rev.,* 14, 269, 1993.

70. **Lerea, L. S., Butler, L. S., and McNamara, J. O.,** NMDA and non-NMDA receptor-mediated increase of c-fos mRNA in dentate gyrus neurons involves calcium influx via different routes, *J. Neurosci.,* 12, 2973, 1992.

71. **Brostrom, C. O. and Brostrom, M. A.,** Calcium dependent regulation of protein synthesis in intact mammalian cells, *Annu. Rev. Physiol.,* 52, 577, 1990.

72. **Catterall, W. A.,** Structure and function of voltage-gated ion channels, *Science,* 242, 50, 1988.

73. **Jan, L. Y. and Jan, Y. N.,** Voltage-sensitive ion channels, *Cell,* 56, 13, 1989.

74. **Hille, B.,** *Ionic Channels of Excitable Membranes,* Sinauer, Sunderland, MA, 1992.

75. **Tsien, R. W., Ellinor, P. T., and Horne, W. A.,** Molecular diversity of voltage-dependent Ca^{2+} channels, *Trends Pharmacol. Sci.,* 12, 349, 1991.

76. **Bertolini, M. and Llinas, R.,** The central role of voltage-activated and receptor operated calcium channels in neuronal cells, *Annu. Rev. Pharmacol. Toxicol.,* 32, 399, 1992.

77. **Mintz, I. M., Adams, M. E., and Bean, B. P.,** P-Type calcium channels in rat central and peripheral neurons, *Neuron,* 9, 85, 1992.

78. **Mintz, I. M., Veneman, V. J., Swiderek, K. M., Lee, T. D., Bean, B. P., and Adams, M. E.,** P-Type calcium channels blocked by the spider toxin ω-Aga-IVa, *Nature,* 355, 827, 1992.

79. **Llinas, R., Sugimori, M., and Cherksey, B.,** Blocking and isolation of a calcium channel from neurons in mammals and cephalopods utilizing a toxin fraction (FTX) from funnel-web spider poison, *Proc. Natl. Acad. Sci. U.S.A.,* 86, 1689, 1989.

80. **Lang, D. G. and Ritchie, A. K.,** Tetraethylammonium sensitivity of a 35-pS Ca^{2+}-activated K^+ channel in GH3 cells that is activated by thyrotropin-releasing hormone, *Pflügers Arch.,* 416, 704, 1990.

81. **Allbritton, N. L., Meyer, T., and Stryer, L.,** Range of messenger action of calcium ion and inositol 1,4,5-trisphophate, *Science,* 258, 1812, 1992.

82. **Brandl, C. L., Green, N. M., Korczak, B., and MacLennan, D. H.,** Two Ca^{2+} ATPase genes: homologies and mechanistic implications of deduced amino acid sequences, *Cell,* 44, 597, 1986.

83. **McNaughton, P. A.,** Fundamental properties of the Na^+-Ca^{2+} exchange. An overview, *Ann. N.Y. Acad. Sci.,* 639, 2, 1991.

84. **Unwin, N.,** Neurotransmitter action: opening of ligand gated ion channels, *Cell,* 72(Suppl.), 31, 1993.

85. **Klepper, M., Hans, M., and Takeda, K.,** Nicotinic modulation of voltage-dependent calcium current in bovine adrenal chromaffin cells, *J. Physiol.,* 428, 545, 1990.

86. **Clapham, D. E.,** Direct G protein activation of ion channels? *Annu. Rev. Neurosci.,* 17, 441, 1994.
87. **Armstrong, D. and Eckert, R.,** Voltage-activated calcium channels that must be phosphorylated to respond to membrane depolarization, *Proc. Natl. Acad. Sci. U.S.A.,* 84, 2518, 1987.
88. **White, R. E., Schonbrunn, A., and Armstrong, D. L.,** Somatostatin stimulates Ca^{2+} activated K^+ channels through protein dephosphorylation, *Nature,* 351, 570, 1991.
89. **Levitan, I. B.,** Modulation of ion channels by protein phosphorylation and dephosphorylation, *Annu. Rev. Physiol.,* 56, 193, 1994.
90. **Schuman, E. M. and Madison, D. V.,** Nitric oxide and synaptic function, *Annu. Rev. Neurosci.,* 17, 153, 1994.
91. **Kato, M.,** Involvement of nitric oxide in growth hormone (GH)-releasing hormone-induced GH secretion in rat pituitary cells, *Endocrinology,* 131, 2133, 1992.
92. **Ceccatelli, S., Hulting, A. L., Zhang, X., Gustafson, L., Villar, M., and Hokfelt, T.,** Nitric oxide synthase in the rat anterior pituitary gland and the role of nitric oxide in regulation of luteinizing hormone secretion, *Proc. Natl. Acad. Sci. U.S.A.,* 90, 11292, 1993.
93. **Zhou, Z. and Misler, S.,** Single vesicle release is coupled to action potentials in rat adrenal chromaffin cells, *Biophys. J.,* 66, A54, 1994.
94. **Ámmälä, C., Eliasson, L., Bokvist, K., Larsson, O., Ashcroft, F. M., and Rorsman, P.,** Exocytosis elicited by action potentials and voltage-clamp calcium currents in individual mouse pancreatic B-cells, *J. Physiol.,* 472, 665, 1993.
95. **Roberts, W. M., Jacobs, R. A., and Hudspeth, A. J.,** Colocalization of ion channels involved in frequency selectivity and synaptic transmission at presynaptic active zones of hair cells, *J. Neurosci.,* 10, 3664, 1990.
96. **Neher, E. and Zucker, R. S.,** Multiple calcium-dependent processes related to secretion in bovine chromaffin cells, *Neuron,* 10, 21, 1993.
97. **Katz, B. and Miledi, R.,** The effect of temperature on the synaptic delay at the neuromuscular junction, *J. Physiol.,* 181, 656, 1965.
98. **Peters, A., Palay, S. L., and Webster, H. D.,** *The Fine Structure of the Nervous System,* Oxford University Press, New York, 1991, 1–494.
99. **Stanley, E. F.,** Single calcium channels and acetylcholine release at a presynaptic nerve terminal, *Neuron,* 11, 1007, 1993.
100. **von Rüden, L. and Neher, E.,** A Ca-dependent early step in the release of catecholamines from adrenal chromaffin cells, *Science,* 262, 1061, 1993.
101. **Burgoyne, R. D. and Morgan, A.,** Regulated exocytosis, *Biochem. J.,* 293, 305, 1993.
102. **Artalejo, C. R., Adams, M. E., and Fox, A. P.,** Three types of Ca^{2+} channel trigger secretion with different efficacies in chromaffin cells, *Nature,* 367, 72, 1994.
103. **Trifaró, J.-M. and Vitale, M. L.,** Cytoskeleton dynamics during neurotransmitter release, *Trends Neurosci.,* 16, 466, 1993.
104. **Thomas, P., Wong, J. G., Lee, A. K., and Almers, W.,** A low affinity Ca^{2+} receptor controls the final steps in peptide secretion from pituitary melonotrophs, *Neuron,* 11, 93, 1993.
105. **Bennett, M. K. and Scheller, R. H.,** The molecular machinery for secretion is conserved from yeast to neurons, *Proc. Natl. Acad. Sci. U.S.A.,* 90, 2559, 1993.
106. **Jahn, R. and Südhof, T. C.,** Synaptic vesicles and exocytosis, *Annu. Rev. Neurosci.,* 17, 219, 1994.
107. **Tsien, R. W., Lipscombe, D., Madison, D. V., Bley, K. R., and Fox, A. P.,** Multiple types of neuronal calcium channels and their selective modulation, *Trends Neurosci.,* 11, 451, 1988.

Chapter 4

ELECTRICAL ACTIVITY AND INTERCELLULAR COMMUNICATION IN NEUROENDOCRINE CELLS

Eduardo Rojas, David Mears, Rosa M. Santos, and Illani Atwater

CONTENTS

1. SUMMARY

The hormonal secretions of the neuroendocrine system (NES) mediate strict temporal control over the behavior of tissues throughout an organism. While the process of stimulus-secretion coupling has been studied extensively in most endocrine organs at the cellular level, little is known about the role of direct cell-to-cell communication in shaping the behavior of the entire organ. This review focuses on recent advances in the characterization of the biophysical properties of intercellular junctions (i.e., gap junctions) in NES and other systems.

The chapter begins with a discussion of the various criteria which have been used to classify tissues as NES. Recently, the use of new immunohistochemical techniques for tracing cell lineage have led to a much broader view of which tissues should be considered NES, such that most peripheral endocrine glands are now included along with the neuroendocrine structures of NES. However, to date it has not been suggested that endothelial cells, which are present in every endocrine tissue in the form of blood vessels, be considered an integral part of NES. We believe that such an omission yields an incomplete picture of the functional NES. Indeed, endothelial cells are not only present in each NES gland, but their secretion in response to a variety of blood-borne signals have dramatic influence over the behavior of hormone-secreting cells in the gland. Furthermore, the endothelial cells often communicate directly with the secreting cells via gap junctions, an observation with clear functional implications. To illustrate this point further, a recent study on cultured adrenal medullary endothelial and chromaffin cells has shown that endothelial cells can inhibit cholinergic agonist-evoked catecholamine secretion by a mechanism involving the L-arginine/nitric oxide/cGMP pathway.[1] We propose that in order to fully understand the behavior of an endocrine organ, the role of communication between secretory cells and the ubiquitous endothelial cells of the gland must be deciphered.

The last two sections of Part 2 discuss the structure and composition of gap junction channels, the types of gap junctions expressed by various NES and non-NES tissues, and regulation of gap junction properties at the genetic and molecular levels. Despite the wide range of tissues which have been studied, a very limited number of gap junction types have been characterized. The biophysical properties of the intercellular junctions in a given tissue are determined by genetic regulation of the gap junction type expressed as well as cell-specific, post-translational modifications to the protein(s) making up the channel. Functional gap junction channels are further influenced by the activity of a variety of second messengers such as Ca^{2+} and cAMP.

There are two reasons for including non-NES tissues in this discussion. First, as pointed out above, endocrine glands contain many cell types (generally considered non-NES), and intracellular communication between these cells and the secretory cells may be an important aspect of NES function. Second, while information on direct cell-to-cell coupling in the NES is quite limited, these tissues contain gap junction types which have been studied extensively in other tissues. In theory, the gap junction properties of NES tissue can be predicted once the gap junction type has been established. However, because the anatomical make-up of each NES gland includes a variety of hormone-secreting cells as well as endothelial cells, the presence of small amounts of a particular gap junction protein in the extract of a gland does not necessarily imply the protein forms functional channels in the tissue of interest. The expression of a connexin type in an immortalized cell line provides more compelling evidence that the protein functions in the cells of interest, although it may always be argued that the immortalization or culturing process can alter protein expression in the tissue. Therefore, caution must be used when attempting to infer gap junction properties based on connexin expression alone.

The third portion of the chapter summarizes the current state of knowledge of gap junction communication in NES itself. While both homologous and heterologous cell-to-cell communication has been demonstrated in the majority of NES tissues studied, the properties of the junctional channels in these cells are yet to be characterized. The noted exception is the endocrine pancreas, which has been studied extensively by morphological as well as dye and electrical coupling techniques. The results show that electrical coupling in the organ is quite extensive, although the number of functional channels and channel conductances are lower than in other tissues expressing the same junctional type. Dye coupling is quite restricted in these cells. There is strong evidence that the extent of cell-to-cell coupling in the islet of Langerhans is directly related to the secretory state of the gland.

The extent of coupling and the junctional properties in other NES tissues are less well understood. Most importantly, little is known about the relationship between coupling and function in these tissues. However, since gap junctions are present in most of these tissues, cell-to-cell communication likely plays an important role in shaping the behavior of endocrine organs.

The final section discusses endogenous tissue clocks which function in many NES tissues. In particular, the gonadotropin-releasing hormone (GnRH)-evoked oscillations in cytosolic Ca^{2+} in pituitary gonadotrophs are discussed in detail. While these clocks, which consist of both membrane and cytosolic components, are endogenous to the individual cells, cell-to-cell transfer of electrolytes and second messengers via gap junctions likely plays a crucial role in synchronizing and setting the frequency of these oscillations. Since oscillations are required for proper NES function, we conclude that gap junctions play a crucial role in shaping NES tissue responsiveness. The challenge for researchers in the future is to elucidate the properties of NES gap junctions and determine exactly how intercellular communication affects tissue behavior, thus completing the picture of NES function at the whole organ level.

2. INTRODUCTION

Intercellular junctions were first observed between homologous[2] and heterologous cell types[3] 3 decades ago. Despite the plethora of detailed information accumulated during these 3 decades on the structural, molecular, and biochemical aspects of the channels responsible for direct cell-to-cell coupling, i.e., the gap junctions, the role of direct coupling on the overall tissue physiology[4,5] is still not well understood. Although we will often refer to the seminal work published during the 1960s and 1970s by the pioneers in this field, the focus of this review will be on the developments occurring since the comprehensive examination of the status of knowledge on direct cell-to-cell coupling in endocrine glands by Meda et al.[6]

2.1. THE NEUROENDOCRINE SYSTEM

As the central theme of this review is direct cell-to-cell communication in the NES, it is necessary to provide a criterion to select the tissues to be considered. This section outlines the various criteria which have been used to define tissues as NES. Historically, the nervous and endocrine systems were considered separate functional and anatomical entities. However, recent studies have provided evidence that the biochemical elements of these systems have a common phylogenetic origin. Thus, despite their anatomical diversity, the two major systems of intercellular communication are biochemically similar.[7] Embryologically, all secretory tissues derived from the neural plate (7.5 d mouse embryo) were considered to be part of NES. By this criterion the entire neurohypophysis, consisting of the pars eminens (part of the hypothalamic infundibulum) and the pars nervosa (part of the hypophysis proper), and the ganglionic cell layer of the retina were included.[8] However, a long-standing controversy in endocrinology concerns the developmental lineage of other endocrine and paracrine tissues classically included in NES.[9]

Over the past 20 years, highly sensitive and specific immunohistochemical techniques to establish cell lineage have become available. The markers used to immunize and induce antibodies include different polypeptide hormones found in the secretory vesicles and in various organelles along the pathways for biosynthesis and secretion. Based on the presence of these markers, one might include in NES not only the peripheral endocrine organs classically included, i.e., the adrenal gland and the endocrine pancreas, but many other cell systems distributed throughout the body. Among other nonhormonal markers used to determine neuronal origin, chromogranin, a 75-kDa acidic protein found in adrenal chromaffin cell secretory granules, has been detected in cells of the so-called "diffuse neuroendocrine system", which includes putative mechanoreceptor epidermal Merkel cells[10,11] (see Akaike and Yamashita, Chapter 14 of this volume), enterochromaffin cells, thyroid C cells, parathyroid cells, thymic epithelial cells, the inner segment of rods and cones, and submandibular gland cells.[12] Another important marker is the neuron-specific enolase[13] which is present in four basic islet cell types of the human fetal pancreas (10–14 weeks of gestational age) and in the pancreatic duct epithelium.

Monoclonal antibodies HISL-5, HISL-9, and HISL-14, generated by immunization of mice with human pancreatic islet cell preparations, recognize a differentiation antigen expressed by the pancreatic islet cells and have also been used to identify NES markers. These monoclonal antibodies react strongly with all endocrine cell subtypes of human pancreatic islets, but minimally if at all with the exocrine acinar cells, vascular cells, and stromal connective tissue cells of the pancreas. Antigenic determinants of these three antibodies are also expressed by thyroid follicular cells, parathyroid chief cells, and anterior pituitary cells. The status of the thymus as part of NES is less clear. However, since it is well established that cranial crest cells differentiate into mesenchyme and, later on, contribute to connective tissue, muscular and nervous components of cephalic organs, and proper morphogenesis of the thymus requires specific interactions between epithelial and mesenchymal components, and appropriate T cell differentiation is not possible if the thymic epithelium is not associated with mesenchyme, this endocrine tissue has also been included.[14]

Another cell marker is a polypeptide detected by the monoclonal antibody HISL-19, which is also generated after immunization of mice with human pancreatic islet cell preparations. This marker is a water-soluble acidic protein (67 kDa, 35/32 kDa dimer) that is stored within secretory granules of peptide hormone-producing cells, and released in detectable amounts into the serum of patients bearing neuroendocrine carcinomas. The protein is also expressed by neuronal cells and shares many biochemical and molecular key features with the chromogranin proteins.[15] Therefore, the protein detected by HISL-19 represents a novel component of the soluble compartments of neurosecretory granules, which is distinct from chromogranin A, B, and C. In good agreement with previous finding using different cell markers, the subunit C of aldolase (aldolase C) has been detected in all peripheral neuroendocrine cells examined, including the pituitary gland, thyroid, pancreas, adrenal gland, bronchus, and gastrointestinal tract.[16] Finally, since adrenal steroids acting via specific receptors may influence brain (namely hippocampal) structure and function, affect the survival and death of neurons, and exert a negative feedback control of pituitary-adrenal function,[17] the steroid-producing tissue has also been considered an integral part of NES.

2.2. INTERCELLULAR COMMUNICATION AMONG NEUROENDOCRINE TISSUES

Bidirectional interactions between different cell types of the NES specifically modulate physiological activities as diverse as hypothalamic neural signal transmission and pituitary hormone secretion. For example, to achieve an optimal bolus-type release of GnRH into the portal vessels, GnRH neurons must enter into synchronous and phasic activity. GnRH neuron activity is in turn controlled by GABA neurons in the preoptic/anterior hypothalamic area.[18] Therefore, the relationship among the central nervous, neuroendocrine, and immune systems are bidirectional and constitute an integrated loop. Thus, the intercellular communication among NES tissues appears to be more efficacious if each specific tissue is releasing the hormones in a pulsatile fashion, and the endogenous frequency of the tissue-specific oscillations is controlled by a tissue clock. The overall control of NES involves tissue clock resetting by pulsatile and episodic hormonal release from other tissues in NES and the autonomous nervous system. Cell-to-cell electrical coupling is essential to synchronize the cells within each specific tissue and thus, electrical coupling might be a crucial requirement to set the pace of the clock at the NES control centers. From this perspective, one challenge for us in this review was to delineate the specific role of cell-to-cell electrical coupling in the function of each specific NES tissue.

NES cells and tissues often communicate by means of periodic signals. Periodic signaling prevents the phenomenon of desensitization brought about by constant stimuli.[19] Intercellular communication through aqueous intercellular channels, known as gap junctions, has been postulated to provide an important mechanism for coordinating the rapid and synchronous responses of several endocrine tissues. Gap junction channels provide a pathway for exchange of ions and small molecules between coupled cells, and this exchange is believed to be critical for normal tissue growth and development.[20,21]

2.3. THE MOLECULAR BASIS FOR DIRECT CELL-TO-CELL COMMUNICATION

It is well established that gap junctions provide means of direct intercellular communication in various tissues. Gap junction channels (or connexons) are formed by protein subunits called connexins, with each communicating cell contributing six such subunits to the channel. The connexins form a family of membrane-spanning proteins with less than a dozen isoforms. As will became apparent in the following sections, there are as many types of gap junctions as connexin isoforms. Furthermore, structural studies suggest that each type of gap junction consists of only one connexin type. Thus, functional characteristics of gap junctions in a given tissue can be inferred from data acquired from other tissues provided the building block of the gap junction, namely the connexin, is the same. Based on this evidence, and because very little information is available about gap junction coupling in NES tissues (with the exception of the endocrine pancreas), much of this review will focus on data from tissues in which powerful biophysical methods have been employed to explore gap junction coupling. The following two sections are intended to provide an overview of the molecular properties and functional implications of gap junction-mediated direct cell-to-cell coupling in tissues other than NES.

2.3.1 Early Developments

Early studies on intercellular communication were carried out on epithelial cells from the giant salivary gland cells in *Chironomus* and in *Drosophila*.[2,22] The majority of these investigations involved one or both of two techniques: analysis of spread of intracellularly microinjected fluorescent dye (dye coupling) and measurement of flow of intracellularly microinjected electrical current (electrical coupling). These studies followed from work carried out in the 1950s which established the existence of electrical communication between cardiac muscle fibers and at certain crayfish neuronal synapses.[23-25] Loewenstein and Kanno[2,26] reported the presence of an electrical communication pathway between the giant salivary gland cells. Modeling the gland as a cable conductor, they elucidated the electrical properties of the system, showing that the transjunctional resistivity was similar in magnitude to that of the cytoplasm and several orders of magnitude smaller than the resistivity of the nonjunctional membrane. They also established the linearity of the current-voltage relationship (I-V curve) of the intercellular pathway.

Initial dye-coupling experiments involved the tracer fluorescein, but later a variety of fluorescent molecules of different molecular weights (or sizes) were used.[27] A gross overestimate of the maximum size of the permeant species was made at this time, but later work showed that the junctional pathway in insects was permeable to molecules of up to 1.7 kDa, the tracer being in the form of an unbranched peptide labeled with fluorescein isothiocyanate.[28] The maximum size of permeant molecule for a mammalian system was later shown to be around half of that for the insect.[29]

Fluorescein had, meanwhile, been used to establish the presence of intercellular pathways between many cells in both homo- and heterocellular systems.[30-35] Thus, the existence of an intercellular junction which seemed to be almost universally present throughout the animal kingdom had been fairly well established by electrical and fluorescent tracer studies by the mid 1960s.

The localization and morphological description of intercellular junctions were carried out primarily by electron microscopy. The techniques of ultra-thin sectioning, fast freeze-fracture, electron-dense staining and metal shadowing were all employed for tissue preparation. Four types of membrane structure with properties of association between adjacent cells were discovered, these being **gap, tight**, and **septate junctions**, as well as **desmosomes**.

From preliminary studies it seemed that the gap junction was most likely the low-resistance intercellular pathway. This proposal was considerably reinforced when Ravel and Karnovsky[36] produced highly resolved electron micrographs of the junction, both in cross section and *en face*. The cross section showed a septilaminar structure of closely opposed membranes with cross bridges. The central lamina was readily penetrated with colloidal lanthanum hydroxide, but not horseradish peroxidase, indicating hindrance to extracellular movement parallel to the membrane for only large molecules. The *en face* view revealed a regular polygonal lattice of subunits about 8 nm in diameter, seen in poor definition in goldfish Mauthner cells previously,[37] later to be named connexons.[38] Each subunit was seen to have a central area susceptible to lanthanum hydroxide staining, which has since been associated with the intercellular pathway.[39] The freeze-fracture technique has been employed to show the cylindrical nature of the gap junction subunit, which generally associates with the P, or cytoplasmic, face in mammalian preparations, but the E face in insect preparations.

2.3.2. Expression of Gap Junction Channels in NES and Other Tissues

Over the past 5 years, there has been much progress in identifying and characterizing a multigene family that is responsible for producing polypeptides which form gap junction channel oligomers between cells. The products of these genes, connexins, and the multigene family can be categorized into two classes, the alpha class and the beta class, based on their primary sequence and overall predicted topological organization. The gap junction genes map to different chromosomes in both mice and humans, and these genes are utilized on a cell-specific basis. Furthermore, these genes are developmentally regulated, and multiple genes can be coexpressed simultaneously by the same cell type. Efforts to understand the precise structure-function relationship of the products of these genes are now being made using various expression systems.[40,41]

Today, with the sequence of each individual connexin and the polymerase chain reaction amplification technique at hand, researchers have a unique opportunity to screen RNA encoding for specific connexins extracted from NES tissues. In addition, new highly selective antibodies against distinct connexin epitopes for Western blot analysis facilitates the identification of the connexin type in freshly dissociated tissues. However, owing to the unavoidable presence of endothelial cells in endocrine glands and different endocrine cell types in each organ (such as the islet of Langerhans), the method should be used on RNA obtained from highly purified homogeneous NES cell preparations or from cell lines derived from NES tissues.

2.3.2.1 *Isolation and Purification of Gap Junction Channels*

Gap junction plaques were isolated[42] and attempts were made to elucidate their structure more precisely. Combined use of electron microscopy and X-ray diffraction showed that the collection of connexins (or connexon) of the mouse liver cell is a tubular structure made up of six subunits.[43,44] The central pore was estimated to have a diameter of about 2 nm, which corresponds well with the connexon pore diameter estimated from tracer studies on mammalian cells.[43] These studies also illustrated that the connexons were arranged in a lattice with overall long-distance order, but that over short ranges the lattice was more disordered. The arrangement of connexins within a gap junction has been correlated with its conducting state. Using a crayfish septate lateral axon preparation[45] and rat stomach and liver preparations,[46] it was shown that the external diameter of the connexon increases from about 15 to 20 nm in changing from the shut to the open state. At the same time the intercellular gap increases from about 2–3 nm to 4–5 nm. The increase in spacing between connexons of the open channel gives rise to short-range lattice disorder, while in the shut state, the connexon lattice tends to increase in regularity.

Unwin and Zampighi[47] proposed a model for the connexon comprised of six subunits arranged radially around the connexon pore, with their long axes running across the membrane. In this model the six subunits gate the channel by tilting about their axes. Three-dimensional maps of the connexons obtained from electron microscopy data gathered at various incident electron beam angles using Fourier synthesis agree well with those of Caspar et al.,[43,44] and there now seems to be general agreement on the hexameric nature of the connexon. The subunit was believed to be a protein of molecular weight 18 to 30 kDa, as this had been the only major protein isolated from gap junction preparations.[38,48,49]

In a recent review, Stauffer et al.[50] report the methods developed in their laboratory to solubilize gap junction channels, or connexons, from isolated gap junctions and to purify them in milligram quantities. Two sources of material are used: rat liver gap junctions and gap junctions produced by infecting insect cells with a baculovirus containing the cDNA for human liver connexin32. Complete solubilization is obtained with long-chain detergents and requires high ionic strength and high pH as well as reducing conditions. The purification also involves chromatography and gel filtration. A homogeneous product is indicated by a single band on a silver-stained gel and a homogeneous population of doughnut-shaped particles with hexameric symmetry under the electron microscope. The purified connexons have a tendency to form in filaments that grow by end-to-end association of connexons, suggesting that the connexons are paired as in the cell-to-cell channel.

Over the last 15 years we have learned a great deal about the membrane topology and quaternary structure of gap junction ion channels, particularly the cardiac gap junction channels containing connexin43. These two aspects were recently examined using antipeptide antibodies directed to seven different sites in the protein sequence, cleavage by an endogenous protease in heart tissue, and electron microscopic image analysis of native and protease-cleaved two-dimensional membrane crystals of isolated cardiac gap junctions.[51] Based on protein-folding criteria, which is predicted by hydropathy analysis, five antibodies were directed to sites in cytoplasmic domains and two antibodies were directed to the two extracellular loop domains. Isolated gap junctions could not be labeled by the two extracellular loop antibodies using thin-section immunogold electron microscopy. This is consistent with the known narrowness of the extracellular gap region that presumably precludes penetration of antibody probes. However, cryosectioning rendered the extracellular domains accessible for immunolabeling. A cytoplasmic loop domain (residues 101 to 142) is readily accessible to peptide antibody labeling.

The native 43-kDa protein can be protease cleaved on the cytoplasmic side of the membrane, resulting in an approximately 30-kDa membrane-bound fragment. Western immunoblots showed that protease cleavage occurs at the carboxyl tail of the protein, and the cleavage site resides between amino acid residues 252–271. Immunoelectron microscopy demonstrated that the 13-kDa (or alpha 1) carboxyl-terminal peptide(s) is released after protease cleavage and does not remain attached to the 30-kDa membrane-bound fragment.

Electron microscopic image analysis of two-dimensional membrane crystals of cardiac gap junctions revealed that the ion channels are formed by a hexagonal arrangement of protein subunits. This quaternary arrangement is not detectably altered by protease cleavage of the alpha 1 polypeptide. Therefore, the 13-kDa carboxyterminal domain is not involved in forming the transmembrane ion channel. The similar hexameric architecture of cardiac (connexin43) and liver (connexin32) gap junction connexon in both tissues indicates conservation in the molecular design of the gap junction channels formed by different connexins.

2.3.2.2. *Pancreatic β Cells Express Connexin43, but Not Connexin32 and Connexin26*

Immunostaining analysis of cryostat section of insulin secreting β cells, Western blot analysis of crude β cell membrane preparations using antibodies against different types of connexins (26, 32, 43), and Northern blot analysis of total β cell RNA using cRNAs for the three connexins showed that β cells express connexin43, but not connexin32 and connexin26. Dual whole-cell patch clamp recording and Lucifer yellow dye coupling experiments indicated that only a small fraction of rat pancreatic β cell pairs were coupled. Pancreatic β cells from glibenclamide-treated rats, which exhibit sustained insulin release *in vivo*, show a twofold increase in the level of connexin43 gene transcripts and enhanced ionic and Lucifer yellow dye coupling. Furthermore, connexin43 gene transcripts and incidence of junctional coupling are modulated in parallel during sustained stimulation of β cell functioning *in vivo*.[52] Taken together, these results indicate that connexin43 is a major component of communicating channels between insulin-producing cells, and that coupling is clearly relevant to the secretory state of the islet.

2.3.2.3 *Connexin43 Is the Predominant Gap Junction Protein in Hypothalamic Astrocytes*

Astrocytes in the brain, both *in situ* and in culture, communicate via gap junctions. Astroglial gap junction channels and their constituent proteins were characterized by immunocytochemical, molecular biological, and physiological techniques. Comparative immunocytochemical labeling utilizing different antibodies specific for liver connexin32 and connexin26 and antibodies specific for the carboxyl-terminal sequences of the heart gap junction protein indicated that the predominant gap junction protein in astrocytes is connexin43. The expression of this connexin was also established by Western and Northern blot analyses. Cultured astrocytes expressed connexin43 mRNA and did not contain detectable levels of the mRNAs encoding connexin32 or connexin26. Further, the cells contained the same primary connexin43 translation product and the same phosphorylated forms as heart. Finally, electrophysiological recordings under voltage-clamp conditions revealed that astrocyte cell pairs were moderately well coupled, with an average junctional conductance of about 13 nS. Single-channel recordings indicated a unitary junctional conductance of about 50–60 pS, which is of the same order as that found in cultured rat cardiac myocytes, where the channel properties of connexin43 were first described. Thus, physiological properties of gap junction channels appear to be determined by the connexin expressed, independent of the tissue type.[53]

Comparing gap junctions in astrocytes derived from two brain regions, Batter et al.[54] found profound functional differences. Indeed, astrocytes cultured from neonatal rat hypothalamus contain a greater number of functional channels than astrocytes from the striatum, a difference reflected in both connexin43 protein and mRNA. Specifically, in hypothalamic astrocytes the level of connexin43 protein was approximately four times that found in comparable cultures from the striatum, as determined by immunoblotting. Complementary results from immunocytochemical experiments using an antibody specific for connexin43 reveal significantly greater fluorescence in astrocytes cultured from the hypothalamus as compared to those from the striatum. Northern blot analysis showed that connexin43 mRNA levels were also approximately fourfold greater in the hypothalamic cultures, consistent with the difference seen by immunoblotting. Finally, dye coupling studies using confluent cultures consistently showed that within 1 min Lucifer yellow injected into striatal astrocytes spread only to immediately adjacent cells, while in hypothalamic astrocytes dye often spread to apparent third- or fourth-order neighbors within the same time period. Thus, the higher level of connexin43 expression seen in hypothalamic astrocytes results in cells with greater numbers of functional channels.

2.3.2.4.   Coexpression of Connexin43 and Connexin26 in a Single Cell Type in the Nervous System

Leptomeningeal cells in intact meninges or dissociated and cultured meninges were found to be Lucifer yellow coupled, and pairs of freshly dissociated leptomeningeal cells were electrically coupled. Junctional conductance was reversibly reduced by halothane, heptanol, and 100% CO_2 and was augmented by 8-bromo cAMP. Unitary conductances of junctional channels were centered about 40 and 90 pS. Both connexin26 and connexin43 were identified between leptomeningeal cells using immunocytochemical methods. Northern blot analyses of total RNA isolated from cultured leptomeningeal cells showed specific hybridization to cDNA encoding connexin26 and connexin43, but not to a cDNA encoding connexin32. These results demonstrate coexpression of two connexins in a single cell type in the nervous system. The biophysical properties of the gap junction-mediated cell-to-cell coupling do not differ significantly from those of astrocytes and cardiac myocytes, which express only connexin43.[55]

2.3.2.5   Connexin32 and Connexin43 Are Expressed in Rat Hepatocytes, but Connexin32 Is the Major Gap Junction Protein

Bai et al.[56] investigated the genetic basis of the transcriptional regulation of the rat connexin32 gene which encodes the major gap junction protein in rat liver. Their data showed that the active promoter responsible for rat connexin32 mRNA transcription is located upstream of the first exon. Also, a basal promoter region was localized to a 50-bp region which formed multiple DNA:protein complexes. The authors concluded that a multiple proximal and distant regulatory elements are involved in the expression of connexin32.

Cell-to-cell communication studies of a rat liver cell line showed that these cells display high levels of gap junctional communication, as assessed functionally and immunologically. These studies showed that intracellularly injected Lucifer yellow diffused extensively and there was rapid recovery of the dye fluorescence after photobleaching. However, expression of connexin43 evaluated by immunocytochemistry of cell monolayers and Western blot analysis of total cell homogenates was relatively high. Western blot analysis revealed multiple forms of connexin43, which presumably correspond to known dephosphorylated and phosphorylated states of this protein.[57]

2.3.2.6   Connexin43 Forms Gap Junctions in Endothelial Cells

Primary cultures and freshly isolated sheets of endothelial cells from aortae and umbilical veins were also used to study the ultrastructure of gap and tight junctions and the cell-to-cell transfer of small molecules. In thin sections and in freeze-fracture replicas, the gap and tight junctions in the freshly isolated cells from both sources appeared similar to those found in the intimal endothelium. Most of the interfaces in replicas had complex arrays of multiple gap junctions either intercalated within tight junction networks or interconnected by linear particle strands. The particle density in the center of most gap junctions was noticeably reduced. Despite the relative simplicity of the junctions, the cell-to-cell transfer of potential changes, Lucifer yellow, and nucleotides was readily detectable in cultures of both endothelial cell types. The extent and rapidity of dye transfer in culture was only slightly less than that in sheets of freshly isolated cells, perhaps reflecting a reduced gap junctional area combined with an increase in cell size *in vitro*.[58]

Vascular endothelial cultures are composed of flat, polygonal monolayer cells which retain many of the growth, metabolic, and physiological characteristics of the intimal endothelium. However, intercellular gap and tight junctions, which are thought to perform important roles in normal intimal physiology, are reduced in complexity and extent in culture. Using electrophysiological techniques to test confluent (3- to 5-d) primary cultures of calf aortic (BAEC) and umbilical cord vein (BVEC) endothelium for junctional transfer of small ions, it was found that both cell types are extensively electrically coupled. The passive electrical properties of the cultured cells were calculated from the decay of induced membrane potential deflections with distance from an intracellular, hyperpolarizing electrode. Data analyses were based on a thin-sheet model for current flow. The generalized space constants λ were 208 pS (BAEC) and 288 pS (BVEC). The nonjunctional (about 0.8 GΩ) and junctional (0.004 GΩ) resistances were similar for the BAEC and BVEC, respectively. *In vivo* ultrastructural studies have suggested that aortic endothelium has more extensive gap junctions than venous endothelium. We have found that these ultrastructural differences are reduced in culture. The lack of any significant difference in electrical coupling capability suggests that cultured BAEC and BVEC have functionally similar junctional characteristics.[59]

Thus, these studies indicate that vascular endothelial cultures, derived from large vessels, retain many of the characteristics of their *in vivo* counterparts. However, the observed reduction in size and complexity of intercellular gap and tight junctions in these cultured cells suggests that important functions, thought to be mediated by these structures, may be altered *in vitro*. In our continuing studies on intercellular communication in vessel wall cells, we have quantitated the extent of junctional transfer not only of Lucifer yellow, but also of other small molecular tracers (tritiated uridine nucleotides) in confluent BAEC and BVEC. Both types of endothelial cells show extensive (and quantitatively equivalent) dye and nucleotide transfer. As an analog of intimal endothelium, we have also tested dye transfer in freshly isolated sheets of endothelium. Transfer in BAEC and BVEC sheets was more rapid, extensive, and homogeneous than in the cultured cells, implying a reduction in molecular coupling as endothelium adapts to culture conditions.[60]

Gap junction-mediated intercellular communication has been studied using an *in vitro* model in which a confluent monolayer of Lucifer yellow loaded, capillary endothelial cells is mechanically wounded. Approximately 40–50% of the cells in a confluent monolayer were coupled in groups of four to five cells (basal level). Basal levels of communication were also observed in sparse cultures, but were reduced in monolayers after they become confluent. Shortly (0.5 h) after wounding, coupling was markedly reduced between cells lining the wound. Communication at the wound was partially reestablished and exceeded basal levels, reaching a maximum after 24 h, at which stage approximately 90% of the cells were coupled in groups of six to seven cells. When the wound had closed (after 8 d), the increase in communication was no longer observed. Induction of wound-associated communication was unaffected by exposure of the cells to the DNA synthesis inhibitor mitomycin C, but was prevented by the protein synthesis inhibitor cycloheximide. The induction of wound-associated communication was also inhibited when migration was prevented by placing the cells at 22°C immediately after wounding or after exposure to cytochalasin D, suggesting that the increase in communication is dependent on cells migrating into the wound area. In contrast, migration was not prevented when coupling was blocked by exposure of the cells to retinoic acid (RA), although this agent did disrupt the characteristic sheet-like pattern of migration typically seen during endothelial repair. These results suggest that junctional communication may play an important role in wound repair, possibly by coordinating capillary endothelial cell migration.[61]

Recently, comparative studies using endothelial cells from the microvasculature and larger vessels showed that junctional communication, as assessed by transfer of Lucifer yellow, was greater in endothelial cells lining large vessels than in microvessels. Furthermore, basal levels of connexin43 protein and mRNA were also higher in endothelial cells from large vessels than from microvessels. When monolayers of microvessel endothelial cells were mechanically injured, junctional communication was increased between migrating cells at the injured edge. This increase in coupling was accompanied by a substantial increase in the protein connexin43 and mRNA encoding connexin43. In contrast, coupling between endothelial cells was not increased and connexin43 expression was unaltered after injuring monolayers of large vessel endothelial cells.[62]

2.3.2.7   *Connexin43 Is the Predominant Gap Junction Protein Expressed in the Heart*

Fishman et al.[63] determined that connexin43 is the predominant gap junction protein expressed in the heart and, by examing the developmental profiles of connexin43, showed a relationship between cardiac maturation and gap junction gene expression. Connexin43 mRNA levels accumulate progressively (eight-fold) during embryonic and early neonatal stages, accompanied by a parallel, but temporally delayed, accumulation of connexin43 protein (15-fold). As the heart matures further, both mRNA and protein levels subsequently decline, to about 50% and 30% of their maximum levels, respectively. These observations suggest that increases in intercellular coupling that characterize cardiac development do not depend solely on modulation of connexin43 gene expression, but rather are likely to involve organization of gap junction channels into the intercalated disc.

2.4.   REGULATION OF GAP JUNCTION CHANNEL EXPRESSION

The complex and overlapping tissue distribution of different members of the gap junctional connexin protein family, the intermixing of different connexins in the building of intercellular channels, and translational and post-translational regulation of gap junctional channels add additional challenges to the interpretation of the possible functions played by gap junction-mediated intercellular communication in NES and other tissues.[64] In view of the recent surge of information dealing with the cellular and molecular biology of gap junctions in nonnervous tissue, as well as current interest in the cell biology of NES, the next few sections are intended to provide an overview of the molecular and functional implications of gap junction-mediated intercellular communication in NES and several other tissues.

2.4.1.   Distribution of Genes for Gap Junction Proteins in Human and Mouse Chromosomes

By probing somatic cell hybrid DNA on Southern filters with rat or human cDNAs or human genomic fragments, Hsieh et al.[65] mapped four functioning gap junction genes (connexins 43, 32, 26, and 46) to different sites on human chromosomes as follows: connexin43 to 6p21.1-q24.1; connexin32 to Xcen-q22; connexin26 and connexin46 to 13. The connexin46 probe also hybridized to a restriction fragment that was mapped to chromosome 1. A connexin43-related pseudogene GJA1P was assigned to chromosome 5. The homologous loci in mouse were assigned to regions of known conserved syntenic groups: connexin43 to chromosome 10; connexin32 to XD-F4 and connexin26 to 14. Of two sites of hybridization with the connexin46 probe, on mouse 14 and 5, we assume that the site on

14 corresponds to the connexin46 locus on human 13. Based on these data, additional members of this family of related genes can be isolated and characterized, and possible human and mouse mutations can be identified.[65]

2.4.2. Developmental Regulation of Gap Junction Gene Expression
2.4.2.1 *Mouse*

The expression of products from three different gap junction genes (connexin43, connexin32, and connexin26) was studied in pre- and post-implantation mouse embryos, during organogenesis, and in cultured embryonic stem (ES) cells. In pre- and postimplantation mouse embryos, connexin43 transcript was the earliest gap junction RNA detected (four cells embryo) and its abundance increased significantly throughout subsequent development. Developmental expression of these three gap junction genes was also observed during the early stages and throughout the late stages of organogenesis. The expression patterns for these genes may be related to differences in gap junctional communication requirements for fetal organ development vs. neonatal and adult organ function, or the utilization of different genes by different cell types during organogenesis. In an ES cell culture line, connexin43 was the only gap junction gene product detected.[66]

Recent studies of the developmental regulation of rat epidermal gap junctions by immunohistochemistry, together with the molecular composition of gap junction plaques by laser scanning confocal microscopy following double immunolabeling with antibodies specific for connexin43 and connexin26, showed that during the early phase of fetal development (embryonic period), gap junctions are already present, forming large junctional plaques consisting of connexins 43 and 26.[67] The first switch in the utilization of connexins 43 and 26 was observed at the onset of epidermal stratification, when connexin26 expression was downregulated in the periderm and in the upper part of the intermedius. The second change in the expression of gap junction components coincided with epidermal differentiation, when connexin26 was preferentially expressed in the differentiated granular and upper spinous layers. An analysis of immuno-double-stained sections by laser scanning confocal microscopy revealed that large gap junction plaques ($>1~\mu m^2$) can contain segregated domains of connexons, which contain a single protein (homooligomer).

2.4.2.2. *Gap Junction Protein Expression and Junctional Permeability During Differentiation*

The diversity of gap junction proteins may exist to form channels that have different permeability properties. Induction of terminal differentiation in mouse primary keratinocytes by calcium results in a specific switch in gap junction protein expression, such that expression of connexin43 and connexin26 gap junction proteins is downmodulated, and that of connexin31 and connexin31.1 proteins is induced. Although both proliferating and differentiating keratinocytes are electrically coupled, there are significant changes in the permeability properties of the junctions to small molecules. The intercellular transfer of the small dyes neurobiotin, carboxyfluorescein, and Lucifer yellow is significantly reduced in differentiating keratinocytes, whereas that of small metabolites, such as nucleotides and amino acids, proceeds unimpeded. Thus, a switch in gap junction protein expression in differentiating keratinocytes is accompanied by selective changes in junctional permeability that may play an important role in the coordinate control of the differentiation process.[68]

2.4.2.3 *Structural Organization of Connexin43 in Gap Junction Channel Occurs after Exit from the Endoplasmic Reticulum*

Musil and Goodenough[69] have recently studied connexin43 by sucrose gradient fractionation and chemical cross linking to characterize the first step in gap junction assembly — oligomerization of connexin43 monomers into connexon channels. In contrast with other plasma membrane proteins, multisubunit assembly of connexin43 was specifically and completely blocked when endoplasmic reticulum (ER)-to-Golgi transport was inhibited by lowering the temperature to 15°C, incubation with carbonyl cyanide m-chloro-phenylhydrazone, or brefeldin A. Additional experiments indicated that connexon assembly occurred intracellularly, most likely in the trans-Golgi network. These results describe a post-ER assembly pathway for integral membrane proteins and have implications for the relationship between membrane protein oligomerization and intracellular transport.

2.4.3. Regulation of Transport of Connexins to the Plasma Membrane

It clear that phenotype expression depends on several factors including translation, post-translational changes, and transport to the target subcellular component which, in the case of the connexins, is the plasma membrane. Musil and Goodenough[70] demonstrated that the gap junction protein connexin43 is translated as a 42-kDa protein (connexin43-NP) that is efficiently phosphorylated to a 46-kDa species (connexin43-P2) in gap junctional communication-competent, but not in communication-deficient, cells. Furthermore, they were able to show that the phosphorylation of connexin43 to the P2 form was accompanied by acquisition of resistance to solubilization by detergent. Immunohistochemical localization of connexin43 in detergent-extracted cells demonstrated that the detergent-insoluble connexin43-P2 was concentrated in gap junctional plaques, whereas the detergent-soluble connexin43-NP was predominantly intracellular. Using either low temperature (20°C) to block intracellular transport or cell-surface protein biotinylation, it was determined that connexin43 was transported to the plasma membrane in the detergent-soluble connexin43 form. Cell-surface biotinylated connexin43-NP was processed to Triton®-insoluble connexin43-P2 at 37°C. Connexin43-NP was also transported to the plasma membrane in communication-defective, gap junction-deficient cells, but was not processed to detergent-insoluble connexin43-P2. Taken together, their results demonstrate that gap junction assembly is regulated after arrival of connexin43 at the plasma membrane and is temporally associated with acquisition of insolubility in detergent and phosphorylation to the connexin43-P2 form.

2.4.4. Functional Interactions Among Connexins 43, 46, and 50

White et al.[71] examined the functional interactions of the three rodent connexins present in the lens, connexin43, connexin46, and connexin50, by expressing them in paired *Xenopus* oocytes. Homotypic channels containing connexin43, connexin46, or connexin50 all developed high conductance. Heterotypic channels composed of connexin46 paired with either connexin43 or connexin50 were also highly conductive, whereas connexin50 did not form functional channels with connexin43. They also examined the functional response of homotypic and heterotypic channels to transjunctional voltage and cytoplasmic pH_i and showed that all lens connexins exhibited pH_i as well as membrane potential dependence. Channel closures of heterotypic channels for a given connexin were dramatically influenced by its partner connexins in the adjacent cell. Based on the observation that connexin43 can discriminate between connexin46 and connexin50, they also investigated the molecular determinants that specify compatibility by constructing chimeric connexins from portions of connexin46 and connexin50 and testing them for their ability to form channels with connexin43. When the second extracellular (E2) domain in connexin46 was replaced with the E2 of connexin50, the resulting chimera could no longer form heterotypic channels with connexin43. A reciprocal chimera, where the E2 of connexin46 was inserted into connexin50, acquired the ability to functionally interact with connexin43. Together, these results demonstrate that formation of intercellular channels is a selective process dependent on the identity of the connexins expressed in adjacent cells, and that the second extracellular domain is a determinant of heterotypic compatibility between connexins.

2.4.5 Tumor-Promoter TPA and the *ras* Oncogene Modulate Expression and Phosphorylation of Gap Junction Proteins

Gap junctional intercellular communication is inhibited in response to tumor promoters and oncogene transformation, suggesting that loss of this function is an important step in tumor formation. To elucidate the molecular mechanisms responsible for this inhibition, Brissette et al.[72] examined the expression of gap junction proteins and mRNA in mouse primary keratinocytes after treatment with the tumor-promoter TPA and/or ras transformation. During normal cell growth, keratinocytes express connexin43 and connexin26 proteins. Within 5 min of TPA treatment, the alpha 1 protein became rapidly phosphorylated on serine residues and its expression was dramatically reduced by 24 h. The beta 2 protein, after an initial increase in expression, was also significantly reduced 24 h after treatment with TPA. Transformation using *ras* caused changes similar to those induced by TPA. The connexin43 protein underwent an increase in serine phosphorylation, although its expression declined only slightly, while connexin26 expression was greatly reduced. The effects of TPA and *ras* on connexin43 expression were additive; treatment of *ras*-transformed cells with TPA resulted in increased alpha 1 phosphorylation, with greatly decreased protein levels, much lower than those generated by either agent alone. These data provide a likely explanation for the similar and synergistic inhibition of gap junctional intercellular communication by phorbol esters and *ras*.[72]

2.5. REGULATION OF GAP JUNCTION CHANNEL FUNCTION IN NES AND OTHER TISSUES

Over the last 15 years, many important bits of information on the regulation of cell-to-cell coupling have been obtained and the emerging picture provides the basis to understand, at the molecular level, the underlying control mechanisms of cell-to-cell communication. Some of the most recent work is summarized below.

2.5.1 Regulation of Gap Junction Molecular Conformation
2.5.1.1 *Ca^{2+}-Activated Calmodulin Regulates the Gap Junction Channel in Lens Fiber Cells*

Girsch and Peracchia[73] observed that lens fiber cells are coupled by communicating junctions that comprise over 50% of their oppositional surfaces. The main intrinsic protein (MIP) of lens fibers is a 28.2-kDa protein that forms large gap junction-like channels in reconstituted systems. Previously, these authors had shown that Ca^{2+}-activated calmodulin (CaM) regulates the permeability of reconstituted MIP26 channels and changes the conformation of MIP, as measured by intrinsic fluorescence and circular dichroism spectroscopy. Furthermore, examination of the MIP amino acid sequence revealed an α-helical segment (Pep C) on the C-terminus with residue distribution similar to that found in other CaM binding proteins. To test the interaction between the amphophilic segment and CaM, the authors synthesized and purified both a 20-mer peptide and trp-substituted fluorescent analog. With this system evidence from spectro-fluorometric titration was obtained indicating that the Pep C binds with CaM in 1:1 stoichiometry and with a K_d of approximately 10 nM. Neither Ca^{2+} nor H$^+$ alone affected the conformation of the Pep C. However, when mixed with CaM the Pep C undergoes both a dramatic blue shift in tryptophan fluorescence emission, indicative of strong hydrophobic interaction, and an increase in circular dichroism absorption in the α-helical region. Additional fluorescence blue shift and α-helical content occur when Ca^{2+} is added to the CaM:Pep C complex.

2.5.2. Regulation of Gap Junction Channel Properties

Ion channel properties are defined in terms of two measurable electrical parameters namely steady-state channel conductance γ and channel kinetics (also referred to as channel gating).

We will now review a selected group of seminal work on regulation of gap junction channel conductance and kinetics.

2.5.2.1   MIP Gap Junction Channel Acquires Voltage-Dependent Gating After Phosphorylation

The properties of the gap junction channel protein MIP from bovine lens were studied by purification of the protein followed by reconstitution in both liposomes and planar lipid bilayer membranes, together with the freeze-fracture analysis of particle formation in liposomes.[74] MIP channels were found to be permeable to both K^+ and Cl^- and have voltage-dependent gating. Indeed, MIP channel open probability p_o decreased progressively as the absolute value of the transjunctional potential was increased ($p_o = 0.5$ at about 70 mV). Histograms of single-channel current amplitudes showed a monomodal Gaussian distribution centered at 10 pA (80 mV; $\gamma = 120$ pS). Frequency distributions of open and closed times could be fit by a single exponential function. With Cs^+ (or Na^+) in place of K^+ the MIP channel conductance (about 100 pS) remained unchanged. Further studies revealed that MIP channels acquire voltage-dependent gating after phosphorylation. Two isomers of MIP from lens fiber cell membranes were found that differ by a single phosphate at a protein kinase A phosphorylation site, and only the phosphorylated isomer produces voltage-dependent channels. Furthermore, direct phosphorylation with protein kinase A converts voltage-independent channels to voltage-dependent channels *in situ*. Analyses of macroscopic and single-channel currents suggest that phosphorylation increases the voltage-dependent closure of MIP channels by increasing closed-channel lifetimes and the rate of channel closure following the application of a voltage step.[75]

2.5.2.2.   Conductance in Stably Expressed Connexin43 Gap Junction Channels Is Modulated by Phosphorylation

Spray et al.[76] studied connexin43 gap junction channels that interconnect cells of the pacemaking, conduction, and contraction elements of the heart and also endothelial and smooth muscle cells of vasculature. When human connexin43 is stably expressed in pairs of a communication-deficient cell line, channels are produced with unitary conductance γ_j, lipophile sensitivity, and voltage-dependent gating similar to those of mammalian systems in which connexin43 is endogenously expressed. They also found that at moderate transjunctional voltages V_j, two γ_j values dominated the recordings, about 60 and 90 pS (CsCl in patch pipet) and that the 60-pS size is favored by phosphorylating and the 90 pS channel, by dephosphorylation. Human connexin43 mutants (truncated at the carboxyl termini) display a change in γ_j while a point mutation in the third transmembrane spanning domain appears to change channel selectivity. Voltage dependence of the human connexin43 channel is marked and kinetic, but steady-state behavior is not affected by phosphorylation state.

Recently, Moreno et al.[77] recorded from single human connexin43 gap junction channel expressed in a transfected communication-deficient cell line. They found two discrete classes of channel events at moderate absolute transmembrane potentials (<60 mV), with unitary conductances of about 65 and 95 pS (internal CsCl solution). They also found that human connexin43 expressed in these cells displays multiple apparent molecular weights of about 41 to 45 kDa, and noted that treatment of connexin43 from these cells with alkaline phosphatase converted all the species into the 41 kDa only. In addition, application of alkaline phosphatase (included in the internal medium) to the cell interior yields channels that are predominantly 95 pS in conductance. The 65-pS conductance corresponds to the most common channel size seen in cultured rat cardiac myocytes. Furthermore, these channels were more frequently observed after treatment with a phosphatase inhibitor, which was also shown to increase phosphorylation of connexin43 in these cells under similar conditions. Exposure of the cells to the protein kinase inhibitor staurosporine resulted in decreased phosphorylation of human connexin43 in seryl residues and shifted the proportion of events towards the largest unitary conductance. The authors concluded that the unitary conductance of human connexin43 gap junction channels changes with the phosphorylation state of the protein.

2.5.2.3. Synergistic-Antagonistic Modulatory Action of $[Ca^{2+}]_i$, $[H^+]_i$, and Neutral Anesthetics on Gap Junction Channels

Studies of the effects of methylxanthines, classical inhibitors of cytosolic phosphodi-esterases, on gap junctional and nonjunctional resistance in crayfish septate axons at different $[Ca^{2+}]_i$ and $[H^+]_i$ revealed that the uncoupling efficiency of neutral anesthetics (heptanol and halothane) is significantly potentiated by these drugs.[78] Heptanol was also found to cause a decrease in $[Ca^{2+}]_i$ and $[H^+]_i$ both in the presence and absence of either caffeine or theophylline. A similar, but transitory decrease in $[Ca^{2+}]_i$ was observed with halothane. Based on these results and negative results obtained in further experiments with different extracellular $[Ca^{2+}]_o$, Ca^{2+}-channel blockers (nisoldipine and Cd^{2+}), and ryanodine (as well as inhibition of the Ca^{2+}-induced Ca^{2+}-release channel), participation of elevated Ca^{2+} in the heptanol or halothane effect was dismissed. Furthermore, additional experiments with negative results using forskolin to activate plasma membrane adenylate cyclase, permeable forms of cAMP, phorbol ester activation of protein kinase C, and H7 kinase inhibitor, indicate that neither the heptanol effects nor their potentiation by methylxanthines were mediated by cyclic nucleotides and kinase C. The data also suggest a direct effect of anesthetics, possibly involving interactions with channel proteins. Recently, Lazrak and Peracchia[79] extended these studies to Novikoff hepatoma cells. Gap junction channel conductance and gating sensitivity to transmembrane potential, $[Ca^{2+}]$, $[H^+]$, and heptanol were studied by the dual whole-cell clamp method in cell pairs. Heptanol (1 mM) was found to readily uncouple the cells allowing detection of single channel events with conductances γ_j of 46 and 97 pS. In addition, both Ca^{2+} loading (EGTA buffered) and H^+ loading (by equilibration with 100% CO_2) caused uncoupling. However, CO_2 was only effective when $[Ca^{2+}]_i$ was buffered with EGTA (a H^+-sensitive Ca^{2+} buffer), but not with BAPTA (a H^+-insensitive Ca buffer), suggesting a Ca^{2+}-mediated H^+ effect on gap junctions. This was tested by monitoring the g_j decay at different pCa_i and pH_i values. The authors concluded that these gap junctions are sensitive to $[Ca^{2+}]_i$ in physiological range. Furthermore, low pH_i, without an increase in $[Ca^{2+}]_i$, does not affect g_j.

Small (<10 µm in diameter) and large (≥20 µm in diameter) keratinocytes cell pairs isolated from normal human epidermis were used to compare Lucifer yellow dye coupling. Under control conditions, dye coupling was found in only 4% of the small pairs tested, whereas it was evident in 75% of the large pairs. In addition, all-transretinoic acid (RA) was found to inhibit cell-to-cell coupling (1–100 µM). Dual whole-cell patch clamp recordings from large keratinocyte pairs showed a macroscopic junctional conductance g_j of about 9 nS, which was abolished by heptanol (mM range) in a fully reversible way. Compared to heptanol, RA abolished keratinocyte g_j more slowly and irreversibly in the micromolar range. By contrast, 1 µM RA had no significant effect on g_j. Single-gap junctional channels were also identified between large keratinocytes. Event histograms revealed three main unitary conductances γ_j of about 45, 78, and 106 pS. The dye coupling results indicate that junctional communication is markedly different in pairs of small and large cells, which showed the phenotype and keratin markers of basal and suprabasal keratinocytes, respectively. In the latter cell type, coupling is ensured by channels of three sizes and is blocked irreversibly by pharmacologic concentrations of RA.[80]

2.5.2.4. Connexin32 Forms Voltage-Sensitive Gap Junction Channels

Further dual whole-cell voltage clamp work in pairs of communication-deficient cells which were stably transfected with connexin32, the major gap junction protein of rat liver, revealed strong sensitivity of junctional conductance g_j to transjunctional voltage V_j of either polarity, with the ratio of minimal to maximal g_j being approximately 0.1 at the highest V_j. Steady-state g_j values as a function of voltages of either polarity were well fit by the Boltzmann equation (at a transition potential V_o of about 27.5 mV and the Boltzmann parameter describing voltage dependence equal to 0.06 (corresponding to an energy difference between states of approximately 1 kCal/mol and to approximately two gating charges moving through the field). Transitions between open and closed states in response to transjunctional voltages of either polarity were found to be well fit by a single exponential. Further studies with this cell line using the whole-cell voltage clamp technique showed that in only 15% of the cell pairs examined, current flow through junctional membranes was detectable (macroscopic junctional conductance g_j ranged from 100 to 600 pS). In the remaining 85% of the cell pairs, junctional conductance was less than the resolution of the recording system (≤ 10 pS). Unitary junctional conductance γ_j determined in the lowest conductance pairs or after reducing conductance with a short exposure to the uncoupling agent halothane was 25–35 pS.[81,82] The authors also found that g_j increased with time after dissociation, from 1 to 9 nS (measured 2 h after plating with the maximal value at 24 h). Two unitary conductance values of gap junction channels between pairs of transfected cells were measured after reducing g_j by exposure to halothane or heptanol, namely 25 and 125 pS. The smaller size corresponded to channels that were occasionally detected in the cells prior to the transfection with connexin32.

Another example of a voltage-gated gap junction channel was found in neonatal Schwann cells.[83] These authors studied and compared the gating properties of macroscopic and microscopic gap junctional currents by applying the dual whole-cell patch clamp technique to pairs of neonatal rat Schwann cells. In response to transjunctional voltage pulses V_j, macroscopic gap junctional currents decayed exponentially with time constants ranging from <1 to <10 s before reaching steady-state levels. The relationship between normalized steady-state junctional conductance and V_j was well described by a Boltzmann relationship with e-fold decay per 10.4 mV, representing an equivalent gating charge of 2.4. At $V_j > 60$ mV, the steady-state conductance was virtually zero, a property that is unique among the gap junctions characterized to date. Determination of opening and closing rate constants for this process indicated that the voltage dependence of macroscopic conductance was governed predominantly by the closing rate constant. In the majority of the experiments, a single population of unitary junctional currents was detected corresponding to a unitary channel conductance of approximately 40 pS. The presence of only a limited number of junctional channels with identical unitary conductances made it possible to analyze their kinetics at the single-channel level. Gating at the single-channel level was further studied using a stochastic model to determine the open probability p_o of individual channels in a multiple-channel preparation. p_o was found to decrease with increasing V_j following a Boltzmann relationship similar to that describing the macroscopic conductance-voltage dependence. These results indicate that, for V_j of a single polarity, the gating of the 40-pS gap junction channels expressed by Schwann cells can be described by a first-order kinetic model of channel transitions between open and closed states.[84]

2.5.2.5   *Ca²⁺ Flux Through Vascular Smooth Muscle Cell Gap Junction Channels*

An interesting observation made by Christ, et al.[85] concerns the use of gap junctional channel for the spreading of intracellular Ca^{2+} signaling. Ratio imaging using the calcium-sensitive probe fura-2 was employed to measure intracellular calcium concentrations and intercellular calcium flux through gap junctions in homogeneous vascular smooth muscle cell cultures derived from the human corpora cavernosa. Microinjection techniques demonstrated that fura-2 free acid was freely diffusible through gap junctions between cultured cells. The resting intracellular calcium level in fura-2-loaded cells was 177 nM. A robust increase in intracellular calcium was seen in response to both phenylephrine and the calcium ionophore A23187. Microinjection of Ca^{2+} into individual smooth muscle cells always resulted in significant, although temporally delayed, increases in intracellular calcium levels in adjacent cells; this intercellular calcium flux was reversibly blocked by inhibition of gap junctional communication with 2 mM heptanol. However, although microinjection of D-myo-inositol 1,4,5-trisphosphate into individual smooth muscle cells always produced significant increases in intracellular calcium levels in the injected cell, the intercellular spread of Ca^{2+} in response to IP_3 was more variable than for Ca^{2+} injections. These studies demonstrate that Ca^{2+}, and perhaps IP_3 as well, can diffuse between smooth muscle cells through gap junction channels.

2.5.2.6   *Size of Gap Junction Channel Conductance Varies with the Length of the Cytoplasmic End of the Constituent Connexin Isoform*

It now clear that conductance and gating properties of the gap junction channels depend upon the constituent connexin isoform. To identify the structural basis for gap junction channel behavior in the human heart, a tissue that expresses connexin43 was used and site-directed mutant cDNA of connexin43 with shortened cytoplasmic domains were generated.[63] Premature stop codons were inserted, resulting in proteins corresponding in length to the mammalian isoforms connexin32 and connexin26, which are expressed primarily in liver. All constructs restore intercellular coupling when they are transfected into a human hepatoma cell line that is communication deficient. Whereas wild-type connexin43 transfectants display two distinct unitary conductance values of about 60 and 100 pS, transfectants expressing the mutant proteins, from which 80 and 138 amino acids have been deleted, exhibit markedly different single-channel properties, with unitary conductance values of about 160 and 50 pS, respectively. Junctional conductance of channels composed of wild-type connexin43 is less voltage sensitive compared with transfectants expressing wild-type connexin32. However, both of the connexin43 truncation mutants retained this relative voltage insensitivity. These results suggest that the cytoplasmic tail domain is an important determinant of the unitary conductance event of gap junction channels, but not their gating. Furthermore, since the mutant connexins are missing several consensus phosphorylation sites, modification of these particular sites may not be required for membrane insertion or assembly of human connexin43 into functional channels.

3. DIRECT CELL-TO-CELL COMMUNICATION IN NES

The islet of Langerhans is by far the neuroendocrine organ most studied using biochemical, biophysical, and molecular biological approaches. For this reason we will review first the evidence for direct cell-to-cell communication in this tissue. Next we will examine the more sparse literature on gap junction-mediated coupling in the hypothalamic-pituitary tissue system.

3.1. ISLET OF LANGERHANS

Depending on the species studied, various proportions of α-cells (glucagon), β cells (about 80% insulin in murine rodents), δ cells (somatostatin), PP cells (pancreatic polypeptide), and endothelial cells have been identified in the islet (Figure 1).[86] The distribution of different cell types within the islet,[87] i.e., β cells are centrally located in the islet and α cells are more abundant in the superficial cell layers of the islet (Figure 1), together with the vectorial nature of the microvessel blood circulation,[88] support the concept that this organ is a functional unit. However, the presence of α cells in close proximity to β cells (< about 100 μm) makes it possible, at least in principle, for α cell-released glucagon molecules to reach β-cell glucagon receptors by diffusion and stimulate cAMP production via the glucagon receptor-adenylate cyclase pathway (for the insulin paracrine action see Muruyama et al.).[89] Thus, paracrine glucagon action might prime the β cell response to a glucose load.

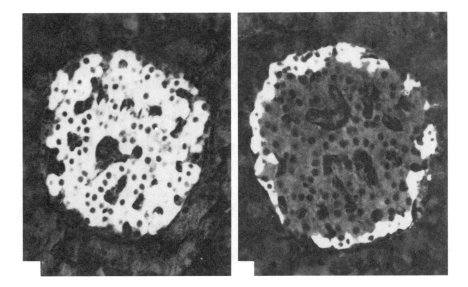

FIGURE 1. Topological distribution of different cell types within the islet. Consecutive serial sections of a rat islet of Langerhans. Indirect immunofluorescence staining with anti-insulin (left) and antiglucagon (right) antisera. Notice the central core of β-cells surrounded by peripheral α-cells. Magnification ×860. (From Orci, L., *Diabetes,* 31, 538, 1982. With permission.)

Despite the lack of experimental data on humoral heterologous cell communication within the islet tissue and, therefore, paracrine action within the islet, it is certain that the different hormones released into the lumen of the islet microvessels in response to changes in blood glucose is perfectly adequate to control plasma glucose *in vivo*.

3.1.1. Gap Junctions in the Islet Tissue

Using freeze-fracture technique, Orci et al.,[90] 20 years ago, found gap junctions connecting different cell types within the islet of Langerhans. This work was followed by the demonstration of cell-to-cell communication between homologous and heterologous cell types within the islet.[91] The development of gap junctions between pancreatic β cells was assessed in freeze-fracture replicas of isolated rat islets under various conditions of insulin secretion. Stimulation of insulin secretion by glucose *in vitro* and by glibenclamide *in vivo* evoked a marked increase in the number of gap junctions between β cells. In addition, the appearance of gap junctions in control and stimulated β cells showed a packing of particles different from that observed under experimental conditions inducing functional uncoupling. These early results suggested a possible role for gap junctions (and probably coupling) in the secretory activity of β cells.[92,93]

The numerical and spatial distribution of gap junctions between insulin-containing β cells analyzed in freeze-fracture replicas of isolated rat islets of Langerhans showed that the β cells located at the periphery of the islet have twice as many gap junctions per unit membrane area as the β cells situated in the islet center. In both locations, gap junctions assumed a nonrandom clustering on the β cell membranes. During stimulation of insulin secretion, the number of gap junctions was found increased and the degree of their clustering was modified in both regions. The latter change depended both on the location of the gap junctions in the islet and on the type of stimulation used, i.e., elevated glucose or glibenclamide.[94,95]

These and further studies revealed that gap junctions between β cells are clustered, occupying only a minute fraction of the cell membrane surface (≤0.4%).[86,94] The β cell plaque is made up of an average of about 26 particles (Figure 2; from Meda et al.),[86] in contrast to those in other secreting cells, such as hepatocytes, which are made up of a few thousand particles.

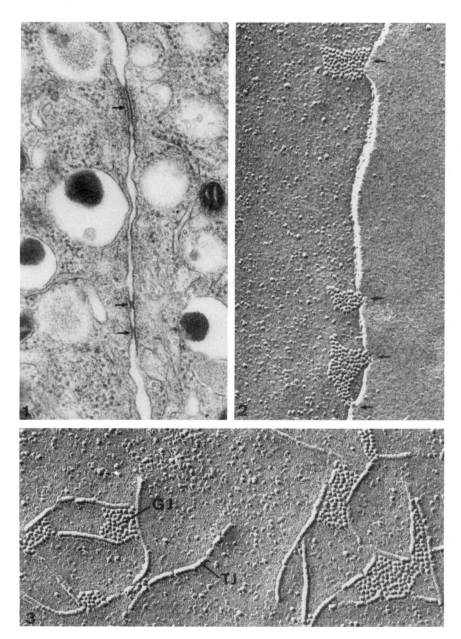

FIGURE 2. Intercellular contacts in a β cell pair (1), gap junctional contacts (2), and junctional particle aggregates (3). (1) Intercellular contacts between two β cells (arrows). (2) Freeze-fracture replica of two juxtaposed plasma membranes at sites of junctional contacts. Arrows indicate that the step between the P face of one cell (to the left) and the E face of the neighboring cell (to the right) is lowered. These sites of intercellular narrowing are bridged with the gap junctional (GJ) particle aggregates seen on the P face (magnification ×85,000). (3) Freeze-fracture replica of an islet cell membrane (P face) showing the association of gap junction particles into aggregates with tight junction (TJ) fibrils (magnification ×91,000). From Meda, P., Perrelet, A., and Orci, L., in *Modern Cell Biology,* Satir, B. H. Ed., Alan R. Liss, New York, 3, 131, 1984. With permission.)

We have already examined evidence showing that the pancreatic β cell expresses connexin43, but not connexins 32 and 26.[52] However, owing to the small size of the gap junction plaques in β cells (Figure 2), the precise localization of connexin43 within the islet cells could not be unambiguously determined by the immunofluorescent methods used to localize the protein.

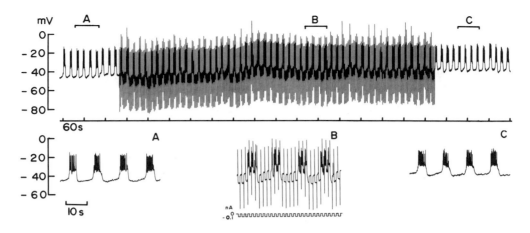

FIGURE 3.   Mouse pancreatic β cell bursting prior to (A), during (B), and after (C) the injection of Lucifer yellow. Upper record depicts the burst pattern of electrical activity obtained during a typical Lucifer yellow dye-coupling experiment to identify by immunocytochemistry the injected cell as well as the dye-coupled neighboring cells. Three segments of the record taken prior to (A), during (B), and after (C) at the times indicated above the upper record are displayed beneath on an expanded time base. (From Santos, R.M., Ph.D. thesis, University of East Anglia, Norwich, U.K.)

3.1.2.   Homologous and Heterologous Dye Coupling

The identification of gap junctions between islet cells indicates that the islet secretory response may be coordinated by the bidirectional flow of signals across gap junctions. We will now review some of the evidence for direct cell-to-cell communication among homologous and heterologous cells within the islet.

Lucifer yellow dye coupling studies were initially carried out on isolated rat islets of Langerhans in physiological medium containing glucose.[96] Intracellularly injected dye was rapidly transferred from the injected cell to neighboring cells regardless of the immunocytochemical identity of either the dye donor cell or the dye recipient cells. The transfer of dye between the islet cells (types α, β, and δ) demonstrates homologous and heterologous cell coupling in a system where the normal proportions and relationships of the cell types are maintained.[96,97] Our own studies of dye coupling among microdissected mouse islet cells not only confirmed these earlier observations, but added the electrophysiological identification of the cells injected with the dye *in situ*.

Figure 3 depicts a segment of a continuous membrane potential record from a mouse islet β cell. The islet was perifused with Krebs bicarbonate buffer containing glucose (11 mM) at 37°C. The intracellular recording microelectrode contained, in addition to the standard high [K^+] filling solution, the fluorescent dye Lucifer yellow for intracellular microinjection. After impalement and identification of the β cell by the classical electrical response to glucose (A), the negatively charged Lucifer yellow molecules were injected by passing negative current into the β cell under current clamp conditions (B). The burst pattern of electrical activity after dye injection (C) remained unchanged. After injection of the dye the islet was rapidly removed from the chamber and fixed in a 4% solution of paraformaldehyde in phosphate buffer as described elsewhere.[98]

To positively identify the cells by successive immunolabeling with specific antihormone serum against one of the three main islet hormones, two to three serial sections cut (1 μm thickness) at a 4-μm distance were taken from each islet. Figure 4 shows examples of dye transfer among homologous islet cells. The territory labeled by the injected dye Lucifer yellow corresponds to the bright yellow fluorescence (left side). The photograph in the middle corresponds to the same section shown on the left using a rhodamine filter to detect the fluorescence from the anti-insulin serum. Right side corresponds to the adjacent islet section

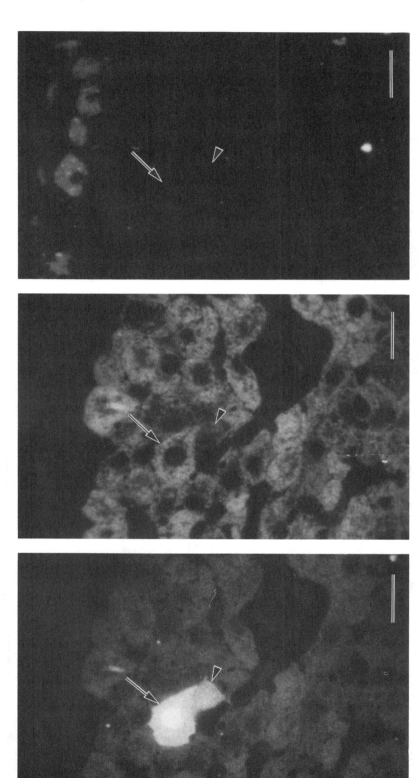

FIGURE 4. Homologous Lucifer yellow coupling in a centrally dye-injected islet β-cell. Shown on the left side is an islet section illuminated to show Lucifer yellow fluorescence. Arrow indicates the injected cell and the arrowhead shows a dye-coupled neighboring cell. The photograph in the middle depicts the rhodamine fluorescence of the same islet section shown on the left after incubation with anti-insulin serum. Another consecutive section from the same islet processed with antiglucagon serum is shown on the right side. Calibration bars represent 10 μm. (From Santos, R.M., Ph.D. thesis, University of East Anglia, Norwich, U.K.)

processed this time with antiglucagon serum to detect glucagon containing α cells. Arrows point to the injected cells and arrow heads to the dye-coupled cell(s). While the photograph on the left indicates a two-cell dye territory, the center one identifies both cells as insulin-containing β cells, and the photo on the right side shows no labeling for glucagon. In conclusion, these data clearly demonstrate that both homologous and heterologous (not shown) coupling exist in the endocrine pancreas.

The number of gap junctions and the extent of dye coupling between insulin-producing β cells was further studied on islets of Langerhans isolated from adult rats treated with either glibenclamide, diazoxide, a blocker of insulin release, or a combination of the two drugs. Glibenclamide treatment was associated with a marked depletion of the islet insulin content, an effect which was blocked by pretreatment of the rats with diazoxide. The study showed that under control conditions, β cells are connected by minute gap junctions (as evaluated on freeze-fracture replicas like those in Figure 2) and exhibit a nonuniform, apparently restricted, dye coupling. Each of the three treatments tested significantly increased the relative and absolute gap junction area of the β cells and the number of detectable, dye-coupled β cells per microinjection. After treatment with glibenclamide alone or with diazoxide plus glibenclamide, an about 1.7-fold increase in gap junction area and an about 3.3-fold increase in the number of dye-coupled β cells were observed. In contrast, following treatment with diazoxide alone, gap junctions and dye coupling were found increased 1.8 and 8.7 times, respectively, as compared with control values.[99]

Previous connexin expression studies already reviewed here revealed that the gap junction channels exhibit different permeabilities to Lucifer yellow and the total number of channels, responsible for the macroscopic g_j will determine the temporal course of detection of Lucifer yellow spreading. Our studies using Lucifer yellow clearly showed very small islet domains which were weakly dye coupled. These data are in good agreement with other published results.

3.1.3. Cell-to-Cell Electrical Coupling

Normal insulin secretion in response to glucose is obtained from undamaged islets isolated either by collagenase digestion from rat pancreas or by microdissection from mouse pancreas. Although cultured islet cells spontaneously form cell aggregates, glucose-induced insulin secretion from these cell aggregates is also diminished. These and other electrophysiological observations suggest that structural integrity is required for full functional capacity. Until now the microdissected mouse pancreatic islet, under mammalian physiological conditions, including the temperature of 37°C and bicarbonate Krebs buffer, is the only islet system that functions as a syncytium. Indeed, stimulatory levels of glucose (>5 mM) evoke synchronized oscillations in: (1) electrical activity in nearby islet cells (bursting),[86,98,100-103] (2) intercellular $[K^+]_e$ and bursting,[104] (3) intercellular $[Ca^{2+}]_e$ and bursting,[105] (4) intracellular $[Ca^{2+}]_i$ and bursting,[106-108] and (5) insulin release and bursting.[109,110]

The oscillatory electrical response to glucose is exquisitely dependent on glucose concentration and, shortly after step changes in glucose level (5–33 mM), β cell oscillations became phase synchronized at different sites throughout the islet.[86,101,107]

3.1.3.1. Synchronized Bursting Among Mouse Pancreatic Islet β Cells

Meissner[100] first demonstrated that in microdissected mouse pancreatic islet cells glucose-evoked bursts of electrical activity were phase synchronized by using pairs of microelectrodes to simultaneously impale two cells within the same islet. Figure 5 illustrates such synchronized activity from our own work. Two membrane potential records were made in the presence of glucose (11 mM) from two nearby cells (labeled Cell-1 and Cell-2 above the records). The diagram on the left side illustrates the islet with the two microelectrode tips 20 μm apart. The similarity of the membrane potential oscillations in both cells is remarkable.

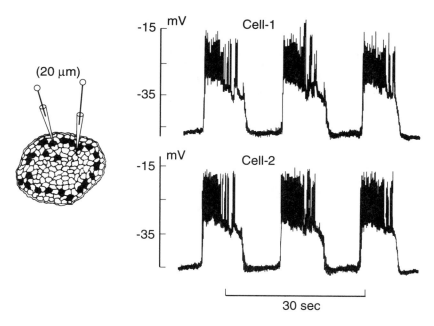

FIGURE 5. Electrical activity from nearby mouse pancreatic islet cells. Diagram on the left represents an intact mouse pancreatic islet of Langerhans. Two microelectrodes were impaled into different islet cells responding to glucose (11 m*M*) with bursting electrical activity. The microelectrodes are positioned and held by two independent micromanipulators. Recording labeled **cell-1** was made under zero current clamp conditions. The other record labeled **cell-2** was made simultaneously using an electrometer amplifier (10^{13} Ω input resistance). Islet was continuously perifused with physiological bicarbonate Krebs buffer at 37°C. (Eddlestone, G. T., Ph.D. thesis, University of East Anglia, Norwich, U.K.)

The maximum tip separation at which a V_m response could be detected in cell-2 during the application of intracellular current pulses in cell-1 ranged between 30 and 40 μm[101] The results from one experiment with the two microelectrode tips 20 μm apart are illustrated in Figure 6A (see diagram on the left). Current injection in cell-1 induced a change in V_m in cell-2 which, when the current pulses were in the depolarizing direction and exceeded threshold, induced a train of action potentials in cell-2 (Figure 6B, bottom record on the right).

Assuming that the current injected into one of the two impaled cells flows radially into a sphere of islet tissue surrounding the microelectrode impaling cell-1, a mathematical relationship, derived by Jack et al.,[111] in terms of membrane resistance (Ωcm^2) and the surface-to-volume ratio for the islet tissue (cm^{-1}) can be used to estimate junctional conductance from the known microinjected current ($i_{c,1}$) and measurement of the membrane potential changes $V_{m,2}$ induced by $i_{c,1}$ in a nearby (≤20 μm) cell.[101]

A least squares regression subroutine was used to calculate the parameters for the best fit of the experimental points (Figure 6B). As illustrated in Figure 6, the spatial distribution of current spread through the islet was mapped and shown to be fit by an equation for radial diffusion in a sphere of coupled β cells. The calculated space constant for the current decay was estimated as 32 μm, (or about 3 cell diameters). Although the estimated junctional conductance g_j was rather low, about 130 pS,[101] this study suggested that most cells in the mouse islet are electrically coupled. In good agreement with these data, the incidence of dye coupling among mouse islet cells was found to be high, with only a third of the cells not dye coupled[98] to any surrounding cells.

3.1.3.2 *Intercellular [K⁺]ₑ and [Ca²⁺]ₑ Oscillations and Bursting*

Action potentials during glucose-induced bursting result from the activation of voltage-gated Ca^{2+} and K$^+$ channels[112] (see Ashcroft and Rorsman, Chapter 14 of this volume). The

A

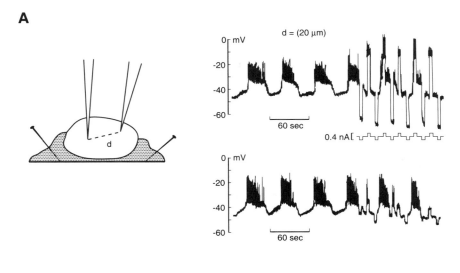

B

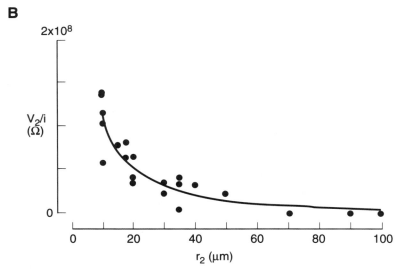

FIGURE 6.   Demonstration of direct cell-to-cell electrical coupling among islet cell pairs. (A) Diagram on the left side shows the relative position of the acinar tissue, islet of Langerhans, and microelectrodes. Three to four entomological pins are used to hold the acinar tissue to the silicone bottom of the chamber (volume 50 μl), avoiding penetration of the islet tissue. The electrical activity from each impaled cell is depicted on the right side (same format as for Figure 5). Rectangular pulses of current were injected into cell 1 (upper record). The polarity of the pulses was alternated. Note that depolarizing pulses during the silent phase of the bursts were able to elicit action potentials. This property is a trait for excitable cells. Also notice the deflections in the membrane potential record of cell 2. We can measure the size of each deflection in membrane potential of cell 2 (V_2) and divide that by the amplitude of the current pulse injected (i) into cell 1. (B) The relationship between V_2/i and distance (d) between the microelectrode tips (see diagram in panel A). Each symbol represents the mean value from at least five deflections per experiment. As explained elsewhere,[101] we used a model developed by Jack et al.[111] to estimate the volume space constant of decay of the electrotonic spread of the injected current. The solid curve represents the best fit of the model to the data. (From Eddlestone, G.T., et al., *J. Membr. Biol.,* 77, 1, 1984. With permission.)

data reviewed so far clearly demonstrate that the endocrine cells of mouse islets are electrically coupled,[101] and for the most part, glucose-induced bursting is synchronous throughout the islet.[86] Since the intercellular space within the islet is rather confined, it is likely that synchronous activity causes the extracellular concentration of the ions involved in the electrical activity undergo fluctuations concomitant with the burst pattern. Figure 7 demonstrates that this is precisely the case for both [K$^+$] and [Ca^{2+}] concentrations. Panel A depicts the burst pattern of glucose (11 mM)-evoked electrical activity from a microelectrode recording V$_m$ (upper record), together with the simultaneously recorded signal from a K$^+$-sensitive liquid membrane resin microelectrode (about 1 μm tip) inserted into the extracellular space within the islet (lower record). The records show that [K$^+$]$_e$ rises to a maximum level of about 6 mM during the active phase. The return to basal [K$^+$]$_e$ commences towards the end of the active phase. Panel B illustrates that the changes in [Ca^{2+}]$_e$ occurred in the opposite direction, i.e., [Ca^{2+}]$_e$ drops to 2.1 mM along the active phase.[104] These oscillatory decreases in [Ca^{2+}]$_e$ must result from Ca^{2+} influx into the β cells lining the intercellular space and thus, the intracellular [Ca^{2+}]$_i$ should exhibit periodic elevations. This interpretation was proven to be correct by simultaneous measurements of [Ca^{2+}]$_i$ and V$_m$ in the same islet. Figure 7C, taken with minor modifications from Santos et al.,[107] demonstrates that [Ca^{2+}]$_i$ is indeed oscillating in perfect phase with the bursts of electrical activity.

The possibility of K$^+$ accumulation in the intercellular spaces between excitable cells was first considered by Frankenhaeuser and Hodgkin[113] to explain changes in equilibrium potential for K$^+$ as a result of repetitive electrical activity in the giant axon. Today, K$^+$ accumulation is considered a common phenomenon in the central nervous system[114] and in the heart.[115-117] Recently, Stokes and Rinzel[118] considered the possibility that localized periodic [K$^+$]$_e$ elevations might spread throughout the islet and may serve to synchronize bursting. By modeling the islet as a spherical array of homogenous β cells and using Hodgkin-Huxley-type equations to calculate the bursts of electrical activity, they generated synchronized electrical activity and periodic [K$^+$]$_e$ rises which were concomitant with the active phase of the cells in the model islet, even in the absence of direct electrical coupling. Therefore, these calculations support the idea that K$^+$ accumulation might synchronize electrical activity at least among β cells forming part of discrete islet domains. It should be noticed that cultured single isolated cells exposed to glucose (11 mM) exhibit chaotic electrical activity.[119] Furthermore, synchronization of glucose-induced electrical activity in a cluster of electrically coupled β cells is only achieved in clusters of 50 or more cells.[120,121]

3.1.3.3 Intracellular [Ca^{2+}]$_i$ Oscillations and Bursting

The influx of Ca^{2+} required for secretion in the pancreatic β cell is provided by a highly synchronized oscillatory electrical activity occurring in the form of bursts of Ca^{2+} action potentials. The previous observation that islet intercellular free Ca^{2+} levels undergo spontaneous oscillations in the presence of glucose, together with the fact that islet cells are coupled through gap junctions, hinted at a highly effective coordination between individual islet cells. Through the use of simultaneous recordings of intracellular calcium and membrane potential it is now reported that the islet calcium waves are synchronized with the β cell bursting electrical activity.[107] This observation suggests that each calcium wave is due to Ca^{2+} entering the cells during a depolarized phase of electrical activity. Moreover, fura-2 fluorescence image analysis indicates that calcium oscillations occur synchronously across the whole islet tissue. The maximal phase shift between oscillations occurring in different islet cells is estimated as 2 s.[107] This highly coordinated oscillatory calcium signaling system may underlie pulsatile insulin secretion and the islet behavior as a secretory syncytium. Since increasing glucose concentration lengthens calcium wave and burst duration without significantly affecting wave amplitude, we further propose that it is the fractional time at an enhanced Ca^{2+} level, rather than its amplitude, that encodes for the primary response of insulin-secreting cells to fuel secretagogues.

3.1.3.4. Pulsatile Insulin Release and Bursting

The demonstration of phase-synchronized electrical activity within the islet of Langerhans suggested that insulin release from islet β cells might also be synchronized and pulsatile, with the maximum occurring during the active phase of the bursts. Our simultaneous measurements

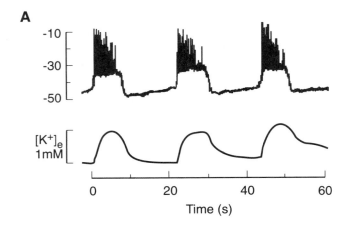

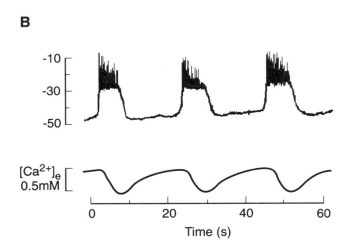

FIGURE 7. Oscillatory behavior of $[K^+]_e$ (A), $[Ca^{2+}]_e$ (B), $[Ca^{2+}]_i$ (C), and insulin secretion (D) in the mouse islet of Langerhans. A and B were reproduced with minor modifications from Perez-Armendariz et al.;[105] C was reproduced with minor modifications from Santos et al.[107] D was reproduced from Rosario et al.[110] with minor changes. See text for explanations.

of insulin release and electrical activity from single normal mouse islets confirmed this hypothesis.[109] To improve the time resolution of the insulin measurements, we repeated these experiment using the larger ob/ob mouse islets which display slower burst frequency.[110] Figure 7D depicts the results from one such experiments. With the ob/ob islet preparation we were able to obtain a better time resolution by collecting the perifusion solution every 10 s. As expected, insulin release was pulsatile and each insulin bolus from the entire islet was found to be released during the active phase of the bursting electrical activity recorded from one impaled cell.[110,122]

3.1.4. Characteristics of Gap Junction Channels in Cultured Mouse Islet Cells

Using the dual whole-cell recording protocol, Perez-Armendariz et al.[123] studied direct coupling between pairs of cultured mouse pancreatic islet cells. The majority of the cell pairs examined were coupled (67%) through a mean junctional conductance of 215 pS. Electrical coupling could be maintained for long-lasting periods (up to 60 min) and, during this time, the junctional conductance remained insensitive to transjunctional potential. Interestingly, the

C

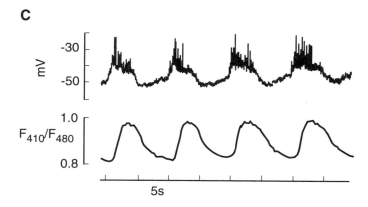

D

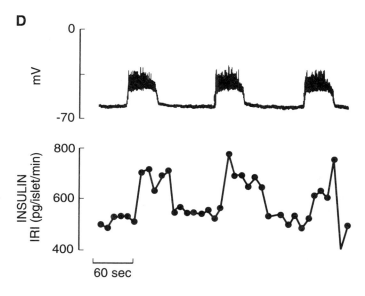

FIGURE 7. (continued)

size of the junctional conductance between cultured cell pairs is similar to that estimated by Eddlestone et al.[101] from measurements in electrophysiologically identified β cell pairs *in situ* under physiological conditions of bicarbonate Krebs buffer at 37°C. Thus, low junctional conductance values between cultured islet cell pairs appears to be a trait of the mouse islet cell-to-cell coupling. The authors also noted that electrically coupled pairs were not necessarily Lucifer yellow dye coupled. Since connexin43 gap junction channel conductance has been found to range between 40 and 150 pS in other cell systems, the group expected to detect discrete transitions including single-channel events as previously reported for other connexin43 gap junction channels. However, no single-channel events were ever apparent in the records. The authors conclude that the single gap junction channel conductance in islet β cells is less than 20 pS, the resolution of their recording system. This value is significantly lower than single-channel events in other tissues expressing connexin43, and may account, in part, for the lower G_j in β cells as opposed to other connexin43-coupled tissues.

Despite the apparent wide range of electrical coupling, Lucifer yellow coupling studies, both at room and physiological temperature of 37°C, conclusively show the dye coupling to be restricted to two to five cell islet domains.[96,98] We have reviewed some of the evidence showing that connexin43 gap junction channel conductance depends on the phosphorylation state of the protein. It is possible that the process of phosphorylation, by addition of a negative charge to the phosphorylation domain of the channel protein, may differentially affect cation and anion permeabilities. By this mechanism Lucifer yellow flux through the open channel might be impeded, while cation flux might be augmented. Furthermore, g_i and dye coupling measurements were carried out simultaneously in cultured β cells by delivering Lucifer yellow through the patch pipet into one cell of the pair in a dual whole-cell recording configuration. While electrical coupling was rapidly detected and persisted up to 15 min, no dye transfer into the second cell could be observed. Thus, absence of observed dye transfer between β cells does not preclude the existence of electrical coupling.[123] It should be mentioned here that in other cell systems which exhibit both dye-coupled as well as electrically coupled domains, almost no overlapping of these territories is observed.[124-127]

3.1.5. Modulation of Cell-to-Cell Coupling Among β Cells *in Situ*

The dual whole-cell voltage-clamp technique provides a measure of the junctional conductance between a pair of cells by allowing one to selectively change the potential across the junction and observe the change in coupling current directly. However, due to the nature of the patch clamp technique, these experiments generally must be carried out at low temperatures (below 30°C) with nonphysiological buffers. Furthermore, all information about the number of neighbors with which a given cell communicates is lost when the tissue is dispersed into cell pairs.

The development of an intracellular voltage-clamp technique to measure membrane ionic currents from β cells within the intact islet of Langerhans[128] led Sherman et al.[129] to predict that gap junction electrical conductance measurements could be made *in situ* under near-physiological conditions. The method, described in detail below, resembles the dual whole-cell voltage-clamp technique in that the potential of one cell is held at a constant level to measure changes in coupling current. Here, however, the glucose-induced oscillations in membrane potential of the surrounding cells provide the necessary changes in potential across the cell-to-cell junctions, and the conductance measured is the total, parallel conductance between the clamped cell and each neighbor to which it is directly coupled.

Figure 8A depicts the experimental setup used to voltage clamp β cells within the intact islet of Langerhans. Microdissected islets of Langerhans are pinned to the sample chamber and continuously perifused at 37°C with bicarbonate-buffered Krebs solution supplemented with glucose (11 mM), conditions which stimulate both glucose-induced bursting electrical activity and pulsatile insulin release from β cells *in situ*. β cells are impaled with high-resistance (about 150 MΩ) glass microelectrodes, and either voltage or current clamp recordings are made using a conventional patch clamp amplifier.

The inset at the top right of Figure 8A shows an expanded view of the vicinity of an impaled β cell. When the amplifier is set to voltage clamp mode, the membrane potential of the impaled cell is held constant at the command level, V_h, but the surrounding cells remain free to burst. As the figure illustrates, the holding current is given by

$$I_h = I_m + I_c \tag{1}$$

where I_m is the current through the cell membrane to ground, and I_c is the coupling component into neighboring cells through gap junctions. Since nearby β cells have nearly identical membrane potential[100-102] and β cell gap junctions are ohmic,[123] the holding current can be expressed as

$$I_h = I_m + G_j(V_n - V_h) \tag{2}$$

where V_n is the potential of the neighboring cells and G_j is the parallel conductance

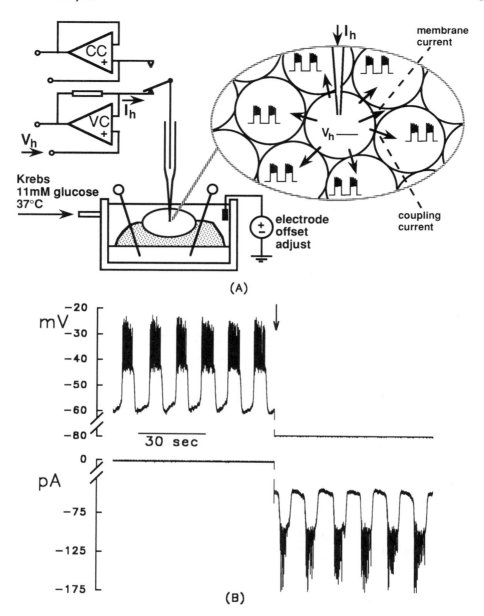

FIGURE 8. Junctional conductance measurements among β-cells *in situ*. (A) Schematic diagram of the experimental arrangement used for voltage clamping single β cells *in situ* within the intact islet of Langerhans. A manual switch on the amplifier allows the choice of voltage or current clamp recording. An expanded view of the vicinity of the impaled β cell (inset at right of A) during voltage clamp recording shows the membrane potential activity of the neighboring cells and the various components of the holding current. (B) Simultaneous recordings of β cell membrane potential (top trace) and holding current (bottom trace) during an *in situ* coupling conductance measurement. Arrow indicates the switching of the amplifier from current clamp to voltage clamp mode. Burst magnitude is defined as the difference between the silent phase level and the plateau of the active phase.

between the impaled cell and its nearest neighbors. Since I_h is directly proportional to V_n, the holding current record will consist of "bursts" of current reflecting the electrical activity of neighboring cells. If the active and silent phase potentials of the neighboring cells are given by V_a and V_s, respectively, then substituting for V_n in Equation 2 gives

$$I_{h,a} = I_m + G_j(V_a - V_h) \qquad (3)$$

as the holding current during the active phase, and

$$I_{h,s} = I_m + G_j(V_s - V_h) \tag{4}$$

as the holding current during the silent phase. Subtracting these two equations and rearranging yields an expression for the parallel coupling conductance G_j:

$$G_j = \frac{\left(I_{h,s} - I_{h,a}\right)}{\left(V_s - V_a\right)} \tag{5}$$

In an actual experiment, the electrical behavior of the surrounding cells is unknown, but can be estimated by recording a few voltage bursts from the impaled cell in current clamp mode.[101,102] After switching to voltage clamp, the magnitude of the current bursts can be divided by that of the previously obtained voltage bursts to estimate G_j. Using a mathematical model of coupled β cells, Sherman et al.[129] determined that for reasonable cell-to-cell conductance values (based on dual whole-cell voltage-clamp data), the parallel gap junction conductance could be measured to within 10–25%.

Figure 8B shows the results of an *in situ* gap junction conductance experiment, with membrane potential on the upper trace and holding current on the lower trace. In current clamp mode ($I_h = 0$), glucose-induced bursts of membrane potential were observed in the impaled cell. Upon switching to voltage clamp mode ($V_h = -80$ mV), as indicated by the arrow in Figure 8B, bursts of inward current were obtained, with nearly identical shape to the previously observed voltage bursts. Dividing the magnitude of the current bursts (47 pA) by that of the voltage bursts (19 mV) yields a coupling conductance of 2.47 nS between the impaled cell and its nearest neighbors.

Each of over 25 cells tested to date has been found to be coupled to at least one neighboring cell, as indicated by the occurrence of bursts in the holding current record during voltage clamp. The mean coupling conductance from this group of cells was close to 3.0 nS. This number agrees reasonably well with that which would be expected based on the study of mouse β cell pairs by Perez-Armendariz et al.[123]. While the measurements provide no information about the size of individual junctional conductances or the total number of neighbors with which each β cell communicates, the 100% incidence of coupling and large mean junctional conductance provide evidence that electrical coupling within the islet of Langerhans is quite extensive, and is sufficient to synchronize glucose-induced bursting.[121] Furthermore, the method can be used to study changes in cell-to-cell coupling brought about by various stimuli, provided that the islet continues to burst. Coupling in other systems which exhibit spontaneous membrane potential oscillations, such as pituitary gonadotrophs, could also be investigated using the single microelectrode voltage clamp technique.

3.2. HYPOTHALAMUS
3.2.1. Dye Coupling

Vertebrate reproduction is dependent on the operation of a central signal generator that directs the episodic release of GnRH, a neuropeptide that stimulates secretion of the pituitary gonadotropic hormones and, thereby, controls gonadal function.[130] The electrophysiological correlates of this pulse generator are characterized by abrupt increases in hypothalamic multiunit electrical activity invariably associated with the initiation of secretory episodes of luteinizing hormone.

Most magnocellular neurosecretory cells that terminate in the posterior pituitary secrete either vasopressin, oxytocin, or enkephalin. Intracellular injection of the fluorescent dye Lucifer yellow into single magnocellular neurons in slices of rat hypothalamus resulted in dye transfer between these cells. Freeze-fracture replicas of these cells occasionally revealed gap junctions, which presumably contain channels that mediate the dye coupling. These two independent techniques strongly suggest that some mammalian neuropeptidergic cells are electrotonically coupled, providing a possible means for recruitment and synchronization of their electrical activity.[131,132]

The ependyma lines the ventricular system of the vertebrate brain and spinal cord. Although its embryology and morphology have been studied extensively, little is known of its physiological properties, particularly in mammals. Tanycytes are modified ependymal cells that are found predominantly lining the floor of the third ventricle, overlying the median eminence. Their processes accompany and enwrap neuroendocrine axons that course from hypothalamic nuclei to terminals in the median eminence, but the significance of this interaction is not yet understood. Intracellular recording and injection techniques were used to study ependymal cells and tanycytes of the rat in the hypothalamic slice preparation after differentiating their respective regions morphologically. With physiologic extracellular $[K^+]_o$, the mean membrane potential for both common ependyma and tanycytes was about −80 mV and both cell types exhibit a near-Nernstian response to changes in extracellular in $[K^+]_o$ between 3 and 20 mM. Input resistances (Rin) were extraordinarily low (<1 MΩ). Single-cell injection of Lucifer yellow revealed dye coupling among 2–70 ependymal cells and 5–48 tanycytes. In both freeze-fractured replicas and thin sections, large numbers of gap junctions were found between adjacent ependymal cells and between adjacent tanycytes. These observations demonstrate that both populations are strongly coupled networks. Electrical stimulation of the arcuate nucleus did not elicit any detectable synaptic response in impaled tanycytes, so that the functional significance of synaptoid contacts between neuroendocrine neurons and the postsynaptic tanycytes is not yet apparent. The high negative Nernstian resting potential, extremely low input resistance, extensive dye coupling, and absence of spontaneous electrical excitability demonstrate that ependymal cells possess numerous glial characteristics and may therefore have similar functions. In the hypothalamus, ependyma probably take up K^+ released from adjacent endocrine neurons and shunt it to the ventricular space.[133]

3.2.2. GT1-7 Interconnections

An immortalized hypothalamic cell line has recently been developed by genetically targeting LHRH neurons for tumorigenesis. One subclone, the GT1-7 cells, has been characterized at both the light and electron microscopic levels to study the cellular and subcellular organization of these cells, particularly as they relate to biosynthesis, processing, and secretion. LHRH and GnRH-associated peptide (GAP) immunoreactivities were detected by immunocytochemistry using colloidal gold labeling in cells fixed onto slides 18–36 h after plating. While all of the cells immunostained for both LHRH and GAP, GAP immunoreactivity was always more pronounced than that for LHRH. These cultured cells exhibited the classical neuronal appearance of LHRH neurons, and they established numerous interconnections. Neighboring neurons were coupled by tight junctions, while more distant cells were interconnected with neural axon-like processes and collaterals. This cellular organization is suggestive of a neural network where neuronal activity is coordinated.[134]

3.3. HIPPOCAMPUS
3.3.1. Dye Coupling Among Pyramidal Hippocampal Cells

Intracellular injections of Lucifer yellow into CA1 pyramidal cells were made in rat hippocampal slices with the goal of using dye transfer between neurons as evidence that these cells are electrotonically coupled. Extensive control procedures were performed which substantially reduced inadvertent staining. Over half of the neurons were dye-coupled after injections in stratus pyramidale. Dye coupling occurred even when spike amplitudes were greater than or equal to 70 mV throughout the impalement and was still present after chemical synapses were blocked with a low Ca^{2+} solution containing Mn^{2+}. Somata of dye-coupled cells were usually located within 35 μm (postfixation) of the injected cell and showed no preferred orientation. Fast prepotentials and dye coupling occurred independently. Intradendritic injections of Lucifer yellow in stratum radiatum also yielded dye coupling between CA1 pyramidal cells, although the dye coupling was less frequent. Thus, injection of Lucifer yellow into the soma or dendrite of a single CA1 pyramidal cell often resulted in multiple staining, and in many ensembles the somata were well spaced. Control experiments suggested that such dye transfer is not by an extracellular route. This implied that some CA1 cells are electrotonically coupled. Within stratum radiatum, neither extracellular ejections nor intracellular injections of interneurons were associated with multiple staining. Similarly, neurons in superior cervical ganglia, which were sliced and injected using similar procedures, showed no dye coupling. Further electrophysiological and morphological studies are required to resolve the discrepancies among various techniques used to evaluate the amount of coupling in the hippocampus.[135]

3.4. HYPOPHYSIS
3.4.1. Anterior Pituitary Gland
3.4.1.1. Direct Cell-to-Cell Coupling Among Gonadotropes
Nearly 20 years ago, extensive examination of the anterior pituitary tissue revealed that some endocrine cells were joined by gap junctions.[136] Mammotropes (cells producing prolactin) were the prevalent cell type joined by gap junctions while such contacts were less frequent between somatotropes (cells secreting growth hormone), gonadotropes (releasing either luteinizing hormone [LH] or folicle-stimulating hormone) or thyrotropes (secreting thyroid-stimulating hormone). Usually the junction occurred between the body of one cell and a deeply invaginated process of a nonidentifiable partner. Whether or not these cell contacts involved folliculo-stellate cell corticotropes (producing adreno-corticotrophic hormone, ACTH) was uncertain. Further analysis using freeze-fracture of the gland showed small macular gap junctions. Mammotropes and gonadotropes could be readily distinguished and differentiated from nonendocrine follicle cells. Electrophysiological studies of cultured pituitary endocrine (containing numerous secretory granules) cells showed them to be electrically coupled. The authors proposed that "homologous endocrine cells of the anterior pituitary gland when joined by gap junctions can act as a functional syncytium".[136]

Immunohistochemical methods were employed to investigate the cellular and ultrastructural localization of the gap junction protein connexin43 in rat pituitary (also see Meda et al.).[137] Western blots of pituitary homogenates probed with anti-connexin43 antibodies showed the presence of the peptide in both anterior and posterior pituitary lobes. By light microscopy, connexin43-immunoreactive puncta were found in all areas of the posterior lobe, but at greater concentrations in peripheral regions of this structure. By electron microscopy with immunogold labeling, connexin43 was seen at gap junctions between thin cytoplasmic processes of pituicytes. No immunoreactivity was detected in the intermediate lobe. The anterior lobe contained puncta similar to, but more sparsely scattered than those in the posterior lobe, and by electron microscopic analysis these were demonstrated to correspond to labeled gap junctions between stellate cells. In addition, anti-connexin43 antibodies produced intracellular labeling in a small percentage of endocrine cells which were distributed throughout the anterior lobe and determined by double immunostaining methods to contain LH. Labelling within these cells was associated with predominantly large secretory granules and other loosely organized organelles. The results indicate that gap junctions in the pituitary are composed of connexin43 and that this or a related protein may have a novel intracellular function within gonadotrophs.[138,139]

The pulsatile pattern of GnRH release from the hypothalamus is driven by a functionally interconnected and synchronized network of GnRH neurons termed the GnRH pulse generator. Several recent observations have revealed that immortalized GnRH neurons can generate an episodic pattern of GnRH release when cultured in the absence of other cell types. The *in vitro* operation of the pulse generator depends on the development of synaptic contacts among GnRH neurons, the electrical properties of individual GnRH cells, and the GnRH-induced modulation of its secretory mechanism. The expression of several other receptors by GnRH neurons provides the means for integrated regulation of pulse generator activity from without the network by agonists including glutamate, GABA, endothelin, and catecholamines.[140]

3.4.1.2. *Cell-to-Cell Coupling Between Folliculo-Stellate Cells*

In a series of recent studies, Soji et al.[139,140] and Yamamoto et al.[140] followed the postnatal development of gap junctions and cell-to-cell communication in male rat anterior pituitary cells (age 10–40 d). The junctions initially appeared between adjacent folliculo-stellate cells (day 20). The number of junctions increased to a level similar to that found in adults (ca. 40 d). The ontogeny of these junctions was examined in rats treated with LH releasing hormone or testosterone. The appearance of gap junctions was accelerated in a similar fashion by both treatments. The results suggest a role for the gonadal steroid hormones in the formation of gap junctions. Similar postnatal gap junction development was observed in the female rat anterior pituitary. In addition, a correlation was evident between the number of gap junctions and stages of the estrous cycle, where they were most numerous during either proestrus or estrus. Taken together, these results suggest that gap junction formation within the female rat hypophysis is in part modulated by both gonadal steroid hormones as well as prolactin[141] (also see Reference 137).

3.4.1.3. *Lack of Cell-to-Cell Coupling During the Early Phase of Human Pituitary Development*

The ultrastructure of cell and tissue contacts in the developing pituitary gland in human embryos (3 to 5 weeks of age) revealed that at an early stage, the pituitary anlage, Rathke's pouch (RP) of the pharyngeal gut wall, had no direct contact with the diencephalon. At later stages, several close cell-to-cell contact sites were observed between the two tissues, on the upper anterior aspect of RP. However, even at these contact sites a thin layer of extracellular material always remained between the tissues, which were elsewhere covered by a well-developed basement membrane. Rathke's pouch was, in addition, surrounded by closely adjacent mesenchymal cells during the time studied.[142]

3.5. PINEAL GLAND
3.5.1. Direct Coupling Among Pinealocytes

Freeze-fracture and conventional electron microscopic studies of the intercellular junctions between the pinealocytes of male rats revealed that the intercellular contacts between pineal cells, formerly described as zonulae adhaerentes or zonulae occludentes, are in fact gap junctions. These gap junctions are difficult to characterize in thin sections due to their peculiar geometrical arrangement, which is in the form of fenestrated communicating zonules.[143] The authors propose that the arrangement of these communicating zonules around rudimentary lumina of pineal clusters and rare transitions between tight and gap junctions may point to phylogenetic transformations of occluding into communicating zonules, corresponding to the change of the pineal gland from a sensory to a secretory organ. Alternatively, these tight-to-gap junctional transitions may reflect the periodic (circadian or seasonal) activity of the pineal gland.

In an earlier study, Huang et al.[144] used combined thin-section/freeze-fracture techniques on the superficial pineal gland of the golden hamster, comparing the parenchymal and interstitial cells of this animal with those previously investigated in rats. In contrast to rats, no gap junctions and gap/tight junction combinations could be found between pineal parenchymal cells of the hamster. Furthermore, the interstitial cells of the hamster pineal gland were found to have large flat cytoplasmic processes, which abut over large areas equipped with tight junctions. In thin sections, profiles of interstitial cell processes were seen to surround groups of pinealocytes. Interstitial cells and their sheet-like, tight junction-sealed processes thus appear to delimit lobule-like compartments of the hamster pineal gland. Because the typification of the interstitial cells is uncertain, the expression of several markers characteristic of mature and immature astrocytes and astrocyte subpopulations were also studied by immunohistology. Many of the nonneuronal elements in the pineal gland are vimentin-positive glial cells, subpopulations of which express glial fibrillary acidic protein and C1 antigen. The astroglial character of these cells is supported by the lack of expression of markers for neuronal, meningeal, and endothelial cells. M1 antigen-positive cells have not been detected.

In accordance with previous results in rats, additional studies[145] showed belt-like arrangements of fenestrated gap junctions around the collicular segments of pineal cells in the guinea pig. In addition, macular interpinealocyte gap junctions have been observed in this species.

3.6. ADRENAL GLAND
3.6.1. Medulla

Grynszpan-Wynograd and Nicolas[146] examined freeze-fractured specimens and found intercellular junctions in adrenal medulla which appear to be different in nature and number according to species. Only gap junctions of diverse size exhibiting characteristic loop-like configurations were found in hamster chromaffin cells. In addition to such gap junctions, polymorphic focal tight junctions occasionally combined with particle clusters or small gap junctions were found in guinea pig. So far, no intercellular junctions were found in rat. Discussion is focused on the possible function of these junctions, in keeping with their presumably high lability.[147] A recent differential expression analysis by immunostaining cryostat sections with antibodies against connexins 26, 32, and 43 has identified connexin43 in the rat adrenal medulla.[137]

3.6.2. Cortex

Meda et al.[137] also identified by immunofluorescence labeling connexin43 in cells from zonas glomerulosa, fasciculata, and reticularis. Freeze-fracture replicas of human adrenocortical cells revealed *en face* fractures (both P and E faces) of the cell surface which could be grouped into three patterns: (1) fracture face with no microvillous projections; (2) fracture face mainly with short microvillous projections which arise horizontally from the edges of the small cavity on the cell surface to form a flower bud-like configuration on the P face, and (3) fracture face mainly with varying-sized, longer microvillous projections which arise almost vertically from the cell surface and show no specific organization. The former two fracture faces were supposed to correspond to the plasma membranes facing the interstitium including pericapillary spaces, since they had no gap junctions and frequently appeared along sinusoidal capillaries. On the other hand, the latter fracture face seemed to be equivalent to the plasma membranes facing intercellular spaces, since it frequently showed gap junctional specialization of intramembranous particles and usually both the P and E fracture faces of plasma membrane on the same surface. *En face* fractures of these gap junctions were various in shape and size. Tight junctions and their complex form with gap junctions were not encountered. Cell organelles and the nucleus had more numerous intramembranous particles on the P fracture face than on the E fracture face.[148]

Furthermore, studies using ACTH-sensitive and ACTH-insensitive Y-1 adrenal cortical tumor cell lines suggested a relationship between responsiveness to ACTH and the presence of gap junctions. An ACTH-sensitive clone of Y-1 cells was shown to possess gap junctions, and these junctions appear to enlarge with ACTH treatment. Gap junctions have not been observed, however, in an ACTH-insensitive clone of Y-1 tumor cells even when stimulated to produce cAMP and steroids with cholera toxin.[149,150] Although gap junctions have been detected recently in the adrenal cortex,[151] direct cell-to-cell electrical and dye-coupling studies have not yet been carried out on these cells. However, gap junction protein expression, evaluated by immunostaining cryostatsections with antibodies against connexins 26, 32, and 43, showed that the adrenal gland expresses as connexin43, variable levels of connexin26, and no connexin32.[137]

4. DIRECT CELL-TO-CELL COMMUNICATION IN THE DIFFUSE NES

4.1. EPIDERMAL MERKEL CELLS

Merkel cells are specialized epidermal cells that make synaptic contacts with nerve endings[152,153] (see Akaike and Yamashita, Chapte 14). Whether or not these intercellular communications are mediated by gap junctions remains to be elucidated.

4.2. THYROID

Connexin43 has been detected in thyroid follicles and within the interfollicular regions of the rat thyroid by differential immunofluorescence labeling. In addition to connexin43, the follicles showed strong labeling with the connexin32 antibody.[137]

The acute influence of thyroid-stimulating hormone (TSH) on the $[Ca^{2+}]_o$ and cytoskeletal dependence of the thyroid epithelial occluding barrier was investigated in thyrocyte monolayers cultured on filter support. Acute stimulation with TSH protects the barrier function of thyrocyte monolayers from the otherwise deleterious influence of $[Ca^{2+}]$-chelating and microfilament-disruptive agents. This effect of TSH is mediated by cAMP. The data indicate that the integrity of the thyroid occluding barrier not only remains but is strengthened in TSH-stimulated cultures. The possible involvement of cell adhesion molecules in this response to TSH is discussed.[154,155]

4.2.1. Cell-to-Cell Coupling Is Modulated by Phosphorylation

The ability of thyroid cells in primary culture to reconstitute gap junctions and the effects of extracellular signals on the functional activity of these junctions were studied by Lucifer yellow cell-to-cell transfer. Isolated thyrocytes in culture either formed monolayers or reorganized in follicular structures in the presence of the glycoprotein hormone TSH. In both systems dye-coupled cells were evident after a few days. The communication between cells forming a reconstituted thyroid follicle was maintained for up to 9 d. In contrast, the dye coupling between cells in monolayers progressively decreased with time. The number of dye-coupled cells in thyroid cell monolayers was augmented by TSH in a time- and concentration-dependent manner. The TSH action was not related to *de novo* protein synthesis. Furthermore, cAMP exhibited stimulatory effects similar, in terms of time course and amplitude of action, to those of TSH. The phorbol ester TPA rapidly inhibited both basal and TSH- or cAMP-activated cell-to-cell communication. The dye coupling of cells in reconstituted follicles was also blocked by a short TPA treatment in both the presence and absence of TSH. The data show that thyroid cells in culture, regardless of the full expression of the differentiated phenotype, rapidly reestablish intercellular gap junctions. The functional activity of gap junctions appears to be regulated positively by a hormone, TSH, probably acting via the cAMP and protein kinase A pathway, and negatively by phorbol esters through the activation of protein kinase C, the two regulatory pathways being interdependent.[156]

Further dye coupling studies on monolayer cultures of rat thyrocytes revealed that cell-to-cell coupling among thyrocytes gradually developed during the culture. Intercellular communication was demonstrated in approximately 60% of the cells cultured for 8 d, while it could not be detected in the remaining 40% of the cells even after longer culture. When thyrocytes were cultured in the presence of thyroid-stimulating hormone the fluorescence recovery after photobleaching was not affected.[157]

Additional work showed that thyroid cells, cultured in the presence of TSH, reorganized within 36–48 h into follicular structures, the *in vitro* reconstituted thyroid follicles. By microinjection of fluorescent probes either into the neoformed intrafollicular lumen or into cells forming the follicles, the development and functional properties of cell-cell contacts involved in the formation of the thyroid follicular lumen and the communication between thyrocytes within the follicle could be further determined. Microinjection of Lucifer yellow in a cell involved in the follicle structure led to the rapid labeling of the other cells forming the follicle, but the dye did not penetrate the intraluminal space. Unlike Lucifer yellow, the higher-molecular weight FITC-Dextran injected into cells delimiting the lumen remained within the microinjected cells. Alterations of medium or intralumenal Ca^{2+} concentration which caused the opening of the intraluminal space did not affect the cell-to-cell dye transfer. From these studies it was concluded that the structure of the thyroid follicles appears to be under the control of both extracellular and intraluminal Ca^{2+} concentrations.[158]

Porcine thyroid follicles, when isolated by enzymatic digestion and suspended in a medium containing 10% fetal calf serum, undergo inversion of cellular polarity. After isolation, the strands for the tight junctions (*zonulae occludentes*) between follicle cells begin to move towards the side of the medium and gather at this side of the lateral plasma membrane during 24 h of incubation. Around this time, microvilli of many follicular cells protrude into the culture medium. The elements of the Golgi apparatus are located at the luminal as well as the culture medium side of the cytoplasm, and also at the lateral side of the nucleus after 24 h of suspension culture, and by 94 h of incubation almost all elements of this organelle, as well as lysosomes and the central cilium, have migrated to the side of the medium. The migration of the zonulae occludentes is considered to be the initial change in the reversal of the polarity of this cell.[159]

The organization of tight junctional complexes has been studied also in cultured porcine thyroid cells during the inversion of polarity induced by collagen embedding of inside-out follicles, using freeze-fracture replicas and lanthanum penetration. During the early steps of polarity reversal, freeze fractures showed that tight junctions generally persisted. They increased in width and progressively branched out into the basolateral surfaces, towards the basal pole. Later, the number of tight junction strands decreased and gap junctions inserted within tight junction networks were found between cells in reversed follicles, in the same manner as in typically polarized follicles, embedded in collagen or in suspension.[160]

In the thyroid follicles and blood of goats with congenital hypothyroidism and goiter, abnormal iodoproteins (e.g., iodoalbumin) are found. To study the mechanism involved in the passage of these proteins between the follicles and the blood, the morphology of tight junctions in goiters and normal thyroids of goats was studied by means of freeze fracturing. The mean numbers of strands composing the tight junctions were negatively correlated with TSH levels. Hence, the tight junctions of the glands of hypothyroid goats are narrower and are composed of fewer strands than those of normal thyroids. This reduction in tight junction complexity may provide an explanation for the leakage of proteins into the follicles of goitrous glands.[161]

4.3. PARATHYROID

Gap junction protein expression studies and differential tissue distribution by immuno-fluorescence labeling revealed the presence of connexin43 and variable levels of connexin26, and no connexin32 in the parathyroid gland.[137] Previous analysis of the fine structure of the intercellular junctions of the hen parathyroid gland using freeze-fracture replicas and thin sections showed the presence of desmosomes, intermediate junctions (maculae adherentes), and gap junctions. In the lanthanum-fixed sections, tight junctions (maculae occludentes) were demonstrated as well. In the freeze-fracture replicas, desmosomes, gap junctions, tight junctions, and combination forms of gap and tight junctions occurred, but intermediate junctions were not identified. The gap junctions varied in size and shape, ranging from minute assemblages of particles to large aggregations of a round or elliptic outline. The tight junctions and the combination forms of gap and tight junctions also exhibited a variety of shape and dimension, and, depending on the form of the tight junctional strands, they were classified into three types: type I consisted of a simple line of strands; type II consisted of a closed network of strands; and type III consisted of an open network of strands. The combination forms were more numerous than the typical tight junctions.[162]

4.4. THYMUS

Analysis of thymic lymphocytes isolated from mice has revealed a minority population able to form permeable, intercellular (gap) junctions. This population is largest in mice aged between 3 and 6 weeks, much smaller in fetal and newborn mice and undetectable in mice aged 12 weeks or more. Fractionation of the thymocytes on Percoll gradients or with peanut agglutinin shows the cells able to form junctions are enriched in lower density fractions and agglutinated by PNA, suggesting they are among the most immature. Fractionation by complement-mediated cytotoxicity and by fluorescence-activated cell sorting using monoclonal antibodies to specific cell surface determinants shows the junction forming cells are Lyt-1+/Lyt-2- and that the phenotype is associated with both high and low Thy-1 and H-2K epitope densities.[163-165]

5. DIRECT CELL-TO-CELL COUPLING AND BIOLOGICAL CLOCKS

5.1. ENDOGENOUS TISSUE CLOCKS

Most cells in NES are endowed with endogenous oscillators which, at least in principle, could set the pace for a specific cellular process and therefore constitute a biological clock. Classical examples are hypothalamic neurons and pituitary gonadotrophs. At the cellular level, the tissue specificity of the plasma membrane oscillator is conferred by the membrane resident cell receptors, channels, and enzymes. Similarly, for the cytosolic oscillator the cytoplasmic enzymes together with the intracellular organelles, including the ER, also confer tissue specificity.

Cells often use the same biochemical components to activate or inactivate either the membrane or the cytosolic oscillator. In addition, recent work show that the plasma membrane not only sends messenger molecules (IP_3, DAG) to intracellular organelles (or cytosolic enzyme systems), but the organelles also send a new class of messenger molecules (CIF, cADP-R) to the plasma membrane. These reciprocal interactions (messenger coupling) between the two oscillators requires specific receptor-effector protein systems on both, the plasma membrane and the cytoplasm including the organelles.[166-173] There is now a vast body of evidence to indicate that direct cell-to-cell coupling serves to synchronize endogenous cell clocks within a tissue via exchange of second messengers. While the importance of such synchronization has not been completely determined, it may lie in the recruitment of more cells, or in producing a pulsatile response, which may help prevent the phenomenon of desensitization of target tissues.

5.2. CYTOSOLIC CA^{2+} OSCILLATORS IN NES TISSUES

5.2.1. Receptor-Operated, InsP$_3$-Mediated [Ca^{2+}]$_i$ Oscillators

Inositol 1,4,5-triphosphate (IP$_3$)-mediated intracellular Ca^{2+} release and the generation of distinct temporal and spatial [Ca^{2+}]$_i$ patterns arising from this mobilization are widespread modes of signal transduction in many cell types.[174] The mechanisms of InsP$_3$-mediated Ca^{2+} release have been studied mainly in electrically nonexcitable cells, which has contributed to their consideration as phenomena independent of plasma membrane potential (V$_m$), a major regulator of [Ca^{2+}]$_i$ in excitable cells.[175,176] However, electrical excitability and InsP$_3$-mediated intracellular responses coexist in NES neurones, pituitary cells, pancreatic islet cells, and other cells from the diffuse NES, which raises the certain possibility of interaction between these two modes of signal transduction. Thus, it is possible that voltage-activated Ca^{2+} channels may provide a pathway to refill the InsP$_3$-sensitive compartments in excitable cells or that the coordinate activation of Ca^{2+} fluxes through the plasma and ER membranes could determine the pattern of the [Ca^{2+}]$_i$ transients. Models of generation of repetitive [Ca^{2+}]$_i$ transients based on fluctuating InsP$_3$ concentrations propose the activation of phospholipase C by Ca^{2+} increases via voltage-gated channels as a link between V$_m$ and Ca^{2+} mobilization.[177] On the other hand, in models that consider a constant InsP$_3$ level for the generation of oscillations, the modulatory role of Ca^{2+} on its own release provides a mechanism of modulation of InsP$_3$-induced Ca^{2+} release by Ca^{2+} entry through voltage-gated channels. This mechanism has been suggested experimentally at different levels[178-180] and incorporated in theoretical models of [Ca^{2+}]$_i$ oscillations.[181]

5.2.2. GnRH-Evoked Cytosolic [Ca^{2+}]$_i$ Oscillations in Pituitary Gonadotrophs

The response of gonadotrophs of the anterior hypophysis to the hypothalamic GnRH involves the generation of InsP$_3$, which in turn initiates robust [Ca^{2+}]$_i$ oscillations.[182,183] Ca^{2+} mobilization from intracellular stores and Ca^{2+} entry through voltage-gated channels are usually considered as independent mechanisms of regulation of [Ca^{2+}]$_i$. However, in gonadotrophs maintenance of GnRH-initiated repetitive [Ca^{2+}]$_i$ oscillations[184,185] requires the activation of depolarization-gated Ca^{2+} channels.[186] These channels constitute the main pathway for the replenishment of the intracellular InsP$_3$-sensitive Ca^{2+} pool, as demonstrated by the steep voltage dependence of the amplitude of the sustained [Ca^{2+}]$_i$ oscillations.[187] In addition, Ca^{2+} entry during action potentials can command the frequency of the intrinsic cycles of Ca^{2+} release from intracellular compartments through a mechanism resistant to ryanodine and compatible with the potentiation by Ca^{2+} of InsP$_3$-triggered release. Thus, plasma membrane potential in excitable cells can directly modulate the pattern of InsP$_3$-controlled Ca^{2+} signals, providing means of integrating stimuli that use different signal transduction pathways.

The control of gonadotropin secretion is tightly linked to this [Ca^{2+}]$_i$ response.[188,189] In addition to the second messenger-mediated response, gonadotrophs also display spontaneous firing of action potentials. The firing pattern is modified by GnRH, which initiates periodic hyperpolarization of the membrane, as a consequence of the activation of apamin-sensitive K$^+$ channels by [Ca^{2+}]$_i$.[190] The role of V$_m$ in the overall response of the gonadotrophs is unclear, but the dependence of the maintained secretory response on extracellular Ca^{2+} and its sensitivity to blockade of voltage-gated Ca^{2+} channels suggests that the channel participates in the Ca^{2+} entry required for the sustained InsP$_3$ response. Several types of Ca^{2+} channels have been described in gonadotrophs[191,189] and their role as modulators of the [Ca^{2+}]$_i$ oscillations has been proposed.[192]

Data obtained by measuring [Ca^{2+}]$_i$ under voltage-clamp conditions in rat gonadotrophs indicate that by activation of voltage-gated Ca^{2+} channels, by depolarization of the plasma membrane, determines the amplitude of sustained GnRH-induced [Ca^{2+}]$_i$ oscillations. It is clear from these studies that the interaction between V$_m$ and Ca^{2+} mobilization from intracellular stores involves the activation of voltage-gated L-type Ca^{2+} channels (VDCC), although other types of Ca^{2+} channels[189,191] are not excluded, and therefore these Ca^{2+} channels constitute the main route for replenishment of the $InsP_3$-sensitive compartment.

Several mechanisms of Ca^{2+} entry other than voltage-gated Ca^{2+} channels have been proposed for nonexcitable cells, for example, inositol phosphates and depletion-activated pathways.[179,193-196] Electrophysiological measurement of an intracellular Ca^{2+} depletion-activated current in mast cells revealed its activity at very negative V_m. Contrasting with this observation, the marked voltage dependence of the amplitude of the $[Ca^{2+}]_i$ transients in gonadotrophs suggests that a different mechanism operates in these excitable cells. Thus, at negative V_m, at which Ca^{2+} entry through voltage-independent pathways should not occur, $[Ca^{2+}]_i$ oscillations decrease in amplitude and eventually disappear. These observations indicate that voltage-independent pathways are not expressed or that they are insufficient for the replenishment of the intracellular Ca^{2+} pool. In accord with this, we were unable to detect the activation of sustained inward currents associated to the depletion of the Ca^{2+} pool in gonadotrophs.

ABBREVIATIONS USED

BAPTA:	(1,2-bis(2-aminophenoxy)ethane-*N,N,N′,N′*-tetraacetic acid)
cAMP:	3′,5′-cyclic adenosine monophosphate
cGMP:	2′,3′-cyclic guanosine monophosphate
CIF:	calcium influx factor
cADP-R:	cyclic adenosine diphosphate-ribose
DAG:	1,2-diacylglycerol
EGTA:	ethylene glycol-bis (β-aminoethyl ether)*N,N,N′,N′*-tetraacetic acid
GABA:	γ-aminobutiric acid
GnRH:	gonadotropin-releasing hormone
IP_3:	D-myo-inositol 1,4,5-trisphosphate
LHRH:	luteinizing hormone-releasing hormone
RP:	Rathke's pouch
TPA:	12-*0*-tetradecanoyl-phorbol-13-acetate
TSH:	thyroid-stimulating hormone

REFERENCES

1. Torres, M., Ceballos, G., and Rubio, R., Possible role of nitric oxide in catecholamine secretion by chromaffin cells in the presence and absence of cultured endothelial cells, *J. Neurochem.*, 63, 988, 1194.
2. Loewenstein, W.R. and Kanno, Y., Studies on an epithelial (gland) cell junction. I. Modification of surface membrane permeability, J. Cell Biol., 22, 565, 1964.
3. Kuffler, S. W. and Potter, D.D, Gliain the leech central nervous system: physiological properties and neuron-glia relationships, *J. Neurophysiol.*, 27, 290, 1964
4. Sheridan, J.D., and Atkinson, M.M., Physiological roles of permeable junctions: some possibilities, *Annu. Rev. Physiol.*, 47, 337, 1985.
5. Bruzzone, R. and Meda, P., The gap junction: a channel for multiple functions? *Eur. J. Clin. Invest.*, 18, 444, 1988.
6. Meda, P., Perrelet, A., and Orci, L., Gap junctions and cell-to-cell coupling in endocrine glands, in *Modern Cell Biology*, Vol. III, Satir, B. H., Ed., Alan R. Liss, New York, 131, 1984.
7. Le Roith, D., Shemer, J., and Roberts, C.T., Evolutionary origins of intercellular communication systems: implications for mammalian biology, *Horm. Res.*, 38, 1, 1992.
8. Malven, P. V., *Mammalian Neuroendocrinology,* CRC Press, Boca Raton, FL, 1993.
9. Le Douarin, N.M., On the origin of pancreatic endocrine cells, *Cell*, 53, 169, 1988.

10. Gauweiler, B., Weihe, E., Hartschuh, W., and Yanaihara, N., Presence and coexistence of chromogranin A and multiple neuropeptides in Merkel cells of mammalian oral mucosa, *Neurosci. Lett.,* 89, 121, 1988.
11. Hartschuh, W. and Weihe, E., Multiple messenger candidates and marker substances in the mammalian Merkel cell-axon complex: a light and electron microscopic immunohistochemical study, *Prog. Brain Res.,* 74, 181, 1988.
12. Nolan, J.A., Trojanowski, J.Q., and Hogue-Angeletti, R., Neurons and neuroendocrine cells contains chromogranin: detection of the molecule in normal bovine tissue by immunochemical methods, *J. Histochem. Cytochem.,* 33, 791, 1985.
13. Von Dorsche, H.H., Falt, K., Hahn, H.J., Reiher, H., and Krisch, K., Neuron-specific enolase (NSE) as a neuroendocrine cell marker in the human fetal pancreas, *Acta Histochem.,* 85, 227, 1989.
14. Le Douarin, N.M. and Joterreau, F.V., Tracing of cells of the avian thymus through embryonic life in interspecific chimeras, *J. Exp. Med.,* 1942, 17, 1975.
15. Horvat, G., Krisch, I., Wengler, G., Alibeik, H., Neuhold, N., Ulrich, W., Braun, O., and Hochmeister, M., Immunochemical characterization of a novel secretory protein (defined by monoclonal antibody HISL-19) of peptide hormone producing cells which is distinct from chromogranin A, B, and C, *Lab. Invest.,* 58, 411, 1988.
16. Inagaki, H., Haimoto, H., Hosoda, S., and Kato, K., Aldolase C is localized in neuroendocrine cells, *Experientia,* 44, 749, 1988.
17. Mc Ewen, B.S., Brinton, R.E., Chao, H.M., Corrini, H., Gannon, M.N., Gould, E., O'Callaghan, J., Spencer, R.L., Sakai, R.R., and Wolley, C.S., The hippocampus: a site for modulatory interactions between steroid hormones, neurotransmitters and neuropeptides, in *Neuroendocrine Perspectives,* Müller, E.E. and MacLead, R.M., Elsevier, New York, 1993, 93.
18. Jarry, H., Leonhardt, S., and Wuttke, W., Gamma-aminobutyric acid neurons in the preoptic/anterior hypothalamic area synchronize the phasic activity of the gonadotropin-releasing hormone pulse generator in ovariectomized rats. *Neuroendocrinology,* 53, 261, 1991.
19. Li, Y-X. and Goldbeter, A., Frequency specificity in intercellular communication influence of patterns of periodic signaling on target responsiveness, *Biophys. J.,* 55, 125, 1989.
20. Eghbali, B., Kessler, J.A., Reid L.M., Roy, C., and Spray, D.C., Involvement of gap junctions in tumorigenesis: transfection of tumor cells with connexin 32 cDNA retards growth in vivo, *Proc. Natl. Acad. Sci. U.S.A.,* 88, 10701, 1991.
21. Bennett, M.V., Barrio, L.C., Bargiello, T.A., Spray, D.C., Hertzberg, E., and Saez J.C., Gap junctions: new tools, new answers, new questions, *Neuron,* 6, 305, 1991.
22. Loewenstein, W.R., Socolar, S. J., Higashino, S., Kanno, Y. and Davidson, N., Intercellular communication: renal urinary bladder, sensory and salivary gland cells, *Science,* 149, 295, 1965.
23. Weidmann, S., The electrical constant of Purkinje fibres, *J. Physiol.,* 118, 348, 1952.
24. Furshpan, E. J. and Potter, D.D., Mechanism of nerve-impulse transmission at a crayfish synapse, *Nature,* 180, 342, 1957.
25. Woodbury, J.W. and Crill, W.E., On the problem of impulse conduction in the atrium, in *Nervous Inhibition,* Florey E., Ed., Pergamon Press, New York, 1961, 124.
26. Loewenstein, W.R., Permeability of membrane junctions, *Ann. N.Y. Acad. Sci.,* 137, 441, 1966.
27. Stewart, W.W., Functional connections between cells as revealed by dye-coupling with a highly fluorescent naphthalimide tracer, *Cell,* 14, 741, 1978.
28. Simpson, I., Rose., B., and Loewenstein, W.R., Size limit of molecules permeating the junctional membrane channels, *Science,* 195, 294, 1977.
29. Flagg-Newton, J., Simpson, I., and Loewenstein, W. R., Permeability of the cell-to-cell membrane channels in mammalian cell junction, *Science,* 205, 404, 1979.
30. Pappas, G.D., and Bennet, M.V.L., Specialized junctions involved in electrical transmission between neurons, *Ann. NY Acad. Sci.,* 137, 495, 1966.
31. Furshpan, E.J., and Potter, D.D., Low resistance junctions between cells in embryos and tissue culture, *Curr. Top., Dev. Biol.,* 3, 95, 1968.
32. Bennet, M.V.L., and Trinkaus, J.P., Electrical coupling between embryo cells by way of extracellular space and specialized junctions, *J. Cell Biol.,* 44, 592, 1970.
33. Sheridan, J., Dye movement and low resistance junctions between reaggregated embryonic cells, *Dev. Biol.,* 26, 627, 1971.
34. Azarnia, R. and Loewenstein, W.R., Intercellular communication and tissue growth. V. A cancer cell strain that fails to make permeable membrane junctions with normal cells, *J. Membr. Biol.,* 6, 368, 1971.
35. Michalke, W. and Loewenstein, W.R., Communication between cells of different types, *Nature,* 232, 121, 1971.
36. Ravel, J.P. and Karnovsky, M.J., Hexagonal array of subunits in intercellular junctions of the mouse heart and liver, *J. Cell Biol.,* 33, C7, 1967.
37. Robertson, J.D., The occurrence of a sub-unit pattern in the unit membrane of club endings in Mauthner cell synapses in goldfish brains, *J. Cell Biol.,* 19, 201, 1963.

38. Goodenough, D.A., Methods for the isolation and structural characterization of hepatocyte gap junctions, in *Methods in Membrane Biology,* Vol. III, Karn, E.D., Ed., Plenum Press, New York, 1975.
39. Goodenough, D. A., *In vitro* formation of gap junction vesicles, *J. Cell Biol.,* 68, 220, 1976.
40. Gilula, N.B., Isolation of liver gap junctions and characterization of the polypeptides, *J. Cell Biol.,* 64, 114, 1974.
41. Kumar N.M. and Gilula, N.B., Molecular biology and genetics of gap junction channels, *Semin. Cell. Biol.,* 3, 3, 1992.
42. Goodenough, D.A. and Stockenius, W., The isolation of mouse hepatocyte gap junctions, *J. Cell Biol.,* 54, 646, 1972.
43. Caspar, D.L.D., Goodenough, D.A., Makowski, L., and Philips, W.C., Gap junction structures. I. Correlated electron microscopy and X-ray diffraction, *J. Cell Biol.,* 74, 605, 1977.
44. Caspar, D.L.D., Goodenough, D.A., Makowski, L., and Philips, W.C., Gap junction structures. II. *J. Cell Biol.,* 74, 629, 1977b.
45. Peracchia, C. and Dulhunty, A.F., Low resistance junction in crayfish: Structural changes with functional uncoupling, *J. Cell Biol.,* 70, 419, 1976.
46. Peracchia, C., Gap junctions: structural changes after uncoupling procedures, *J. Cell Biol.,* 72, 628, 1977.
47. Unwin, P.N.T. and Zampighi, G., Structure of the junctions between communicating cells. *Nature,* 283, 545, 1980.
48. Hertzberg, E.L. and Gilula, N.B., Isolation and characterization of gap junctions from rat liver. *J. Biol. Chem.,* 254, 2138, 1979.
49. Zampighi, G., and Unwin, P.N.T., Two forms of isolated gap junctions, *J. Mol. Biol.,* 135, 451, 1979.
50. Stauffer, K.A., Kumar, N.M., Gilula, N.B., and Unwin, N., Isolation and purification of gap junction channels, *J. Cell. Biol.,* 115, 141, 1991.
51. Yeager, M. and Gilula, N.B., Membrane topology and quaternary structure of cardiac gap junction ion channels, *J. Mol. Biol.,* 223, 929, 1992.
52. Meda, P., Chanson, M., Pepper, M., Giordano, E., Bosco, D., Traub, O., Willecke, K., el Aoumari, A., Gros, D., and Beyer, E.C., In vivo modulation of connexin 43 gene expression and junctional coupling of pancreatic β-cells, *Exp. Cell Res.,* 192, 469, 1991.
53. Dermietzel, R., Hertberg, E.L., Kessler, J.A., and Spray, D.C., Gap junctions between cultured astrocytes: immunocytochemical, molecular, and electrophysiological analysis, *J. Neurosci.,* 11, 1421, 1991.
54. Batter, D.K., Corpina, R.A., Roy, C., Spray, D.C., Hertzberg, E.L., and Kessler, J.A., Heterogeneity in gap junction expression in astrocytes cultured from different brain regions, *Glia,* 6, 213, 1992.
55. Spray, D.C., Moreno, A.P., Kessler, J.A., and Dermietzel, R., Characterization of gap junctions between cultured leptomeningeal cells, *Brain Res.* 568, 1, 1991.
56. Bai, S., Spray, D.C., and Burk, R.D., Identification of proximal and distal regulatory elements of the rat connexin32 gene, *Biochim. Biophys. Acta,* 1216, 197, 1993.
57. Budunova, I.V., Williams, G.M., and Spray, D.C., Effect of tumor promoting stimuli on gap junction permeability and connexin43 expression in ARL18 rat liver cell line, *Arch. Toxicol.,* 67, 565, 1993.
58. Larson, D.M. and Sheridan, J.D., Intercellular junctions and transfer of small molecules in primary vascular endothelial cultures, *J. Cell Biol.,* 92, 183, 1982.
59. Larson, D.M., Kam, E.Y., and Sheridan, J.D., Junctional transfer in cultured vascular endothelium. I. Electrical coupling, *J. Membr. Biol.,* 74, 103, 1983.
60. Larson, D.M. and Sheridan, J.D., Junctional transfer in cultured vascular endothelium. II. Dye and nucleotide transfer, *J. Membr. Biol.,* 83, 157, 1985.
61. Pepper, M.S., Spray, D.C., Chanson, M., Montesano, R., Orci, L., and Meda, P., Junctional communication is induced in migrating capillary endothelial cells, *J. Cell Biol.,* 109, 3027, 1989.
62. Pepper, M.S., Montesano, R., el Aoumari, A., Gros, D., Orci, L., and Meda, P., Coupling and connexin 43 expression in microvascular and large vessel endothelial cells, *Am. J. Physiol.,* 262, C1246, 1992.
63. Fishman, G.I., Hertzberg, E.L., Spray, D.C., and Leinwand, L.A., Expression of connexin43 in the developing rat heart, *Circ. Res.,* 68, 782, 1991.
64. Goodenough, D.A., and Musil, L.S., Gap junctions and tissue business: problems and strategies for developing specific functional reagents, *J. Cell. Sci.,* 17, 133, 1993.
65. Hsieh, C.L., Kumar, N.M., Gilula, N.B. and Francke, U., Distribution of genes for gap junction membrane channel proteins on human and mouse chromosomes, *Somat. Cell. Mol. Genet.,* 17, 191, 1991.
66. Nishi, M., Kumar, N.M., and Gilula, N.B., Developmental regulation of gap junction gene expression during mouse embryonic development, *Dev. Biol.,* 146, 117, 1991.
67. Risek, B., Klier, F.G., and Gilula, N.B., Developmental regulation and structural organization of connexins in epidermal gap junctions, *Dev. Biol.,* 164, 183, 1994.
68. Brissette, J.L., Kumar, N.M., Gilula, N.B., Hall, J.E., and Dotto, G.P., Switch in gap junction protein expression is associated with selective changes in junctional permeability during keratinocyte differentiation, *Proc. Natl. Acad. Sci. U.S.A.,* 91, 6453, 1994.

69. Musil, L.S. and Goodenough, D.A., Multisubunit assembly of an integral plasma membrane channel protein, gap junction connexin43, occurs after exit from the ER, *Cell,* 74, 1065, 1993.
70. Musil, L.S. and Goodenough, D.A., Biochemical analysis of connexin43 intracellular transport, phosphorylation, and assembly into gap junctional plaques, *J. Cell Biol.,* 115, 1357, 1991.
71. White, T.W., Bruzzone, R., Wolfram, S., Paul, D.L., and Goodenough, D.A., Selective interactions among the multiple connexin proteins expressed in the vertebrate lens: the second extracellular domain is a determinant of compatibility between connexins, *J. Cell Biol.,* 125, 879, 1994.
72. Brissette, J.L., Kumar, N.M., Gilula, N.B., and Dotto, G.P., The tumor promoter 12-O-tetradecanoylphorbol-13-acetate and the ras oncogene modulate expression and phosphorylation of gap junction proteins, *Mol. Cell. Biol.,* 11, 5364, 1991.
73. Girsch, S.J. and Peracchia, C., Calmodulin interacts with a C-terminus peptide from the lens membrane protein MIP26, *Curr. Eye Res.,* 10, 839, 1991.
74. Shen, L., Shcrager, P., Girsch, S.J., Donaldson, P.J., and Peracchia, C., Channel reconstitution in liposomes and planar bilayers with HPLC-purified MIP26 of bovine lens, *J. Membr. Biol.,* 124, 21, 1991.
75. Ehring, G.R., Lagos, N., Zampighi, G.A., and Hall, J.E., Phosphorylation modulates the voltage dependence of channels reconstituted from the major intrinsic protein of lens fiber membranes, *J. Membr. Biol.,* 126, 75, 1992.
76. Spray, D.C., Moreno, A.P., and Campos-de-Carvalho, A.C., Biophysical properties of the human cardiac gap junction channel, *Braz. J. Med. Biol. Res.,* 26, 541, 1993.
77. Moreno, A.P., Saez, J.C., Fishman, G.I., and Spray, D.C., Human connexin43 gap junction channels. Regulation of unitary conductances by phosphorylation, *Circ. Res.,* 74, 1050, 1994.
78. Peracchia, C., Effects of the anesthetics heptanol, halothane and isoflurane on gap junction conductance in crayfish septate axons: a calcium- and hydrogen-independent phenomenon potentiated by caffeine and theophylline, and inhibited by 4-aminopyridine, *J. Membr. Biol.,* 121, 67, 1991.
79. Lazrak, A. and Peracchia, C., Gap junction gating sensitivity to physiological internal calcium regardless of pH in Novikoff hepatoma cells, *Biophys. J.,* 65, 2002, 1993.
80. Salomon, D., Chanson, M., Vischer, S., Masgrau, E., Vozzi, C., Saurat, J.H., Spray, D.C., and Meda, P., Gap junctional communication of primary human keratinocytes: characterization by dual voltage clamp and dye transfer, *Exp. Cell Res.,* 201, 452, 1992.
81. Moreno, A.P., Eghbali, B., and Spray, D.C., Connexin32 gap junction channels in stably transfected cells. Equilibrium and kinetic properties, *Biophys. J.,* 60, 1267, 1991a.
82. Moreno, A.P., Eghbali, B., and Spray, D.C., Connexin32 gap junction channels in stably transfected cells: unitary conductance, *Biophys. J.,* 60, 1254, 1991.
83. Chanson, M., Chandross, K.J., Rook, M.B., Kessler, J.A., and Spray, D.C., Gating characteristics of a steeply voltage-dependent gap junction channel in rat Schwann cells, *J. Gen. Physiol.,* 102, 925, 1993.
84. Rojas, E., Gating mechanism for the activation of the sodium conductance in nerve membranes, *Cold Spring Harbor Symp. Quant. Biol.,* XL, 305, 1976.
85. Christ, G., Moreno, A.P., Melman, A., and Spray, D.C., Gap junction-mediated intercellular diffusion of Ca^{2+} in cultured human corporal smooth muscle cells, *Am. J. Physiol.,* 263, C373, 1992.
86. Meda, P., Perrelet, A., and Orci, L., Gap junctions and cell-to-cell coupling in endocrine glands, in *Modern Cell Biology,* Vol. III, Satir, B.H., Ed., Alan R. Liss, New York, 131, 1984.
87. Orci, L., Macro- and micro-domains in the endocrine pancreas, *Diabetes,* 31, 538, 1982.
88. Bonner-Weir, S. and Orci, L., New perspectives on the microvasculature of the islets of Langerhans in the rat, *Diabetes,* 31, 883, 1982.
89. Muruyama, H., Hisatomi, A., Orci, L., Grodsky, G.M., and Unger, R.H., Insulin within islets is a physiologic glucagon release inhibitor, *J. Clin. Invest.,* 74, 2296, 1984.
90. Orci, L., Unger, R.H., and Renold, A.E., Structural coupling between pancreatic islet cells, *Experientia,* 29, 1015, 1973.
91. Orci, L., Malaisse-Lagae, F., Ravazzola, M., Rouiller, D., Renold, A.E., Perrelet, A., and Unger, R., A morphological basis for intercellular communication between β and β-cells in the endocrine pancreas, *J. Clin. Invest.,* 56, 1066, 1975.
92. Meda, P., Perrelet, A., and Orci, L., Gap junctions and β-cell function, *Horm. Metab. Res.,* 10, 157, 1980.
93. Meda, P., Halban, P., Perrelet, A., Renold, A.E., and Orci, L., Gap junction development is correlated with insulin content in the pancreatic β-cell, *Science,* 209, 1026, 1980.
94. Meda, P., Perrelet, A., and Orci, L., Increase of gap junctions between pancreatic β-cells during stimulation of insulin secretion, *J. Cell Biol.,* 82, 441, 1979.
95. Meda, P., Denef, J.F., Perrelet, A., and Orci, L., Nonrandom distribution of gap junctions between pancreatic β-cells, *Am. J. Physiol.,* 238, C114, 1980.
96. Michaels, R.L., and Sheridan, J.D., Islets of Langerhans: dye coupling among immunocytochemically distinct cell types, *Science,* 214, 801, 1981.
97. Meda, P., Kohen, E., Kohen, C., Rabinovitch, A., and Orci, L., Direct communication of homologous and heterologous endocrine islet cells in culture, *J. Cell Biol.,* 92, 221, 1982.

98. Meda, P., Santos, R.M., and Atwater, I., Direct identification of electrophysiologically monitored cells within intact mouse islets of Langerhans, *Diabetes,* 35, 232, 1986.
99. Meda, P., Michaels, R.L., Halban, P.A., Orci, L., Sheridan, J.D., In vivo modulation of gap junctions and dye coupling between B cells of the intact pancreatic islet, *Diabetes,* 32, 858, 1983.
100. Meissner, H.P., Electrophysiological evidence for coupling between β-cells of pancreatic islets, *Nature,* 262, 502, 1976.
101. Eddlestone, G.T., Gonçalves, A.A., Bangham, J.A., and Rojas, E., Electrical coupling between cells in islets of Langerhans from mouse, *J. Membrane Biol.,* 77, 1, 1984.
102. Santos, R.M. and Rojas, E., Evidence for modulation of cell-to-cell electrical coupling by cAMP in mouse islets of Langerhans, *FEBS Lett.,* 220, 342, 1987.
103. Chanson, M., Bruzzone, R., Spray, D.C., Regazzi, R., and Meda, P., Cell uncoupling and protein kinase C: correlation in a cell line but not in a differentiated tissue, *Am. J. Physiol.,* 255, C699, 1988.
104. Perez-Armendariz, E.M., Atwater, I., and Rojas, E., Glucose-induced oscillatory changes in extracellular ionized potassium concentration in mouse islets of Langerhans, *Biophys. J.,* 48, 741, 1985.
105. Perez-Armendariz, E.M. and Atwater, I., Glucose-evoked changes in $[K^+]$ and $[Ca^{2+}]$ in the intercellular spaces of the mouse islet of Langerhans, *Adv. Exp. Med. Biol.,* 211, 31, 1985.
106. Valdeomillos, M., Santos, R.M., Contreras, D., Soria, B., and Rosario, L.M., Glucose-induced oscillations of intracellular Ca^{2+} concentration resembling bursting electrical activity in single mouse islets of Langerhans, *FEBS Lett.,* 259, 19, 1989.
107. Santos, R.M., Rosario, L.M., Nadal, A., Garcia-Sancho, J., Soria, B., and Valdeomillos, M., Widespread synchronous $[Ca^{2+}]_i$ oscillations due to bursting electrical activity in single pancreatic islets, *Pfluegers Arch.,* 418, 417, 1991.
108. Rosario, L.M., Barbosa, R.M., Antunes, C.M., Silva, A.M., Abrunhosa, A.J., and Santos, R.M., Bursting electrical activity in pancreatic beta-cells: evidence that the channel underlying the burst is sensitive to Ca^{2+} influx through L-type Ca^{2+}-channels, *Pfluegers Arch.,* 424, 439, 1993.
109. Atwater, I., Rojas, E., and Scott, A., Simultaneous measurements of insulin release and electrical activity from single microdissected mouse islets of Langerhans, *J. Physiol. (London),* 291, 57P, 1979.
110. Rosario, L.M., Atwater I., and Scott, A.M., Pulsatile insulin release and electrical activity from single ob/ob mouse islets of Langerhans, *Adv. Exp. Med. Biol.,* 211, 413, 1985.
111. Jack, J.J.B, Noble, D., and Tsien, R.W., *Electric Current Flow in Excitable Cells,* Clarendon Press, Oxford, 1975.
112. Atwater, I., Carroll, P., and Li, M-X., Electrophysiology of the pancreatic B cell, in *Molecular and Cellular Biology of Diabetes Mellitus,* I. Insulin secretion, Draznin, B., Melmed, S. and Le Roith, D., Eds., Alan R. Liss, New York, 1989, 49.
113. Frankenhaeuser, B. and Hodgkin, A.L., The after-effects of impulses in the giant nerve fibres of *Loligo, J. Physiol. (London),* 131, 341, 1956.
114. Sykova, E., Extracellular K^+ accumulation in the central nervous system. *Prog. Biophys. Mol. Biol.,* 42, 135, 1983.
115. Cleemann, L. and Morad, M., Extracellular potassium accumulation in voltage-clamped frog ventricular muscle, *J. Physiol. (London),* 286, 83, 1979.
116. Morad, M., Physiological implications of K accumulation in heart muscle, *Fed. Proc., Fed. Am Soc. Exp. Biol.,* 39, 1533, 1980.
117. Cleemann, L., Pizarro, G., and Morad, M., Optical measurements of extracellular calcium depletion during a single heartbeat, *Science,* 226, 174, 1984.
118. Stokes, C. and Rinzel, J., Diffusion of extracellular K^+ can synchronize bursting oscillations in a model islet of Langerhans, *Biophys. J.,* 65, 597, 1993.
119. Smith, P.A., Ashcroft, F.M., and Rorsman, P., Simultaneous recordings of glucose dependent electrical activity and ATP-regulated K^+-currents in isolated mouse pancreatic beta-cells, *FEBS Lett.,* 261, 187, 1990.
120. Sherman, A. and Rinzel, J., Model for synchronization of pancreatic β-cells by gap junction coupling. Biophys. J. 59, 547, 1991.
121. Smolen, P., Rinzel, J., and Sherman, A., Why pancreatic islets burst but single β-cells do not: the heterogeneity hypothesis, *Biophys. J.,* 64, 1668, 1993.
122. Scott, A.M., Atwater, I., and Rojas, E., A method for the simultaneous measurement of insulin release and B cell membrane potential in single mouse islets of Langerhans, *Diabetologia,* 21, 470, 1981.
123. Perez-Armendariz, E.M., Roy, C., Spray, D.C., and Bennett, M.V.L., Biophysical properties of gap junctions between freshly dispersed pairs of mouse pancreatic beta cells, *Biophys. J.,* 59, 59, 1991.
124. Lo, C.W. and Gilula, N.B., Gap junctional communication in the post-implantation mouse embryo, *Cell,* 18, 411, 1979.
125. Lo, C.W. and Gilula, N.B., PCC4azal teratocarcinoma stem cell differentiation in culture. II. Morphological characterization, *Dev. Biol.,* 75, 93, 1980.
126. Serras, F., and van den Biggelaar, J.A., Is a mosaic embryo also a mosaic of communication compartments? *Dev. Biol.,* 120, 132, 1987.

127. Serras, F., Buultjens, T.E., and Finbow, M.E., Inhibition of dye-coupling in Patella (Mollusca) embryos by microinjection of antiserum against nephrops (Arthropoda) gap junctions, *Exp. Cell Res.,* 179, 282, 1988.

128. Rojas E., Stokes, C.L., Mears, D., and Atwater, I., Single-microelectrode voltage-clamp measurements of pancreatic β-cell membrane ionic currents *in situ, J. Membr. Biol.,* 143, 65, 1995.

129. Sherman, A. Xu, L., and Stokes, C.L., Estimating and eliminating junctional current in coupled cell populations. A computational study, *J. Membr. Biol.,* 143, 78, 1995.

130. Cardenas, H., Ordog, T., O'Byrne, K. T., and Knobil, E., Single unit components of the hypothalamic multiunit electrical activity associated with the central signal generator that directs the pulsatile secretion of gonadotropic hormones, *Proc. Natl. Acad. Sci. U.S.A.,* 90, 9630, 1993.

131. Andrew, R.D., MacVicar, B.A., Dudek, F.E., and Hatton, G.I., Dye transfer through gap junctions between neuroendocrine cells of rat hypothalamus, *Science,* 211, 1187, 1981.

132. Dermietzel, R., and Spray, D.C., Gap junctions in the brain: where, what type, how many and why? *Trends Neurosci.,* 16, 186, 1993.

133. Jarvis, C.R. and Andrew, R.D., Correlated electrophysiology and morphology of the ependyma in rat hypothalamus, *J. Neurosci.,* 8, 3691, 1988.

134. Liposits, Z., Merchenthaler, I., Wetsel, W.C., Reid, J.J., Mellon, P.L., Weiner, R.I., and Negro-Vilar, A., Morphological characterization of immortalized hypothalamic neurons synthesizing luteinizing hormone-releasing hormone, *Endocrinology,* 129, 1575, 1991.

135. Andrew, R.D., Taylor, C.P., Snow, R.W., and Dudek, F.E., Coupling in rat hippocampal slices: dye transfer between CA1 pyramidal cells, *Brain Res. Bull.,* 8, 211, 1982.

136. Fletcher, W.H., Anderson, N.C., Jr., and Everett, J.W., Intercellular communication in the rat anterior pituitary gland. An in vivo and in vitro study, *J. Cell Biol.,* 67, 469, 1975.

137. Meda, P., Pepper, M.S., Traub, O., Willecke, K., Gros, D., Beyer, E., Nicholson, B., Paul, D., and Orci, L., Differential expression of gap junction connexins in endocrine and exocrine glands, *Endocrinology,* 133, 2371, 1993.

138. Soji, T., Yashiro, T., and Herbert, D.C., Intercellular communication within the rat anterior pituitary gland. I. Postnatal development and changes after injection of luteinizing hormone-releasing hormone (LH-RH) or testosterone, *Anat. Rec.,* 226, 337, 1990.

139. Soji, T., Nishizono, H., Yashiro, T., and Herbert, D.C., Intercellular communication within the rat anterior pituitary gland. III. Postnatal development and periodic changes of cell-to-cell communications in female rats, *Anat. Rec.,* 231, 351, 1991.

140. Yamamoto, T., Hossain, M.Z., Hertzberg, E.L., Uemura, H., Murphy, L.J., and Nagy, J.I., Connexin43 in rat pituitary: localization at pituicyte and stellate cell gap junctions and within gonadotrophs, *Histochemistry,* 100, 53, 1993.

141. Stefanovic, V., Saraga-Babic, M., and Wartiovaara, J., Cell contacts in early human pituitary development, *Acta Anat.,* 148, 169, 1993.

142. Schwartz, J. and Cherny, R., Intercellular communication within the anterior pituitary influencing the secretion of hypophysial hormones, *Endocr. Rev.,* 13, 453, 1992.

143. Taugner, R., Schiller, A., and Rix, E., Gap junctions between pinealocytes. A freeze-fracture study of the pineal gland in rats, *Cell Tissue Res.,* 218, 303, 1981.

144. Huang, S.K., Nobiling, R., Schachner, M., and Taugner, R., Interstitial and parenchymal cells in the pineal gland of the golden hamster. A combined thin-section, freeze-fracture and immunofluorescence study, *Cell Tissue Res.,* 235, 327, 1984.

145. Huang, S.K., and Taugner, R., Gap junctions between guinea-pig pinealocytes, *Cell Tissue Res.,* 235, 137, 1984.

146. Grynszpan-Wynograd, O. and Nicolas G., Intercellular junctions in the adrenal medulla: a comparative freeze-fracture study, *Tissue Cell,* 12, 661, 1980.

147. Grynszpan-Winograd, O., Close relationship of mitochondria with intercellular junctions in the adrenaline cells of the mouse adrenal gland, *Experientia,* 38, 270, 1982.

148. Setoguti, T. and Inoue, Y., Freeze-fracture replica studies of the human adrenal cortex, with special references to microvillous projections, *Acta Anat.,* 111, 207, 1982.

149. Decker, R.S., Donta, S.T., Larsen, W.J., and Murray S.A., Gap junctions and ACTH sensitivity in Y-1 adrenal tumor cells, *J. Supramol. Struct.,* 9, 497, 1978.

150. Decker, R.S., Gap junctions and steroidogenesis in the fetal mammalian adrenal cortex, *Dev. Biol.,* 82, 20, 1981.

151. Usadel, H., Bornstein, S.R., Ehrhart-Bornstein, M., Kreysch, H.G., and Scherbaum, W.A., Gap junctions in the adrenal cortex, *Horm. Metab. Res.,* 25, 653, 1993.

152. Mihara, M., Hashimoto, K., Ueda, K., and Kumakiri, M., The specialized junctions between Merkel cell and neurite: an electron microscopic study, *J. Invest. Dermatol.,* 73, 325, 1979.

153. Hartschunh, W. and Weihe, E., Fine structural analysis of the synaptic junction of Merkel cell-axon-complexes, *J. Invest. Dermatol.,* 75, 159, 1980.

154. Nilsson, M., Integrity of the occluding barrier in high-resistant thyroid follicular epithelium in culture. I. Dependence of extracellular Ca^{2+} is polarized, *Eur. J. Cell Biol.,* 56, 295, 1991.

155. Nilsson, M., Molne, J., and Ericson, L.E., Integrity of the occluding barrier in high-resistant thyroid follicular epithelium in culture. II. Immediate protective effect of TSH on paracellular leakage induced by Ca^{2+} removal and cytochalasin B, *Eur. J. Cell Biol.,* 56, 308, 1991.

156. Munari-Silem, Y., Audebet, C., and Rousset, B., Hormonal control of cell to cell communication: regulation by thyrotropin of the gap junction-mediated dye transfer between thyroid cells, *Endocrinology,* 128, 3299, 1991.

157. Asakawa, H., Yamasaki, H., Hanafusa, T., Kono, N., and Tarui, S., Cell-to-cell communication in cultured rat thyroid monolayer cells is inhibited dose-dependently by methimazole, *Res. Commun. Chem. Pathol. Pharmacol.,* 77, 131, 1992.

158. Munari-Silem, Y., Mesnil, M., Selmi, S., Bernier-Valentin, F., Rabilloud, R., and Rousset, B., Cell-cell interactions in the process of differentiation of thyroid epithelial cells into follicles: a study by microinjection and fluorescence microscopy on in vitro reconstituted thyroid follicles, *J. Cell Physiol.,* 145, 414, 1990.

159. Kitajima, K., Yamashita, K., and Fujita, H., Fine structural aspects of the shift of zonula occludens and cytoorganelles during the inversion of cell polarity in cultured porcine thyroid follicles, *Cell Tissue Res.,* 242, 221, 1985.

160. Barriere, H., Chambard, M., Selzner, J.P., Mauchamp, J., and Gabrion, J., Polarity reversal of inside-out thyroid follicles cultured within collagen gel: structure of the junctions assessed by freeze-fracture and lanthanum permeability, *Biol. Cell,* 62, 133, 1988.

161. van Uigen, A.J., van Dijk, J.E., Koch, C.A., and de Vijlder, J.J., Freeze fracture morphology of thyroid tight junctions in goats with different thyrotropin stimulation, *Endocrinology,* 117, 114, 1985.

162. Setoguti, T., Inoue, Y., and Suematsu, T., Intercellular junctions of the hen parathyroid gland. A freeze-fracture study, *J. Anat.,* 135, 395, 1982.

163. Carolan, E.J., and Pitts, J.D., Some murine thymic lymphocytes can form gap junctions, *Immunol. Lett.,* 13, 255, 1986.

164. Tamir, M., Rozenszajn, L.A., Malik, Z., and Zipori, D., Thymus-derived stromal cell lines, *Int. J. Cell. Cloning,* 5, 289, 1985.

165. Leceta, J., Villena, A., Razquin, B., Fonfria, J., and Zapata, A., Interdigitating cells in the thymus of the turtle Mauremys caspica. Possible relationships to macrophages, *Cell Tissue Res.,* 238, 381, 1984.

166. Galione, A., Cyclic ADP-ribose: a new way to control calcium, *Science,* 259, 325, 1993.

167. Takasawa, S., Nata, K., Yonekura, H., and Okamoto, H., Cyclic ADP-ribose in insulin secretion from pancreatic β-cells, *Science,* 259, 370, 1993.

168. Putney, J.W., and Bird, G.St., The signal for capacitative calcium entry, *Cell,* 75, 199, 1993.

169. Fasolato, C., Horth, M., and Penner, R., A GTP-dependent step in the activation mechanism of capacitative calcium influx, *J. Biol. Chem.,* 268, 20737, 1993.

170. Worley, J.F., III, McIntyre, M.S., Spencer, B., Mertz, R.J., Roe, M.W., and Dukes, I.D., Endoplasmic reticulum calcium store regulates membrane potential in mouse islet β-cells, *J. Biol. Chem.,* 269, 14359, 1994.

171. Dukes, I.D., McIntyre, M.S., Mertz, R.J., Philipson, L.H., Roe, M.W., Spencer B., and Worley, J.F., III, Dependence on NDH produced during glycolisis for β-cell glucose signaling, *J. Biol. Chem.,* 269, 10979, 1994.

172. Ehrlich, B.E., Kaftan, E., Bezprozvannaya, S., and Bezprozvanny, I., The pharmacology of intracellular Ca^{2+}-release channels, *TIPS,* 15, 145, 1994.

173. Fasolato, C., Innocenti, B., and Pozzan, T., Receptor-activated Ca^{2+}-influx: how many mechanisms for how many channels? *TIPS,* 15, 77, 1994.

174. Berridge, M.J., Inositol trisphosphate and calcium signaling, *Nature,* 361, 315, 1993.

175. Berridge, M.J., Calcium oscillations, *J. Biol. Chem.,* 265, 9583, 1990.

176. Tsien, R.W., and Tsien, R.Y., Calcium channels, stores and oscillations, *Annu. Rev. Cell Biol.,* 6, 715, 1990.

177. Harootunian, A.T., Kao, J.P.Y., Paranjape, S., and Tsien, R.Y., Generation of calcium oscillations in fibroblasts by positive feedback between calcium and IP3, *Science,* 251, 75, 1991.

178. Bezprozvanny, I., Watras, J., and Ehrlich, B.E., Bell-shaped calcium-response curves of Ins(1,4,5)P3- and calcium-gated channels from endoplasmic reticulum of cerebellum, *Nature,* 351, 751, 1991.

179. Finch, E.A., Turner, T.J., and Goldin, S.M., Calcium as a coagonist of inositol 1,4,5-trisphosphate-induced calcium release, *Science,* 252, 443, 1991.

180. Yao, Y. and Parker, I., Potentiation of inositol trisphosphate-induced Ca^{2+} mobilization in Xenopus oocytes by cytosolic Ca^{2+}, *J. Physiol. (London),* 458, 319, 1992.

181. Keizer, J. and De Young, G., Effect of voltage-gated plasma membrane Ca^{2+} fluxes on IP_3-linked Ca^{2+} oscillations, *Biophys. J.,* 61, 649, 1992.

182. Shangold, G.A., Murphy, S.N., and Miller, R.J., Gonadotropin-releasing hormone-induced Ca^{2+} transients in single identified gonadotropes require both intracellular Ca^{2+} mobilization and Ca^{2+} influx, *Proc. Natl. Acad. Sci. U.S.A.,* 85, 6566, 1988.

183. Iida, T., Stojilković, S.S., Izumi, S.I., and Catt, K.J., Spontaneous and agonist-induced calcium oscillations and secretory responses in pituitary gonadotrophs, *Mol. Endocrinol.,* 5, 949, 1991.

184. Stojilković, S.S., Torsello, A., Iida, T., Rojas, E., and Catt, K.J., Calcium signaling and secretory responses in agonist-stimulated pituitary gonadotrophs, *J. Steroid Biochem. Mol. Biol.,* 41, 453, 1992.

185. Stojilković, S.S., Kukuljan, M., Tomi_, M., Rojas, E., and Catt, K.J., Mechanism of agonist-induced $[Ca^{2+}]_i$ oscillations in pituitary gonadotrophs, *J. Biol. Chem.,* 268, 7713, 1993.

186. Stojilković, S.S., Iida, T., Virmani, M.A., Izumi, S-I., Rojas, E., and Catt, K.J., Dependence of hormone secretion on activation-inactivation kinetics of voltage-sensitive Ca^{2+} channels in pituitary gonadotrophs, *Proc. Natl. Acad. Sci. U.S.A.,* 87, 8855, 1990.

187. Kukuljan, M., Rojas, E., Catt, K.J. and Stojilković, S.S., Membrane potential regulates inositol 1,4,5-trisphosphate-controlled cytoplasmic Ca^{2+} oscillations in pituitary gonadotrophs, *J. Biol. Chem.,* 269, 4860, 1994.

188. Tse, A., Tse, F.W., Almers, W., and Hille, B., Rhythmic exocytosis stimulated by GnRH-induced calcium oscillations in rat gonadotropes, *Science,* 260, 82, 1993.

189. Tse, A. and Hille, B., GnRH-induced Ca^{2+} oscillations and rythmic hyperpolarization of pituitary gonadotrophs, *Science,* 255, 462, 1992.

190. Kukuljan, M., Stojilković, S.S., Rojas, E., and Catt, K.J., Apamin-sensitive potassium channels mediate agonist-induced oscillations of membrane potential in pituitary gonadotrophs, *FEBS Lett.,* 301, 19, 1992.

191. Stutzin, A., Stojilković, S.S., Catt, K.C., and Rojas, E., Characteristics of two types of calcium channels in rat pituitary gonadotrophs, *Am. J. Physiol.,* 257, C865, 1989.

192. Stojilković, S.S., Kukuljan, M., Iida, T., Rojas, E., and Catt, K.J., Integration of cytoplasmic calcium and membrane potential oscillations maintains calcium signaling in pituitary gonadotrophs, *Proc. Natl. Acad. Sci. U.S.A.,* 89, 4081, 1992.

193. Takemura, H., Hughes, A.R., Thastrup, O., and Putney, J.W., Activation of calcium entry by the tumor promote thapsigargin in parotid acinar cells. Evidence that an intracellular calcium pool and not an inositol phosphate regulates calcium fluxes at the plasma membrane, *J. Biol. Chem.,* 264, 12266, 1989.

194. Bird, G.S., Rossier, M.F., Hughes, A.R., Shears, S.B., Armstrong, D.L., and Putney, J.W., Activation of Ca^{2+} entry into acinar cells by a non-phosphorylable inositol trisphosphate, *Nature,* 352, 162, 1991.

195. Lückhoff, A. and Clapham, D.E., Inositol 1,3,4,5-tetrakisphosphate activates an endothelial Ca(2+)-permeable channel, *Nature,* 355, 356, 1992.

196. Hoth, M. and Penner, R., Depletion of intracellular calcium stores activates a calcium current in mast cells, *Nature,* 355, 353, 1992.

Section III
Hypothalamus and Pituitary Gland

Chapter 5

ELECTRICAL ACTIVITY, OSMOSENSITIVITY, AND NEUROMODULATION OF HYPOTHALAMIC SUPRAOPTIC AND PARAVENTRICULAR NEUROSECRETORY CELLS

Leo P. Renaud and Bin Hu

CONTENTS

1. INTRODUCTION

The hypothalamus is the central nervous system (CNS) site in mammalian brain for regulation of neuroendocrine homeostasis and autonomic functions. This chapter is focused on the "classical" magnocellular neurosecretory cells (MNC) of the supraoptic (SON) and paraventricular nuclei (PVN). These are neurons that synthesize the precursor molecules prepropressophysin and preprooxytocin (which contain the sequences for the biologically active hormones vasopressin [VP] and oxytocin [OXY] respectively) and transport them in neurosecretory vesicles to local release sites in dendrites and systemic release sites in posterior pituitary axon terminals.

0-8493-2477-7/95/$0.00+$.50

Investigations to define the parameters that evoke hormone release in this neuronal system have benefited from the fact that their neurohypophysial nerve endings can be studied in an isolated neural lobe preparation. Such experiments have repeatedly demonstrated that hormone release involves a stimulus-secretion coupled exocytotic process that is activity dependent in terms of both the frequency and pattern of impulses invading the axon terminals. Thus, increasing the frequency of stimulation up to ~40 Hz enhances hormone release *per pulse* and, for a given stimulation frequency, hormone release is greater when the stimuli are clustered into bursts than when applied continuously.[1,2] Consequently, excitability and firing patterns, which are generated at the level of the somata of MNC and then propagated into the neurohypophysis, are important factors in the regulation of neurosecretion.

There has been substantial new knowledge in this area recently, and this chapter will elaborate on some of the extrinsic and intrinsic factors that are presently known to influence the excitability and the activity patterns in MNC. For supplementary information on synaptic connections, immunocytochemistry, electrophysiology, neuropharmacology, cell biology, and plasticity, readers are referred to recent reviews.[3-7] Readers are also referred elsewhere[8,9] for information on local conditions at the level of neurohypophysial axon terminals which can influence stimulus-secretion coupling and hormone release (e.g., voltage- and ligand-gated ion channels, coexisting and coreleased amines and peptides, receptors on axon terminal and glial membranes).

2. ELECTRICAL PROPERTIES

2.1. DISTINCT ACTIVITY PATTERNS OF OXY-SECRETING AND VP-SECRETING MNC

Axons of MNC are sparsely branched and virtually all MNC project their axons to the neurohypophysis, a property that permits ready identification through antidromic activation. In addition, MNC in the rat (and perhaps other species) display activity and physiological response profiles that permit an "on-line" distinction between recordings from "OXY-secreting" and "VP-secreting" MNC (see References 3 and 10 for details).

Recordings from **putative OXY-secreting MNC** can be recognized by several features. First, their spontaneous activity pattern is usually slow and irregular, and is not altered by stimuli that activate peripheral chemoreceptors or baroreceptors. Second, in the lactating female rat, the stimulus from the suckling pups provokes intermittent high-frequency (up to 50 Hz), 2- to 4-s bursts of action potentials that occur synchronously[11] and specifically in putative OXY-secreting MNC, resulting in an intermittent release of OXY from the neurohypophysis and milk ejection approximately 10 s later as the circulating hormone reaches and contracts the myoepithelial tissue in the breast. Third, and contrasting with the preceding patterns of activity, another powerful and selective stimulus that activates OXY-secreting MNC in the rat is systemically administered cholecystokinin octapeptide (CCK8) which, through its action at peripheral (gastrointestinal) CCK A-type receptors, triggers a vagally mediated and brainstem-relayed input to OXY-secreting MNC. This stimulus induces a prolonged 50-200% increase in continuous firing from the OXY-secreting MNC, as does exposure to an osmotic challenge (a nonspecific stimulus), and thereby achieves an elevation in plasma OXY.

Recordings from **putative VP-secreting MNC** demonstrate a different activity profile. First, they may be electrically silent, but when they are spontaneously active, their firing pattern may be either continuous (at 12-20 Hz) or "phasic", i.e., silent intervals alternating with bursts of continuous firing that last for variable time intervals. Second, their spontaneous activity can be initiated by activation of peripheral chemoreceptors, or abruptly and transiently depressed following activation of peripheral baroreceptors. Third, these cells demonstrate an ability to adapt their firing patterns in response to stimuli that demand increased hormonal release, so that a "phasic" bursting and/or fast continuous activity pattern emerges as a response to conditions simulating hemorrhage, dehydration, or induced elevations in plasma osmotic pressure. Fourth, phasic bursting patterns seldom correlate with milk ejection in the lactating rate or with the large increase in plasma OXY levels that follow systemic injections of CCK8 in the rat.

2.2. INTRINSIC MEMBRANE CONDUCTANCES

MNC possess a variety of intrinsic membrane conductances that contribute to action potential configuration, discharge frequency, and firing patterns and contribute to the richness of their response repertoire to environmental and synaptic perturbations. Somatically generated action potentials are carried by an influx of both Na and Ca ions, and display a conspicuous, frequency-dependent "shoulder" on the repolarization phase, due largely to activation of high voltage-activated Ca^{2+} channels (Figure 1Aa). Action potential repolarization involves a slowly inactivating delayed rectifying outward potassium current, I_K. Each action potential is followed by a 50- to 100-ms hyperpolarizing afterpotential (HAP) resulting from the activation of a *fast* Ca^{2+}-dependent K^+ conductance, most likely a transient Ca^{2+}-dependent K^+ current (I_{to}) with characteristics similar to I_A;[12] the HAP functions as an "intrinsic" inhibition to set an upper limit on the maximal firing frequency that can be achieved during a depolarizing stimulus (Figure 1Aa). A direct current stimulus-evoked burst (>10 Hz) of action potentials is followed by a prominent, but distinct afterhyperpolarization, or AHP, whose amplitude and duration are proportional to the number of impulses elicited, due to the activation of a *slow* Ca^{2+}-dependent K^+ conductance or I_{AHP}[13] (Figure 1Ab). The I_{AHP} serves to stabilize the steady-state firing **frequency** of continuously active MNC and possibly also their firing **patterns**, since this conductance can shunt or mask the occurrence of subsequent depolarizing afterpotentials, or DAP (see below). The distinction between postspike HAP and the postburst AHP is justified by their distinct ionic currents and pharmacology; HAP are blocked by 4-aminopyridine, which blocks I_{to}[12], whereas AHP and the I_{AHP} are blocked by apamin,[14] a peptide component of bee venom which has a high density of binding sites within the rat SON. Modulation of either I_{to} and/or I_{AHP} (e.g., by neurotransmitters) can be expected to alter not only cell excitability, but also spontaneous firing patterns.

A majority of VP-secreting MNC demonstrate "phasic" firing and many, but not all, phasic-firing MNC display a voltage- and Ca^{2+}-dependent inward membrane current (I_{DAP}) that underlies a slow DAP which may summate during successive action potentials to form a small (<10 mV) depolarizing plateau[15] (Figure 1Ab). While activation of I_{DAP} by an action potential always causes a postspike depolarization (i.e., a DAP), its ability to promote and sustain phasic bursting activity strongly depends on the prevailing membrane potential. Activation of I_{DAP} induces a region of negative slope resistance which crosses spike threshold; when the cell voltage enters this region of negative slope following an action potential, a suprathreshold regenerative depolarization will occur, ensuring that repetitive firing will be sustained. However, MNC with relatively hyperpolarized resting potentials will display a region of net outward current in the voltage range immediately negative to the zone of negative resistance, from where the postspike DAP will not be regenerative and repetitive firing will not be sustained.[16] Thus the onset and maintenance of a phasic discharge can be explained by the properties of the inward postspike aftercurrent I_{DAP}. The termination of a phasic discharge may coincide with an extrinsic event, e.g., arrival of a barrage of postsynaptic inhibitory potentials, but the more typical process is first the cessation of action potential generation followed by collapse of the plateau potential, a sequence that involves intrinsic mechanisms not yet understood. Current-clamp observations suggest that a time-dependent inactivation of the conductance underlying the formation of the plateau potential (i.e., I_{DAP}) is a factor, with recovery occurring slowly during the silent interval.[15] Indeed, the inactivation of prolonged plateau depolarizations in adult MNC has been shown to be Ca^{2+}-dependent.[17] Neurotransmitters

A

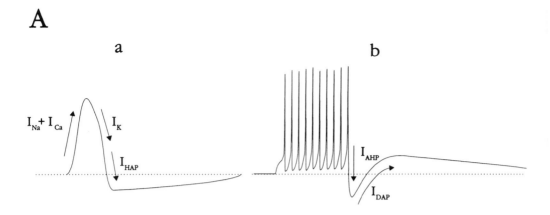

B

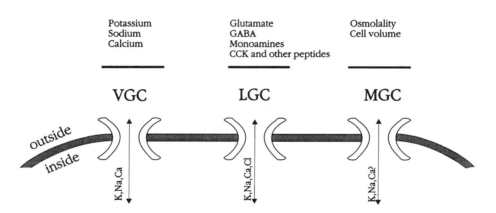

FIGURE 1. Schematic drawings of basic electrical membrane properties and ion channel families that are known to exist in MNC. (Aa) Single action potential waveform and the underlying membrane conductances. (Ab) Burst of spikes triggered by direct current injection are followed by an AHP and a DAP. (B) Summary of channel families so far identified in MNC and their respective ion permeability. Abbreviations: VGC, voltage-gated channels; LGC, ligand-gated channels; MGC, mechano-gated channels.

affecting intracellular $[Ca^{2+}]$, or Ca^{2+} sequestration or buffering, may therefore alter firing patterns or induce subtle changes in periodicity in these MNC. I_{DAP} may also be a select target for neurotransmitter modulation. For example, the excitatory actions of histamine can be partly explained by an H_1 receptor-mediated enhancement of DAP in SON VP-secreting neurons.[18]

2.3. INTRINSIC OSMOSENSITIVITY/OSMORECEPTOR COMPLEX

The pioneering efforts of Verney and Jewell established the importance of the area of the hypothalamic supraoptic nucleus in central osmoreception and hydromineral regulation. It is now thought that hormonal (i.e., VP secretion) and behavioral (i.e., drinking) responses to an osmotic challenge involve a central "osmoreceptor (regulatory) complex" of neurons that incorporates both an "intrinsic" osmoresponsiveness of MNC together with synaptic inputs from other osmosensitive and nonosmosensitive neurons located along the lamina terminalis and area rostroventral to the third ventricle. This notion follows from observations obtained during intracellular recordings from MNC in the SON of brain slice preparations where exposure to hyperosmotic media was associated with both membrane depolarization *and* an increase in the frequency of postsynaptic excitatory potentials.[19] Further support for an osmoreceptor "complex" comes from observations of an intrinsic osmosensitivity in both MNC[20] and in neurons to which they are synaptically connected; some of the latter are located in the rostroventral periventricular regions (e.g., the organum vasculosum lamina terminalis, or OVLT).[20,21] While synaptic inputs may contribute to the normal sensitivity of the whole-animal response to an osmotic challenge,[23] the important feature at the cellular level is the interaction of "exogenous" inputs with "intrinsic" conductances. In the SON, MNC behave as intrinsic osmoreceptors in the absence of synaptic transmission. Data from current clamp recordings indicate that this intrinsic osmosensitivity results from the activation of a voltage-independent and nonselective cationic conductance (I_O).[20] Some clarification as to how MNC transduce osmolality into electrical signals comes from recent studies on acutely isolated MNC where their exposure to hyper- or hypoosmotic media produces membrane depolarization or hyperpolarization, respectively. Unitary current measurements support macroscopic data and lead to the proposal that osmotically induced changes in cell volume modulate the activity of stretch-inactivated mechanosensitive cation channels.[24] This "intrinsic" osmosensitivity, while unique to MNC, may also be a feature of other neurons that participate in the response to an osmotic challenge. For example, cells in circumventricular organs (structures that lack a blood-brain barrier) such as OVLT and the subfornical organ would be strategically positioned to inform the CNS about peripheral events (e.g., plasma osmolality or circulating hormones). Information of an osmotic nature does appear to be conveyed from the lamina terminalis to MNC since (1) injections of hyperosmotic media into the OVLT induce spike discharges and glutamate-mediated excitatory postsynaptic potentials in MNC in SON[25] and (2) animals bearing lesions in the rostral third ventricle area lack the appropriate hormonal (VP release) or behavioral (drinking) responses to a hyperosmotic challenge, and their MNC lose their sensitivity to hyperosmotic stimuli.[26]

3. SYNAPTIC INPUTS: EXCITATORY AMINO ACIDS

Although glutamate has been shown previously to have a modulating influence on both the excitability and activity patterns of MNC,[27] it has only recently been confirmed that MNC in SON express functional postsynaptic excitatory amino acid receptors of both NMDA and non-NMDA subtypes.[28] The former are characterized by a voltage-sensitive cationic conductance with negative resistance properties below spike threshold, are reversibly blocked by micromolar concentrations of aminophosphonovaleric acid, and contribute to rhythmic bursting activity in MNC.[29] Non-NMDA receptors are activated by micromolar concentrations of kainate or quisqualate, and induce a voltage-insensitive cationic conductance. *In situ* hybridization data from SON support the presence of a third or metabotropic receptor subtype,[30] although these remain to be studied in detail.

Recent electrophysiological observations *in vivo* indicate that glutamate, acting at non-NMDA and NMDA receptor sites, is a major contributor to the spontaneous activity of VP-secreting MNC and thus a likely determinant of their threshold for firing in response to osmotic or other excitatory inputs. In anesthetized rats, intravenous administration of ketamine, an NMDA receptor channel blocker, is associated with an immediate cessation in spontaneous phasic firing that persists for seconds to tens of minutes, depending on the dose.[31] This effect is highly reproducible and specific to NMDA receptor antagonism rather than general anesthetic depression. This observation has been further confirmed with local drug application techniques[32] where a rapid, sustained, and selective blockade of ongoing phasic activity consistently follows applications of both NMDA and non-NMDA receptor antagonists. Interestingly, NMDA receptor ligands do not seem to affect the spontaneous activity of a majority of putative OXY-secreting MNC. Moreover, whereas α_1 noradrenergic receptors also induce membrane depolarization and firing in VP-secreting MNC, pharmacological blockade of noradrenergic receptors or destruction of noradrenergic afferents has little influence on spontaneous phasic firing patterns of supraoptic MNC.[33] Hence, the maintenance of spontaneous phasic firing *in vivo* in SON VP-secreting MNC seems to require tonic synaptic activation involving both NMDA and non-NMDA glutamate receptors. Therefore, glutamatergic inputs appear to be in a unique position to influence the genesis of electrical activity in VP-secreting NMC.

4. SYNAPTIC INPUTS: INHIBITORY AMINO ACIDS

Whereas GABAergic input acting at $GABA_A$ (and possibly $GABA_B$) receptors can modulate hormone release from neurohypophysial axon terminals, immunocytochemical and electrophysiological data attest to a pronounced GABAergic input on MNC within the SON and PVN.[34] A local GABAergic innervation, synaptically driven via an ascending pathway that involves neurons in the diagonal band of Broca, is considered to mediate a $GABA_A$ receptor inhibition of VP-secreting MNC that follows activation of peripheral baroreceptors.[35] *In vivo* studies suggest additional sources of GABAergic input to MNC arising from the median preoptic nucleus[36] and possibly other neurons located along the lamina terminals.[22] Some perinuclear zone GABAergic neurons also contain estrogen receptors,[37] raising speculation that they may be part of a steroid-sensitive neural circuitry. Indeed, a remarkable structural plasticity is detected in GABAergic synapses, specifically on OXY-secreting neurons, during lactation.[38]

The functional consequences of this prominent GABAergic innervation can be appreciated during intracellular recordings in SON MNC whether obtained *in vivo*[39] or *in vitro*.[40] With potassium acetate-filled electrodes, SON neurons demonstrate spontaneous inhibitory postsynaptic potentials (IPSP) and prominent compound IPSP of 60-100 ms duration, typically evoked by electrical stimulation in the diagonal band of Broca[40] and in OVLT.[22] Both spontaneous and evoked IPSP and responses to exogenous GABA reflect an enhanced permeability to chloride ions, blocked by bicuculline and potentiation by barbiturates,[41] i.e., features of a $GABA_A$ receptor-mediated inhibition. In SON, MNC show no response to applications of baclofen, implying a lack of postsynaptic $GABA_B$ receptors. However, baclofen can reduce OVLT-evoked IPSP, presumably through an action on presynaptic $GABA_B$ receptors (unpublished observations).

5. SYNAPTIC INPUTS: ACETYLCHOLINE AND MONOAMINES

5.1. ACETYLCHOLINE

Acetylcholine is a stimulant for the release of neurohypophysial hormones, and cholinergic drugs do appear to activate MNC.[27] However, since most studies report little or no innervation of SON by choline acetyltransferase-positive fibers, this issue requires more study. Interestingly, rat SON neurons exposed to *d*-tubocurarine demonstrate block of I_{AHP}, although this appears unrelated to the cholinoceptive properties of the cell.[42]

5.2. HISTAMINE

Histamine, long implicated in the release of neurohypophysial hormones, may be a transmitter in histidine decarboxylase-containing fibers that project to MNC from the tuberomammillary nuclei and supramammillary mesencephalic area.[43] In the SON, MNC are depolarized by histamine acting at H_1 receptors; among phasically firing MNC, at least a part of its action appears directed to an effect on I_{DAP}.[18]

5.3. DOPAMINE

MNC receive a specific dopamine innervation, in part arising from the A11 cells within the periventricular dopaminergic system and/or the A13-15 cells in the anterior and dorsal hypothalamus.[44] Intracellular recordings from SON neurons in hypothalamic explants indicate that a majority of MNC are depolarized following exogenous application of dopamine, acting at a D_2 receptor that is coupled with a nonselective cationic channel.[45]

5.4. NOREPINEPHRINE

The SON and PVN contain a dense catecholamine terminal innervation that arises primarily from the A1 and A2 noradrenergic cells in the caudal ventrolateral medulla and nucleus tractus solitarius, respectively, and are more closely associated with the somata of VP- rather than OXY-secreting MNC.[46] While the nature of the role of norepinephrine in regulating hormone release from the neurohypophysis has generated considerable controversy, exogenous drug application studies indicate that MNC possess adrenergic receptors that permit both stimulatory and depressant functions: a facilitatory role, observed with micromolar concentrations of agonist, is mediated through α_1 postsynaptic receptors whose activation results in gradual membrane depolarization and bursting activity that is, in part, the result of a suppression of I_{to}; a depressant action, requiring application in millimolar concentrations, is mediated through α_2 and β adrenoreceptors.[3] However, the role of norepinephrine as a synaptic mediator of information ascending from the medulla to magnocellular neurons remains unclear. Whereas VP-secreting MNC neurons appear to be selectively activated by discrete electrical stimulation in both the caudal ventrolateral medulla (VLM) and the caudal nucleus tractus solitarius (NTS), a response that can be abolished after 6-hydroxydopamine depletion of catecholamines at the level of the hypothalamic termination of these fibers,[33] it is curious that the evoked response is resistant to local and systemic adrenergic antagonists, even when present in sufficient concentration to block responses to exogenous norepinephrine.[47] This has raised suspicion that a coexisting nonadrenergic transmitter (possibly ATP) may mediate the evoked short latency synaptic responses to VLM and NTS stimulation, whereas norepinephrine itself may serve to alter MNC cell firing patterns so that MNC can provide a more efficient release of hormone (VP) from their pituitary axon terminals during stress (such as might follow systemic hypovolemia and hypotension).

6. NEUROMODULATORY PEPTIDES

6.1. CHOLECYSTOKININ (CCK)

CCK, predominantly in the form of its octapeptide CCK-8, is widely distributed within the peripheral and central nervous systems. In the neurohypophysial system, CCK is colocalized within a portion of the VP- and OXY-synthesizing neurons[7] and can influence their function at two levels. First, CCK acts at CCK_B-type receptors on MNC to induce a tetrodotoxin (TTX)-resistant membrane depolarization that involves nonselective cationic channels.[48] Second, CCK can evoke a TTX-resistant release of both VP and OXY from the isolated neural lobe.[49] Both effects persist in the absence of extracellular calcium, and may involve the release of Ca^{2+} from intraterminal stores as part of the signal transduction cascade. The CCK-induced release of hormones in the isolated neural lobe preparation[49] is suppressed by staurosporine, a protein kinase inhibitor, suggesting that CCK receptors in the neural lobe may be functionally linked to a second-messenger system involving protein kinase C.

6.2. ANGIOTENSIN

Angiotensin (Ang) that is present as a circulating hormone *indirectly* affects CNS function through an action at AT_1 receptors on neurons in certain circumventricular organs (e.g., subfornical organ and area postrema), resulting in drinking behavior, a rise in arterial blood pressure, and the release of neurohypophysial hormones.[3] In addition, CNS regulation of cardiovascular and hydromineral homeostasis engages an intrinsic renin-angiotensin neuronal system. This system includes subfornical organ neurons, and they can act *directly* to facilitate firing in MNC, mediated via AT_1 receptors that ultimately induce a nonselective cationic conductance.[50,51] Ang pathways and Ang receptors are also a prominent feature along the lamina terminalis; this peptide may therefore have a transmitter role in pathways that contribute to the "osmoregulatory complex". In addition, immunocytochemical data demonstrate Ang-like immunoreactivity within MNC and their neurohypophysial projections, indicating a possible role for this peptide in the neural lobe.[52]

6.3. VASOPRESSIN AND OXYTOCIN

In addition to their role as circulating hormones, VP and OXY and certain fragments of these molecules are likely participants in several CNS functions (e.g., thermoregulation, cardiovascular regulation, and learning behavior), perhaps serving as neurotransmitters in an extensive, but selective neural network in brain (see Reference 3 for review). Within the magnocellular neurohypophysial axis, functional[53] and structural[54] observations support the notion that VP and OXY (and possibly other molecules endogenous to MNC) may be released by exocytosis from somatic, dendritic, or axonal sites to influence their excitability. Centrally released OXY, but not VP, has been linked to a marked reduction in glial separation of magnocellular OXY-secreting neurons and terminals during lactation and dehydration.[55] Concordant with a possible transmitter function for neurohypophysial peptides is their binding to select brain regions, including SON and PVN,[56] where there are indications that OXY may act at excitatory receptors on OXY-secreting MNC.[57]

6.4. OPIOID PEPTIDES

Early reports of a morphine-induced antidiuresis and suppression of milk letdown have resulted in numerous studies as to the possible mechanisms whereby opiates exert their influence on the hypothalamic-neurohypophysial axis. It is now apparent that MNC may be innervated by opioidergic pathways and synthesize opioid peptides.[7] Both the somata and terminals of MNC are targets for opiate actions. Exogenous application of morphine, opioid peptides, and opiate receptor agonists is usually accompanied by a naloxone-sensitive depression in the excitability of MNC, acting at μ and κ receptors.[58] Evidently, MNC do not receive a tonically active opioidergic input, since there is little or nonsignificant change in their firing rate following administration of an opiate receptor antagonist (naloxone) to urethane-anesthetized male or female rats in control or osmotically stimulated states. However, towards the end of pregnancy in the rat there is a reduction in the density of μ receptors in the SON, coincident with evidence that the firing of OXY-secreting MNC can be modulated by endogenous opioids.[59] Also, in lactating rats rendered morphine dependent, naloxone administration precipitates a pronounced and selective increase in the firing of OXY-secreting MNC, likely due to an alteration of μ receptors on the somata of neurosecretory cells.[60] An analysis of peptide release from the electrically or chemically stimulated isolated neural lobe from morphine-naive rats reveals that exposure to naloxone results in a preferential enhancement in OXY release, presumably due in part to dynorphin (released from VP-secreting terminals) acting at κ-type receptors to depress hormone release from adjacent OXY-secreting terminals.

7. CONCLUDING COMMENTS

Multidisciplinary *in vivo* and *in vitro* electrophysiological and pharmacological investigations accomplished in the past several years have led to significant progress in our understanding of the neuroendocrine mechanisms whereby hypothalamic MNC regulate body homeostasis and cardiovascular function. These studies support the model in which MNC act both as a biological osmometer and final hormone output pathway whose effectiveness is closely regulated by three distinct ion channel families (Figure 1B). Interactive changes in membrane ion permeation brought about by voltage and ligand channels as well as by mechanosensitive conductance therefore provide a functionally coherent machinery of cell regulation in these neuroendocrine neurons.

REFERENCES

1. **Cazalis, M., Dayanithi, G., and Nordmann, J. J.,** The role of patterned burst and interburst interval on the excitation coupling mechanism in the isolated rat neural lobe, *J. Physiol.,* 369, 45, 1985.
2. **Bicknell, R. J.,** Endogenous opioid peptides and hypothalamic neuroendocrine neurons, *J. Endocrinol.,* 107, 437, 1988.
3. **Renaud, L. P. and Bourque, C. W.,** Neurophysiology and neuropharmacology of hypothalamic magnocellular neurons secreting vasopressin and oxytocin, *Prog. Neurobiol.,* 36, 131, 1991.
4. **Hatton, G. I.,** Emerging concepts of structure-function dynamics in adult brain: the hypothalamo-neurohypophysial system, *Prog. Neurobiol.,* 34, 437, 1990.
5. **Theodosis, D. T. and Poulain, D. A.,** Activity-dependent neuronal-glial and synaptic plasticity in the adult mammalian hypothalamus, *Neuroscience,* 57, 501, 1993.
6. **Gainer, H., Altstein, M., Whitnall, M. H., and Wray, S.,** The biosynthesis and secretion of oxytocin and vasopressin, in *The Physiology of Reproduction,* Knobil, E. and Neill, J., Eds., Raven Press, New York, 1988, chap. 57.
7. **Meister, B.,** Gene expression and chemical diversity in hypothalamic neurosecretory neurons, *Mol. Neurobiol.,* 7, 87, 1993.
8. **Wang, X., Treistman, S. N., Wilson, A., Nordmann, J. J., and Lemos, J. R.,** Ca^{2+} channels and peptide release from neurosecretory terminals, *News Physiol. Sci.,* 8, 64, 1993.
9. **Bielefeldt, K. and Jackson, M. B.,** A calcium-activated potassium channel causes frequency-dependent action potential failures in a mammalian nerve terminal, *J. Neurophysiol.,* 70, 284, 1993.
10. **Poulain, D. A., Wakerley, J. B., and Dyball, R. E. J.,** Electrophysiological differentiation of oxytocin- and vasopressin-secreting neurons, *Proc. R. Soc. London Ser. B.,* 196, 367, 1977.
11. **Wakerley, J. B. and Ingram, C. D.,** Synchronisation of bursting in hypothalamic oxytocin neurones: possible coordinating mechanisms, *News Physiol. Sci.,* 8, 129, 1993.
12. **Bourque, C. W.,** Transient calcium-dependent potassium current in magnocellular neurosecretory cells of the rat supraoptic nucleus, *J. Physiol.,* 397, 331, 1988.
13. **Bourque, C. W., Randle, J. C. R., and Renaud, L. P.,** Calcium dependent potassium conductance in rat supraoptic nucleus neurosecretory neurons, *J. Neurophysiol.,* 54, 1375, 1985.
14. **Bourque, C. W. and Brown, D. A.,** Apamin and d-tubocurarine block the afterhyperpolarization of rat supraoptic neurosecretory neurons, *Neurosci. Lett.,* 185, 1987.
15. **Andrew, R. D. and Dudek, F. E.,** Analysis of intracellularly recorded phasic bursting by mammalian neuroendocrine cells, *J. Neurophysiol.,* 51, 552, 1984.
16. **Bourque, C. W.,** Intrinsic features and control of phasic burst onset in magnocellular neurosecretory cells, in *Organization of the Autonomic Nervous System: Central and Peripheral Mechanisms,* Ciriello, J., Calaresu, F. R., Renaud, L. P., and Polosa, C., Eds., Alan R. Liss, New York, 1987, 387-396.
17. **Bourque, C. W., Brown, D. A., and Renaud, L. P.,** Barium ions induce prolonged plateau depolarizations in neurosecretory neurones of the adult supraoptic nucleus, *J. Physiol.,* 375, 573, 1986.
18. **Smith, B. N. and Armstrong, W. E.,** Histamine enhances the depolarizing afterpotential of immunohistochemically identified vasopressin neurons in the rat supraoptic nucleus via H_1-receptor activation, *Neuroscience,* 53, 855, 1993.
19. **Mason, W. T.,** Supraoptic neurones of rat hypothalamus are osmosensitive, *Nature,* 287, 154, 1980.
20. **Bourque, C. W.,** Ionic basis for the intrinsic activation of rat supraoptic neurones by hyperosmotic stimuli, *J. Physiol.,* 417, 263, 1989.
21. **Nissen, R., Bourque, C. W., and Renaud, L. P.,** Membrane properties of organum vasculosum lamina terminalis neurons recorded in vitro, *Am. J. Physiol.,* 264, R811, 1993.
22. **Yang, C. R., Senatorov, V. V., and Renaud, L. P.,** Organum vasculosum lamina terminalis-evoked postsynaptic responses in rat supraoptic neurones *in vitro, J. Physiol.,* 477, 59, 1994.
23. **Sladek, C. D. and Armstrong, W. E.,** Osmotic control of vasopressin release, *Trends Neurosci.,* 8, 166, 1985.
24. **Oliet, S. H. R. and Bourque, C. W.,** Mechanosensitive channels transduce osmosensitivity in supraoptic neurons, *Nature,* 364, 341, 1993.

25. **Richard, D. and Bourque, C. W.,** Synaptic activation of rat supraoptic neurones by osmotic stimulation of the organum vasculosum lamina terminalis, *Neuroendocrinology,* 55, 609, 1992.
26. **Leng, G., Dyball, R. E. J., and Russell, J. A.,** Neurophysiology of body fluid homeostasis, *Comp. Biochem. Physiol.,* 90A, 781, 1988.
27. **Arnauld, E., Cirino, M., Layton, B. S., and Renaud, L. P.,** Contrasting actions of amino acids, acetylcholine, noradrenaline and leucine enkephalin on the excitability of supraoptic vasopressin-secreting neurons, *Neuroendocrinology,* 36, 187, 1983.
28. **Hu, B. and Bourque, C. W.,** Functional N-methyl-D-aspartate and non N-methyl-D-aspartate receptors are expressed by rat supraoptic neurosecretory cells *in vitro, J. Neuroendocrinol.,* 3, 509, 1991.
29. **Hu, B. and Bourque, C. W.,** NMDA receptor-mediated rhythmic bursting activity in rat supraoptic nucleus neurones *in vitro, J. Physiol.,* 458, 667, 1992.
30. **Tanabe, Y., Nomura, A., Masu, M., Shigemoto, R., Mizuno, N., and Nakanishi, S.,** Signal transduction, pharmacological properties, and expression patterns of two rat metabotropic glutamate receptors, mGluR3 and mGluR4, *J. Neurosci.,* 13, 1372, 1993.
31. **Nissen, R., Hu, B., and Renaud, L. P.,** N-Methyl-D-aspartate receptor antagonist ketamine selectively attenuates spontaneous phasic activity of supraoptic vasopressin neurons *in vivo, Neuroscience,* 59, 115, 1994.
32. **Nissen, R., Hu, B., and Renaud, L. P.,** Glutamate receptors regulate spontaneous phasic firing of rat supraoptic vasopressin neurones in vivo, *J. Physiol.,* 484, 415, 1995.
33. **Day, T. A., Ferguson, A. V., and Renaud, L. P.,** Facilitatory influence of noradrenergic afferents on the excitability of rat paraventricular nucleus neurosecretory cells, *J. Physiol.,* 355, 237, 1984.
34. **Decavel, C. and Van Den Pol, A. N.,** Converging GABA- and glutamate-immunoreactive axons make synaptic contact with identified hypothalamic neurosecretory neurons, *J. Comp. Neurol.,* 316, 104, 1992.
35. **Nissen, R., Cunningham, J. T., and Renaud, L. P.,** Lateral hypothalamic lesions alter baroreceptor-evoked inhibition of rat supraoptic vasopressin neurones, *J. Physiol.,* 470, 751, 1993.
36. **Nissen, R. and Renaud, L. P.,** GABA receptor mediation of median preoptic nucleus-evoked inhibition of supraoptic neurosecretory neurones in rat, *J. Physiol.,* 479, 207, 1994.
37. **Herbison, A. E.,** Immunocytochemical evidence for oestrogen receptors within GABA neurones located in the perinuclear zone of the supraoptic nucleus and GABA$_A$ receptor β_2/β_3 subunits on supraoptic oxytocin neurones, *J. Neuroendocrinol.,* 6, 5, 1994.
38. **Gies, U. and Theodosis, D. T.,** Synaptic plasticity in the rat supraoptic nucleus during lactation involves GABA innervation and oxytocin neurons: a quantitative immunocytochemical analysis, *J. Neurosci.,* 14, 2861, 1994.
39. **Bourque, C. W. and Renaud, L. P.,** Membrane properties of rat magnocellular neuroendocrine cells in vivo, *Brain Res.,* 540, 349, 1991.
40. **Randle, J. C. R., Bourque, C. W., and Renaud, L. P.,** Characterization of spontaneous and evoked inhibitory postsynaptic potentials in rat supraoptic neurosecretory neurons in vitro, *J. Neurophysiol.,* 56, 1703, 1986.
41. **Randle, J. C. R. and Renaud, L. P.,** Actions of gamma-aminobutyric acid in rat supraoptic nucleus neurosecretory neurones in vitro, *J. Physiol.,* 387, 629, 1987.
42. **Bourque, C. W. and Brown, D. A.,** Apamin and D-tubocurarine block the after-hyperpolarization of rat supraoptic neurosecretory neurons, *Neurosci. Lett.,* 185, 1987.
43. **Panula, P., Yang, H.-Y. T., and Costa, E.,** Histamine-containing neurons in the rat hypothalamus, *Proc. Natl. Acad. Sci. U.S.A.,* 81, 2572, 1984.
44. **Decavel, C., Geffard, M., and Calas, A.,** Comparative study of dopamine- and noradrenaline-immunoreactive terminals in the paraventricular and supraoptic nuclei of the rat, *Neurosci. Lett.,* 7, 149, 1987.
45. **Yang, C. R., Bourque, C. W., and Renaud, L. P.,** Dopamine D_2 receptor activation depolarizes rat supraoptic neurones in hypothalamic explants, *J. Physiol.,* 443, 405, 1991.
46. **Swanson, L. W. and Sawchenko, P. E.,** Hypothalamic integration: organization of the paraventricular and supraoptic nucleus, *Annu. Rev. Neurosci.,* 6, 269, 1983.
47. **Day, T. A., Renaud, L. P., and Sibbald, J. R.,** Excitation of supraoptic vasopressin cells by stimulation of the A1 noradrenaline cell group: failure to demonstrate role for established adrenergic or amino acid receptors, *Brain Res.,* 516, 91, 1990.
48. **Jarvis, C. R., Bourque, C. W., and Renaud, L. P.,** Depolarizing action of cholecystokinin on rat supraoptic neurones in vitro, *J. Physiol.,* 458, 621, 1992.
49. **Bondy, C. A., Jensen, R. T., Brady, L. S., and Gainer, H.,** Cholecystokinin evokes secretion of oxytocin and vasopressin from rat neural lobe independent of external calcium, *Proc. Natl. Acad. Sci. U.S.A.,* 861, 5198, 1989.
50. **Jhamandas, J. H., Lind, R. W., and Renaud, L. P.,** Angiotensin II may mediate excitatory neurotransmission from the subfornical organ to the hypothalamic supraoptic nucleus: an anatomical and electrophysiological study in the rat, *Brain Res.,* 487, 52, 1989.

51. **Yang, C. R., Phillips, M. I., and Renaud, L. P.,** Angiotensin II receptor activation depolarizes rat supraoptic neurons in vitro, *Am. J. Physiol.,* 263, R1333, 1992.

52. **Imboden, H. and Felix, D.,** An immunocytochemical comparison of the angiotensin and vasopressin hypothalamo-neurohypophysial systems in normotensive rats, *Reg. Peptides,* 36, 197, 1991.

53. **Moos, F. and Richard, P. H.,** Paraventricular and supraoptic bursting oxytocin cells in rat are locally regulated by oxytocin and functionally related, *J. Physiol.,* 408, 1, 1989.

54. **Pow, D. V. and Morris, J. F.,** Dendrites of hypothalamic magnocellular neurons release neurohypophysial peptides by exocytosis, *Neuroscience,* 32, 435, 1989.

55. **Chapman, D. B., Morris, J. F., Theodosis, D. T., Montagnese, C., and Poulain, D. A.,** Osmotic stimulation causes structural plasticity of neurone-glia relationships of the oxytocin- but not the vasopressin-secreting neurones in the hypothalamic supraoptic nucleus, *Neuroscience,* 17, 679, 1986.

56. **Tribollet, E., Barberis, C., Jard, S., Dubois-Dauphine, M., and Dreifuss, J. J.,** Localization and pharmacological characterization of high affinity binding sites for vasopressin and oxytocin in the rat brain by light microscopic autoradiography, *Brain Res.,* 442, 105, 1988.

57. **Freund-Mercier, J. J., Moos, F., Poulain, D. A., Richard, P., Rodriguez, F., Theodosis, D. T., and Vincent, J.-D.,** Role of central oxytocin in the control of the milk ejection reflex, *Brain Res. Bull.,* 20, 737, 1988.

58. **Leng, G. and Russell, J. A.,** Opioids, oxytocin and parturition, in *Brain Opioid Systems in Reproduction,* Dyer, R. G. and Bicknell, R. J., Eds., Oxford University Press, London, 1989, 231-256.

59. **Bicknell, R. J., Leng, G., Russell, J. A., Dyer, R. G., Mansfield, S., and Zhao, B.-G.,** Hypothalamic opioid mechanisms controlling oxytocin neurons during parturition, *Brain Res. Bull.,* 20, 743, 1988.

60. **Bicknell, R. J., Leng, G., Lincoln, D. W., and Russell, J. A.,** Naloxone excites oxytocin neurones in the supraoptic nucleus of lactating rats after chronic morphine treatment, *J. Physiol.,* 396, 297, 1988.

Chapter 6

ELECTROPHYSIOLOGY OF ANTERIOR PITUITARY CELLS

Brian J. Corrette, Christiane K. Bauer, and Jürgen R. Schwarz

CONTENTS

1. SUMMARY

The anterior pituitary contains 6 major types of cells which secrete different hormones. Previous work has shown that secretion in these cells is largely controlled by voltage-gated Ca channels which adjust the influx of Ca^{2+} into the cell based on membrane potential. A large body of additional knowledge has accumulated concerning the electrophysiology of these cells since the establishment of the patch clamp technique and the development of quantitative microfluorimetry to measure free intracellular Ca^{2+}. A number of other methods, such as the reverse hemolytic plaque assay and various cell separation methods based on sedimentation, have allowed identification and isolation of defined groups of cells from the heterogeneous population of the anterior pituitary, and these methods are briefly described. Experimental results relating to the electrophysiology of particular cell types within the anterior pituitary are presented. Clonal GH cell lines have served as a model pituitary cell system. The basic electrophysiology of growth hormone (GH) cells has been intensively studied, and recent experiments concerning several signaling pathways, in particular those of the thyrotropin-releasing hormone (TRH) response, are summarized. Results of electrophysiological experiments on somatotroph, lactotroph, gonadotroph, corticotroph, and melanotroph cells of the anterior pituitary are reviewed. For each of these cell types, information concerning the basic ionic currents and the electrical activity is summarized. The electrophysiological responses induced by substances known to stimulate or inhibit secretion are discussed. Finally, experiments

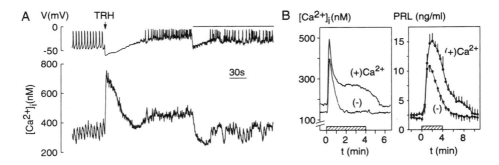

FIGURE 1. Comparison of the TRH-induced biphasic changes in membrane potential, in cytosolic Ca^{2+} concentration and in secretion of PRL in GH_4C_1 cells. (A) Simultaneous measurement of membrane potential (top) and $[Ca^{2+}]_i$ (bottom) during continuous stimulation with 100 nM TRH, starting from the time indicated by the arrow. A similar time course is apparent for the changes in membrane potential (whole-cell current clamp) and Ca concentration. Hyperpolarizing current, injected during the second phase as indicated by the horizontal bar, reversed the increased firing frequency during the second phase and abolished the Ca plateau. (From Iijima, T., Sand, O., Sekiguchi, T., and Matsumoto, G., *Acta Physiol. Scand.*, 140, 269, 1990. With permission.) (B) Effect of the presence (+) and absence (–) of external Ca on the TRH-induced (hatched bars denote 10 nM TRH) changes in $[Ca^{2+}]_i$ (left) and in secretion of prolactin (PRL, right). Absence of the second phase plateau of $[Ca^{2+}]_i$ results in a large reduction in the secretion of PRL. Note the difference in time scales. (From Sato, N., Wang, X. B., Greer, M. A., *Mol. Cell. Endocrinol.*, 77, 193, 1991. With permission.)

are described in which modulation of ionic currents and changes in responsiveness to these substances have been shown to occur as part of the short- and long-term regulation of secretion.

2. INTRODUCTION

The electrophysiology of the various cell types of the anterior pituitary was last covered in two major reviews by Douglas and Taraskevich in 1985[1] and by Ozawa and Sand in 1986.[2] These papers summarize about 2 decades of work beginning with the formulation of the calcium hypothesis of stimulus-secretion coupling by Douglas and Rubin in the early 1960s[3] and culminating in a rapid growth in electrophysiological studies, which was prompted by Kidokoro's[4] demonstration that clonal pituitary cells generate spontaneous action potentials whose frequency is modulated during stimulation of secretion. This initial work clearly showed that secretion is in large part controlled by voltage-gated Ca channels which adjust the influx of Ca^{2+} into the cell based on membrane potential.

A large body of additional knowledge has accumulated since 1986 concerning the electrophysiology of the various cell types of the anterior pituitary. The growth in this field has come mainly from the routine use of the patch clamp method, particularly in its whole-cell configuration. In addition, quantitative fluorescent microscopy using fluorescent dyes such as fura-2 has allowed measurements of the changes in free cytosolic Ca^{2+} concentration ($[Ca^{2+}]_i$) in individual endocrine cells. In combination with a video imaging system, this method makes it possible to monitor $[Ca^{2+}]_i$ for a number of cells simultaneously, and has proved useful for the examination of Ca-dependent secretory processes.[5] Hormone secretion, itself, can be monitored using the well-established method of radioimmunology.

As shown in the examples of Figure 1, by combining these methods it has been possible to directly correlate action potential activity in a pituitary cell with changes in $[Ca^{2+}]_i$ (Figure 1A), and the same changes in $[Ca^{2+}]_i$ can be related to hormone secretion (Figure 1B). Anterior pituitary cells are small and lack the processes of nerve cells that complicate the control of membrane voltage in patch clamp experiments. These cells are therefore ideal objects for studies of electrical activity involved in hormone secretion, including the currents and ionic channels which underlie it. By taking advantage of the dialysis which accompanies whole-cell recording, it has also been possible to exactly define the intracellular milieu during experiments, and thus to advance our knowledge of the multitude of signaling pathways connecting receptors with effectors.

Based on its embryonic development, the pituitary gland can be divided into two major parts: a posterior neurohypophysis and an anterior adenohypophysis.[1] The adenohypophysis or anterior pituitary contains six types of cells, each of which secretes different hormones. The secretion of hormones by these different classes of cells is controlled by releasing or inhibiting factors, which are produced by neurons in the hypothalamus with nerve endings in the hypophysial stalk. From the hypophysial stalk, these factors are secreted into the portal blood system and are transported to the anterior pituitary. The neurohypophysis or neural lobe mainly contains the axon terminals of hormone-producing cells located in hypothalamic nuclei and need not be further considered, with the exception that a fraction of its nerve fibers innervates melanotrophs within the neurointermediate lobe at the boundary of the anterior with the posterior pituitary.

A large number of studies have made use of the patch clamp method to characterize the basic electrical properties of five of the six cell types found in the anterior pituitary. Although the patch clamp technique has made detailed electrophysiological studies of individual pituitary cells into a routine laboratory procedure, adaptation of a number of other methodologies has been necessary to help isolate and identify well-defined groups of cells within the heterogeneous population making up the anterior pituitary. From work with defined groups of cells, initial experiments performed by several laboratories have shown that many external factors impinge upon these cells. The results of these studies give examples of how these factors can modulate the basic electrical properties of pituitary cells, and thereby regulate their secretion, depending on the exact physiological context.

Given the importance of these methods to separate and identify cells, they are described briefly in Section 3 of this review. In Section 4, we have tried to chronicle the new developments for each of the major cell types. First we discuss the most recent results obtained with GH cell lines. These model cells have been most intensively studied and therefore provide a good introduction. Each of the following parts deals with one of the major cell types of the anterior pituitary including somatotrophs, lactotrophs, gonadotrophs, corticotrophs, and melanotrophs. Both thyrotrophs and the nonhormone-producing folliculo-stellate cells have not been included, since little information is available on the electrophysiology of these cells.

3. METHODS OF ISOLATING AND IDENTIFYING PITUITARY CELLS

The simplest method of identifying anterior pituitary cells in electrophysiological experiments is immunocytochemical staining using antibodies for particular peptide hormones. Immunocytochemical staining can be performed either before (avidin-fluorescein method)[6] or after (peroxidase-antiperoxidase method) an experiment.

Neill and Frawley[7] introduced the use of another technique, the reverse hemolytic plaque assay (RHPA), to measure hormones released by individual pituitary cells in culture. The pituitary cells are plated together with a high density of erythrocytes which have been coupled to an antibody that binds the secreted hormone. A complement-mediated lysis of the erythrocytes produces a clear zone or plaque in the vicinity of pituitary cells producing the particular hormone. These plaques identify those cells which are actively secreting a particular hormone, and the size of each plaque reflects the amount of hormone released by single cells during the incubation period. The RHPA has made it possible to study subpopulations of cells which differ in either basal or stimulated secretion of a hormone.

A few studies have begun to make use of the cell immunoblot assay (CIBA) to measure small peptides secreted by single endocrine cells.[8,9] This technique brings the cells in contact with a transfer membrane that has a high protein-binding capacity and is not toxic to living cells. Hormone secreted by these cells binds directly to this membrane and can be stained with hormone-specific antibody by immunocytochemical procedures. With the use of hormone standards on the same membrane and quantitative image analysis, the CIBA offers a reliable alternative to RHPA. CIBA also offers the advantage that very small amounts of substances can be measured from single cells and high dilutions of antibodies can be used.

A number of improved methods have also been introduced to produce cultures which are enriched in specific subtypes of cells. These methods make use of differences in cell characteristics such as sedimentation velocity in gradients of either bovine serum albumin (BSA)[10] or Percoll.[11] Cells can also be identified based on a cell-surface reaction with particular antibodies, and then isolated by fluorescence-activated cell sorting.[12]

Two cell types, corticotrophs and melanotrophs, can be identified purely based on anatomy or cell morphology. Based on their unique morphology, corticotrophs can be identified in mixed pituitary cultures.[13] Melanotrophs are the only hormone-producing cells in the intermediate lobe of the pituitary and have therefore been well studied.

In addition to dispersed cells, clonal cell lines, such as the various GH cells derived from a rat pituitary adenoma,[14] have continued to provide readily available populations of homogeneous cells for studying particular details of the control of secretion.

4. ELECTROPHYSIOLOGY OF ANTERIOR PITUITARY CELLS

4.1. GH CELL LINES
4.1.1. Characterization of Cell Lines
4.1.1.1. *GH Cell Lines*

The GH strains of pituitary cell lines are derived from an original primary culture of pituitary tumor cells isolated by Tashjian et al.[14] in 1965 from two female Wistar/Furth rats bearing pituitary tumor MtT/W5. Of the various GH cell lines, only the GH_3 (GH_3/B_6) and GH_4C_1 cell lines are electrophysiologically and biochemically well characterized. Both of these lines secrete growth hormone (GH) as well as prolactin (PRL), but they differ in the amount and relation of the secreted hormones. In the GH_3 cell line there exist more GH- than PRL-secreting cells,[15] whereas more PRL- than GH-secreting cells have been found in the GH_4C_1 cell line.[16]

The GH cells have often been used as a model for lactotrophs because they produce PRL and they possess functional TRH, vasoactive intestinal peptide (VIP), and cholecystokinin (CCK) receptors, which all act to increase PRL secretion. Unlike lactotrophs, though, no functional high-affinity dopamine (DA) D_2 receptors are present in GH cells.[17] Similar to somatotrophs, GH cells have functional somatostatin (SST) receptors which inhibit hormone secretion. The intracellular pathways linked to the SST receptor are similar to those used by DA receptors in lactotrophs (See Sections 4.1.4 and 4.3.3 and compare Figures 6 and 8). GH cells have been less often used as a model for somatotrophs, though, due to the inability of GH-releasing factor (GhRF) to increase hormone production or hormone mRNA levels.[18]

4.1.1.2. *Other Related Pituitary Cell Lines*

MMQ cells developed from a rat pituitary tumor secrete only PRL, no GH.[19] They could therefore serve as a model for lactotrophs, especially as they have functional D_2 receptors. They also possess VIP, SST, and muscarinic acetylcholine receptors, but they lack TRH, bombesin, angiotensin II, and neurotensin receptors. The MMQ cells and several other lines like GH_1, P_0 or G3 cells are only marginally or not at all electrophysiologically characterized; therefore, we have concentrated on GH cells in this review.

4.1.2. Membrane Currents and Electrical Excitability

Although current densities have been shown to differ between the GH_3 and GH_4C_1 cell lines, as described, for example, for Na and A-like K current,[20] no ionic current has been demonstrated to exclusively exist in only one or the other of these two cell lines. Therefore, a description of ionic channels is given for GH cells in general, without reference to the cell line.

4.1.2.1. *Membrane Currents*

GH cells exhibit a tetrodotoxin (TTX)-sensitive Na current[20] and possess T- and L-type Ca channels.[21,22] Three different Ca-dependent K channels have been described: the high-conductance BK channel which is also extremely voltage dependent, the small-conductance SK channel, and the intermediate-conductance IK channel.[23,24] In addition, another Ca-dependent K channel with 4 pS unit conductance has been mentioned.[23] A Ca-dependent Cl current is also present in GH cells.[25] Several voltage-dependent K currents have been described: the A-like fast-inactivating current,[26,27] a slow-inactivating delayed rectifier-like current,[28] and an inward-rectifying K current (K_{IR}) which inactivates at potentials negative to -60 mV in a potential-dependent and Na-independent manner.[29] In addition, K channels with 55-pS unit conductance can be activated by SST and the muscarinic ligands, acetylcholine and carbachol.[30]

4.1.2.2. *Electrical Activity*

The cells of the GH strains are electrically excitable and the majority of cells measured with the perforated-patch method exhibit spontaneous action potentials. The mean frequencies of unstimulated cells are between 0.2 and 0.5 Hz,[31-33] and the "resting" membrane potential between action potentials lies in the range between -60 mV, reached during afterhyperpolarization, and about -40 mV, which is the threshold for initiation of action potentials.[32]

The action potentials are mainly due to a flow of Ca^{2+} and, to a variable degree, Na^+ ions.[4,34] An evidence of the tight coupling between Ca^{2+} influx and the depolarizing phase of the action potentials was given by simultaneous measurements of $[Ca^{2+}]_i$ using microfluorimetry, and membrane potential using the patch clamp method [35] (see also Figure 1A.)[36]

4.1.2.3. *Ionic Currents Involved in the Action Potential*

The upstroke velocity of an action potential depends mainly on the amount of Na current involved.[34] The peak amplitude depends mainly on Ca currents and can be influenced by changing extra- and intracellular Ca concentrations.[37] Repolarization of an action potential is mediated by voltage-dependent K channels (A-like and delayed rectifier-like) and Ca-dependent K channels, SK, IK, and in particular the BK channel, which is also voltage dependent.[38] The duration of action potentials therefore depends strongly on the amount of Ca^{2+} influx and on the level of $[Ca^{2+}]_i$. The afterhyperpolarization is mediated by voltage-dependent K channels and the Ca-dependent K channels, SK and IK.[24]

4.1.2.4. *Ionic Currents Contributing to the Resting Membrane Potential*

Na channels are suggested not to be involved in the maintenance of the resting membrane potential, since TTX has not been found to affect this potential.[39] T-type Ca currents are most probably absent at the resting potential, in spite of their negative activation threshold (-53 mV),[40] since when these currents are activated by depolarizing potentials to values near the resting potential (-60 to -40 mV) they completely inactivate.[41] In addition, steady-state inactivation is already complete at -40 mV. Although the activation threshold for L-type Ca currents is more positive (-20 mV)[40] than the resting potential, small steady-state currents of a few pA flow through L-type Ca channels at the resting potential, as has been demonstrated.[42] These steady-state currents act to depolarize the membrane potential and hold it close to the threshold for eliciting action potentials.

The delayed rectifier-like K current is not involved in maintenance of the resting potential as it activates at relatively positive potentials around −10 mV.[28] It is unclear if the A-like fast inactivating K current contributes to the maintenance of the resting potential. Reported values for activation differ slightly, although this might be explained by the different holding potentials used. With a holding potential of −60 mV, which lies just within the normal range of resting potentials, this current activates at potentials positive to −40 mV,[27] suggesting no contribution of this current to the resting potential. With a more negative holding potential of −86 mV, below the normal resting potential range, the current already activates at potentials positive to −50 mV.[28] Values for 50% steady-state inactivation differ considerably: Oxford and Wagoner[27] found a 50% steady-state inactivation at −40 mV, whereas a 50% steady-state inactivation at −55 mV and an almost complete inactivation at −45 mV is reported by Simasko.[28]

Inward rectifiers are classical contributors to the resting membrane potential[43] and the inward-rectifying K current, K_{IR}, of GH cells is also suggested to play a crucial role in the maintenance and modulation of the membrane potential.[29,44] It has been demonstrated that a considerable amount of K_{IR} current is present in the range from −60 to −40 mV.[29,44] At potentials negative to −60 mV, the K_{IR} starts to inactivate in a time- and voltage-dependent manner. This fact could be functionally correlated with the absence of more negative resting potentials in these cells and could also account, in part, for action potentials which are elicited following cessation of hyperpolarizing pulses. Since all experiments on the K_{IR} have been performed in increased external K solutions to increase K_{IR} current amplitude, the magnitude of its contribution to the resting potential is still unknown.

The Ca-dependent BK channel does not contribute to the resting potential due to its very low Ca sensitivity over this potential range.[23] In contrast, the Ca sensitivity of both the SK and IK channels is high in the range of the resting potential[24] and they are therefore able to contribute to resting potential under conditions where $[Ca^{2+}]_i$ is increased with respect to resting levels (slightly below 100 nM).[33] Ca transients due to Ca^{2+} influx during single action potentials have been shown to last more than 10 s.[35] During such a long period, the open probability of SK and IK channels decays progressively,[24] resulting in a continuous slow depolarization of the membrane potential following the afterhyperpolarization. This slow depolarization might be important for the pacemaker mechanism in these cells. Since the voltage-dependent A-like K current has deactivation time constants of less than 10 ms,[27] it is probably not involved in such a pacemaker mechanism as it would already be deactivated during the initial phase of the afterhyperpolarization.

4.1.3. Effects of PRL- and GH-Releasing Factors

Secretion-stimulating factors differ in the intracellular pathways they use to produce an increase in secretion. Clear electrophysiological responses are produced by factors which induce secretion mainly via an increase in $[Ca^{2+}]_i$, which can occur either by liberation of Ca^{2+} from intracellular pools or by an increased Ca^{2+} influx (see Figure 1A,[36] 1B).[45] TRH is an example of this kind, whereas VIP is an example of a hormone which increases secretion without consistent effects on firing activity[46] and $[Ca^{2+}]_i$.[47-49] VIP acts to increase cAMP via a strong activation of adenylate cyclase.[50] This increase in cAMP can, itself, induce substantial secretion independent of Ca, although Ca can augment cAMP-induced secretion.[51] Drastic pharmacological elevation of cAMP levels by forskolin or membrane-permeable cAMP analogs has been reported to induce a small (less than 100-nM) sustained increase in $[Ca^{2+}]_i$ due to Ca^{2+} influx through L-type Ca channels[51] and also to be able to induce or increase spiking activity.[47] Stimulation of adenylate cyclase activity by VIP does not result in such high levels of cAMP,[47] and this fact might account for the lack of a membrane response to VIP.

TRH, the "classical" releasing hormone for PRL in GH cells, increases secretion mainly through elevation of $[Ca^{2+}]_i$ (see Figure 1B).[45] The TRH-induced increase in $[Ca^{2+}]_i$ is biphasic, consisting of a transient first phase in which Ca^{2+} is liberated from intracellular pools, and a sustained second phase in which Ca^{2+} influx is increased. Other releasing factors like bombesin[52]

FIGURE 2. An increased duration of action potentials (APs) is apparent during the second phase of the TRH response in GH$_3$ cells, similar to that observed after a TEA-induced block of the high-conductance, Ca-dependent BK channel. (A) Biphasic response to 1 μM TRH recorded in a perforated-patch, current-clamped cell. A first phase hyperpolarization and a second-phase increase in AP frequency can be seen. Two APs before and after TRH are shown in expanded time scale at the bottom. Horizontal lines indicate 0 mV. (From Barros, F., Delgado, L. M., Macia, C., and De la Pena, P., *FEBS Lett.*, 279, 33, 1991. With permission.) (B) Effect of 1 mM TEA on AP duration in whole-cell current clamp. (From Ritchie, A. K. and Lang, D. G., in *Neural Control of Reproductive Function*, Lakoski, J. M., Perez-Polo, J. R., and Rassin, D. K., Eds., Alan R. Liss, New York, 1989, 436. With permission from ©Wiley-Liss, a subsidiary of John Wiley & Sons, Inc.)

or epidermal growth factor (EGF)[53] induce similar biphasic Ca responses, whereas another class of substances elicit only a single-phase Ca response. CCK8, for example, induces only the first phase,[54] in contrast to platelet-activating factor[55] and galanin,[56] which stimulate only Ca^{2+} influx. Under certain conditions, GH cells also respond to growth hormone-releasing hormone (GhRH) with a short increase in the frequency of action potentials due to increased Ca conductance.[57-59] The mechanism by which other releasing factors act to increase Ca^{2+} influx remains unknown, but they might share second messenger pathways involved in the signal cascade triggered by TRH.

4.1.3.1. *TRH Signal Cascade*

Elements of the TRH signal cascade mediating the biphasic rise in [Ca^{2+}]$_i$ and the increase in secretion are indicated in the model shown in Figure 5. The biphasic Ca response to TRH occurs in parallel with a biphasic electrical response. During the first phase the cell becomes hyperpolarized, and in the second phase the frequency and duration of action potentials increase (see Figure 2A),[31] and act synergistically to increase Ca^{2+} influx. In perforated-patch experiments, 1 μM TRH was found to increase action potential frequency by about 0.1 Hz and action potential duration by about 50%.[31-33]

It is now well established that the transient first-phase increase of [Ca^{2+}]$_i$ is the result of release of Ca^{2+} from inositol trisphosphate (IP$_3$)- and thapsigargin-sensitive, caffeine-insensitive intracellular pools.[60] This release is mediated by a G protein of the G$_{q/11}$ family coupled to phospholipase C (PLC)[61] which act to increase IP$_3$ generated together with diacylglycerol (DG) by phosphatidylinositol bisphosphate (PIP$_2$) hydrolysis.[62] The hyperpolarization results from the first-phase Ca transient and is mediated by opening of the Ca-dependent IK and SK channels.[24]

It has long been assumed that the electrophysiological changes associated with the second phase are mediated by a decrease in K conductance.[2] Up to now, consistent effects of TRH have been found on two K currents: the Ca-dependent K current (see Figure 3)[31,32] and the inward-rectifying K current K$_{IR}$ (see Figure 4).[29,33,44] It is likely that TRH acts on the BK channel, as the effects of TRH on outward K current were able to be eliminated by addition of 5 mM TEA (Figure 3),[31] a concentration which completely blocks BK. The SK and IK channels are less sensitive to TEA, although 5 mM also considerably reduces SK.[24] Therefore, an effect of TRH on SK and IK cannot be excluded. The inhibitory actions of TRH on BK and K$_{IR}$ would be sufficient to explain the observed effects on action potential duration and action potential frequency, respectively. A quite selective block of BK by 1 mM TEA results in a

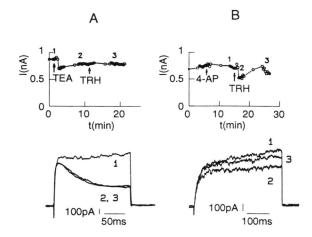

FIGURE 3. Effect of TRH on the TEA-insensitive, voltage-dependent K currents (A) and Ca-dependent K currents (B) in perforated-patch recordings from GH_3 cells. Upper row: time course of maximum outward currents evoked by depolarizing steps to +40 mV (A) and +50 mV (B) from a holding potential of –70 mV. Lower row: representative current traces obtained at the times indicated by numbers in the upper panels. Concentration of TEA and 4-AP 5 mM, of TRH 100 nM. (From Barros, F., Delgado, L. M., Maciá, C., and de la Peña, P., *FEBS Lett.*, 279, 33, 1991. With permission.)

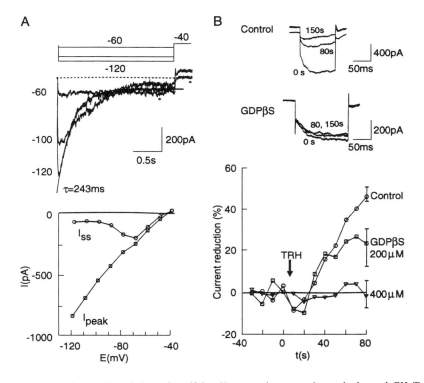

FIGURE 4. (A) A voltage-activated, inward-rectifying K current is present in patch-clamped GH_3/B_6 cells in isotonic KCl. Top: inward K currents elicited by the hyperpolarizing pulses shown in the pulse program at the top. Holding potential: –40 mV. Bottom: current-voltage relationship for the experiment shown above. Current amplitude is shown as the difference from the holding current at –40 mV. (B) TRH produces a G protein-mediated block of this current. Top: current traces elicited by hyperpolarizing steps from –40 to –100 mV just before (0 s) and after (80 s, 150 s) TRH application without (control) and with 400 μM GDPβS in the patch pipette. Bottom: time course of the reduction of inward-rectifying current produced by TRH without (control) and with 200 or 400 μM GDPβS in the patch pipette. (From Bauer, C. K., Meyerhof, W., and Schwarz, J. R., *J. Physiol.*, 429, 169, 1990. With permission.)

pronounced increase in action potential duration (Figure 2B). [38] As K_{IR} is most probably involved in the maintenance of the resting membrane potential, an inhibition of K_{IR} should result in membrane depolarization, leading to an increase in the frequency or an onset of action potentials. [29]

The intracellular mechanisms of the TRH-induced inhibition of BK and K_{IR} are only partially known. In outside-out patches and in cell-attached patches, the BK channel has been found to be blocked by either cAMP, itself, or a membrane-permeable analog. [63] TRH is suggested to elevate cAMP via at least three different pathways: a direct activation of adenylate cyclase via the G_s protein or a G_s-like protein, [64] an inhibition of the adenylate cyclase-inhibiting G_i protein via activation of protein kinase C (PKC), [50] and a PKC-mediated increase in the percentage of high activity state adenylate cyclase. [65] Nevertheless, additional mechanisms by which TRH inhibits Ca-dependent K current ("?" in the model of Figure 5) must be assumed since the TRH-induced increases in adenylate cyclase activity are small compared to the action of VIP, [66] and cAMP-mediated effects of VIP on spiking activity have not been observed (see paragraph on VIP). One additional mechanism is probably the indirect modulation of Ca-dependent K currents via Ca. Ca-dependent inhibition of the L-type Ca current [67] is produced by the TRH-induced release of Ca from intracellular pools. In perforated-patch experiments the inhibition of L-type Ca current was maximal (about 35%) shortly after TRH application and slowly reversed within the next several minutes. [32,68] This inhibition of L-type Ca current could contribute to the TRH-induced lengthening of action potentials and produce the observed decrease in their upstroke velocity and peak amplitude. [32,37] The TRH-induced increase in cAMP seems to be insufficient to compensate for the Ca-dependent inhibition of the L-type Ca current. [67]

TRH induces a voltage-dependent reduction of K_{IR} which can be partially antagonized by preceding depolarizations. [29] This inhibitory effect of TRH on K_{IR} is mediated by a G protein (Figure 4B) [29] and an involvement of a cholera toxin (CTX)-sensitive G protein different from G_s has been suggested. [33,68] The TRH effect requires a minimum of $[Ca^{2+}]_i$, [29] and probably involves a phosphorylation event, since prevention of protein phosphatase-mediated dephosphorylation renders the TRH-induced K_{IR} inhibition irreversible, as also occurs during whole-cell patch clamp recordings. [44] The phosphorylation of K_{IR} suggested to underlie the inhibitory effect of TRH is not mediated by either cAMP or by PKC. [29,33,44]

For a long period, PKC has been suggested to play a crucial role in mediating the second phase of the TRH response. [2] This was mainly based on experiments in which PKC-activating phorbol esters at least partially mimicked the second phase of the TRH response. [69,70] Experiments in which PKC was downmodulated by phorbol esters have shown that phorbol ester-sensitive PKC is essential for the TRH-induced secretory event, [71,72] but that PKC is not required for the biphasic increase in Ca. [72,73] Considerable cross talk between various second-messenger pathways as shown between PKC and adenylate cyclase (see above) may be one of several reasons for unphysiological or PKC-nonspecific actions of phorbol esters. [74] Given the variety of TRH-induced effects which could play a role in the second-phase Ca^{2+} influx like changes in pH, [75] in the activity of membrane pumps, [76] and various types of protein kinases, [77,78] this model of action of TRH (Figure 5) still needs to be investigated, and might represent an enormous simplification.

On the single-cell level, the TRH response is not always biphasic. In electrophysiological experiments using the intracellular recording technique [79] or the perforated-patch configuration, [31] it has been found that about two thirds of the TRH-reponsive cells exhibit a biphasic response, and the other one third is almost equally divided between cells which show only a first or a second phase supporting the idea that the two phases are mediated by distinct pathways.

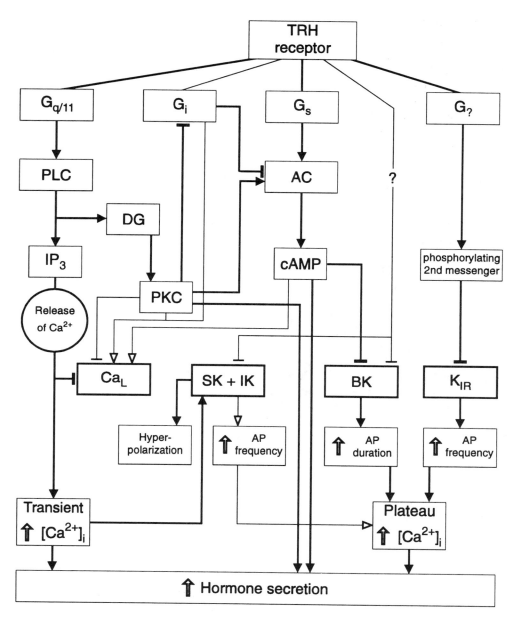

FIGURE 5. Proposed mechanism underlying the TRH-induced biphasic increase in cytosolic Ca concentration and hormone release. At least four different G proteins ($G_{q/11}$, G_i, G_s and $G_?$) are suggested to be involved in this response. The first phase is a transient release of stored Ca^{2+} into the cytosol with concomitant hyperpolarization as seen toward the left of the diagram. The second phase involves reductions in K conductances which depolarize the cell and alter Ca-dependent action potentials to produce a plateau increase in Ca concentration (right side), increasing secretion. Ca-independent effects of cAMP and PKC on hormone secretion have also been measured (middle). Thick lines indicate direct observations from TRH experiments on GH_3 or GH_4C_1 cells; thin lines represent other possible effects. Abbreviations: AC, adenylate cyclase; BK, high-conductance K channel; Ca_L, L-type Ca channel; DG, diacylglycerol; IP_3, inositol 1,4,5-trisphosphate; K_{IR}, inward-rectifying K channel; PKC, protein kinase C; PLC, phospholipase C; SK + IK, small- and intermediate-conductance, Ca-dependent K channels. Arrowheads and bars at the end of the connecting lines denote stimulatory and inhibitory effects, respectively. For further details see text.

4.1.3.2. Variations in Ca Responses to TRH

There exist two types of biphasic increases of $[Ca^{2+}]_i$ induced by TRH: (1) the first-phase Ca transient passes directly over to the second-phase Ca plateau (see Figure 1B) or (2) the first and a delayed second phase are separated by a Ca "nadir" during which $[Ca^{2+}]_i$ is very low.[80] Both types of TRH responses occur in GH_3 and GH_4C_1 cells, and it is not clear if one response type predominantly occurs in connection with particular experimental procedures or cell culture conditions. One explanation of the cellular mechanism underlying the "nadir" is given by Drummond.[81] He demonstrated that the magnitude of the nadir was concentration dependent (insignificant at 1 and 10 nM, but obvious at 100 nM TRH) and suggested that it is mediated by activation of PKC, since it could be mimicked by active phorbol esters and PLC. Martin et al.[82] found that TRH activates PKC only transiently for less than 1 min with maximal effect at 1 μM and half-maximal effect at 40 nM. In studies using 100 nM TRH, though, a Ca "nadir" was not consistently present in the TRH Ca response.[33,36,83] This is not surprising, since the results of application of TRH or pharmacological PKC activators on Ca currents are also not consistent. In perforated-patch experiments on GH_3 cells, TRH was found to inhibit L-type Ca currents using concentrations of 100 nM[68] or 1 μM[32] (see above). Under certain conditions, stimulating effects of TRH on Ca channels have also been observed as, for example, a transient activation of T-type channels, which occurred when 20 μM GTPγS was added to the whole-cell patch pipette solution,[84] or, in complete absence of Ca, a transient increase in Na current through Ca channels, involving the G_{i2} protein and PKC.[85,86]

Using pharmacological PKC activators, it was found, for example, that oleoyl-acetyl glycerol (OAG) inhibits T-type and L-type Ca currents of GH_3 cells,[87] that the phorbol ester TPA inhibits L-type Ca current via activation of PKC in the majority of GH_3 cells, but that some cells are resistant to TPA,[88] that phorbol esters, but not dioctanoyl glycerol (DOG), inhibit depolarization-induced Ca^{2+} influx through L-type channels in GH_3 cells,[89] or that the phorbol ester PMA does not alter T- or L-type currents in GH_4C_1 cells.[90]

Taken together, these results suggest that Ca currents of GH cells are able to be modulated by PKC, but also that the Ca currents in these cells can be resistant to PKC activation. It would be very interesting to find the reasons for this heterogeneity.

4.1.4. Inhibiting Factors

Several factors have been found to inhibit secretion in GH cells. Based on analysis of the underlying cellular events, these inhibiting factors seem to act via the same mechanism: they decrease cAMP levels and reduce $[Ca^{2+}]_i$, and their action is abolished by pretreatment of the cells with pertussis toxin (PTX). These effects have been demonstrated for acetylcholine,[30,91,92] adenosine,[93,94] gonadotropin-releasing hormone-associated peptide[95,96] and SST, whose actions are summarized in Figure 6 and will be described in more detail.

In spontaneously active cells, SST induces a more or less pronounced hyperpolarization associated with an elimination of action potentials or a decrease in their frequency. This inhibitory response is completely blocked by pretreatment of the cells with PTX.[97] The inhibitory effect of SST on hormone secretion is also abolished by PTX preincubation,[98] demonstrating that SST effects are exclusively mediated by the PTX-sensitive $G_{i/o}$ proteins. Ca-independent effects of SST are due to inhibition of adenylate cyclase, probably via G_i,[99] which results in a reduction of cAMP levels or at least an inhibition of increased cAMP levels and thereby a decrease in secretion.

The Ca-dependent action of SST on hormone secretion is mediated by stimulatory effects on K channels and inhibitory effects on Ca channels. SST activates a 55-pS K channel directly coupled to a G_i-type G protein.[30] This results in the hyperpolarization of the membrane observed in current-clamp experiments. SST decreases L-type, but not T-type Ca currents. This effect is also independent of cAMP and is suggested to be mediated via a G_o protein either directly or by an as yet unidentified component.[100]

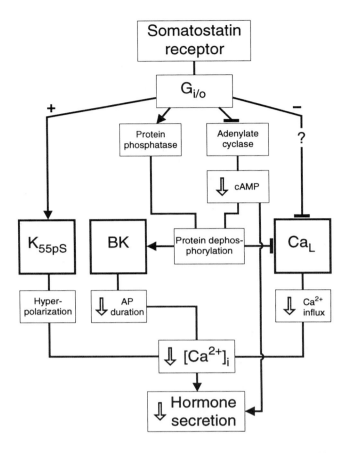

FIGURE 6. Schema showing the ionic channels and $G_{i/o}$-mediated signaling pathways which have been found to be involved in the SST response of GH_3 and GH_4C_1 cells. A "direct" G protein activation of a 55-pS K channel hyperpolarizes the membrane to voltage levels where the probability of Ca channel opening is reduced, decreasing Ca^{2+} influx, intracellular Ca concentration, and thereby inhibiting hormone secretion. The L-type Ca channel is inhibited via an unknown cAMP-independent pathway. Stimulation of protein phosphatase and inhibition of adenylate cyclase change the level of phosphorylation of the high-conductance potassium channel (BK) adding to the decrease in hormone secretion through changes in duration of Ca action potentials and a reduction in Ca^{2+} influx. Since L-type Ca channels are known to require phosphorylation to respond to depolarization, a dephosphorylation of these channels will also reduce Ca^{2+} influx. A Ca-independent effect of cAMP on hormone secretion is well documented. Arrowheads and bars at the end of the connecting lines denote stimulatory and inhibitory effects, respectively.

In addition to these cAMP-independent effects on ionic channels, stimulation of a protein phosphatase is suggested to mediate the SST-induced activation of the high-conductance BK channel,[63] acting to repolarize the membrane. Since dephosphorylation activates the BK channel and cAMP blocks this channel,[63] the inhibitory effect of SST on cAMP should support the activation of BK. Phosphorylation is suggested to be a prerequisite of the response of L-type Ca channels to depolarization.[101] An increased dephosphorylation of these channels would therefore also reduce the Ca^{2+} influx during depolarization. In addition, low cAMP levels increase the Ca-dependent inhibition of L-type Ca channels.[67] All these effects would act synergistically to reduce Ca^{2+} influx during action potentials.

4.1.5. Heterogeneity of GH Cells and Long-Term Modulation

Even within the "clonal" cell lines GH_3 and GH_4C_1, there exists a considerable heterogeneity of the cells with respect to the hormones secreted. About 50% of the GH_3 cells secrete

GH and 20–25% secrete PRL.[15] All of the PRL-secreting cells were suggested to also secrete GH, since about 50% of the GH_3 cells secrete neither GH nor PRL. In the GH_4C_1 cell line, only 25% of the cells secrete GH and 45% of cells secrete PRL.[16] Long-term treatment with TRH, 17β-estradiol, and cortisol has been demonstrated to shift the percentage of cells in these groups.[15,102] The existence of such functional "subpopulations" in the GH cell lines might be correlated with the fact that a variety of substances act only on a proportion of GH cells. This heterogeneity and plasticity of GH cells suggest the importance of classifying individual GH cells used in experiments.

Chronic effects on GH cells have been demonstrated not only for releasing factors and steroids, but also for inhibiting factors, growth factors, PKC- or protein kinase A (PKA)-activating substances, and vitamins. These factors modulate the level of a wide variety of mRNA coding for proteins of the signal cascade. This has been demonstrated for mRNA coding for receptors (TRH receptor,[103,104] induction of D_2 receptors[105,106]), G proteins,[107,108] hormones,[109,110] and ionic channels (K channels,[111,112] L-type Ca channels,[113,114] T-type Ca channels[115]). The cell culture media for GH cells contain high levels of sera. These sera are well-known sources of large numbers of undefined substances with long-term effects at unknown concentrations. Uncontrolled culture conditions could underlie discrepancies in the results obtained by different research groups.

4.2. SOMATOTROPH CELLS

The secretion of GH by somatotrophs of the anterior pituitary occurs in a pulsatile manner in all species which have been examined.[116] This pulsatile release of GH is controlled by an interplay between two hypothalamic peptides, GhRH and SST. GhRH stimulates GH release,[117] while SST inhibits basal and GhRH-induced GH release.[118]

4.2.1. Membrane Currents and Electrical Excitability

A number of studies have examined the basic ionic currents and electrical activity which underlie the regulation of GH release by somatotrophs. Both T- and L-type voltage-activated Ca currents have been found to be present in most somatotrophs where they have been measured (bovine,[119] adult rat,[120] male rat,[121] female rat,[122] human GH-secreting adenoma[123]), although in some cases only L-type currents were apparent (young male rat,[124] goldfish[125]). A TTX-sensitive Na current (goldfish,[125] young male rat,[124] male rat,[129] bovine[119], human GH-secreting adenoma)[130] is also present. Two voltage-activated K currents: a transient, 4-aminopyridine (4-AP)-sensitive current (I_A) and a slowly inactivating, TEA-sensitive current (I_K) (bovine,[119] ovine,[126] male rat,[127] young male rat,[124] human GH-secreting adenoma,[128] goldfish[125]) have been regularly observed in somatotrophs. A Ca^{2+}-activated K current has been reported for bovine,[119] and young male rat[124] somatotrophs. In adult female rats, an ATP-sensitive K current has been shown to be involved in the control of GH secretion.[131] A voltage-activated inward-rectifying K current, similar to that found in lactotrophs[132] and GH_3/B_6 cells[29] (see sections on lactotrophs and GH cell lines) has also been measured in ovine somatotrophs.[126] As discussed below, SST activates an inward-rectifying K current in freshly dispersed rat somatotrophs[124] and in human GH-secreting adenoma cells.[130]

The resting membrane potential measured for somatotrophs varies considerably, but much of this variability is likely related to the differences in culture conditions and in recording methods between studies. Values obtained with intracellular recordings (rat: –43 mV,[133] –54 mV[134]) tend to be less negative than those from whole-cell patch clamp measurements (rat: –78 mV[127], bovine: –83 mV).[119] In measurements obtained from perforated-patch recordings of freshly dispersed somatotrophs, which are likely to be the most accurate, an intermediate value of about –60 mV was found for rat somatotrophs.[124]

Nearly all somatotrophs examined, independent of the physiological state or type of animal, produced similar bursts of action potentials in response to injected depolarizing current.[119,124,133,135] These bursts have been shown to consist of one or more fast Na action potentials followed by several slower Ca potentials and assumed to be important for both basal and GhRH-stimulated secretion of GH.[129] Supporting this assumption, a number of studies[122,124,129] have shown that a large percentage of somatotrophs spontaneously fire bursts of action potentials at low frequency during the first few days in culture and, as discussed below, GhRH has also been shown to be able to induce low-frequency rhythmic oscillations in silent somatotrophs.[136]

4.2.2. Effects of GhRH

In initial whole-cell voltage clamp recordings from bovine somatotrophs, application of GhRH had no effect on any ionic current, giving a first suggestion that cytoplasmic compounds are involved in GhRH activation of somatotrophs.[119] Nussinovitch[136] used the cell-attached configuration of the patch clamp technique to examine the changes in electrical activity associated with the action of GhRH on pituitary cells obtained from adult male rats during days 1–3 in culture. In the absence of dialysis of the cytoplasm, GhRH produced rhythmic hyperpolarizing currents, attributed to rhythmic hyperpolarizations in individual anterior pituitary cells, presumed to be somatotrophs. This GhRH response was found to require the presence of extracellular Ca^{2+} and appeared to be due to periodic increases in K conductance, based on the measured reversal potential of the GhRH-induced currents.

The effects of GhRH on the electrical properties of cells from somatotroph-enriched cultures from adult male rat pituitaries were also examined by Chen et al.[134] using intracellular recording. They observed that GhRH induces a dose-dependent membrane depolarization associated with an increase in conductance due in part to activation of Ca current, since Ca channel blockers could inhibit it. Replacement of Na^+ with choline was without significant effect, but other ionic currents were suggested to be involved since the reversal potential of this depolarization was found to be –40 mV.

In contrast, the depolarization produced by GhRH was found to be Na dependent in experiments where membrane potential was monitored with voltage-sensitive fluorescent dyes,[137] but this effect of Na did not rely on voltage-activated Na channels, since GhRH-induced GH release was not blocked by TTX.[138] Responses of male rat somatotrophs to application of GhRH were further examined using fluorescence measurements of changes in $[Ca^{2+}]_i$.[139] A continuous application of 10 nM GhRH was found to produce a rapid rise in internal Ca^{2+} which was sustained in most somatotrophs. This GhRH-induced rise in $[Ca^{2+}]_i$ was found to be inhibited by Co^{2+} and suppressed by substitution of mannitol for Na^+, yet it was not affected by TTX. Elevation of intracellular cAMP levels using dibutyryl-cAMP caused a similar rise in $[Ca^{2+}]_i$ to that produced by GhRH with the same ionic dependence. Based on these results, Kato and co-workers[137-139] have proposed that GhRH activates TTX-insensitive Na (or nonselective cation) channels via cAMP which leads to depolarization and subsequent activation of Ca channels and secretion of GH.

4.2.3. Effects of Somatostatin (SST)

Chen and Clark[140] have examined the modulation of ionic currents which underlie the inhibitory effect of SST on GH release. To eliminate hormonal state of the animal as a potential source of variability, somatotroph-enriched primary cultures from adult male rats were exclusively used and electrophysiological measurements were made after 7–15 d in culture, when evoked responses were first able to be recorded.[133] SST was found to induce a dose-dependent hyperpolarizing response in somatotrophs, which results from activation of a TEA- and 4-AP-insensitive K current.[141] In addition, SST reduced voltage-dependent Ca currents and increased voltage-dependent K currents.[127] The L-type Ca current was reduced without changes in its kinetics, whereas the reduction of the T-type current was accompanied by an acceleration of time-dependent inactivation and a hyperpolarizing shift in voltage-dependent inactivation. The inhibitory effect of SST on Ca currents and on the release of GH has been shown to be mediated by G proteins of the $G_{i/o}$ family.[127]

Sims et al. [124] examined the SST response of freshly dispersed somatotrophs from young male rats. Spontaneous rhythmic bursts of action potentials were observed in perforated-patch recordings from these cells, and the hyperpolarization induced by SST was able to suppress these bursts. Voltage-clamp experiments showed the presence of an inward-rectifying K channel which is activated by SST and underlies the SST-induced membrane hyperpolarization. A ligand-gated inward-rectifying K conductance similar to that in the rat was also found in human GH-producing pituitary adenoma cells.[128] In these tumor cells, the induction of the inward-rectifying K conductance by SST was inhibited by PTX, suggesting that the coupling is also via a $G_{i/o}$ family protein.

4.2.4.   Effects of Pituitary Adenylate Cyclase-Activating Polypeptide (PACAP)

Although GH secretion is still assumed to be mainly regulated by the two hypothalamic hormones GhRH and SST, recent evidence suggests that other factors may be involved. In particular, evidence has accumulated to suggest that PACAP acts as a hypophysiotropic regulator of GH secretion.[142] As suggested by its name, PACAP was isolated from the anterior pituitary based on its ability to stimulate cAMP production in pituitary cells.[143] PACAP immunoreactivity is present in the median eminence[144] and high-affinity binding sites for PACAP exist in pituitary membranes.[145] PACAP has been shown to produce one of two Ca responses in about half of the somatotrophs from adult male rats, consisting of either a slow rise in $[Ca^{2+}]_i$ or several repeated transients.[142] Both responses involve cAMP and require extracellular Ca^{2+},[146] and produce a dose-dependent increase in GH secretion which can be inhibited by SST and which occurs by means of a receptor distinct from that for GhRH.[147] These findings suggest that a greater complexity of the regulation of GH secretion at the level of the somatotroph is to be expected.

4.2.5.   Subpopulations

Somatotrophs obtained from adult male rats and separated on a continuous-density BSA gradient were found to be heterogeneous as regards the responsiveness of their membranes to factors controlling GH release.[134] About half of the somatotrophs in these cultures were hyperpolarized by SST and depolarized by GhRH, whereas the other half responded to SST, but not to GhRH. Using cultures of bovine somatotrophs, Kineman et al.[148] made a combined analysis of the proportion of GH secretors and the relative amount of hormone released per cell measured using the RHPA. Their results indicated that a substantial fraction of GH secreted after stimulation with GhRH is attributable to recruitment of "silent" cells, which are immunoreactive for GH, but show no basal GH secretion.[148] Although the physiological significance of these two examples of subtypes of somatotrophs is unknown, they suggest that a functional heterogeneity of somatotrophs forms a part of the system regulating GH secretion.

4.3.   LACTOTROPH CELLS

PRL is secreted by a group of the acidophilic cells within the anterior pituitary, the lactotrophs. In contrast to other pituitary hormones, PRL has been shown to be involved in a large variety of physiological processes in many organs,[149] but the only well-studied function of PRL remains its decisive action in the control of lactation in mammals.[150,151] As might be expected from the diverse functions of PRL, lactotrophs have been found to be extremely heterogeneous with regard to their ultrastructure,[152] their distribution within the anterior pituitary,[153] their responsiveness to releasing factors,[154,155] and the rate of release.[156]

Studies of the ontogeny of the anterior pituitary have shown that the appearance of GH-containing cells considerably precedes that of PRL-containing cells in many species. Using sequential RHPA and double immunocytochemistry at the electron microscopic level, it has been possible to show the existence of mammosomatotrophic cells, which release both GH and PRL, in normal fetal pituitaries of several species.[157] Mammosomatotrophs are also present in the adult pituitary gland, although in the male rat, for example, the percentage of these cells found by investigators varied between 33% and none at all.[157] The presence of mammosomatotrophs in fetal and in early neonatal life which persists into adulthood is an important aspect of the regulation of both GH and PRL secretion.

The available physiological data suggest that somatotrophs and lactotrophs share a common developmental lineage and that they are functionally interconvertible in response to various stimuli in the adult (for review see Reference 158) and that this plasticity presumably underlies the anatomical and functional heterogeneity of adult lactotrophs. Although this heterogeneity makes studies of lactotrophs difficult, the development of improved separation techniques for pituitary cells[122,159,160] and the adaptation of the RHPA to identify and measure the rate of secretion by individual cells[7] have permitted electrophysiological studies of well-defined subpopulations of lactotrophs and given some first insights into how the various ionic currents of a cell are modulated to produce the particular pattern of hormone secretion necessary for diverse functions.

4.3.1. Membrane Currents and Electrical Excitability

The basic ionic currents found in lactotrophs are similar to those found in other pituitary cells and neurons. Inward currents consist of two types of Ca currents: T-type and L-type (ovariectomized rat,[161] adult rat,[120] bovine,[162] lactating rat,[163] male rat,[164] female rat),[122] and a TTX-sensitive Na current (bovine,[135] adult male rat).[165] Outward currents are mainly K currents including a transient, 4-AP-sensitive current (I_A), a slowly inactivating, TEA-sensitive current (I_K), and a Ca-activated K current ($I_{K(Ca)}$, ovariectomized rat[161], lactating rat).[166] In bovine lactotrophs, mainly I_K and $I_{K(Ca)}$ are present.[135] A Ca^{2+}-activated Cl conductance is present in lactotrophs of female rats[167] and Cl channels activated by γ-aminobutyric acid (GABA) have been found in bovine lactotrophs.[168] As discussed in detail below, a DA-activated K current has been characterized in female rats.[169,170] A voltage-dependent inward-rectifying K current is present in lactotrophs and could contribute to the resting potential[132] (see section on GH cell lines above).

Both the membrane potential and the electrical activity of lactotrophs have been found to be quite variable in primary culture cells. In an early study, Israel et al.[171] made intracellular recordings from human prolactinoma cells in culture to examine the electrical properties of lactotrophs. Their results indicated that these cells have a resting potential between –60 and –55 mV and, though not spontaneously active, could produce bursts of action potentials in response to depolarizing current. These action potentials were sensitive to Ca current blockers and prolonged by TEA, but were insensitive to TTX. Based on intracellular recording, Ingram et al.[172] observed two classes of normal bovine lactotrophs based on membrane potential and spontaneous activity. Silent cells lacking spontaneous action potentials had a more negative membrane potential (–59 mV) than active cells which showed spontaneous firing of Ca-dependent action potentials at an average frequency of 3.4 Hz and had a reduced membrane potential of –38 mV. All cells showed an active response to depolarizing current in the form of either fast regenerative action potentials or slow depolarizations. No changes in these properties were observed with time in culture. In early studies using the whole-cell mode of patch clamp recording, Lingle et al.[161] found that lactotrophs isolated from ovariectomized rats had an average resting membrane potential of –43 mV and none of these cells showed spontaneous activity during the first week in culture. With improvements in recording and culture techniques, a substantial percentage of lactotrophs have been found to produce spontaneous action potential activity in studies using both the whole-cell configuration of the patch clamp technique (female rat: 20%,[173] 55%,[122] 85-95%;[174] bovine: 11%)[135] and intracellular recording (lactating rat: 15%,[169] 60%).[166] In all these studies, depolarizing current pulses could elicit action potentials in lactotrophs not showing spontaneous activity. In particular lactotrophs from lactating rats, action potentials produced by cessation of hyperpolarization, called "off" potentials, were also readily elicited.[175]

4.3.2. Lactotroph Heterogeneity
4.3.2.1. Lactotrophs Differ in the Rate of Basal Secretion

Subpopulations of functionally different lactotrophs have been identified using the RHPA (see Section 3).[7] In adult rats, lactotrophs in culture can be classified as small-plaque (SP) or large-plaque (LP) forming cells according to the relative rate of basal PRL secretion.[154,164] Basal PRL release is presumably set by the cytoplasmic Ca concentration maintained by Ca^{2+} influx through plasma membrane ionic channels which transiently open during spontaneous electrical activity of the lactotrophs.[35,169,172,173,176] As might be expected, therefore, rat lactotrophs could also be classified into two groups based on the dynamics of cytosolic Ca^{2+}. Optical measurements of $[Ca^{2+}]_i$ in single lactotrophs from adult female rats, for example, have identified two distinct cell groups which differ in $[Ca^{2+}]_i$ homeostasis. Some lactotrophs present low and stable levels of $[Ca^{2+}]_i$, whereas others exhibit large $[Ca^{2+}]_i$ fluctuations probably caused by spontaneous action potentials.[122,177]

4.3.2.2. Subpopulations of Lactotrophs in Male Development

Cota et al.[164] went on to show that voltage-gated Ca channels in SP and LP lactotrophs are differentially expressed in adult male rats. LP-forming cells which have a high basal rate of secretion also have larger amplitudes of high voltage-activated, L-type Ca currents in response to membrane depolarization than SP cells. By contrast, the amplitude of low voltage-activated, T-type Ca currents did not differ between the two lactotroph subtypes. In other experiments,[165] LP-forming cells were also found to have relatively large TTX-sensitive Na currents in response to depolarizations. Normalized for membrane capacitance, the Na currents in LP cells were about sixfold larger than those in SP-forming cells. These results suggest that the density of high voltage-activated Ca channels and of Na channels in the plasma membrane is a major factor that determines the basal rate of secretion in individual cells and contributes to the heterogeneity of lactotrophs.

The development of these differences in basal secretion was investigated by comparing the number of cells making up the two subpopulations of lactotrophs from neonatal and adult male rats.[178] SP cells represented 6% of all pars distalis cells in both neonatal and adult male rats. LP lactotrophs, on the other hand, were scarce in neonatal cultures, but represented 13% of the cells in adult cultures. The adult LP lactotrophs were also characterized by a much higher density of Ba^{2+} current through Ca channels. Simultaneous plaque assays for PRL and GH hormone further showed that the number of SP and LP cells was equal to the number of dual hormone secretors (mammosomatotrophs) and PRL secretors (classical lactotrophs), respectively. Taken together, these results suggest that the SP neonatal lactotrophs are mammosomatotrophs and persist into the adult. The appearance of LP secretors, which may correspond to classical lactotrophs, represents a major developmental change in the population of lactotrophs.

4.3.2.3. Lactotroph Heterogeneity Associated with Lactation

In anterior pituitaries from lactating rats, two types of lactotrophs could be separated on a continuous density gradient of BSA at unit gravity. Distinct subpopulations of lactotrophs (immunoreactive for PRL) were found among "light" fractions, 3–5, and among "heavy" fractions, 7–9.[169] Lactotrophs were not present in the heavy fractions of cells prepared from adult males[179] or from virgin females.[169] However, treatment of males with 17β-estradiol led to the appearance of a band of lactotrophs in the heavy fractions.[179]

Electrophysiological studies of these two subpopulations in lactating rats showed that the resting membrane potential and membrane capacitance values were lower in light than in heavy fraction cells.[180] Based on the analysis of tail currents, fast (FD or L-type) and slowly deactivating (SD or T-type) components of Ca current were found to be present with different proportions in each subpopulation. The ratio of the current components, SD/FD, was 2.42 in light fraction cells and 1.17 in heavy fraction cells. The proportion of T-type to L-type Ca current correlates with the level of basal PRL secretion. A majority of light fraction cells produced LP in the RHPA, whereas in heavy fraction cells SP predominated. Furthermore, it was shown in perifusion experiments that basal PRL release was higher in light than in heavy fraction cells.

Electrophysiological responses of these two groups of lactotrophs, their levels of $[Ca^{2+}]_i$, and their release of PRL in response to DA and TRH were also shown to be distinctive.[180] The DA-induced inhibition of PRL release was greater in light fraction cells than in heavy fraction cells, whereas TRH-induced increase of PRL release was significantly smaller in light fraction than in heavy fraction cells. The majority of the cells in the light fraction was hyperpolarized by DA and did not respond to TRH. In the heavy fractions most of the cells responded to TRH application, while only a minority was DA sensitive.

In light fraction cells the $[Ca^{2+}]_i$ exhibited an unstable level between 100 and 200 nM. Application of DA resulted in the disappearance of these fluctuations and in an accompanying decrease in $[Ca^{2+}]_i$. Cells from the heavy fraction showed a lower basal $[Ca^{2+}]_i$ level of about 60 nM. Many of these cells were quiescent, exhibited a $[Ca^{2+}]_i$ increase in response to TRH, but were DA insensitive.[180] In perifusion experiments, Lledo et al.[180] also showed that DA induced a greater inhibition of basal PRL release in light than in heavy fraction cells, whereas the TRH-induced increase of PRL release was significantly smaller in light relative to heavy fraction cells.

Taken together, these results suggest that lactotrophs exist as two subpopulations: the light fraction cells present a high level of basal PRL release associated with spontaneous electrical activity and a basal $[Ca^{2+}]_i$ level that is high and oscillating. The heavy fraction cells exhibit low basal PRL release accompanied by a high resting membrane potential and a low and stable $[Ca^{2+}]_i$ level. DA may preferentially suppress the activity of the lactotroph subpopulation with a high basal release, while TRH appears to be effective on the subpopulation characterized by a low basal release.[180] Another study[177] has shown that DA does not significantly affect the $[Ca^{2+}]_i$ level of silent cells, but induces a marked and rapid decrease of $[Ca^{2+}]_i$, with disappearance of oscillations in spontaneously active cells. Silent and active subtypes of lactotrophs might therefore correspond to SP- and LP-secreting cells, respectively.

As suggested by this example from the lactating rat, hormonal regulation at the level of the pituitary can involve the development of a new subpopulation of cells with appropriate responsiveness to releasing and inhibiting factors and with release characteristics tailored to a particular physiological task, such as lactation.

4.3.3. Factors Controlling PRL Secretion

In the absence of normal hypothalamic and hypophysiotropic factors, lactotrophs secrete PRL at a high rate.[150,157] DA tonically inhibits the secretion of PRL *in vitro* and is considered to be the main inhibitory regulator *in vivo*.[181] The importance of factors which stimulate PRL release, though, remains unclear. In rats, the acute stimulation of PRL release occurring during late proestrus and during suckling seems to involve TRH-induced stimulation. Under other conditions, VIP or low concentrations of DA can also increase the release of PRL (see below). Good evidence exists for the presence of at least two additional unknown stimulatory factors released from the posterior and intermediate lobes of the pituitary.[150] Much progress has been made in characterizing the ionic currents and intracellular signaling pathways associated with the actions of DA. The currents involved in TRH responses have also been described, but its mode of intracellular signaling has only been examined in GH cell lines and not in lactotrophs. These actions of TRH and DA are discussed below.

4.3.3.1. Effects of TRH

The effect of TRH has been mainly characterized from studies of GH clonal pituitary cells. In these cells, TRH produces a biphasic response consisting of a brief hyperpolarization and an increase in action potential frequency (see previous section on GH cell lines and model in Figure 5). Similar responses to TRH have been observed for bovine[172] and human lactotrophs.[182] Sartor et al. [174] examined TRH responses in lactotrophs of diestrus and lactating rats. In this study, TRH induced only the first-phase transient hyperpolarization and the responses of cells from lactating rats were larger and faster when compared to those from diestrus animals. TRH was found to mobilize a Ca-dependent Cl conductance in lactotrophs, which is not activated in GH_3 cells.[167] Distinct responses to TRH were also found in lactotrophs identified from primary cultures of adult male rats.[155] In this study, TRH was found to raise $[Ca^{2+}]_i$ in most lactotrophs examined, but only a first transient phase was observed in some cells, and the kinetics of both phases varied considerably from cell to cell.

Israel et al. [175] made a detailed study of the TRH response of lactotrophs of lactating rats taken from the heavy fractions of a continuous BSA gradient. These cells had been previously shown to release large amounts of PRL in response to TRH, but not to respond to DA.[169] This population of cells characterized by a high mean resting potential was found to respond to TRH with a plateau potential (see Figure 7), consisting of an initial slow depolarization to a threshold which then evokes a second rapid depolarization leading to the plateau. Plateau responses also occurred with the onset of brief large depolarizing pulses or following hyperpolarizing pulses. Local application of TEA during the plateau was without effect on these responses, but Ca channel blockers could shorten or block the plateau. In spite of the fact that no effect of DA could be observed at resting potential, plateau responses induced by current injection or TRH could be terminated by application of DA.

A voltage-dependent inward-rectifying K current has also been shown to be present in lactotrophs from lactating rats and to be able to be reduced by TRH.[132] This K current is suggested to contribute to the resting potential of lactotrophs and its reduction by TRH could therefore underlie the second-phase depolarization leading to increased firing activity in lactotrophs showing a biphasic response to TRH and also the slow depolarization which releases the plateau response in lactotrophs with high resting potential. In addition, the TRH-stimulated release of PRL differs for lactotrophs from the light fractions of a BSA gradient obtained from male and female rats.[183] The above examples suggest that the TRH response of lactotrophs varies considerably in form, depending on the hormonal environment.

4.3.3.2. Effects of DA

The inhibitory action of DA is mediated by several transduction mechanisms. Binding of DA to D_2 receptors has been shown to inhibit both adenylate cyclase activity[184,185] and inositol phospholipid metabolism.[186-189] D_2 receptor activation also leads to modulation of K and Ca channels, resulting in electrophysiological changes (see schema in Figure 8). DA induces a membrane hyperpolarization[169,177] which is caused by the activation of a nonvoltage-dependent K current.[170] A 50-pS K channel was found to underlie this response[190] and to be coupled to the D_2 receptor by a PTX-sensitive G protein via a direct, membrane-delimited pathway.[191] In addition, two voltage-activated K currents have been shown to be increased and two voltage-activated Ca currents to be decreased by D_2 receptor activation.[163,166] All of these effects of DA are blocked by preincubation of pituitary cells with PTX, which ADP-ribosylates several mammalian G proteins, including G_o and G_i.[192] Recently it has been shown that $G_{o\alpha}$

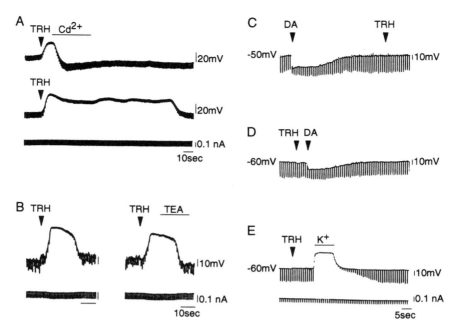

FIGURE 7. Typical responses to TRH and DA of cells from two populations of lactotrophs from lactating rats, isolated on a continuous BSA gradient. (A) In cells with a high resting potential (HRP cells), TRH (100 nM) induced a plateau potential (second trace from top). The resting potential of HRP cells was not altered by DA, but the plateau could be abolished by application of DA or, as shown here, by Cd^{2+} (1 mM, top trace). (B) The TRH-induced plateau was not influenced by TEA (30 mM). (C,D) Ejection of DA (10^{-7} M, 50 ms) onto low-resting potential (LRP) cells induced a hyperpolarization of membrane potential, but TRH (10^{-6} M, 400 ms) was without effect (C) and did not abolish the DA response (D). (E) For comparison, high-K^+ medium (56 mM) was able to induce a depolarization in a LRP cell in which TRH was without effect. Decreases in membrane resistance are indicated in all traces by an attenuation of deflections induced by hyperpolarizing currents (0.1 nA, 100 ms, 1 Hz; lower traces in A, B, and E). (From Israel, J. M., Kukstas, L. A., and Vincent, J. D., Neuroendocrinology, 51, 113, 1990. With permission from S. Karger AG, Basel.)

is involved in the reduction of both Ca currents and $G_{i3\alpha}$ in mediating the increase of the K currents.[193,194] Therefore, in lactotrophs the D_2 receptor responsible for the inhibitory action of DA is linked to at least two different G proteins.

As mentioned above, the majority of lactotrophs obtained from the heavy fraction of a BSA gradient of pituitary cells from lactating rats do not show an inhibitory (type 1) response to DA at the resting potential, but if the membrane has been previously depolarized, DA induces a repolarization associated with a decrease in Ca conductance (type 2 response). Two subtypes of D_2 receptor ($D_2$415 and $D_2$444) result from differential splicing of the premessenger RNA coding for the D_2 receptor.[195,196] These subtypes differ in the region which is believed to be responsible for the binding of G proteins and could therefore activate different second-messenger systems. Kukstas et al.[197] compared groups of lactotrophs in lactating female rats which show either the type 1 or type 2 DA response and found that these subpopulations differentially splice D_2 receptor pre-mRNA to produce significantly different ratios of $D_2$415/$D_2$444. Furthermore, this ratio could also be changed by treatment of primary culture cells with progesterone or testosterone. Estrogen, on the other hand, diminished the total amount of cDNA without changing the ratio. As pointed out by the authors, this type of regulation of differential splicing under the influence of sex steroids could provide a mechanism for modifying lactotroph responsiveness in different physiological situations in order to produce characteristic patterns of PRL release.

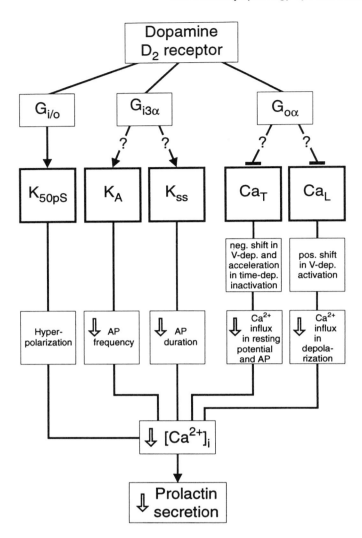

FIGURE 8.   Schema summarizing the ionic channels and signaling pathways involved in the response to activation of DA D_2 receptors in lactotrophs. The major effect of DA is an activation of a "direct" G protein-coupled 50-pS K channel which hyperpolarizes the membrane to voltage levels where the probability of Ca channel opening is reduced, decreasing Ca^{2+} influx, $[Ca^{2+}]_i$, and, thereby, inhibiting PRL secretion. This receptor is also coupled through $G_{i3\alpha}$ to an increase in the conductance of voltage-activated K currents (K_A: transient, 4-AP sensitive, and K_{ss}: slowly inactivating TEA sensitive) and through $G_{3\alpha}$ to a decrease in conductance and change in the kinetics of Ca (T- and L-type) currents, and can hence decrease Ca^{2+} influx resulting from spontaneous action potentials. The pathways by which these voltage-activated channels are modulated are unknown. Arrowheads and bars at the end of the connecting lines denote stimulatory and inhibitory effects, respectively.

It is now well documented that at extremely low concentrations, DA has the ability to stimulate PRL secretion in a particular subpopulation of lactotrophs.[198-201] This paradoxical stimulatory effect of DA involves a different DA receptor subtype and a different class of G proteins (CTX sensitive) than the DA-induced inhibition.[201] Recent results have suggested that both D_1 and D_5 DA receptors are capable of mediating the stimulatory effects of DA in manipulated subclones of GH_4C_1 cells stably expressing these receptors.[202] Stimulatory doses of DA do not affect inositol phosphates nor cAMP, and Ca^{2+}, itself, has been suggested to be the likely second messenger.[203]

DA exerts a tonic inhibition on PRL secretion *in vivo*, but almost without exception, all electrophysiological studies of the effects of DA on lactotrophs have examined only short-term effects. Lledo and co-workers[204] used low-resting potential lactotrophs from lactating rats to assess the effects of chronic DA exposure. Their data suggest that chronic treatment with DA can prevent the increase in T- and L-type Ca current observed to occur over time in culture. This chronic effect of DA was found to be mediated by a cAMP-independent, PTX-sensitive G protein pathway. Such a long-term limit preventing increases in voltage-activated Ca currents could be an important aspect of the mechanism by which DA tonically inhibits PRL release *in vivo*.

4.3.4. Influences of Cell-to-Cell Interactions

A large number of studies have provided evidence that normal regulation of PRL secretion involves both paracrine interactions with neighboring cells in the pituitary and autocrine controls of the lactotroph, and this literature has been extensively reviewed.[150,205]

4.3.4.1. Galanin

An important example of the subtlety of this type of cell-to-cell interaction has been provided in recent work on the role of the neuropeptide galanin in lactotroph function. Galanin is known to be stored and synthesized in lactotrophs[206,207] and can directly stimulate the release of PRL,[208,209] but is predominantly localized to the lactotrophs of female rats.[210] Wynick and co-workers[8] have shown that galanin plays a dynamic role in the regulation of PRL secretion by lactotrophs. Only 9% of all female lactotrophs are galanin secretors. This small galanin-secreting population releases PRL normally in response to TRH, but is unresponsive to galanin. Addition of VIP produces secretion of galanin and not PRL from these cells. The nongalanin-secreting population, on the other hand, was found to have a reduced basal rate of secretion and to release large amounts of PRL in response to galanin. Hyperestrogenization drastically increases the number of galanin-secreting lactotrophs to 39%, and the effects of estrogen treatment on basal PRL secretion and lactotroph proliferation can be eliminated by coincubation with galanin antisera.

These findings give direct evidence for paracrine regulation of basal PRL secretion by galanin and suggest that estrogen-stimulated changes in lactotrophs also make use of this local mediator. As underscored by this example, the differential release of PRL underlying its diversified biological functions results from a complex mesh of endocrine, paracrine, and autocrine interactions which make appropriate adjustments relative to the momentary physiological state. Extreme care must be taken in the design and interpretation of future studies of PRL release to take this complexity into account.

4.4. GONADOTROPH CELLS AND αT3-1 CELLS

Unlike other anterior pituitary cells, gonadotrophs secrete two hormones: luteinizing hormone (LH) and follicle-stimulating hormone (FSH), which are co-localized and co-released from the same cell, following stimulation with gonadotropin-releasing hormone (GnRH). GnRH is the only hypothalamic stimulating hormone for pituitary gonadotrophs, acting on both the synthesis of the polypeptide chains of LH and FSH and on their secretion.[152]

4.4.1. Electrical Excitability of Gonadotrophs

Only about 5–15% of the cells in the anterior pituitary are gonadotrophs.[211] For electro-physiological recordings, the fluorescence immuncytochemical staining of living cells[13] or the RHPA[7] has been used to identify gonadotrophs.[212,213] The resting potential of female rat gonadotrophs varied between –44 mV[13] and –60 mV.[214] About 60% of gonadotrophs of adult female rats exhibited spontaneous action potentials with a frequency ranging between 0.1 and

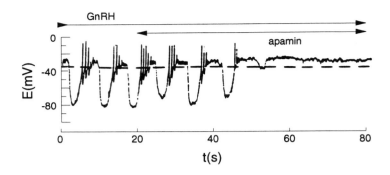

FIGURE 9. Electrophysiological actions of GnRH. Membrane potential changes in a male rat gonadotroph recorded in the continuous presence of 1 nM GnRH in the bath. Apamin (1 μM) was bath-applied during the time indicated by an arrow. Dashed line denotes the resting potential (–35 mV) of this cell before GnRH was applied. (From Tse, A. and Hille, B., Endocrinology, 132, 1475, 1993. With permission from ©The Endocrine Society.)

1.5 Hz.[215] Spontaneous activity could be blocked by 2 mM Co^{2+}, but remained unchanged in Na-free solutions.[13,214] In contrast, gonadotrophs of male rats are usually silent because of their low resting potential of about –30 mV at which Na and transient Ca channels are inactivated. After GnRH application rhythmic membrane potential changes occur, comprising a strong hyperpolarization followed by a burst of action potentials (see Figure 9). [212]

4.4.2. Membrane Currents

In ovine[216] and male rat[213] gonadotrophs, TTX-sensitive, voltage-dependent Na currents have been recorded. In female rat gonadotrophs only small-amplitude Na currents were present.[13] In male rat gonadotrophs at the normal low resting potential the majority of Na channels are inactivated. Recovery from inactivation is fast enough, however, to remove Na inactivation of almost all Na channels during the 1- to 3-s periodic hyperpolarization to –90 mV.[213] These properties of Na channels are different from those reported for Na channels of gonadotrophs from ovine[216] and female rat gonadotrophs,[13,214] as well as from the gonadotroph clonal cell line αT3-1.[217] In gonadotrophs of these species the resting potential is negative enough that Na channels are not totally inactivated and spontaneous activity can be initiated.

T- and L-type Ca currents are present in female rat gonadotrophs.[214] In male rat gonadotrophs, about 40% of the high voltage-activated (HVA) Ca channels have been identified as L-type.[213] The other Ca channels could not be identified. ω-Conotoxin, a specific blocker of N-type Ca channels, had no effect. The identity of the other Ca channels is not clear. Perhaps they are similar to the P-type channel first described in the cerebellar Purkinje cell[218] and which seems also to exist in rat melanotrophs.[219]

Two types of outward-rectifying voltage-activated K currents, the transient K current and the sustained K current, have been described in female rat gonadotrophs.[214] The relative amplitude of the transient and the sustained K current varied from cell to cell, e.g., in female rat gonadotrophs the sustained outward K current was predominantly present.[13] Both K currents are supposed to participate in action potential repolarization. In addition, there exists a Ca-dependent K current.[214]

4.4.3. Effects of GnRH on Gonadotrophs

In male rat gonadotrophs, the action of GnRH induces rhythmic elevations of [Ca^{2+}]$_i$ accompanied by rhythmic membrane hyperpolarizations, which are followed by intermittent bursts of action potentials. Binding of GnRH to its plasma membrane receptor induces a rhythmic release of Ca^{2+} from intracellular stores presumably due to activation of PLC and generation of IP$_3$. The increase in [Ca^{2+}]$_i$ activates apamine-sensitive, Ca-dependent K channels

(SK-type), inducing a transient hyperpolarization which leads to removal of steady-state inactivation of voltage-gated Na and transient Ca channels.[213,215] Opening of these channels during depolarization generates the firing of one to four Na-dependent action potentials superimposed on a slow Ca-dependent potential which leads to an influx of Ca^{2+} into the cytoplasm. These action potentials are characterized by a small overshoot of about 10 mV. TTX reduced the overshoot and the rate of rise of the action potentials, but did not block them. Block of action potentials could be induced by additional application of Cd^{2+}, demonstrating the contribution of high voltage-activated Ca channels.[213] GnRH decreased the Ca current in female rat gonadotrophs, but at the same time induced a negative shift of the activation curve by 10 mV, and therefore increased the activity of Ca channels for potentials close to the resting potential.[220]

In gonadotrophs the hypothalamic peptide PACAP increases cAMP concentration and $[Ca^{2+}]_i$. Two forms of response of $[Ca^{2+}]_i$ were observed. The majority of cells (about 72%) responded with a Ca spike independent of the extracellar Ca concentration, and 28% responded with a Ca plateau.[146] To date, the underlying electrophysiological changes produced by PACAP are not known.

4.4.4. αT3-1 Cells

αT3-1 cells were derived from a mouse tumor. This cell line synthesizes and secretes the α chain but not the β chain, of the LH molecule.[221] The electrophysiological properties of αT3-1 cells have been described by Bosma and Hille.[217] αT3-1 cells contain TTX-sensitive Na channels and T- and L-type Ca currents. The two types of Ca channel are differentially expressed in individual cells, and in a few cells only the transient or the sustained current was found. Transient and sustained outward K current components upon depolarization were also measured. As was the case with Ca currents, the relative sizes of both K current components varied from cell to cell, with the sustained level ranging from 25–90% of the peak current recorded at +52 mV; 0.5–1 mM 4-AP blocked more than 75% of the transient current, whereas the sustained component was blocked by less than 25%. Both components were strongly blocked by 10 mM TEA. A variable amount of the outward K currents, about 20%, was due to Ca-dependent K currents. The Ca-dependent K currents could be blocked with apamine only in a few cells, indicating that only a minority of the cells express the small conductance apamine-sensitive type of K channel.

4.4.5. Effects of GnRH on αT3-1 Cells

In αT3-1 cells, two phases of GnRH-stimulated LH release occur. The first is independent of extracellular Ca, lasts 2 to 3 min, and is correlated with the activation of PLC, production of IP_3, and a transient rise in $[Ca^{2+}]_i$ due to release of Ca^{2+} from intracellular stores.[222] The second phase lasts for about 10 min or longer and occurs with continuous application of GnRH. It corresponds to a slower and smaller elevation of $[Ca^{2+}]_i$ and is abolished by removal of external Ca or application of dihydropyridine Ca channel antagonists.[222,223] This nonoscillatory, amplitude-modulated signal is in contrast to normal rat pituitary gonadotrophs, in which GnRH induces a frequency-modulated oscillatory $[Ca^{2+}]_i$ response.[272]

As a consequence of PLC activation, activation of the GnRH receptor in αT3-1 cells causes a translocation of PKC from the cytoplasm to the membrane, indicating that phospholipase C-mediated hydrolysis of phosphoinositides with the production of IP_3 and DG is necessary for secretion from gonadotrophs.[222-224] However, the contribution of PKC activation to secretion seems to be negligible, since PKC depletion does not affect levels of GnRH-stimulated LH release.[211] In αT3-1 cells, the current through L-type Ca channels is augmented, but probably not induced by PKC, because phorbol esters exert an inhibitory modulation of L-type Ca channels in gonadotrophs.[223]

4.5. CORTICOTROPH CELLS AND AtT-20/D16-16 CELLS

Corticotrophs secrete adreno-corticotrophic hormone (ACTH) and β-endorphin. The main stimulus for secretion is the corticotrophin-releasing factor (CRF). Other secretagogues are vasopressin (VP), oxytocin (OXY), EGF, and angiotensin II. Inhibition of ACTH secretion is mediated by glucocorticoids.

Corticotrophs have a characteristic morphology and appear as small ovoid cells with short processes.[6,225] About 10% of the cells in a normal rat anterior pituitary (pars distalis) are corticotrophs as determined from immunocytochemistry.[225] Because of the small number of cells, it had been difficult to enrich these cells, and prior to the application of immunocytochemical methods it was not possible to identify these cells unequivocally. This may be the reason why so little electrophysiological work has been done in corticotrophs. For cell identification, an avidin-fluorescein stain of cells previously exposed to biotinylated analogs of CRF was employed,[13] or corticotrophs were first chosen by their morphological appearance and then positively identified by immunocytochemistry after having done the electrophysiological experiment.[225,226]

4.5.1. Electrical Excitability

In identified rat corticotrophs, Marchetti et al.[13] measured resting membrane potentials of about -40 mV and showed that part of the cells fire spontaneous action potentials. They described TTX-sensitive Na currents, T- and L-type Ca currents, and an inactivating as well as a sustained outward-rectifying K current. Except for the presence of a TTX-sensitive Na current, similar observations were made in human corticotroph tumor cells.[225] These authors performed simultaneous measurements of $[Ca^{2+}]_i$ and patch clamp recordings and showed that single action potentials induced transient increases in $[Ca^{2+}]_i$ which could be blocked by Cd^{2+} or a specific DHP-blocker (PN200-110), indicating that Ca^{2+} influx occurred mainly through L-type Ca channels.

4.5.2. Effect of CRF

CRF induces a characteristic cell response comprised of a slight membrane depolarization and the initiation of a train of action potentials accompanied by a series of Ca transients.[225] The increase in $[Ca^{2+}]_i$ is due to an influx of Ca^{2+} through Ca channels in the plasma membrane and not due to release of Ca^{2+} from internal stores.[225,226] Although CRF increases cAMP, it is not known whether or not this second messenger modulates Ca channels or whether or not they are changed indirectly through other pathways.[146]

4.5.3. Effect of VP

The intracellular signal cascade activated by VP is different from that induced by CRF.[226] The responses of corticotrophs to VP were heterogeneous.[226] In some cells, VP induces a large transient increase in $[Ca^{2+}]_i$ accompanied by a transient hyperpolarization due to activation of apamin-sensitive, Ca-dependent K channels. In other cells it evokes a long-lasting increase in $[Ca^{2+}]_i$ due to superposition of Ca oscillations induced by Ca^{2+} influx during high-frequency action potentials.[226] Another type of cell response to VP consisted of sinusoidal-like oscillations. These oscillations as well as the short spike-like increase in $[Ca^{2+}]_i$ were due to release of Ca^{2+} from internal stores attributed to the production of IP_3 and DG induced by activation of PLC.[227] The usage of two different second-messenger pathways to initiate an increase in secretion could enable corticotrophs to potentiate the CRF-induced cell response by VP. Different effects of VP were observed in cells in which application of this substance did not induce action potentials. In these cells, VP induced a maintained depolarization within the range of -55 to -35 mV, during which a sustained DHP-sensitive Ca^{2+} influx occurred. It is not known whether or not this Ca current is an L-type Ca channel similar to that recently described in GH_3 cells by Scherübl and Hescheler.[42]

In addition to CRF and VP, ACTH secretion is also stimulated by EGF, OXY, and angiotensin II. CRF and EGF act *in vitro* by increasing the percentage of cells that store ACTH or express mRNA for proopiomelanocortin (POMC). VP and angiotensin II increase the percentage of cells that bind CRF.[228]

4.5.4. AtT-20/D16-16 Cells

AtT-20/D16-16 cells, which were derived from a radiation-induced murine anterior pituitary tumor, synthesize, store, and release the ACTH/β-endorphin family of peptides.[229,230] The principal stimulator of secretion is CRF[229] and secretion is mainly inhibited by SST.[231]

4.5.5. Electrical Excitability

The mean resting membrane potential of these cells was –53 mV (ranging from –30 to –72 mV) and more than 95% of the cells exibited spontaneous bursting of action potentials with a frequency of about 2 Hz. The bursts consisted of one or a few fast Na-dependent action potentials superimposed on a slow Ca-dependent depolarization and an afterhyperpolarization due to a Ca-dependent K conductance.[229] In AtT-20/D16-16 cells, voltage-dependent Na and Ca currents[229] as well as a Ca-dependent K current[229,232] and a Ca-dependent Cl current[233] have been described. The action potential bursts occurring in AtT-20/D16-16 cells are an important source for Ca^{2+} influx, by which the $[Ca^{2+}]_i$ and the regulation of secretion are mainly determined.[230] Ca^{2+} influx can be changed directly by modulating voltage-dependent Ca channels and indirectly by mechanisms which are able to change the frequency as well as the amplitude and duration of action potentials. In addition to the modulating effect of voltage-dependent and Ca-dependent K channels, the Ca-dependent Cl current is important in regulating action potential duration in AtT-20/D16-16 cells.[233] Several other mechanisms within the plasma membrane such as the Ca^{2+}-ATPase and the Na^+/Ca^{2+} transporter have also been described as important regulators of $[Ca^{2+}]_i$.[234]

Two types of inward-rectifying K currents have been characterized in AtT-20/D16-16 cells. A voltage-dependent, inward-rectifying K current can be activated with hyperpolarizing potential steps more negative than the K^+ reversal potential and is closed at the resting potential. This K current can be completely blocked with 1 mM Ba^{2+}. Inactivation of this current is voltage dependent and can be largely removed by decreasing the external Na concentration.[235] These and other properties suggest that this current is similar to that found in myocytes.[236] In addition, there is a noninactivating, inward-rectifying K current, which is not active at the resting potential, but can be activated by β receptor agonists like isoproterenol.[237]

4.5.6. Effects of CRF

Secretion of ACTH/β-endorphin is mainly stimulated by CRF.[230] Other stimulating agents are norepinephrine, isoproterenol, VIP, 8-bromo-cAMP, and forskolin.[238] The principal intracellular mechanism activated by all of these secretagogues seems to be an increase in cAMP with subsequent activation of PKA, phosphorylation of voltage-dependent Ca channels, and increase in $[Ca^{2+}]_i$.[230] The increase in $[Ca^{2+}]_i$ is brought about by an increase in firing of Ca-dependent action potentials due to a small depolarization of about 7 mV.[230] It is not known which types of Ca channels are involved in this effect. Binding of isoproterenol to the β receptor leads to activation of the noninactivating, inward-rectifying K current which seems to be directly coupled to a CTX-sensitive G protein.[237] This pathway is independent of an increase in $[Ca^{2+}]_i$ and cAMP.[237] However, activation of a K conductance would lead to hyperpolarization and by itself would inhibit instead of activate secretion.

4.5.7. Effects of SST

Inhibition of secretion is mediated by SST in AtT-20/D16-16 cells.[231] The intracellular mechanisms inducing this effect are similar to those described in other pituitary cells (see

Figure 6). The main effects of SST are a decrease in the cAMP concentration[231] and a decrease in resting $[Ca^{2+}]_i$. This decrease in $[Ca^{2+}]_i$ is brought about by blocking of voltage-dependent Ca currents[239] and activation of an inward-rectifying K current, inducing a small hyperpolarization with an ensuing decrease in action potential frequency.[235] However, the duration and amplitude of the action potentials after SST application were not changed,[230] although this should be expected if the amplitude of voltage-dependent Ca currents in AtT-20/D16-16 cells is reduced by SST.[231] In both reactions a PTX-sensitive G protein is involved.[231,235]

There might be additional Ca-independent mechanisms leading to inhibition of secretion. Increase in $[Ca^{2+}]_i$ induced by a high external KCl-mediated depolarization was not decreased by SST, although elevated external KCl reduced β-endorphin secretion.[230] Similarly, it had been reported that SST can inhibit ionomycin-evoked ACTH secretion from AtT-20/D16-16 cells.[238] Ionomycin induces Ca^{2+} influx independent of voltage-activated Ca channels and SST does not block Ca^{2+} entry induced by ionomycin.[231] The same signal cascade as that of SST is used by acetylcholine.[235]

4.6. MELANOTROPH CELLS

Melanotrophs are located in the intermedial lobe of the pituitary (pars intermedia). They synthesize POMC from which melanocyte-stimulating hormone (α-MSH), β-endorphin, and related peptides are generated. These hormones are stored in and secreted from secretory granules on an equimolar basis.[240] α-MSH controls pigment dispersion in dermal melanophores in amphibians and mammals. In amphibians it is one of the major factors for background adaptation. When the clawed toad, *Xenopus laevis,* is placed on a dark background, the melanotrophs secrete α-MSH, leading to a darkening of the skin.[241]

In contrast to the cells in the other major part of the anterior pituitary (pars distalis), melanotrophs are synaptically innervated by hypothalamic nerve fibers.[242] The synaptic terminals are enlarged, forming varicosities.[243] These varicosities contain DA, GABA, and neuropeptide Y (NPY).[244] In amphibians they also contain TRH and melanostatin, the amphibian analog of NPY.[245]

The predominant effect of the neurotransmitters located in the synaptic terminals is a strong inhibition of the spontaneous secretory activity.[246] This inhibition is mainly brought about by DA,[247-249] GABA,[250,251] and NPY/melanostatin.[245] In frog melanotrophs secretion of α-MSH is stimulated by TRH.[3,252]

The synaptic nature of the GABAergic transmission in melanotrophs was first shown by intracellular recordings performed in isolated pituitary preparations.[253] Stimulation of the pituitary stalk evoked inhibitory postsynaptic potentials mediated by activation of the GABA$_A$ receptors. Experiments in a thin slice preparation combined with patch clamp recordings from visually identified cells showed that the synaptic contacts between neurones and rat pars intermedia melanotrophs are in many respects similar to synaptic contacts between central neurones and could also be well described by a "classical" quantal model.[250] In contrast, *in situ* recordings from *Xenopus* melanotrophs showed that the amplitudes of the spontaneous inhibitory postsynoptic currents (IPSC) varied considerably.[240]

Cells located in the rat pars intermedia are not homogeneous. Recordings in a thin slice preparation showed that most of the cells (about 90%) contained a uniform pattern of Na, Ca, and K currents and in all cases responded to GABA. The remaining group of cells was not excitable, i.e., they did not respond to GABA and only a slowly activating K current could be recorded upon depolarizations.[254] In contrast, the population of hormone-secreting melanotrophs in the frog could be separated into two different cell groups with different contents of α-MSH, different rates of secretion, and different sensitivity to TRH using a Percoll density gradient.[255]

4.6.1. Electrical Excitability and Membrane Currents

A high percentage of melanotrophs exhibit spontaneous firing of action potentials,[256] which occurs at a frequency of about 1.5 Hz and was seen in 82% of melanotrophs investigated with intracellular recording and in 64% when whole-cell recording was used.[249] This percentage is in good agreement with measurements of $[Ca^{2+}]_i$ in *Xenopus* melanotrophs, in which 77% of melanotrophs displayed spontaneous oscillatory changes in $[Ca^{2+}]_i$.[257] The average resting potential of spontaneously active melanotrophs was -43 mV with intracellular recording and -54 mV with whole-cell recording.[249] In rat and frog melanotrophs action potentials are mainly the result of an inward Na current and an additional small Ca current.[256,258] Repolarization takes place due to activation of transient and sustained K currents.[259] TTX abolished action potentials, but small-amplitude membrane potential oscillations persisted which could be blocked by Co^{2+}, Ni^{2+}, and Cd^{2+}.[249]

In melanotrophs there are TTX-sensitive Na channels which have almost the same properties as Na channels in other preparations. This is true for rat,[259] porcine[260] and frog[258] melanotrophs. Na currents provide the main component in generating action potentials and presumably induce the strong depolarization required to trigger the influx of Ca^{2+} through high voltage-activated Ca channels.

Voltage-dependent, outward-rectifying K currents in rat and frog melanotrophs consist of a transient and a sustained component.[261] The transient K current is TEA insensitive[261] and can be blocked by 4-AP.[254,261] The properties of this current are similar to that of the A-current described in GH_3 cells.[26] Similar voltage-dependent K currents have been described in frog melanotrophs.[258] In addition, there is a K channel which is normally closed at the resting potential, is activated by DA, and has a single channel conductance of about 100 pS.[248]

It is still uncertain whether there are two or more different types of voltage-activated Ca currents in melanotrophs. At least two different Ca channels have been characterized, a low voltage-activated (LVA) and a HVA Ca channel,[260,262-265] which have similar properties as those encountered in other preparations.[266] The LVA Ca current is unequivocally classified as a T-type current by several groups,[260,262,263] it has a single channel conductance of 8.1 pS,[267] and is activated at potentials positive to -60 mV. However, in frog melanotrophs no T-type Ca currents were found.[245,258] In contrast, the components of the HVA Ca current have been classified differently.[249,260,263,264] In rat melanotrophs, the HVA Ca current seems to be composed of a transient and a sustained component, which led to the interpretation that two L-type Ca currents are present,[263] both with a single channel conductance of 25 pS.[267] In addition, there is also a rapidly inactivating, high-threshold current component which was classified as an N-type current.[264,268] Three types of Ca current (T-, L-, and N-type) were also found in porcine melanotrophs.[260] However, the existence of N-type Ca channels has been doubted, since only a transient block[249] of a Ca current component by ω-conotoxin was reported. In contrast, the observation of a 15-pS Ca channel is in favor of the existence of an N-type Ca channel in frog melanotrophs.[248] Recently, a fourth type of Ca channel was suggested with the demonstration that an HVA Ca current component could be inhibited with funnel web spider toxin, indicating the existence of a P-type current.[219] This Ca current component had also been shown to be depressed by synaptically released DA.[219]

It has been assumed that Ca entry in melanotrophs takes place only through Ca channels opened by the depolarizing phase of the action potential. However, it had been shown that secretion persists even after block of action potentials with TTX.[269] This indicated that Ca entry into the cytoplasm can occur through Ca channels open at the normal resting potential. In addition to fluctuations of the basal Ca concentrations, spontaneous elevations of $[Ca^{2+}]_i$ occur at intervals of several minutes and are due to Ca^{2+} influx through Ca channels which open phasically.[246] They can be blocked in *Xenopus* melanotrophs by DA, GABA, and NPY and stimulated by TRH and CRF.[269] These oscillations of $[Ca^{2+}]_i$ measured in *Xenopus* melanotrophs were totally dependent on extracellular Ca^{2+} and could be blocked with ω-conotoxin, but not by nifedipine, indicating that Ca^{2+} entered the cell via N-type rather than L-type Ca channels.[257]

In porcine melanotrophs a small-conductance (2 to 3-pS) Cl channel has been described which is activated by micromolar concentrations of internal Ca.[270] A similar Ca^{2+}-activated Cl

FIGURE 10. Typical membrane potential response to DA of a cultured frog melanotroph recorded in the whole-cell configuration. DA (1 μ*M*), ejected in the vicinity of the cell for 500 ms starting at the dot, hyperpolarized the cell and inhibited spontaneous action potentials. (Reprinted from Valentijn, J. A., Louiset, E., Vaudry, H., and Cazin, L., Dopamine regulates the electrical activity of frog melanotrophs through a G-protein-mediated mechanism, *Neuroscience*, 44, 85–95, Copyright 1991. With kind permission from Elsevier Science Ltd., The Boulevard, Langford Lane, Kidlington OX5 1GB, UK.)

channel has been described in oocytes of *X. laevis*.[271] Opening of these channels would enhance Cl⁻ efflux and by this induce fluid secretion, causing a temporary increase of the extracellular space. This could facilitate extrusion of secreted peptides which indeed accumulate in the colloid.[272,273]

4.6.2. Effects of Releasing and Inhibiting Factors
4.6.2.1. Effects of DA

DA and GABA are the main inhibiting factors mediating a powerful tonic inhibition of hormone release.[247-250] Both receptors, GABA$_A$ and GABA$_B$, are present in the plasma membrane of melanotrophs.[274] Melanostatin in the frog, and NPY in mammals, provide an additional inhibition of α-MSH secretion.[245,275] NPY or melanostatin are colocalized in the nerve terminals with DA and GABA.[3,252]

Binding of DA as well as of the D$_2$ agonists bromocryptine and quinpirole to D$_2$ dopaminergic receptors in melanotrophs activates several second-messenger pathways which finally leads to an effective inhibition of hormone release.[3] One of these pathways decreases adenylate cyclase activity through an inhibitory G protein[276] which does not involve a modulating effect on ionic channels. Electrophysiologically, in melanotrophs DA induces a membrane hyperpolarization of 5–28 mV (Figure 10) and blocks or slows spike discharge.[248,249] This response is similar to that in lactotrophs. The hyperpolarization is due to an activation of a K channel which has a single channel conductance of 100 pS.[248] In frog melanotrophs, DA decreases Na as well as L- and N-type Ca currents,[248,277] whereas in rat melanotrophs DA blocks the T- and N-type Ca current as well as the L$_1$-subtype sparing the L$_2$-subtype Ca current.[264] Obviously there are species-related differences in the mode of action of DA on Ca currents. In addition, DA activates an outward-rectifying K current.[248] All of these effects are abolished by the specific D$_2$ receptor antagonist sulpiride.[248,249] As in rat lactotrophs,[193,194] a direct coupling between channel proteins and G proteins is likely to occur. The G proteins involved in activation of the K currents and inhibition of Na and Ca currents are PTX-sensitive.[247]

When melanotrophs are removed from the pituitary and placed in culture, there is a marked increase of the current amplitude through Ca channels as a function of time in culture.[262] HVA and LVA Ca channels are affected differently by denervation; the density of HVA Ca channels increases more than that of LVA Ca channels with time (up to 16 d). The voltage dependence and kinetics of both channel types do not change in culture. This increase in Ca channel density seems to be due to an increased mRNA and protein synthesis. Chronic stimulation of DA receptors prevents or even reverts the development of Ca channel activity.[278] On the other hand, a faster development of Ca channel activity is obtained by chronic stimulation of the adrenergic receptors.

4.6.2.2. Effects of GABA

The inhibitory effect exerted by GABA on frog[251] and rat[250] melanotrophs is mediated by activation of GABA$_A$ receptors, leading to an increase in membrane Cl conductance which can be blocked by bicuculline.

4.6.2.3. Effects of NPY and Melanostatin

NPY and its amphibian counterpart melanostatin inhibit α-MSH release from frog melanotrophs.[275] Melanostatin inhibits electrical activity by inducing a strong hyperpolarization due to an activation of K channels via activation of a G protein.[245] The experiments by Valentijn et al.[245] suggest that melanostatin and DA may activate the same signal pathways, i.e., melanostatin activates two types of K current and inhibits Na currents and N- and L-type Ca currents. The depression of Ca currents was attenuated when the cells were pretreated with pertussis toxin. Whether the inhibition of Ca currents by DA and melanostatin is mediated by the same G protein or by two distinct G proteins is not known. As well, the physiological significance of these coexisting neurotransmitters is not known. If they are coreleased, the two endocrine signals could exert additive effects on the target cell. Alternatively, DA and melanostatin may be released differently, depending on whether the hypothalamic neurones fire with low or high frequency, as previously shown for acetylcholine/VIP-containing neurones in the submandibular gland.[279] A similar inhibition of α-MSH secretion is mediated via binding of adrenaline to an α_2-adrenergic receptor subtype, which leads to hyperpolarization due to activation of a K conductance. This effect could be blocked by yohimbine.[280]

4.6.2.4. Effects of TRH

In frog melanotrophs, TRH induces a transient hyperpolarization as a result of the opening of Ca-dependent K channels and a subsequent depolarization accompanied by an enhanced frequency of action potentials,[252] presumably due to a decrease in a voltage-dependent K conductance induced by TRH. These electrophysiological changes correlate with the finding that TRH stimulates a biphasic secretion of α-MSH.[252] The effects of TRH on melanotrophs are phenomenologically similar to those induced by TRH in GH$_3$/B$_6$ cells.

4.6.2.5. Effects of Isoproterenol

Activation of the β-adrenergic receptor with isoproterenol stimulates secretion of α-MSH which is not blocked by TTX.[249] Stimulation of the β receptor increased the intracellular cAMP level mediated by the stimulatory GTP-binding protein (G$_s$).[276] In addition, isoproterenol increases the amplitude of the L-type and not the T-type Ca current. This increase in the L-type Ca current amplitude is mimicked by 8-bromo-cAMP.

5. CONCLUSIONS

The combined work reviewed here on the electrophysiology of five of the major cell types found in the anterior pituitary has shown that a variety of ionic currents is present in all of these cells. In each particular cell type, the combination of these ionic currents seems to be correlated with the particular secretory task of this cell.

With few exceptions, all electrophysiological studies have concentrated on acute experiments to examine the effects of the factors which stimulate or inhibit hormone release. These experiments have elucidated details of the underlying signal cascades for a few substances, as for example the action of TRH in GH cells. A model system, such as GH cells, can help in the design and interpretation of experiments in less well-studied cells, but to be useful, such a model system needs to be based on studies of homogeneous cells.

As many examples described in this review make clear, though, cells producing the same hormone can be heterogeneous and exist in a number of subpopulations which depend on the exact physiological state of the donor. Heterogeneity occurs at all levels in the anterior pituitary. Individual cells differ in terms of the hormones secreted, the mode of release, the types of receptors and the set of associated G proteins, the signal pathways, and the combination of ionic currents. The examples presented in this review suggest that this heterogeneity is directly related to function. Separation of enriched populations of cells based on anatomical differences have been shown to result in groups of cells with more homogeneous electrophysiological properties. The degree of the differences between subpopulations have been most apparent when related to a particular physiological context (e.g., sexual development or lactation in rats).

Even clonal cell lines, such as the GH cells, are not necessarily homogeneous. Also, as shown in GH cells, the subpopulations of cells change, depending on which factors are present in the culture medium. These changes can occur within hours, similar to the rapid changes observed in primary cultured cells during the first period in cell culture. Since changes in cell culture conditions or separation of a native cell from its normal physiological context can quickly result in a change of its secretory behavior, the control cells in electrophysiological experiments must be carefully defined.

Given the extensive heterogeneity shown both in cell lines and in primary culture cells, a future task will be the development of refined separation methods based more directly on functional differences to yield highly homogeneous cell groups in which particular mechanisms can be studied. The goal remains to understand the common regulatory principles underlying coordinated activity of the anterior pituitary within the entire organism. In addition to simple acute experiments, in future studies more attempts should be made to recreate the endocrine, paracrine, and autocrine factors of the extracellular milieu which define secretion.

ACKNOWLEDGMENTS

We would like to thank C. Reißmann and M. Leoni for preparing the figures and K. Klapper for help in organizing the literature and preparing the manuscript. Thanks also to the Deutsche Forschungsgemeinschaft for continuing support of our research projects.

REFERENCES

1. Douglas, W. W. and Taraskevich, P. S., The electrophysiology of adenohypophyseal cells, in *The Electrophysiology of the Secretory Cell*, Poisner, A. M., and Trifaro, J. M., Eds., Elsevier, Amsterdam, 1985, Chap. 4.
2. Ozawa, S. and Sand, O., Electrophysiology of excitable endocrine cells, *Physiol. Rev.*, 66, 887, 1986.
3. Douglas, W. W., Calcium, stimulus-secretion coupling and exocytosis — glancing back 30 years — and recent lessons on spontaneous secretion from melanotrophs revealing calcium-driven autonomous secretion, independent of action potential discharge, and its uncoupling by the secretacurbins — dopamine, GABA and neuropeptide-Y, *Biomed. Res.*, 14, 9, 1993.
4. Kidokoro, Y., Spontaneous calcium action potentials in a clonal pituitary cell line and their relationship to prolactin secretion, *Nature*, 258, 741, 1975.
5. Mason, W. T., Hoyland, J., Davison, I., Carew, M., Somasundaram, B., Tregear, R., Zorec, R., Lledo, P. M., Shankar, G., and Horton, M., Quantitative real-time imaging of optical probes in living cells, in *Fluorescent and Luminescent Probes for Biological Activity*, Mason, W. T., Ed., Academic Press, London, 1993, chap. 12.
6. Childs, G. V., Unabia, G., Burke, J. A., and Marchetti, C., Secretion from corticotropes after avidin-fluorescein stains for biotinylated ligands (CRF or AVP), *American J. Physiol.*, 252, E347, 1987.
7. Neill, J. D. and Frawley, L. S., Detection of hormone release from individual cells in a mixed population using a reverse hemolytic plaque assay, *Endocrinology*, 112, 1135, 1983.
8. Wynick, D., Hammond, P. J., Akinsanya, K. O., and Bloom, S. R., Galanin regulates basal and oestrogen-stimulated lactotroph function, *Nature*, 364, 529, 1993.

9. Arita, J., Analysis of the secretion from single anterior pituitary cells by cell immunoblot assay, *Endocr. J.*, 40, 1, 1993.

10. Hymer, W. C. and Hatfield, J. M., Separation of cells from the rat anterior pituitary gland, in *Cell Separation: Methods and Selected Applications*, Vol. 8, Academic Press, New York, 1984, 163.

11. Hall, M., Howell, S. L., Schulster, D., and Wallis, M., A procedure for the purification of somatotrophs isolated from rat anterior pituitary glands using Percoll density gradient, *J. Endocrinol.*, 94, 257, 1982.

12. St. John, P. A., Dufy-Barbe, L., and Barker, J. L., Anti-prolactin cell-surface immunoreactivity identifies a subpopulation of lactotrophs from the rat anterior pituitary, *Endocrinology*, 119, 2783, 1986.

13. Marchetti, C., Childs, G. V., and Brown, A. M., Membrane currents of identified isolated rat corticotropes and gonadotropes, *Am. J. Physiol.*, 252, E340, 1987.

14. Tashjian, A. H., Yasumura, Y., Levine, L., Sato, G. H., and Parker, M. L., Establishment of clonal strains of rat pituitary tumor cells that secrete growth hormone, *Endocrinology*, 82, 342, 1968.

15. Boockfor, F. R. and Schwarz, L. K., Cultures of GH_3 cells contain both single and dual hormone secretors, *Endocrinology*, 122, 762, 1988.

16. Kineman, R. D. and Frawley, L. S., Secretory characteristics and phenotypic plasticity of growth hormone- and prolactin-producing cell lines, *J. Endocrinol.*, 140, 455, 1994.

17. Cronin, M. J., Faure, N., Martial, J. A., and Weiner, R. I., Absence of high affinity dopamine receptors in GH_3 cells: a prolactin-secreting clone resistant to the inhibitory action of dopamine, *Endocrinology*, 106, 718, 1980.

18. Zeytin, F., Gick, G. G., Brazeau, P., Ling, N., McLaughlin, M., and Bancroft, C., Growth hormone (GH)- releasing factor does not regulate GH release or GH mRNA levels in GH_3 cells, *Endocrinology*, 114, 2054, 1984.

19. Judd, A. M., Login, I. S., Kovacs, K., Ross, P. C., Spangelo, B. L., Jarvis, W. D., and MacLeod, R., Characterization of the MMQ cell, a prolactin-secreting clonal cell line that is responsive to dopamine, *Endocrinology*, 123, 2341, 1988.

20. Dubinsky, J. M. and Oxford, G. S., Ionic currents in two strains of rat anterior pituitary tumor cells, *J. Gen.Physiol.*, 83, 309, 1984.

21. Armstrong, C. M. and Matteson, D. R., Two distinct populations of calcium channels in a clonal line of pituitary cells, *Science*, 227, 65, 1985.

22. Simasko, S. M., Weiland, G. A., and Oswald, R. E., Pharmacological characterization of two calcium currents in GH_3 cells, *Am. J. Physiol.*, 245, E328, 1988.

23. Lang, E. G. and Ritchie, A. K., Large and small conductance calcium-activated potassium channels in the GH_3 anterior pituitary cell line, *Pflügers Arch.*, 410, 614, 1987.

24. Lang, D. G. and Ritchie, A. K., Tetraethylammonium ion sensitivity of a 35-pS Ca^{2+}-activated K^+ channel in GH_3 cells that is activated by thyrotropin-releasing hormone, *Pflügers Arch.*, 416, 704, 1990.

25. Rogawski, M. A., Inoue, K., and Barker, J. L., A slow calcium-dependent chloride conductance in clonal anterior pituitary cells, *J. Neurophysiol.*, 59, 1854, 1988.

26. Rogawski, M. A., Transient outward current (I_A) in clonal anterior pituitary cells: blockade by aminopyridine analogs, *Naunyn-Schmiedeberg's Arch. Pharmacol.*, 338, 125, 1988.

27. Oxford, G. S. and Wagoner, P. K., The inactivating K^+ current in GH_3 pituitary cells and its modification by chemical reagents, *J. Physiol.*, 410, 587, 1989.

28. Simasko, S. M., Evidence for a delayed rectifier-like potassium current in the clonal rat pituitary cell line GH_3, *Am. J. Physiol.*, 261, E66, 1991.

29. Bauer, C. K., Meyerhof, W., and Schwarz, J. R., An inward-rectifying K^+ current in clonal rat pituitary cells and its modulation by thyrotropin-releasing hormone, *J. Physiol.*, 429, 169, 1990.

30. Yatani, A., Codina, J., Sekura, R. D., Birnbaumer, L., and Brown, A. M., Reconstitution of somatostatin and muscarinic receptor mediated stimulation of K^+ channels by isolated G_K protein in clonal rat anterior pituitary cell membranes, *Mol. Endocrinol.*, 1, 283, 1987.

31. Barros, F., Delgado, L. M., Maciá, C., and de la Peña, P., Effects of hypothalamic peptides on electrical activity and membrane currents of perforated patch clamped GH_3 anterior pituitary cells, *FEBS Lett.*, 279, 33, 1991.

32. Simasko, S. M., Reevaluation of the electrophysiological actions of thyrotropin-releasing hormone in a rat pituitary cell line (GH_3), *Endocrinology*, 128, 2015, 1991.

33. Bauer, C. K., Davison, I., Kubasov, I., Schwarz, J. R., and Mason, W. T., Different G proteins are involved in the biphasic response of clonal rat pituitary cells to thyrotropin-releasing hormone, *Pflügers Arch.*, 428, 17, 1994.

34. Biales, B., Dichter, M. A., and Tischler A., Sodium and calcium action potential in pituitary cells, *Nature*, 267, 172, 1977.

35. Schlegel, W., Winiger, B. P., Mollard, P., Vacher, P., Wuarin, F., Zahnd, G. R., Wollheim, C. B., and Dufy, B., Oscillations of cytosolic Ca^+ in pituitary cells due to action potentials, *Nature*, 329, 719, 1987.

36. Iijima, T., Sand, O., Sekiguchi, T., and Matsumoto, G., Simultaneous recordings of cytosolic Ca^{2+} level and membrane potential and current during the response to thyroliberin in clonal rat anterior pituitary cells, *Acta Physiol. Scand.*, 140, 269, 1990.

37. Simasko, S. M., Effect of calcium on membrane potential behavior in a rat pituitary cell line (GH_3), *Mol. Cell. Endocrinol.*, 78, 79, 1991.

38. Ritchie, A. K., and Lang, D. G., Activation of single ion channels that regulate membrane excitability in GH_3 anterior pituitary cells, in *Neural Control of Reproductive Function*, Lakoski, J. M., Perez-Polo, J. R., and Rassin, D. K., Eds., Alan R. Liss, New York, 1989, 463.

39. DuPont, J. S., Triiodothyronine affects the electrical properties of GH_3 cells, *Acta Endocrinol.*, 123, 51, 1990.

40. Kunze, D. L. and Ritchie, A. K., Multiple conductance levels of the dihydropyridine-sensitive calcium channel in GH_3 cells, *J. Membr. Biol.*, 118, 171, 1990.

41. Herrington, J. and Lingle, C. J., Kinetic and pharmacological properties of low voltage-activated Ca^{2+} current in rat clonal (GH_3) pituitary cells, *J. Neurophysiol.*, 68, 213, 1992.

42. Scherübl, H. and Hescheler, J., Steady-state currents through voltage-dependent, dihydropyridine-sensitive Ca^{2+} channels in GH_3 pituitary cells, *Proc. R. Soc. London Ser. B*, 245, 127, 1991.

43. Hille, B., *Ionic Channels of Excitable Membranes*, 2nd ed., Sinauer Associates, Sunderland, MA, 1992.

44. Barros, F., Delgado, L. M., Del Camino, D., and De la Pena, P., Characteristics and modulation by thyrotropin-releasing hormone of an inwardly rectifying K^+ current in patch-perforated GH_3 anterior pituitary cells, *Pflügers Arch.*, 422, 31, 1992.

45. Sato, N., Wang, X. B., and Greer, M. A., Medium hyperosmolarity depresses thyrotropin-releasing hormone-induced Ca^{2+} influx and prolactin secretion in GH_4C_1 cells, *Mol. Cell. Endocrinol.*, 77, 193, 1991.

46. Sand, O., Chen, B., Li, Q., Karlsen, H. E., Bjøro, T., and Haug, E., Vasoactive intestinal peptide (VIP) may reduce the removal rate of cytosolic Ca^{2+} after transient elevations in clonal rat lactotrophs, *Acta Physiol. Scand.*, 137, 113, 1989.

47. Mollard, P., Zhang, Y., Rodman, D., and Cooper, D. M. F., Limited accumulation of cyclic AMP underlies a modest vasoactive-intestinal-peptide-mediated increase in cytosolic Ca^{2+} transients in GH_3 pituitary cells, *Biochem. J.*, 284, 637, 1992.

48. Chiavaroli, C., Vacher, P., and Schlegel, W., Modulation of Ca^{2+} influx by protein phosphorylation in single intact clonal pituitary cells, *Eur. J. Pharmacol.-Mol. Pharmacol. Sect.*, 227, 173, 1992.

49. Bjøro, T., Sand, O., Østberg, B. C., Gordeladze, J. O., Torjesen, P., Gautvik, K. M., and Haug, E., The mechanisms by which vasoactive intestinal peptide (VIP) and thyrotropin releasing hormone (TRH) stimulate prolactin release from pituitary cells, *Biosci. Rep.*, 10, 189, 1990.

50. Gordeladze, J. O., Sletholt, K., Thorn, N. A., and Gautvik, K. M., Hormone-sensitive adenylate cyclase of prolactin-producing rat pituitary adenoma (GH_4C_1) cells: molecular organization, *Eur. J. Biochem.*, 177, 665, 1988.

51. Inukai, T., Wang, X., Greer, S. E., and Greer, M. A., Adenosine 3′,5′-Cyclic monophosphate-mediated prolactin secretion in GH_4C_1 cells involves Ca^{2+} influx through L-type Ca^{2+} channels, *Cell Calcium*, 14, 219, 1993.

52. Törnquist, K., 1,25-Dihydroxycholecalciferol enhances both the bombesin-induced transient in intracellular free Ca^{2+} and the bombesin-induced secretion of prolactin in GH_4C_1 pituitary cells, *Endocrinology*, 128, 2175, 1991.

53. Aanestad, M., Røtnes, J. S., Torjesen, P. A., Haug, E., Sand, O., and Bjøro, T., Epidermal growth factor stimulates the prolactin synthesis and secretion in rat pituitary cells in culture (GH_4C_1 cells) by increasing the intracellular concentration of free calcium, *Acta Endocrinol.*, 128, 361, 1993.

54. Kuwahara, T., Nagase, H., Takamiya, M., Yoshizaki, H., Kudoh, T., Nakano, A., and Arisawa, M., Activation of CCK-B receptors elevates cytosolic Ca^{2+} levels in a pituitary cell line, *Peptides*, 14, 801, 1993.

55. Yang, J. P. and Tashjian, A. H., Platelet-activating factor affects cytosolic free calcium concentration and prolactin secretion in GH_4C_1 rat pituitary cells, *Biochem. Biophys. Res. Commun.*, 174, 424, 1991.

56. Guérineau, N., Drouhault, R., Corcuff, J. B., Vacher, A. M., Vilayleck, N., and Mollard, P., Galanin evokes a cytosolic calcium bursting mode and hormone release in GH_3/B_6 pituitary cells, *FEBS Lett.*, 276, 111, 1990.

57. Bresson, L., Fahmi, M., Sartor, P., Dufy, B., and Dufy-Barbe, L., Growth hormone-releasing factor stimulates calcium entry in the GH_3 pituitary cell line, *Endocrinology*, 129, 2126, 1991.

58. Fahmi, M., Bresson, L., Dufy-Barbe, L., Sartor, P., Taupignon, A., and Dufy, B., Electrophysiological studies of the mechanism of action of growth-hormone releasing-hormone (GH-RH) on GH_3 pituitary cells, *C.R. Soc. Biol.*, 185, 224, 1991.

59. Drouhault, R., Vacher, P., Darret, D., Larrue, J., Vacher, A. M., and Vilayleck, N., Long-term effects induced by only one ketoconazole treatment on rat tumoral pituitary cells (GH_3/B_6), *C.R. Acad. Sci. Paris*, 312, 615, 1991.

60. Law, G. J., Pachter, J. A., Thastrup, O., Hanley, M. R., and Dannies, P. S., Thapsigargin, but not caffeine, blocks the ability of thyrotropin-releasing hormone to release Ca^{2+} from an intracellular store in GH_4C_1 pituitary cells, *Biochem. J.*, 267, 359, 1990.

61. Aragay, A. M., Katz, A., and Simon, M. I., The $G_{\alpha q}$ and $G_{\alpha 11}$ proteins couple the thyrotropin-releasing hormone receptor to phospholipase C in GH_3 rat pituitary cells, *J. Biol. Chem.*, 267, 24983, 1992.

62. Gershengorn, M. C., Mechanism of thyrotropin releasing hormone stimulation of pituitary hormone secretion, *Ann.Rev. Physiol.*, 48, 515, 1986.

63. White, R. E., Schonbrunn, A., and Armstrong, D. L., Somatostatin stimulates Ca^{2+}-activated K^+ channels through protein dephosphorylation, *Nature*, 351, 570, 1991.

64. Paulssen, R. H., Paulssen, E. J., Gautvik, K. M., and Gordeladze, J. O., The thyroliberin receptor interacts directly with a stimulatory guanine-nucleotide-binding protein in the activation of adenylyl cyclase in GH_3 rat pituitary tumour cells — evidence obtained by the use of antisense RNA inhibition and immunoblocking of the stimulatory guanine-nucleotide-binding protein, *Eur. J. Biochem.*, 204, 413, 1992.

65. Summers, S. T., Walker, J. M., Sando, J. J., and Cronin, M. J., Phorbol esters increase adenylate cyclase activity and stability in pituitary membranes, *Biochem. Biophys. Res. Commun.*, 151, 16, 1988.

66. Gourdji, D., Bataille, D., Vauclin, N., Grouselle, D., Rosselin, G., and Tixier-Vidal, A., Vasoactive intestinal peptide (VIP) stimulates prolactin (PRL) release and cAMP production in a rat pituitary cell line (GH_3/B_6) additive effects of VIP and TRH on PRL release, *FEBS Letters*, 104, 165, 1979.

67. Kalman, D., O'Lague, P. H., Erxleben, C., and Armstrong, D. L., Calcium-dependent inactivation of the dihydropyridine-sensitive calcium channels in GH_3 cells, *J. Gen. Physiol.*, 92, 531, 1988.

68. Barros, F., Villalobos, C., Garciasancho, J., Del Camino, D., and De la Pena, P., The role of the inwardly rectifying K^+ current in resting potential and thyrotropin-releasing-hormone-induced changes in cell excitability of GH_3 rat anterior pituitary cells, *Pflügers Arch.*, 426, 221, 1994.

69. Gammon, C. M., Oxford, G. S., Allen, A. C., McCarthy, K. D., and Morell, P., Diacylglycerol modulates action potential frequency in GH_3 pituitary cells: correlative biochemical and electrophysiological studies, *Brain Res.*, 479, 217, 1989.

70. Dufy, B., Jaken, S., and Barker, J. L., Intracellular Ca^{2+}-dependent protein kinase C activation mimics delayed effects of thyrotropin-releasing hormone on clonal pituitary cell excitability, *Endocrinology*, 121, 793, 1987.

71. Cheng, K., Chan, W. W. S., Arias, R., Barreto, A., and Butler, B., PMA-sensitive protein kinase-C is not necessary in TRH-stimulated prolactin release from female rat primary pituitary cells, *Life Sci.*, 51, 1957, 1992.

72. Haymes, A. A. and Hinkle, P. M., Activation of protein kinase-C increases Ca^{2+} sensitivity of secretory response of GH_3 pituitary cells, *Am. J. Physiol.*, 264, C1020, 1993.

73. Bauer, C. K., Carratù, M. R., and Schwarz, J. R., The TRH-induced biphasic increase in intracellular calcium concentration persists following down-regulation of protein kinase C in clonal rat pituitary cells, in *Proceedings of the 22nd Göttingen Neurobiology Conference*, Vol. II, Elsner, N., and Breer, H., Eds., Georg Thieme Verlag, New York, 1994, 739.

74. Wilkinson, S. E. and Hallam, T. J., Protein kinase C — is its pivotal role in cellular activation over-stated ?, *Trends Pharmacol. Sci.*, 15, 53, 1994.

75. Mariot, P., Dufy, B., Audy, M. C., and Sartor, P., Biphasic changes in intracellular pH induced by thyrotropin-releasing hormone in pituitary cells, *Endocrinology*, 132, 846, 1993.

76. Törnquist, K. and Tashjian, A. H., Importance of transients in cytosolic free calcium concentrations on activation of Na^+/H^+ exchange in GH_4C_1 pituitary cells, *Endocrinology*, 128, 242, 1991.

77. Ohmichi, M., Sawada, T., Kanda, Y., Koike, K., Hirota, K., Miyake, A., and Saltiel, A. R., Thyrotropin-releasing hormone stimulates MAP kinase activity in GH_3 cells by divergent pathways — evidence of a role for early tyrosine phosphorylation, *J. Biol. Chem.*, 269, 3783, 1994.

78. Jefferson, A. B., Travis, S. M., and Schulman, H., Activation of multifunctional Ca^{2+}/calmodulin-dependent protein kinase in GH_3 cells, *J. Biol. Chem.*, 266, 1484, 1991.

79. Ozawa, S. and Kimura, N., Membrane potential changes caused by thyrotropin-releasing hormone in the clonal GH_3 cell and their relationship to secretion of pituitary hormone, *Proc. Natl. Acad. Sci. U.S.A.*, 76, 6017, 1979.

80. Albert, P. R. and Tashjian, A. H., Dual actions of phorbol esters on cystolic free Ca^{2+} concentrations and reconstitution with ionomycin of acute thyrotropin-releasing hormone responses, *J. Biol. Chem.*, 260, 8746, 1985.

81. Drummond, A. H., Bidirectional control of cytosolic free calcium by thyrotropin-releasing hormone in pituitary cells, *Nature*, 315, 752, 1985.

82. Martin, T. F. J., Hsieh, K. P., and Porter, B. W., The sustained 2nd-phase of hormone-stimulated diacylglycerol accumulation does not activate protein kinase-C in GH_3 cells, *J. Biol. Chem.*, 265, 7623, 1990.

83. Winiger, B. P. and Schlegel, W., Rapid transient elevations of cytosolic calcium triggered by thyrotropin releasing hormone in individual cells of the pituitary line GH_3/B_6, *Bioch. J.*, 255, 161, 1988.

84. Suzuki, N., Takagi, H., Yoshioka, T., Tanakadate, A., and Kano, M., Augmentation of transient low-threshold Ca^{2+} current induced by GTP-binding protein signal transduction system in GH_3 pituitary cells, *Bioch. Biophys. Res. Commun.*, 187, 529, 1992.

85. Gollasch, M., Haller, H., Schultz, G., and Hescheler, J., Thyrotropin-releasing hormone induces opposite effects on Ca²⁺ channel currents in pituitary cells by two pathways, *Proc. Natl. Acad. Sci. U.S.A.*, 88, 10262, 1991.

86. Gollasch, M., Kleuss, C., Hescheler, J., Wittig, B., and Schultz, G., G(i2) and protein kinase-C are required for thyrotropin-releasing hormone-induced stimulation of voltage-dependent Ca²⁺ channels in rat pituitary GH₃ cells, *Proc. Natl. Acad. Sci. U.S.A.*, 90, 6265, 1993.

87. Marchetti, C. and Brown, A. M., Protein kinase activator 1-oleoyl-2-acetyl-sn-glycerol inhibits two types of calcium currents in GH₃ cells, *Am. J. Physiol.*, 254, C206, 1988.

88. Haymes, A. A., Kwan, Y. W., Arena, J. P., Kass, R. S., and Hinkle, P. M., Activation of protein kinase-C reduces L-type calcium channel activity of GH₃ pituitary cells, *Am. J. Physiol.*, 262, C1211, 1992.

89. MacEwan, D. J., Mitchell, R., Calcium influx through L-type channels into rat anterior pituitary cells can be modulated in two ways by protein kinase-C — (PKC-isoform selectivity of 1,2-dioctanoyl sn-glycerol), *FEBS Lett.*, 291, 79, 1991.

90. Duchemin, A. M., Enyeart, J. A., Biagi, B. A., Foster, D. N., Mlinar, B., and Enyeart, J. J., Ca²⁺ channel modulation and kinase-C activation in a pituitary cell line — induction of immediate early genes and inhibition of proliferation, *Mol. Endocrinol.*, 6, 563, 1992.

91. Yagisawa, H., Simmonds, S. H., and Hawathorne, J. N., The muscarinic receptor of rat pituitary GH₃ cells is coupled with adenylate cyclase inhibition, but not with phosphoinositide turnover, *Biochem. Pharmacol.*, 37, 2675, 1988.

92. Offermanns, S., Gollasch, M., Hescheler, J., Spicher, K., Schmidt, A., Schultz, G., and Rosenthal, W., Inhibition of voltage-dependent Ca²⁺ currents and activation of pertussis toxin-sensitive G-proteins via muscarinic receptors in GH₃ cells, *Mol. Endocrinol.*, 5, 995, 1991.

93. Delahunty, T. M., Cronin, M. J., and Linden, J., Regulation of GH₃-cell function via adenosine A1 receptors, *Biochem. J.*, 255, 69, 1988.

94. Cooper, D. M. F., Caldwell, K. K., Boyajian, C. L., Petcoff, D. W., and Schlegel, W., Adenosine A1 receptors inhibit both adenylate cyclase activity and TRH-activated Ca²⁺ channels by a pertussis toxin-sensitive mechanism in GH₃ cells, *Cell. Signalling*, 1, 85, 1989.

95. Vacher, P., Mariot, P., Dufy-Barbe, L., Nikolics, K., Seeburg, P. H., Kerdelhue, B., and Dufy, B., The gonadotropin-releasing hormone associated peptide reduces calcium entry in prolactin-secreting cells, *Endocrinology*, 128, 285, 1991.

96. Chuoi, M. T. V., Vacher, P., and Dufy, B., GnRH-associated peptide decreases cyclic AMP accumulation in the GH₃ pituitary cell line, *Neuroendocrinology*, 58, 251, 1993.

97. Mollard, P., Vacher, P., Dufy, B., and Barker, J. L., Somatostatin blocks Ca²⁺ action potential activity in prolactin-secreting pituitary tumor cells through coordinate actions on K⁺ and Ca²⁺ conductances, *Endocrinology*, 123, 721, 1988.

98. Yajima, Y., Akita, Y., and Saito, T., Pertussis toxin blocks the inhibitory effects of somatostatin on cAMP-dependent vasoactive intestinal peptide and cAMP-independent thyrotropin releasing hormone-stimulated prolactin secretion in GH₃ cells, *J. Biol. Chem.*, 261, 2684, 1986.

99. Hildebrandt, J. D., Sekura, R. D., Codina, J., Iyengar, R., Manclark, C. R., and Birnbaumer, L., Stimulation and inhibition of adenylyl cyclase is mediated by distinct proteins, *Nature*, 302, 706, 1983.

100. Rosenthal, W., Hescheler, J., Hinsch, K. D., Spicher, K., Trautwein, W., and Schultz, G., Cyclic AMP-independent, dual regulation of voltage-dependent Ca²⁺currents by LHRH and somatostatin in a pituitary cell line, *EMBO J.*, 7, 1627, 1988.

101. Armstrong, D. and Eckert, R., Voltage-activated calcium channels that must be phosphorylated to respond to membrane depolarization, *Proc. Natl. Acad. Sci. U.S.A.*, 84, 2518, 1987.

102. Boockfor, F. R., Hoeffler, J. P., and Frawley, L. S., Cultures of GH₃ cells are funcionally heterogenous, thyrotropin-releasing hormone, estradiol and cortisol cause reciprocal shifts in the proportions of growth hormone and prolactin secretors, *Endocrinology*, 117, 418, 1985.

103. Hinkle, P. M., Shanshala, E. D., and Yan, Z. F., Epidermal growth factor decreases the concentration of thyrotropin-releasing hormone (TRH) receptors and TRH responses in pituitary GH₄C₁ cells, *Endocrinology*, 129, 1283, 1991.

104. Fujimoto, J., Straub, R. E., and Gershengorn, M. C., Thyrotropin-releasing hormone (TRH) and phorbol myristate acetate decrease TRH receptor messenger RNA in rat pituitary GH₃ cells — evidence that protein kinase-C mediates the TRH effect, *Mol. Endocrinol.*, 5, 1527, 1991.

105. Missale, C., Castelletti, L., Boroni, F., Memo, M., and Spano, P., Epidermal growth factor induces the functional expression of dopamine receptors in the GH₃ cell line, *Endocrinology*, 128, 13, 1991.

106. Gardette, R., Rasolonjanahary, R., Kordon, C., and Enjalbert, A., Epidermal growth factor treatment induces D₂ dopamine receptors functionally coupled to delayed outward potassium current (I-K) in GH₄C₁ clonal anterior pituitary cells, *Neuroendocrinology*, 59, 10, 1994.

107. Paulssen, E. J., Paulssen, R. H., Gautvik, K. M., and Gordeladze, J. O., Hypothalamic hormones modulate-G protein levels and 2nd messenger responsiveness in GH₃ rat pituitary tumour cells, *Biochem. Pharmacol.*, 44, 471, 1992.

108. Chiavaroli, C., Cooper, D. M. F., Boyajian, C. L., Murraywhelan, R., Demaurex, N., Spiegel, A. M., and Schlegel, W., Spontaneous intracellular calcium oscillations and $G_{s\alpha}$ subunit expression are inversely correlated with secretory granule content in pituitary cells, *J. Neuroendocrinol.*, 4, 473, 1992.

109. Laverriere, J. N., Richard, J. L., Buisson, N., Martial, J. A., Tixier-Vidal, A., and Gourdji, D., Thyroliberin and dihydropyridines modulate prolactin gene expression through interacting pathways in GH_3 cells, *Neuroendocrinology*, 50, 693, 1989.

110. Gilchrist, C. A. and Shull, J. D., Epidermal growth factor induces prolactin messenger RNA in GH_4C_1 cells via a protein synthesis-dependent pathway, *Mol. Cell. Endocrinol.*, 92, 201, 1993.

111. Levitan, E. S., Hemmick, L. M., Birnberg, N. C., and Kaczmarek, L. K., Dexamethasone increases potassium channel messenger RNA and activity in clonal pituitary cells, *Mol. Endocrinol.*, 5, 1903, 1991.

112. Takimoto, K., Fomina, A. F., Gealy, R., Trimmer, J. S., and Levitan, E. S., Dexamethasone rapidly induces Kv1.5 K^+ channel gene transcription and expression in clonal pituitary cells, *Neuron*, 11, 359, 1993.

113. Törnquist, K. and Tashjian, A. H., Dual action of 1,25-dihydroxycholecalciferol on intracellular Ca^{2+} in GH_4C_1 cells: evidence for effects on voltage-operated Ca^{2+} channels and Na^+/Ca^{2+} exchange, *Endocrinology*, 124, 2765, 1989.

114. Hinkle, P. M., Nelson, E. J., and Haymes, A. A., Regulation of L-type voltage-gated calcium channels by epidermal growth factor, *Endocrinology*, 133, 271, 1993.

115. Ritchie, A. K., Estrogen increases low voltage-activated calcium current density in GH_3 anterior pituitary cells, *Endocrinology*, 132, 1621, 1993.

116. Tannenbaum, G. S., Physiological role of somatostatin in the regulation of pulsatile growth hormone secretion, in *Somatostatin*, Patel, Y. C. and Tannenbaum, G. S., Eds., Plenum Press, New York, 1985, 229.

117. Frohman, L. A. and Janson, J. O., Growth hormone-releasing hormone, *Endocr. Rev.*, 7, 223, 1986.

118. Patel, Y. C. and Srikant, C. B., Somatostatin mediation of adenohypophysial secretion, *Ann. Rev. Physiol.*, 48, 551, 1986.

119. Mason, W. T. and Rawlings, S. R., Whole-cell recordings of ionic currents in bovine somatotrophs and their involvement in growth hormone secretion, *J. Physiol.*, 405, 577, 1988.

120. DeRiemer, S. A. and Sakmann, B., Two calcium currents in normal rat anterior pituitary cells identified by a plaque assay, *Exp. Brain Res.*, 14, 139, 1986.

121. Chen, C., Zhang, J., Vincent, J. D., and Israel, J. M., Two types of voltage-dependent calcium current in rat somatotrophes are reduced by somatostatin, *J. Physiol.*, 425, 29, 1990.

122. Lewis, D. L., Goodman, M. B., St. John, P. A., and Barker, J. L., Calcium currents and fura-2 signals in fluorescence-activated cell sorted lactotrophs and somatotrophs of rat anterior pituitary, *Endocrinology*, 123, 611, 1988.

123. Yamashita, N., Matsunaga, H., Shibuya, N., Teramoto, A., Takakura, K., and Ogata, E., Two types of calcium channels and hormone release in human pituitary tumor cells, *Am. J. Physiol.*, 255, E137, 1988.

124. Sims, S. M., Lussier, B. T., and Kraicer, J., Somatostatin activates an inwardly rectifying K^+ conductance in freshly dispersed rat somatotrophs, *J. Physiol.*, 441, 615, 1991.

125. Price, C. J., Goldberg, J. I., and Chang, J. P., Voltage-activated ionic currents in goldfish pituitary cells, *Gen. Comp. Endocrinol.*, 92, 16, 1993.

126. Chen, C., Heyward, P., Zhang, J., Wu, D. X., and Clarke, I. J., Voltage-dependent potassium currents in ovine somatotrophs and their function in growth hormone secretion, *Neuroendocrinology*, 59, 1, 1994.

127. Chen, C., Zhang, J., Vincent, J. D., and Israel, J. M., Somatostatin increases voltage-dependent potassium currents in rat somatotrophs, *Am. J. Physiol.*, 259, C854, 1990.

128. Yamashita, N., Shibuya, N., and Ogata, E., Requirement of GTP on somatostatin-induced K^+ current in human pituitary tumour cells, *Proc. Natl. Acad. Sci. U.S.A.*, 85, 4924, 1988.

129. Chen, C., Zhang, J., Vincent, J. D., and Israel, J. M., Sodium and calcium currents in action potentials of rat somatotrophs: their possible functions in growth hormone secretion, *Life Sci.*, 46, 983, 1990.

130. Yamashita, N., Shibuya, N., and Ogata, E., Hyperpolarization of the membrane potential caused by somatostatin in dissociated human pituitary adenoma cells that secrete growth hormone, *Proc. Natl. Acad. Sci. U.S.A.*, 83, 6198, 1986.

131. DeWeille, J. R., Fosset, M., Epelbaum, J., and Lazdunski, M., Effectors of ATP-sensitive K^+ channels inhibit the regulatory effects of somatostatin and GH-releasing factor on growth hormone secretion, *Biochem. Biophys. Res. Commun.*, 187, 1007, 1992.

132. Corrette, B. J., Davison, I., and Schwarz, J. R., An inward-rectifying K current which can be blocked by TRH is present in a population of rat anterior pituitary cells, in *Proceedings of the 22nd Göttingen Neurobiology Conference*, Vol. II, Elsner, N., and Breer, H., Eds., Georg Thieme Verlag, Stuttgart, 1994, 741.

133. Israel, J. M., Denef, C., and Vincent, J. D., Electrophysiological properties of normal somatotrophs in culture, *Neuroendocrinology*, 37, 193, 1983.

134. Chen, C., Israel, J. M., and Vincent, J. D., Electrophysiological responses of rat pituitary cells in somatotroph-enriched primary culture to human growth-hormone releasing factor, *Neuroendocrinology*, 50, 679, 1989.

135. Cobbett, P., Ingram, C. D., and Mason, W. T., Sodium and potassium currents involved in action potential propagation in normal bovine lactotrophs, *J. Physiol.*, 392, 273, 1987.

136. Nussinovitch, I., Growth hormone releasing factor evokes rhythmic hyperpolarizing currents in rat anterior pituitary cells, *J. Physiol.*, 395, 303, 1988.

137. Kato, M. and Suzuki, M., Growth hormone releasing factor depolarizes rat pituitary cells in Na^+-dependent mechanism, *Brain Res.*, 476, 145, 1989.

138. Kato, M., Hattori, M. A., and Suzuki, M., Inhibition by extracellular Na^+ replacement of GRF-induced GH secretion from rat pituitary cells, *Am. J. Physiol.*, 254, E476, 1988.

139. Kato, M., Hoyland, J., Sikdar, S. K., and Mason, W. T., Imaging of intracellular calcium in rat anterior pituitary cells in response to growth hormone releasing factor, *J. Physiol.*, 447, 171, 1992.

140. Chen, C. and Clarke, I. J., Minireview — ion channels in the regulation of growth hormone secretion from somatotrophs by somatostatin, *Growth Regul.*, 2, 167, 1992.

141. Chen, C., Israel, J. M., and Vincent, J. D., Electrophysiological responses to somatostatin of rat hypophysial cells in somatotroph-enriched primary cultures, *J. Physiol.*, 408, 493, 1989.

142. Rawlings, S. R., Canny, B. J., and Leong, D. A., Pituitary adenylate cyclase-activating polypeptide regulates cytosolic Ca^{2+} in rat gonadotropes and somatotropes through different intracellular mechanisms, *Endocrinology*, 132, 1447, 1993.

143. Miyata, A., Arimura, A., Dahl, D. H., Minamino, N., Uehara, A., Jiang, L., Culler, M. D., and Coy, D. H., Isolation of a novel 38-residue hypothalamic polypeptide which stimulates adenylate cyclase in pituitary cells, *Biochem. Biophys. Res. Commun.*, 164, 567, 1989.

144. Koves, K., Arimura, A., Somogyvari-Vigh, A., Vigh, S., and Miller, J., Immunohistochemical demonstration of a novel hypothalamic peptide, pituitary adenylate cyclase-activating polypeptide, in the ovine hypthalamus, *Endocrinology*, 127, 264, 1990

145. Gottschall, P. E., Tatsuno, I., Miyata, A., and Arimura, A., Characterization and distribution of binding sites for the hypothalamic peptide, pituitary adenylate cyclase-activating polypeptide, *Endocrinology*, 127, 272, 1990.

146. Canny, B. J., Rawlings, S. R., and Leong, D. A., Pituitary adenylate cyclase-activating polypeptide specifically increases cytosolic calcium ion concentration in rat gonadotropes and somatotropes, *Endocrinology*, 130, 211, 1992.

147. Goth, M. I., Lyons, C. E., Canny, B. J., and Thorner, M. O., Pituitary adenylate cyclase activating polypeptide, growth hormone (GH)-releasing peptide and GH-releasing hormone stimulate GH release through distinct pituitary receptors, *Endocrinology*, 130, 939, 1992.

148. Kineman, R. D., Faught, W. J., and Frawley, L. S., Bovine pituitary cells exhibit a unique form of somatotrope secretory heterogeneity, *Endocrinology*, 127, 2229, 1990.

149. Fluckiger, E., del Pozo, E., and von Werder, K., Prolactin: synthesis, fate and actions, in *Prolactin, Physiology and Clinical Findings*, Fluckiger, E., del Pozo, E., and von Werder, K., Eds., Springer-Verlag, Berlin, 1982, 1.

150. Neill, J. D. and Nagy, G. M., Prolactin secretion and its control, in *Physiology of Reproduction*, 2nd ed., Vols. 1 and 2, Knobil, E. and Neill, J. D., Eds., Raven Press, New York, 1994, Chap. 33.

151. Leong, D. A., Frawley, L. S., and Neill, J. D., Neuroendocrine control of prolactin secretion, *Annu. Rev. Physiol.*, 45, 109, 1983.

152. Tougard, C. and Tixier-Vidal, A., Lactotropes and gonadotropes, in *Physiology of Reproduction*, 2nd ed., Vols. 1 and 2, Knobil, E. and Neill, J. D., Eds., Raven Press, New York, 1994, chap. 29.

153. Boockfor, F. R. and Frawley, L. S., Functional variations among prolactin cells from different pituitary regions, *Endocrinology*, 120, 874, 1987.

154. Luque, E. H., DeToro, M. M., Smith, P. F., and Neill, J. D., Subpopulations of lactotropes detected with the reverse hemolytic plaque assay show differential responsiveness to dopamine, *Endocrinology*, 118, 2120, 1986.

155. Winiger, B. P., Wuarin, F., Zahnd, G. R., Wollheim, C. B., and Schlegel, W., Single cell monitoring of cytosolic calcium reveals subtypes of rat lactotrophs with distinct responses to dopamine and thyrotropin-releasing hormone, *Endocrinology*, 121, 2222, 1987.

156. Boockfor, F. R., Hoeffler, J. P., and Frawley, L. S., Analysis by plaque assays of GH and prolactin release from individual cells in cultures of male pituitaries, *Neuroendocrinology*, 42, 64, 1986.

157. Lamberts, S. W. J. and Macleod, R. M., Regulation of prolactin secretion at the level of the lactotroph, *Physiol. Rev.*, 70, 279, 1990.

158. Frawley, L. S. and Boockfor, F. R., Mammosomatotropes: presence and functions in normal and neoplastic pituitary tissue, *Endocr. Rev.*, 12, 337, 1991.

159. Hymer, W. C., Evans, W. H., Kraicer, J., Mastro, A., Davis, J., and Griswold, E., Enrichment of cell types from the rat adenohypophysis by sedimentation at unit gravity, *Endocrinology*, 92, 275, 1972.

160. Denef, C., Hautekeete, E., DeWolf, A., and Vaderschueren, B., Pituitary basophils from immature male and female rats: distribution of gonadotrophs and thyrotrophs as studied by unit gravity sedimentation, *Endocrinology*, 103, 724, 1978.

161. Lingle, C. J., Sombati, S., and Freeman, M. E., Membrane currents in identified lactotrophs of rat anterior pituitary, *Journal. Neurosci.*, 6, 2995, 1986.

162. Cobbett, P., Ingram, C. D., and Mason, W. T., Voltage-activated currents through calcium channels in normal bovine lactotrophs, *Neuroscience*, 23, 661, 1987.

163. Lledo, P. M., Legendre, P., Israel, J. M., and Vincent, J. D., Dopamine inhibits two characterized voltage-dependent calcium currents in identified rat lactotroph cells, *Endocrinology*, 127, 990, 1990.

164. Cota, G., Hiriart, M., Horta, J., and Torres-Escalante, J. L., Calcium channels and basal prolactin secretion in single male rat lactotropes, *Am. J. Physiol.*, 259, C949, 1990.

165. Horta, J., Hiriart, M., and Cota, G., Differential expression of Na channels in functional subpopulations of rat lactotropes, *Am. J. Physiol.*, 261, C865, 1991.

166. Lledo, P. M., Legendre, P., Zhang, J., Israel, J. M., and Vincent, J. D., Effects of dopamine on voltage-dependent potassium currents in identified rat lactotroph cells, *Neuroendocrinology*, 52, 545, 1990.

167. Sartor, P., Dufy-Barbe, L., Vacher, P., and Dufy, B., Calcium-activated chloride conductance of lactotrophs — comparison of activation in normal and tumoral cells during thyrotropin-releasing-hormone stimulation, *J. Membr. Biol.*, 126, 39, 1992.

168. Inenaga, K. and Mason, W. T., Chloride channels activated by gamma-aminobutyric acid in normal bovine lactotrophs, *Brain Res.*, 405, 159, 1987.

169. Israel, J. M., Kirk, C., and Vincent, J. D., Electrophysiological responses to dopamine of rat hypophysial cells in lactotroph-enriched primary cultures, *J. Physiol.*, 390, 1, 1987.

170. Einhorn, L. C., Gregerson, K. A., and Oxford, G. S., D_2 Dopamine receptor activation of potassium channels in identified rat lactotrophs: whole-cell and single-channel recording, *J. Neurosci.*, 11, 3727, 1991.

171. Israel, J. M., Jaquet, P., and Vincent, J. D., The electrical properties of isolated human prolactin-secreting adenoma cells and their modification by dopamine, *Endocrinology*, 117, 1448, 1985.

172. Ingram, C. D., Bicknell, R. J., and Mason, W. T., Intracellular recordings from bovine anterior pituitary cells: modulation of spontaneous activity by regulators of prolactin secretion, *Endocrinology*, 119, 2508, 1986.

173. Chen, G. G., St. John, P. A., and Barker, J. L., Rat lactotrophs isolated by fluorescence-activated cell sorting are electrically excitable, *Mol. Cell. Endocrinol.*, 51, 201, 1987.

174. Sartor, P., Dufy-Barbe, L., Corcuff, J. B., Taupignon, A., and Dufy, B., Electrophysiological response to thyrotropin-releasing hormone of rat lactotrophs in primary culture, *Am. J. Physiol.*, 258, E311, 1990.

175. Israel, J. M., Kukstas, L. A., and Vincent, J. D., Plateau potentials recorded from lactating rat enriched lactotroph cells are triggered by thyrotropin releasing hormone and shortened by dopamine, *Neuroendocrinology*, 51, 113, 1990.

176. Taraskevich, P. S. and Douglas, W. W., Action potentials occur in cells of the normal anterior pituitary gland and are stimulated by hypophysiotropic peptide thyrotropin-releasing hormone, *Proc. Natl. Academy of Sciences USA*, 74, 4064, 1977.

177. Malgaroli, A., Vallar, L., Elahi, F. R., Pozzan, T., Spada, A., and Meldolesi, J., Dopamine inhibits cytosolic Ca^{2+} increases in rat lactotroph cells. Evidence of a dual mechanism of action, *J. Biol. Chem.*, 262, 13920, 1987.

178. Felix, R., Horta, J., and Cota, G., Comparison of lactotrope subtypes of neonatal and adult male rats — plaque assays and patch-clamp studies, *Am. J. Physiol.*, 265, E120, 1993.

179. Zhang, J., Chen, C., Kukstas, L. A., Verrier, D., Vincent, J. D., and Israel, J. M., In vitro effects of 17β-estradiol on thyrotropin-releasing hormone-induced and dopamine-inhibited prolactin release from adult male rat lactotrophs in primary culture, *J. Neuroendocrinol.*, 2, 276, 1990.

180. Lledo, P. M., Guérineau, N., Mollard, P., Vincent, J. D., and Israel, J. M., Physiological characterization of two functional states in subpopulations of prolactin cells from lactating rats, *J. Physiol.*, 437, 477, 1991.

181. Ben-Jonathan, N., Dopamine: a prolactin-inhibiting hormone, *Endocr. Rev.*, 6, 564, 1985.

182. Dufy, B., Mollard, P., Dufy-Barbe, L., Manciet, G., Guerin, J., and Roger, P., The electrophysiological effects of TRH are similar in human TSH- and prolactin-secreting pituitary cells, *J. Clin. Endocrinol. Metab.*, 67, 1178, 1988.

183. Kukstas, L. A., Verrier, D., Zhang, J., Chen, C., Israel, J. M., and Vincent, J. D., Evidence for a relationship between lactotroph heterogeneity and physiological context, *Neurosci. Lett.*, 120, 84, 1990.

184. Enjalbert, A. and Bockaert, J., Pharmacological characterization of the D_2 dopamine receptor negatively coupled with adenylate cyclase in rat anterior pituitary, *Mol. Pharmacol.*, 23, 576, 1983.

185. Foord, S. M., Peters, J. R., Dieguez, C., Scanlon, M. F., and Hall, R., Dopamine receptors on intact anterior pituitary cells in culture: functional association with the inhibition of prolactin and thyrotropin, *Endocrinology*, 112, 1567, 1983.

186. Canonico, P. L., Valdenegro, C. A., and Macleod, R. M., The inhibition of phosphatidylinositol turnover: a possible post receptor mechanism for the prolactin secretion-inhibiting effect of dopamine, *Endocrinology*, 113, 7, 1983.

187. Enjalbert, A., Sladeczek, F., Guillon, G., Bertrand, P., Shu, C., Epelbaum, J., Garcia-Sanchez, A., Jard, S., Lombard, C., Kordon, C., and Bockaert, J., Angiotensin II and dopamine modulate both cAMP and inositol phosphate production in anterior pituitary cells, *J. Biol. Chem.*, 261, 4071, 1986.

188. Enjalbert, A., Guillon, G., Mouillac, B., Audinot, V., Rasolonjanahary, R., Kordon, C., and Bockaert, J., Dual mechanisms of inhibition by dopamine of basal and thyrotropin-releasing hormone-stimulated inositol phosphate production in anterior pituitary cells — evidence for an inhibition not mediated by voltage-dependent Ca^{2+} channels, *J. Biol. Chem.*, 265, 18816, 1990.

189. Simmonds, S. H. and Strange, P., Inhibition of inositol phospholipid breakdown by D_2 receptors in dissociated bovine anterior pituitary cells, *Neurosci. Lett.*, 60, 267, 1985.

190. Oxford, G. S. and Tse, A., Modulation of ion channels underlying excitation-secretion coupling in identified lactotrophs and gonadotrophs, *Biol. Reprod.*, 48, 1, 1993.

191. Einhorn, L. C. and Oxford, G. S., Guanine nucleotide binding proteins mediate D_2 dopamine receptor activation of a potassium channel in rat lactotrophs, *J. Physiol.*, 462, 563, 1993.

192. Gilman, A. G., G proteins: transducers of receptor-generated signals, *Annu. Rev. Biochem.*, 56, 615, 1987.

193. Lledo, P. M., Homburger, V., Bockaert, J., and Vincent, J. D., Differential G protein-mediated coupling of D_2 dopamine receptors to K^+ and Ca^{2+} currents in rat anterior pituitary cells, *Neuron*, 8, 455, 1992.

194. Baertschi, A. J., Audiger, Y., Lledo, P. M., Israel, J. M., Bockaert, J., and Vincent, J. D., Dialysis of lactotropes with antisense oligonucleotides assigns guanine nucleotide binding protein subtypes to their channel effectors, *Mol. Endocrinol.*, 6, 2257, 1992.

195. Giros, B., Sokologg, P., Martres, M. P., Riou, J. F., Emorine, L. J., and Schwartz, J. C., Alternative splicing directs the expression of two D_2 dopamine receptor isoforms, *Nature*, 342, 923, 1989.

196. Monsma, F. J., McVittie, L., Gerfen, C., Mahan, L., and Sibley, D., Multiple D_2 dopamine receptors produced by alternative RNA splicing, *Nature*, 342, 926, 1989.

197. Kukstas, L. A., Domec, C., Bascles, L., Bonnet, J., Verrier, D., Israel, J. M., and Vincent, J. D., Different expression of the two dopaminergic D_2 receptors, D_2415 and D_2444, in two types of lactotroph each characterised by their response to dopamine, and modification of expression by sex steroids, *Endocrinology*, 129, 1101, 1991.

198. Shin, S. H., Dopamine-induced inhibition of prolactin release from cultured adenohypophysial cells: spare receptors for dopamine, *Life Sci.*, 22, 67, 1978.

199. Denef, C., Manet, D., and Dewals, R., Dopaminergic stimulation of prolactin release, *Nature*, 285, 243, 1980.

200. Burris, T. P., Stringer, L. C., and Freeman, M. E., Pharmacologic evidence that a D_2-receptor subtype mediates dopaminergic stimulation of prolactin secretion from the anterior pituitary gland, *Neuroendocrinology*, 54, 175, 1991.

201. Burris, T. P., Nguyen, D. N., Smith, S. G., and Freeman, M. E., The stimulatory and inhibitory effects of dopamine on prolactin secretion involve different G-proteins, *Endocrinology*, 130, 926, 1992.

202. Porter, T. E., Grandy, D., Bunzow, J., Wiles, C. D., Civelli, O., and Frawley, L. S., Evidence that stimulatory dopamine receptors may be involved in the regulation of prolactin secretion, *Endocrinology*, 134, 1263, 1994.

203. Burris, T. P. and Freeman, M. E., Low concentrations of dopamine increase cytosolic calcium in lactotrophs, *Endocrinology*, 133, 63, 1993.

204. Lledo, P. M., Israel, J. M., and Vincent, J. D., Chronic stimulation of D_2 dopamine receptors specifically inhibits calcium but not potassium currents in rat lactotrophs, *Brain Res.*, 558, 231, 1991.

205. Schwartz, J. and Cherny, R., Intercellular communication within the anterior pituitary influencing the secretion of hypophysial hormones, *Endocr. Rev.*, 13, 453, 1992.

206. Kaplan, L. M., Gabriel, S. M., Koenig, J. I., Sunday, M. E., Spindel, E. R., Martin, J. B., and Chin, W. W., Galanin is an estrogen-inducible, secretory product of the rat anterior pituitary, *Proc. Natl. Acad. Sci. U.S.A.*, 85, 7408, 1988.

207. Vrontakis, M. E., Peden, L. M., Duckworth, M. L., and Freisen, H. G., Isolation and characterization of a complementary DNA (galanin) clone from estrogen-induced pituitary tumor messenger RNA, *J. Biol. Chem.*, 262, 16755, 1987.

208. Koshiyama, H., Kato, Y., Inoie, T., Murakami, Y., Ishikawa, Y., Yanaihara, N., and Imura, I., Central galanin stimulates pituitary prolactin secretion in rats: possible involvement of hypothalamic vasoactive intestinal polypeptide, *Neurosci. Lett.*, 75, 49, 1987.

209. Melander, T., Fuxe, K., Harfstrand, A., Eneroth, P., and Hokfelt, T., Effects of intraventricular injections of galanin on neuroendocrine functions in the male rat: possible involement o hypothalamic catecholamine neuronal systems, *Acta Physiol. Scand.*, 131, 25, 1987.

210. Steel, J. H., Gon, G., O'Halloran, D. J., Jones, P. M., Yanaihara, N., and Ishikawas, H., Galanin and vasoactive intestinal polypeptide are co-localized with classical pituitary hormones and show plasticity of expression, *Histochemistry*, 93, 183, 1989.

211. McArdle, C. A., Bunting, R., and Mason, W. T., Dynamic video imaging of cystolic Ca^{2+} in the $\alpha t3$-1, gonadotrope-derived cell line, *Mol. Cell. Neurosci.*, 3, 124, 1992.

212. Tse, A. and Hille, B., GnRH-induced Ca^{2+} oscillations and rhythmic hyperpolarizations of pituitary gonadotropes, *Science*, 255, 462, 1992.

213. Tse, A. and Hille, B., Role of voltage-gated Na^+ and Ca^{2+} channels in gonadotropin-releasing hormone-induced membrane potential changes in identified rat gonadotropes, *Endocrinology*, 132, 1475, 1993.

214. Chen, C., Zhang, J., Dayanithi, G., Vincent, J. D., and Israel, J. M., Cationic currents in identified rat gonadotroph cells maintained in primary culture, *Neurochem. Int.*, 15, 265, 1989.

215. Kukuljan, M., Stojilkovic, S. S., Rojas, E., and Catt, K. J., Apamin-sensitive potassium channels mediate agonist-induced oscillations of membrane potential in pituitary gonadotrophs, *FEBS Lett.*, 301, 19, 1992.

216. Mason, W. T. and Sikdar, S.K., Characterization of voltage-gated sodium channels in ovine gonadotrophs: relationship to hormone secretion, *J. Physiol.*, 399, 493, 1988.

217. Bosma, M. M. and Hille, B., Electrophysiological properties of a cell line of the gonadotrope lineage, *Endocrinology*, 130, 3411, 1992.

218. Llinàs, R., Sugimori, M., Lin, J., and Cherksey, B., Blocking and isolation of a calcium channel from neurons in mammals and cephalopods ultilizing a toxin fraction (FTX) from funnel-web spider poison, *Proc. Natl. Acad. Sci. U.S.A.*, 86, 1689, 1989.

219. Williams, P. J., Pittman, Q. J., and MacVicar, B. A., Blockade by funnel web toxin of a calcium current in the intermediate pituitary of the rat, *Neurosci. Lett.*, 157, 171, 1993.

220. Marchetti, C., Childs, G. V., and Brown, A. M., Voltage-dependent calcium currents in rat gonadotropes separated by centrifugal elutriation, *Am. J. Physiol.*, 258, E589, 1990.

221. Windle, J. J., Weiner, R. I., and Mellon, P. L., Cell lines of the pituitary gonadotrope lineage derived by targeted oncogenesis in transgenic mice, *Mol. Endocrinol.*, 4, 597, 1990.

222. Anderson, L., Hoyland, J., Mason, W. T., and Eidne, K. A., Characterization of the gonadotrophin-releasing hormone calcium response in single alphaT3-1 pituitary gonadotroph cells, *Mol. Cell. Endocrinol.*, 86, 167, 1992.

223. Merelli, F., Stojilkovic, S. S., Iida, T., Krsmanovic, L. Z., Zheng, L., Mellon, P. L., and Catt, K. J., Gonadotropin-releasing hormone- induced calcium signaling in clonal pituitary gonadotrophs, *Endocrinology*, 131, 925, 1992.

224. Ben-Menahem, D. and Naor, Z., Regulation of gonadotropin messenger RNA levels in cultured rat pituitary cells by gonadotropin-releasing hormone (GnRH) — role for Ca^{2+} and protein kinase-C, *Biochemistry*, 33, 3698, 1994.

225. Guérineau, N., Corcuff, J. B., Tabarin, A., and Mollard, P., Spontaneous and corticotropin-releasing factor-induced cytosolic calcium transients in corticotrophs, *Endocrinology*, 129, 409, 1991.

226. Corcuff, J. B., Guérineau, N. C., Mariot, P., Lussier, B. T., and Mollard, P., Multiple cytosolic calcium signals and membrane electrical events evoked in single arginine vasopressin-stimulated corticotrophs, *J. Biol. Chem.*, 268, 22313, 1993.

227. Carvallo, P. and Aguilera, G., Protein kinase-C mediates the effect of vasopressin in pituitary corticotrophs, *Mol. Endocrinol.*, 3, 1935, 1989.

228. Childs, G. V., Structure-function correlates in the corticotropes of the anterior pituitary, *Front. Neuroendocrinol.*, 13, 271, 1992.

229. Adler, M., Wong, B. S., Sabol, S. L., Busis, N., Jackson, M. B., and Weight, F. F., Action potentials and membrane ion channels in clonal anterior pituitary cells, *Proc. Natl. Acad. Sci. U.S.A.*, 80, 2086, 1983.

230. Adler, M., Sabol, S. L., Busis, N., and Pant, H. C., Intracellular calcium and hormone secretion in clonal AtT-20/D16-16 anterior pituitary cells, *Cell Calcium*, 10, 467, 1989.

231. Reisine, T., Cellular mechanisms of somatostatin inhibition of calcium influx in the anterior pituitary cell line AtT-20, *J. Pharmacol. Exp. Ther.*, 254, 646, 1990.

232. Luini, A. and Brown, D. A., Effects of corticotrophin releasing factor, muscarine and somatostatin on rubidium and potassium efflux from mouse AtT-20 pituitary cells, *Eur. J. Neurosci.*, 2, 126, 1990.

233. Korn, S. J., Bolden, A., and Horn, R., Control of action potentials and Ca^{2+} influx by the Ca^{2+}-dependent chloride current in mouse pituitary cells, *J. Physiol.*, 439, 423, 1991.

234. Korn, S. J. and Horn, R., A Na^+-independent, pH-dependent mechanism for reduction of intracellular Ca^{+2} after influx through Ca^{2+} channels in mouse pituitary cells, *J. Gen. Physiol.*, 98, 893, 1991.

235. Dousmanis, A. G. and Pennefather, P. S., Inwardly rectifying potassium conductances in AtT-20 clonal pituitary cells, *Pflügers Arch.*, 422, 98, 1992.

236. Harvey, R. D. and Ten Eick, R. E., Characterization of the inward-rectifying potassium current in ventricular myocytes, *J. Gen. Physiol.*, 91, 593, 1988.

237. Weik, R. and Spiess, J., Isoproterenol enhances a calcium-independent potassium current in mouse anterior pituitary tumor cells, *J. Neurophysiol.*, 68, 117, 1992.

238. Luini, A. and DeMatteis, M. A., Evidence that receptor-linked G-protein inhibits exocytosis by a post-2nd-messenger mechanism in AtT-20 cells, *J. Neurochem.*, 54, 30, 1990.

239. Lewis, D. L., Weight, F. F., and Luini, A., A guanine nucleotide-binding protein mediates the inhibition of voltage-dependent calcium current by somatostatin in a pituitary cell line, *Proc. Natl. Acad. Sci. U.S.A.*, 83, 9035, 1986.

240. Borst, J. G. G., Lodder, J. C., and Kits, K. S., Large amplitude variability of GABAergic IPSCs in melanotropes from *Xenopus laevis* — evidence that quantal size differs between synapses, *J. Neurophysiol.*, 71, 639, 1994.

241. Jenks, B. G., Verburg-van Kemenade, B. M. L., and Martens, G. J. M., Proopiomelanocortin in the amphibian pars intermedia: a neuroendocrine model system, in *The Melanotropic Peptides,* Vol. I, Hadley, M. E., Ed., CRC Press, Boca Raton, FL, 1988, 67.

242. Baumgarten, H. G., Björklund, A., Holstein, A. F., and Nobin A., Organization and ultrastructural identification of the catecholamine nerve terminals in the neural lobe and pars intermedia of the rat pituitary, *Z. Zellforsch.,* 126, 483, 1972.

243. De Rijk, E. P. C. T., Jenks, B. G., Vaudry, H., and Roubos, W. W., GABA and neuropeptide Y coexist in axons innervating the neurointermediate lobe of the pituitary of *Xenopus laevis, Neuroscience,* 38, 495, 1990.

244. De Rijk, E. P. C. T., Van Strien, F. J. C., and Roubos, E. W., Demonstration of coexisting catecholamine (dopamine), aminoacid (GABA) and peptide (NPY) involved in inhibition of melantotrope cell activity in *Xenopus laevis, J. Neurosci.,* 12, 864, 1992.

245. Valentijn, J. A., Vaudry, H., Kloas, W., and Cazin, L., Melanostatin (NPY) inhibited electrical activity in frog melanotrophs through modulation of K, Na and Ca currents, *J. Physiol.,* 475, 185, 1994.

246. Shibuya, I., Douglas and W. W., Spontaneous cytosolic calcium pulsing detected in *Xenopus* melanotrophs — modulation by secreto-inhibitory and stimulant ligands, *Endocrinology,* 132, 2166, 1993.

247. Valentijn, J. A., Louiset, E., Vaudry, H. and Cazin, L., Dopamine regulates the electrical activity of frog melanotrophs through a G-protein-mediated mechanism, *Neuroscience,* 44, 85, 1991.

248. Valentijn, J. A., Louiset, E., Vaudry, H., and Cazin, L., Dopamine-induced inhibition of action potentials in cultured frog pituitary melanotrophs is mediated through activation of potassium channels and inhibition of calcium and sodium channels, *Neuroscience,* 42, 29, 1991.

249. Stack, J. and Surprenant, A., Dopamine actions on calcium currents, potassium currents and hormone release in rat melanotrophs, *J. Physiol.,* 439, 37, 1991.

250. Schneggenburger, R. and Konnerth, A., GABA-mediated synaptic transmission in neuroendocrine cells — a patch-clamp study in a pituitary slice preparation, *Pflügers Arch.,* 421, 364, 1992.

251. Louiset, E., Mei, Y. A., Valentijn, J. A., Vaudry, H., and Cazin, L., Characterization of the GABA-induced current in frog pituitary melanotrophs, *J. Neuroendocrinol.,* 6, 39, 1994.

252. Louiset, E., Cazin, L., Lamacz, M., Tonon, M. C., and Vaudry, H., Dual effects of thyrotrophin-releasing hormone (TRH) on K^+ conductance in frog pituitary melanotrophs — TRH-induced alpha-melanocyte-stimulating hormone release is not mediated through voltage-sensitive K^+ channels, *J. Mol. Endocrinol.,* 3, 207, 1989.

253. Williams, P. J., MacVicar, B. A., and Pittman, Y. J., Identification of a GABA-activated chloride-mediated synaptic potential in rat pars intermedia, *Brain Res.,* 483, 130, 1989.

254. Schneggenburger, R. and Lopez-Barnéo, J., Patch-clamp analysis of voltage-gated currents in intermediate lobe cells from rat pituitary thin slices, *Pflügers Arch.,* 420, 302, 1992.

255. DeAguilar, J. L. G., Gracianavarro, F., Tonon, M. C., Ruiznavarro, A., and Vaudry, H., Morphological and functional heterogeneity of frog melanotrope cells, *Neuroendocrinology,* 59, 176, 1994.

256. Douglas, W. W. and Taraskevich, P. S., Action potentials in gland cells of rat pituitary pars intermedia: inhibition by dopamine, an inhibitor of MSH secretion, *J. Physiol.,* 285, 171, 1978.

257. Scheenen, W. J. J. M., Jenks, B. G., Roubos, E. W., and Willems, P. H. G. M., Spontaneous calcium oscillations in Xenopus laevis melanotrope cells are mediated by ω-conotoxin sensitive calcium channels, *Cell Calcium,* 15, 36, 1994.

258. Louiset, E., Cazin, L., Lamacz, M., Tonon, M. C., and Vaudry, H., Patch-clamp study of the ionic currents underlying action potentials in cultured frog pituitary melanotrophs, *Neuroendocrinology,* 48, 507, 1988.

259. Kehl, S. J., Voltage-clamp analysis of the voltage-gated sodium current of the rat pituitary melanotroph, *Neurosci. Lett.,* 165, 67, 1994.

260. Taleb, O., Trouslard, J., Demeneix, B. A., and Feltz, P., Characterization of calcium and sodium currents in porcine pars intermedia cells, *Neurosci. Lett.,* 66, 55, 1986.

261. Kehl, S. J., Catechol blocks the fast outward potassium current in melanotrophs of the rat pituitary, *Neurosci. Lett.,* 125, 136, 1991.

262. Cota, G., Calcium channel currents in pars intermedia cells of the rat pituitary gland: kinetic properties and washout during intracellular dialysis, *J. Gen. Physiol.,* 88, 83, 1986.

263. Keja, J. A., Stoof, J. C., and Kits, K. S., Voltage-activated currents through calcium channels in rat pituitary melanotrophic cells, *Neuroendocrinology,* 53, 349, 1991.

264. Keja, J. A., Stoof, J. C., and Kits, K. S., Dopamine D_2 receptor stimulation differentially affects voltage-activated calcium channels in rat pituitary melanotropic cells, *J. Physiol.,* 450, 409, 1992.

265. Kocmur, L. and Zorec, R., A new approach to separation of voltage-activated Ca currents in rat melanotrophs, *Pflügers Arch.,* 425, 172, 1993.

266. Bean, B. P., Classes of calcium channels in vertebrate cells, *Annu. Rev. Physiol.,* 51, 367, 1989.

267. Keja, J. A. and Kits, K. S., Single-channel properties of high- and low-voltage-activated calcium channels in rat pituitary melanotropic cells, *J. Neurophysiol.,* 71, 840, 1994.

268. Nussinovitch, I. and Kleinhaus, A. L., Dopamine inhibits voltage-activated calcium channel currents in rat pars-intermedia pituitary cells, *Brain Res.*, 574, 49, 1992.

269. Shibuya, I. and Douglas, W. W., Spontaneous cytosolic calcium pulses in Xenopus melanotrophs are due to calcium influx during phasic increases in the calcium permeability of the cell membrane, *Endocrinology*, 132, 2176, 1993.

270. Taleb, O., Feltz, P., Bossu, J. L., and Feltz, A., Small-conductance chloride channels activated by calcium on cultured endocrine cells from mammalian pars intermedia, *Pflügers Arch.*, 412, 641, 1988.

271. Takahashi, T., Neher, E., and Sakman, B., Rat brain serotonin receptors in Xenopus oocytes are coupled by intracellular calcium to endogenous channels, *Proc. Natl. Acad. Sci. U.S.A.*, 84, 5063, 1987.

272. Saland, L. C., Extracellular spaces of the rat pars intermedia as outlined by lanthanum tracer, *Anat. Rec.*, 196, 355, 1980.

273. Boyd, W. H. and Krogsrud, R., Presence of α-melanocyte-stimulating hormone in bovine pituitary intraglandular colloid of intermediate lobe origin, *Am. J. Anat.*, 178, 81, 1987.

274. Shibuya, I., Kongsamut, S., and Douglas, W. W., Effectiveness of $GABA_B$ antagonists in inhibiting baclofen-induced reductions in cytosolic free Ca concentration in isolated melanotrophs of rat, *Br. J. Pharmacol.*, 105, 893, 1992.

275. Danger, J. M., Lamacz, M., Mauviard, F., Saintpierre, S., Jenks, B. G., Tonon, M. C., and Vaudry, H., Neuropeptide-Y inhibits thyrotropin-releasing hormone-induced stimulation of melanotropin release from the intermediate lobe of the frog pituitary, *Gen. Comp. Endocrinol.*, 77, 143, 1990.

276. Cote, T. E., Grewe, C. W., Tsuruta, K., Stoof, J. C., Eskay, R. L., and Kebabian, J. W., D_2 dopamine receptor-mediated inhibition of adenylate cyclase activity in the intermediate lobe of the rat pituitary gland requires GTP, *Endocrinology*, 110, 812, 1982.

277. Valentijn, J. A., Louiset, E., Vaudry, H., and Cazin, L., Voltage-dependent modulation of calcium current by $GTP_\gamma S$ and dopamine in cultured frog pituitary melanotrophs, *Neurosci. Lett.*, 138, 216, 1992.

278. Cota, G. and Hiriart, M., Hormonal and neurotransmitter regulation of Ca channel activity in cultured adenohypophyseal cells, in *Secretion and Its Control*, Oxford, G. S. and Armstrong, C. M., Eds., Rockefeller University Press, New York, 1989, Chap. 9.

279. Lundberg, J. M. and Hökfelt, T., Multiple co-existence of peptides and classical neurotransmitters in peripheral autonomic and sensory neurons — functional and pharmacological implications, *Prog. Brain Res.*, 68, 241, 1986.

280. Valentijn, J. A., Vaudry, H., and Cazin, L., Adrenaline induces hyperpolarization in frog pituitary melanotrophs through activation of potassium channels, *Neuroendocrinology*, 59, 20, 1994.

Section IV
Thyroid and Parathyroid Glands

Chapter 7

VOLTAGE-DEPENDENT CALCIUM CHANNELS AND EXTRACELLULAR CALCIUM SENSING IN C CELLS OF THE THYROID

F. Raue, J. Hescheler, and H. Scherübl

CONTENTS

1. SUMMARY

An essential function of calcitonin (CT)-secreting, parafollicular cells of the thyroid (C cells) is to monitor the extracellular calcium concentration $[Ca^{2+}]_e$ and to increase CT secretion in response to small increments in $[Ca^{2+}]_e$. CT, in turn, decreases $[Ca^{2+}]_e$ via its effects on bone and kidney, thus maintaining a tightly balanced $[Ca^{2+}]_e$. $[Ca^{2+}]_e$-dependent CT secretion is known to be mediated by corresponding changes in intracellular calcium concentrations $[Ca^{2+}]_i$. The $[Ca^{2+}]_e$ sensing of C cells is mediated by dihydropyridine (DHP)-sensitive, voltage-dependent Ca^{2+} channels which allow Ca^{2+} influx even at the resting membrane potential. An increase of $[Ca^{2+}]_e$ stimulates transmembrane Ca^{2+} influx via DHP-sensitive Ca^{2+} channels, thereby increasing $[Ca^{2+}]_i$ and consequently CT secretion. Moreover, $[Ca^{2+}]_e$- and cAMP-dependent oscillations of $[Ca^{2+}]_i$ are observed in C cells. Various neuropeptides and hormones involved in the control of CT secretion act by regulating Ca^{2+} channel activity via G proteins. One of the endogenous modulators is somatostatin (SST), which is produced by C cells themselves, and tonically inhibits CT secretion by inhibiting voltage-dependent Ca^{2+} channels and adenylyl cyclases (AC).

2. INTRODUCTION

2.1. ANATOMY AND EMBRYOLOGY OF C CELLS

The CT-secreting, parafollicular cells of the thyroid (C cells) constitute only a few percent of the total number of thyroid cells in humans. They are unevenly distributed throughout the gland and are concentrated in the lateral upper two thirds of the gland. C cells are of neural crest origin and derive from the ventral portion of the last branchial pouch, which migrates to and fuses with the thyroid early in embryologic development. In all submammalian species, the ventral portion of the last branchial pouch forms a separate gland known as the ultimobranchial body. The distinctive feature of C cells compared with thyroid follicular cells is the presence of numerous fine secretory granules in the cytoplasm. The granules can be demonstrated to contain CT by means of immunohistochemical techniques.

The conceived physiological role of C cells is to sense changes in $[Ca^{2+}]_e$ and to keep $[Ca^{2+}]_e$ within its physiological range of 1.1–1.3 mM by secreting the $[Ca^{2+}]_e$-lowering hormone CT.

2.2. EXTRACELLULAR CA²⁺ AS A MAIN REGULATOR OF CALCITONIN SECRETION

CT is a polypeptide hormone composed of 32 amino acid residues with a disulfide bridge between positions 1 and 7 and a carboxy terminal proline amide. The concentration of Ca^{2+} ions in plasma and extracellular fluids is the principal physiological stimulus for the secretion of CT by C cells. When serum Ca^{2+} rises acutely, there is a proportional increase in serum CT;[1,2] this holds true for both humans[3] and various species of animals.[4] This Ca^{2+} sensitivity is the basis for using Ca^{2+} infusion as a provocative test for CT release in patients with suspected C cell carcinoma (synonym: medullary thyroid carcinoma [MTC]).[5]

In contrast, the effect of chronic hypercalcemia is controversial, and contradictory results have been reported. Normal,[6] elevated,[7] and decreased[8] basal levels of CT have been observed in patients with hypercalcemia due to primary hyperparathyroidism. During chronic hypercalcemia in the rat, a decrease in the CT response to a Ca^{2+} load and a reversible exhaustion of the CT content of the thyroid was observed, while basal serum CT levels remained unchanged.[9] These results indicated that C cells either have a diminished secretory capacity or a decreased sensitivity for Ca^{2+} during chronic hypercalcemia.

In patients with chronic hypocalcemia, such as occurs in hypoparathyroidism, Ca^{2+} infusion results in a greater increase in serum CT than found in controls.[10] Chronic hypocalcemia enhances CT storage and stimulated secretion in the rat.[11] Thus, CT storage and CT response to short-term stimuli show an inverse relationship to serum Ca^{2+} levels.

2.3. ACTION AND PHYSIOLOGICAL FUNCTION OF CALCITONIN

The major target cell for CT action is the osteoclast. CT inhibits osteoclast activity[12] and thereby suppresses serum Ca^{2+} levels. In addition, CT acts on the kidneys by increasing Ca^{2+} excretion, resulting in a decrease of the serum Ca^{2+} concentration, and stimulating $1,25(OH)_2D_3$ production.[13] On the other hand, receptors for $1,25(OH)_2D_3$ are present on thyroid C cells, suggesting that vitamin D may also have a regulatory function on CT secretion. In a feedback loop, $1,25(OH)_2D_3$ decreases the transcription rate of the CT gene,[14] followed by a decrease in CT mRNA and, consequently, reduced CT content and CT secretion. The following two feedback relationships exist between C cells and the whole organism: first, an acute rise in serum Ca^{2+} is the adequate stimulus for CT secretion, which in turn decreases and thus normalizes serum Ca^{2+} rapidly. Second, CT stimulates $1,25(OH)_2D_3$ synthesis in the kidney. $1,25(OH)_2D_3$ in turn inhibits both CT synthesis and secretion and increases serum Ca^{2+}. The overall physiological function of CT thus appears to be protection of the skeleton from excessive osteoclastic bone resorption, especially during periods of increased calcium flux such as pregnancy, lactation, and childhood growth when serum CT levels are elevated.

Together with parathyroid hormone (PTH) and $1,25(OH)_2D_3$, CT plays an important role in the complex regulatory network[2] that maintains extracellular Ca^{2+} ($[Ca^{2+}]_e$) within a very narrow concentration range. C cells and parathyroid cells perform this function by monitoring the Ca^{2+} levels in the blood. In turn, PTH and CT secretion are regulated by serum Ca^{2+} itself. Extracellular Ca^{2+}-dependent secretion of CT and PTH is known to be mediated by corresponding changes in intracellular calcium ($[Ca^{2+}]_i$). The coupling of $[Ca^{2+}]_e$ to $[Ca^{2+}]_i$ appears to be an essential mechanism in establishing Ca^{2+} sensitivity in C cells and parathyroid cells. In C cells the coupling is established by voltage-dependent Ca^{2+} channels.

Most of the current knowledge of $[Ca^{2+}]_e$-sensing in C cells was obtained from studies of permanent C cell carcinoma cell lines (human TT, rat MTC [rMTC] 6–23, 44–2).[15,16]

3. ELECTRICAL PROPERTIES OF C CELLS

Electrophysiological studies of mammalian C cells have been aggravated by difficulty in isolating C cells. Tischler et al.[17] and Sand et al.[18] were the first to record action potentials in C cells derived from MTC. The action potentials elicited by membrane depolarization had both a Na^+ and a Ca^{2+} component. Based on their studies, Sand et al.[19] hypothesized already in 1986 that an "elevated plasma Ca^{2+} concentration within the physiological range might cause membrane depolarization and resistance increase in normal CT-producing cells, thus leading to elevated excitability and increased secretory activity."

3.1. VOLTAGE-DEPENDENT CA^{2+} CHANNELS IN CA^{2+}-SENSITIVE C CELLS

The establishment of rat and human C cell lines has provided extensively used models of C cell function.[15,16] C cells of the Ca^{2+}-sensitive rMTC cell line rMTC 44–2 express voltage-dependent, dihydropyridine (DHP)-sensitive, slowly inactivating Ca^{2+} currents that are essentially involved in the extracellular Ca^{2+} sensitivity[20] (Figure 1). The outstanding property of these Ca^{2+} channels are their slow inactivation kinetics. Both whole-cell and single-channel recordings by the patch clamp technique[21] demonstrate steady-state ion influx through these voltage-dependent DHP-sensitive Ca^{2+} channels even at physiological membrane potentials of -40 mV. Several conductance states of DHP-sensitive Ca^{2+} channels contribute to the steady-state Ca^{2+} influx necessary for Ca^{2+}-dependent CT release from C cells: single-channel recordings reveal multiple conductance levels of the DHP- sensitive steady-state Ca^{2+} currents.[22] The dependency of Ca^{2+} influx through DHP-sensitive, voltage-dependent Ca^{2+} channels

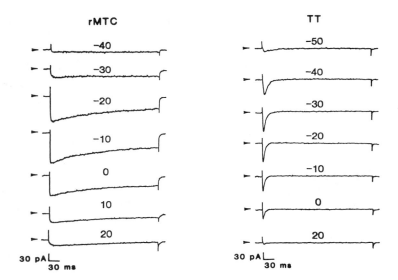

FIGURE 1. Whole-cell recordings of Ca^{2+} currents in C cells. Original current traces of Ca^{2+}-sensitive rMTC and Ca^{2+}-insensitive TT cells are shown. Ca^{2+} currents were elicited by 300 ms-long voltage clamp pulses from −80 mV to the various test potentials indicated by the numbers. The membrane patch under the tip of the pipette was disrupted and free access to the cytoplasm was obtained. Solutions: pipette solution I3 for both rMTC and TT cells; external solutions E2 (1.2 mM Ca^{2+}) for rMTC and E3 (10.8 mM Ca^{2+}) for TT cells. (From Scherübl, H. et al., *FEBS Lett.*, 273, 51–54, 1990. With permission.)

on the $[Ca^{2+}]_e$ is evidenced by whole-cell steady-state recordings. To avoid major disturbances of the cytoplasm and a "run-down" of the Ca^{2+} channel currents,[23] the cytoplasm has to be accessed by the nystatin modification of the patch clamp technique.[24] To suppress steady-state K^+ currents, Ca^{2+} has to be substituted by Ba^{2+} as the divalent charge carrier. Under these conditions, increases in extracellular Ba^{2+} produce a steady-state inward current which is maximal after about 5 s and then slowly decays (Figure 2).[20] The steady-state Ba^{2+} inward current is DHP sensitive[25] and displays a U-shaped voltage dependence with a threshold of about −50 mV. Thus, the steady-state, whole-cell Ca^{2+} channel current enters the cell via DHP-sensitive, voltage-dependent Ca^{2+} channels. These Ca^{2+} channels couple changes in $[Ca^{2+}]_e$ to changes in $[Ca^{2+}]_i$. Thereby, they enable C cells to monitor $[Ca^{2+}]_e$ and to respond to changes in $[Ca^{2+}]_e$ with corresponding changes in CT secretion.

3.2. VOLTAGE-DEPENDENT CA²⁺ CHANNELS IN CA²⁺-INSENSITIVE C CELLS

The concept that DHP-sensitive, voltage-dependent Ca^{2+} channels are essential for the extracellular Ca^{2+} sensitivity of C cells is substantiated by the existence of the "defective", i.e., $[Ca^{2+}]_e$-insensitive C cell line TT (see also Section 4.1.2). The defect in Ca^{2+} signal transduction in TT cells has been localized to the plasma membrane; bypassing the plasma membrane by electropermeabilization or by Ca^{2+} ionophores "restores" the $[Ca^{2+}]_e$ sensitivity of TT cells.[26] Hypothesizing that TT cells might have a "defect" or lack of voltage-dependent Ca^{2+} channels, we performed the first electrophysiological studies. Surprisingly, TT cells displayed DHP-insensitive, voltage-dependent Ca^{2+} currents with a low activation threshold and fast inactivation kinetics.[20] In contrast to the situation in rMTC cells, whose depolarization-induced Ca^{2+} currents show hardly any inactivation around the resting potential, the voltage-dependent Ca^{2+} currents of TT cells quickly inactivated (Figure 1).[20] Steady-state, whole-cell experiments of TT cells have failed to demonstrate any steady-state conductivity of the fast inactivating, so-called T-type Ca^{2+} channels. Increases in extracellular Ba^{2+} did not induce any

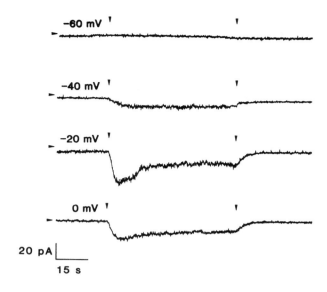

FIGURE 2. Steady-state, whole-cell Ca^{2+} channel currents in a single rMTC cell at various holding potentials: effect of rising the concentration of external Ba^{2+} concentration. To obtain steady-state conditions the (Ca^{2+}-sensitive) rMTC cell was voltage clamped at the respectively indicated potential for 2 min before Ba^{2+} was raised. For the 1-min periods marked in between the vertical arrows, the external Ba^{2+} was respectively raised from 1.2 mM (solution E5) to 10.8 mM (solution E6) and then again lowered to 1.2 mM for 2 min. To avoid a major disturbance of the cytoplasm and to avoid "run-down" of the Ca^{2+} channel current, the cytoplasm was assessed by the nystatin modification of the patch clamp technique. Pipette solution I2 (with 100–200 µg/ml nystatin). Horizontal arrows mark the zero current level. (From Scherübl, H. et al., *FEBS Lett.*, 273, 51–54, 1990. With permission.)

discernible steady-state inward current in TT cells. Consistent with this, T-type Ca^{2+} channels are known to inactivate already at potentials negative to −50 mV,[27] whereas the resting potential of TT cells is −37.9 ± 1.8 mV.

3.3. EXTRACELLULAR CA^{2+} SENSITIVITY AND SPONTANEOUS ELECTRICAL ACTIVITY

Although TT cells can produce an action potential-like wave form in response to membrane depolarization,[28] they do not show any spontaneously occurring action potentials. Changes in $[Ca^{2+}]_e$ within the physiological range do not affect the membrane potential of TT cells (Figure 3).[20] In contrast, in rMTC cells a rise of $[Ca^{2+}]_e$ from 1.2 to 1.8 mM elicits action potentials and/or depolarizes the cells by about 10 mV.[20,29] The effect of high $[Ca^{2+}]_e$ is mimicked by the DHP Ca^{2+} channel opener BAY K 8644, whereas the DHP Ca^{2+} channel blocker isradipine (PN 200–110) reversibly suppresses $[Ca^{2+}]_e$-induced action potentials in rMTC cells. Despite the coincidence of $[Ca^{2+}]_e$-induced action potentials and $[Ca^{2+}]_e$-induced rises of $[Ca^{2+}]_i$, action potentials per se do not appear to be essential for Ca^{2+} influx and Ca^{2+}-dependent CT release. Even with spontaneous electrical activity suppressed by tetrodotoxin, a blocker of voltage-dependent Na^+ channels, CT secretion remains sensitive to changes of $[Ca^{2+}]_e$.[30] Similar observations were reported for the secretion of melanocyte-stimulating hormone from melanocytes.[31] A dissociation of the spontaneous electrical activity and Ca^{2+}-dependent hormone release was already pointed out by Ozawa and Sand in 1986.[32] The function of the electrical activity appears to be to couple individual cells electrically, perhaps thereby synchronizing secretory activity and a variety of other cellular functions.[33]

3.4. 1,4-DIHYDROPYRIDINE BINDING SITES IN C CELLS

By combined electrophysiological and radioligand binding studies, the number of DHP-sensitive Ca^{2+} channels can be calculated to be 2000 to 7000 per rMTC cell.

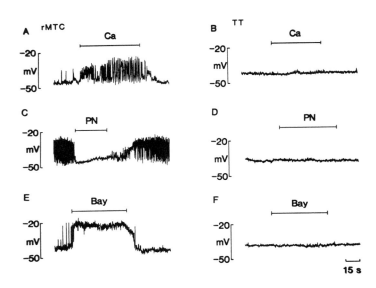

FIGURE 3. Effects of high $[Ca^{2+}]_e$, the Ca^{2+} channel blocker isradipine, and the Ca^{2+} channel agonist BAY K 8644 on membrane potentials in C cells. The effects of 1.8 mM Ca^{2+} (Ca) (A and B), 1 μM isradipine (PN) (C and D), and 1 μM BAY K 8644 (Bay) (E and F) on membrane potentials in Ca^{2+}-sensitive rMTC (left) and Ca^{2+}-insensitive TT (right) cells are shown. The substances were added as indicated by the horizontal lines. Solutions: pipette solution I1 (with 100–200 μg/ml nystatin) and external solution E1 (1.2 mM Ca^{2+}). External solution E1 containing 1.8 mM Ca^{2+} instead of 1.2 mM Ca^{2+} was applied in (A) and (B) as marked by (Ca) and throughout in (C) and (D). To avoid a major disturbance of the cytoplasm, the cytoplasm was accessed by the nystatin modification of the patch clamp technique. (From Scherübl, H. et al., *FEBS Lett.*, 273, 51–54, 1990. With permission.)

Moreover, high-affinity 1,4-DHP binding sites featuring all characteristics of binding to the α_1 subunits of L-type Ca^{2+} channels are detected in membranes of rMTC cells. In contrast, in TT cells, which lack extracellular Ca^{2+} sensitivity, neither Ca^{2+} channel-specific $(+)(-)[^3H]$ PN 200–110 binding, nor DHP-sensitive voltage-dependent Ca^{2+} channels can be detected.[34] The expression of slowly rather than fast-inactivating, voltage-dependent Ca^{2+} channels in rMTC- vs. TT-C cells offers a fascinating opportunity to clarify the molecular diversity of voltage-dependent Ca^{2+} channels and their roles in the regulation of cellular functions.

3.5. SATURATION AND DESENSITIZATION OF EXTRACELLULAR CA^{2+} SENSITIVITY

The extracellular Ca^{2+} sensitivity of rMTC cells "saturates" at $[Ca^{2+}]_e$ of 3–4 mM and even declines with higher $[Ca^{2+}]_e$. Repetitive stimulation pulses with high $[Ca^{2+}]_e$ lead to a (reversible) desensitization of $[Ca^{2+}]_e$-induced CT secretion[35] (see also Section 4.1.5). Although this phenomenon is not fully understood, several mechanisms appear to be involved. First, after an initial membrane depolarization, extreme rises of $[Ca^{2+}]_e$ to 5 or 10 mM cause a long-lasting membrane hyperpolarization to almost −70 mV in rMTC cells. This hyperpolarization most likely results from activation of Ca^{2+}-dependent K^+ channels;[35a] this can be responsible for the declining CT secretion observed at very high $[Ca^{2+}]_e$. Second, the inactivation of Ca^{2+} currents due to high $[Ca^{2+}]_i$[36] may contribute to the declining of the $[Ca^{2+}]_e$-induced secretory response. Third, for chromaffin cells, Augustine and Neher[37] have shown that the secretory response may decline at long-term rises of $[Ca^{2+}]_i$ above 2 or 10 μM. Thus, several mechanisms have to be taken into account, but clearly more detailed studies are required to fully understand the desensitization of CT secretion in response to repetitive $[Ca^{2+}]_e$ pulses.

4. THE ROLE OF INTRACELLULAR CA²⁺ IN THE SECRETION OF CALCITONIN

4.1. CA²⁺-DEPENDENT CALCITONIN SECRETION *IN VITRO*

Early studies with perfused isolated ultimobranchial organs of goose, turkey, and hen demonstrated that acute hypercalcemia increased CT secretion.[4] In order to study the cellular and molecular mechanisms underlying regulation of CT secretion, synthesis, and transcription by Ca^{2+}-dependent mechanisms, *in vitro* systems had to be established. These include ultimobranchial or thyroid gland explants,[38] primary cell cultures of MTC,[39] and permanent C cell carcinoma cell lines (human TT, rMTC 6–23, 44–2).[15,16]

4.1.1. Intracellular Ca²⁺

A rapid elevation of $[Ca^{2+}]_e$ evokes a prompt release of preformed CT, CT gene-related peptide, or neurotensin from C cells of the carcinoma cell lines rMTC 6–23 or rMTC 44–2.[40-43] There is a strong positive relationship between $[Ca^{2+}]_e$ and CT secretion. Increases of $[Ca^{2+}]_e$ similarly cause increases in CT release in rat thyroid explants. This tight linkage between $[Ca^{2+}]_e$ and CT secretion is mediated by changes in $[Ca^{2+}]_i$ as measured with the fluorescent Ca^{2+} indicators quin-2 or fura-2. When $[Ca^{2+}]_e$ is elevated, $[Ca^{2+}]_i$ increases in a biphasic manner, consisting of a rapid spike within 5–8 s and a subsequent steady plateau (Figure 4).[40,44] This plateau shows no evidence of decay for at least 20 min. The pattern of the response to $[Ca^{2+}]_e$ varies from cell to cell; some cells demonstrate a rapid rise with only a small peak followed by a plateau; others exhibit oscillations in $[Ca^{2+}]_i$[45] (see also Section 4.3). A small increase in $[Ca^{2+}]_e$ of as little as 0.1 mM causes a distinct rapid elevation in $[Ca^{2+}]_i$. This degree of responsiveness allows rMTC cells to monitor small physiological changes in $[Ca^{2+}]_e$. Since $[Ca^{2+}]_i$ is the major intracellular messenger for Ca^{2+}-dependent hormone secretion from C cells, the coupling of extra- to intracellular Ca^{2+} appears to be the essential mechanism in establishing Ca^{2+} sensitivity. In general, $[Ca^{2+}]_i$ can be affected by an influx of $[Ca^{2+}]_e$ or by Ca^{2+} mobilization from intracellular stores. Agents which increase $[Ca^{2+}]_i$ by enhancing Ca^{2+} influx or by mobilization of intracellular Ca^{2+} enhance CT release. This has been reported for the Ca^{2+} ionophores ionomycin and A 23187 or for raising $[Ca^{2+}]_i$ by electropermeabilization.[26] These findings again indicate that changes $[Ca^{2+}]_i$ regulate CT secretion in C cells.

4.1.2. Voltage-Dependent Ca²⁺ Channels and Intracellular Ca²⁺

In C cells, coupling of $[Ca^{2+}]_e$ to $[Ca^{2+}]_i$ is accomplished by steady-state Ca^{2+} influx through noninactivating, DHP-sensitive, high-threshold Ca^{2+} channels.[20] The essential role of these voltage-dependent Ca^{2+} channels for $[Ca^{2+}]_e$-dependent CT release could be confirmed by administering Ca^{2+} channel activators (BAY K 8644) or Ca^{2+} channel blockers (nifedipine, verapamil) or by depolarizing the plasma membrane by high external potassium;[35,39,46,47] more direct confirmation has come from electrophysiological studies (see Section 3). BAY K 8644-stimulated CT secretion can be completely inhibited by an equimolar concentration of nifedipine or by chelating $[Ca^{2+}]_e$.[39,46,48] Similarly, the rapid increase in cytosolic Ca^{2+} concentration in response to an elevation in $[Ca^{2+}]_e$ can be completely blocked by nifedipine, nitrendipine, isradipine, or verapamil (Figure 4).[40] In contrast, in the human C cell line TT, which has a defect in the Ca^{2+} signal transduction, a rise of $[Ca^{2+}]_e$ affects neither $[Ca^{2+}]_i$ nor secretion of CT.[26] However, when the plasma membrane of TT cells is bypassed by electropermeabilization, the release of CT can be stimulated by $[Ca^{2+}]_e$.[26] Thus, the $[Ca^{2+}]_e$-insensitive TT-C cells lack the extracellular Ca^{2+} sensitivity, but not the intracellular Ca^{2+} sensitivity (see Section 3.2). They are an excellent model for comparative studies to elucidate how extracellular Ca^{2+} sensitivity works.

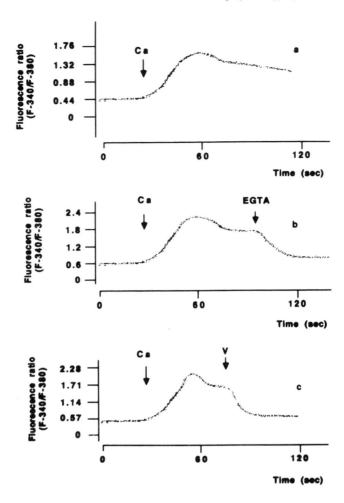

FIGURE 4. Effects of calcium (Ca), EGTA, and verapamil (V) on the intracellular calcium concentration in single rMTC 6–23 cells loaded with fura-2. The traces show relative changes in fluorescence intensity given in the number of counted photons for the 340 nM/380 nM ratio. 3 mM Ca (a), 3 mM Ca and 3 mM EGTA (b) or 3 mM Ca and 10^{-7} M verapamil (c) were directly added to the bath. Basal $[Ca^{2+}]_e$ was 1.0 mM. The arrows indicate the addition of the substances. (From Zink, A. and Raue, F., *Acta Endocrinol.*, 127, 378–384, 1992. With permission.)

4.1.3. Phospholipid Systems

Another mechanism by which $[Ca^{2+}]_e$ and $[Ca^{2+}]_i$ are coupled is through the release of Ca^{2+} from intracellular stores in response to second messengers triggered by $[Ca^{2+}]_e$. In many endocrine cells, the hydrolysis of membrane phospholipids to produce inositol triphosphate (IP_3) and diacylglycerol is an important component of the signaling pathway that mediates secretion.[49] IP_3 is responsible for releasing Ca^{2+} from internal stores, primarily from the endoplasmic reticulum.[50] In C cells, IP_3 production is, however, not stimulated by high $[Ca^{2+}]_e$.[51]

4.1.4. Calmodulin/Protein Kinase C

The anticonvulsant agent phenytoin, which can inhibit Ca^{2+} channel activity and calmodulin (a Ca^{2+}-binding protein), inhibits CT release.[52] Similar actions have been described for other more or less specific calmodulin inhibitors such as W7, trifluoperazine, chlorpromazine, and haloperidol.[53] Calmodulin is activated by Ca^{2+}-binding, resulting in a Ca^{2+}-calmodulin complex which is a functional subunit of several enzymes, e.g., protein kinases C. Activators of protein kinase C such as phorbol ester are highly effective in releasing CT within minutes.[48,54] Thus, Ca^{2+}-dependent secretory pathways of C cells involve plasma membrane Ca^{2+} channels, intracellular Ca^{2+}, calmodulin, and protein kinase C.

4.1.5. Long-Term Elevation of $[Ca^{2+}]_e$

Similar to *in vivo*, *in vitro* long-term high $[Ca^{2+}]_e$ leads to a decline of CT release to unstimulated levels (within 4 h). This decline proves to be reversible by lowering the high $[Ca^{2+}]_e$ concentration to basal levels for 2 h and then increasing $[Ca^{2+}]_e$ again.[35] Thus, the Ca^{2+}-induced desensitization of CT release is not due to an exhaustion of the secretory reserve within the observation period. Also, long-term exposure to BAY K 8644 over 4 d results in decreased CT secretion and CT content in rMTC 6–23 cells, while short-term experiments with BAY K 8644 show a stimulatory effect over 6 h.[55] These results suggest that a reversible modification of Ca^{2+} channels in C cells may cause the desensitization of CT release to repetitive Ca^{2+} stimulation (see Section 3.5). The different time course of desensitization by BAY K 8644 and $[Ca^{2+}]_e$ might be explained by different modifications of the voltage-dependent Ca^{2+} channels.

4.1.6. Extracellular Ca^{2+}-Sensing Receptor

In contrast to the situation in C cells, the Ca^{2+}-sensing in parathyroid cells has been reported to be due to $[Ca^{2+}]_e$-sensing receptors, probably coupled via G proteins to the breakdown of phosphoinositol phosphates and thus intracellular Ca^{2+} release.[56] Recently, in addition to the above described voltage-dependent Ca^{2+} channels, a parathyroid-like Ca^{2+} sensor has been implicated to contribute also to the Ca^{2+} sensing in C cells.[57] Thus, La^{3+} and Ce^{3+} were found to elicit a transient increase of the 340/380-mm fluorescence ratio in primary cultures of human MTC cells at low external Ca^{2+} of 0.5 mM, or even in Ca^{2+}-deficient medium.[57] Unfortunately, analogous experiments were not performed in the $[Ca^{2+}]_e$-insensitive TT-C cells that do not express DHP-sensitive, slowly inactivating Ca^{2+} channels (see Sections 3.2 and 4.1.2) and that do not respond to rises of $[Ca^{2+}]_e$. If a parathyroid-like Ca^{2+} sensor participated in the Ca^{2+} sensing of C cells, one would expect TT-C cells to lack not only DHP-sensitive, slowly inactivating Ca^{2+} channels, but also the parathyroid-like Ca^{2+} sensors. Moreover, unlike the situation in parathyroid cells, rises of $[Ca^{2+}]_e$ fail to produce any discernible rise of intracellular IP_3 production in rMTC C cells.[51] Thus, any $[Ca^{2+}]_e$-sensing receptors of C cells would differ from the described parathyroid-like Ca^{2+} sensor.[56] As concerns C cells, the role of parathyroid-like Ca^{2+} sensors if any is unclear at present; clearly, further detailed studies are needed.

4.2. THE ROLE OF cAMP IN THE SECRETION OF CALCITONIN

There are two major intracellular signals, $[Ca^{2+}]_i$ and cAMP, which are closely interrelated and coordinately regulate cell function in C cells.[58,59] The efficiency of signal transduction may be modified in a number of ways, including modulation of receptor number or receptor function, desensitization of postreceptor coupling systems, and antagonistic or synergistic interactions between different signaling systems.[60,61]

4.2.1. cAMP-Dependent Calcitonin Secretion

Gastrointestinal hormones such as glucagon induce CT secretion by a cAMP-dependent pathway via activation of stimulatory guanine nucleotide-binding proteins (G_s proteins)[59,62] cAMP-dependent CT secretion can be imitated by cAMP analogs (8-bromo cAMP, dibutyryl-cAMP) or forskolin, a diterpene which enhances protein kinase A (PKA) activity by stimulating the catalytic subunit of AC.[48,54,63–65] Other PKA activating substances are cholera toxin, which stimulates ADP ribose transfer to the G_s proteins with subsequent inhibition of their GTPase activity and prolonged activation of AC, or theophylline (or 3-isobutyl-1-methylxanthine, IBMX), which inhibits phosphodiesterases and thereby endogenous degradation of cAMP. Not only gastrointestinal hormones, but also biogenic amines such as norepinephrine (NE)[66,67] and neuropeptides such as GRF[68,69] increase CT secretion via the cAMP-dependent pathway.

4.2.2. Dual Regulation of Adenylyl Cyclase

Coupling to both stimulatory G_s and inhibitory G_i proteins results in a dual regulation of AC. The receptor for the inhibitory agonist is coupled to a G_i protein; the activated G_i α subunit inhibits the catalytic subunit of AC, thereby preventing formation of cAMP from ATP[60,70] In C cells, SST and adenosine (A1) receptors are coupled to G_i proteins, thus inhibiting AC. Glucagon-, GRF-, or NE-stimulated CT secretion can be dose-dependently inhibited by SST or *N*-6-phenylisopropyladenosine, an adenosine analog acting on A1 receptors.[62,65,67,69] These inhibitory effects of SST or adenosine can be partially blocked by pertussis toxin, a bacterial toxin which is known to uncouple G_i from its associated receptor by ADP ribosylating the GTP-binding α subunit. The physiological role of SST or adenosine in C cells remains unclear. Since SST is secreted from C cells themselves[71,72] it is an attractive hypothesis that SST modulates CT secretion in a tonic inhibitory way, possibly in an autocrine or paracrine fashion.

4.2.3. Cross Talk Between cAMP- and Intracellular Ca^{2+}-Dependent Pathways

The combination of the two secretagogues $[Ca^{2+}]_e$ and glucagon results in higher CT release than would be expected from the sum of their effects.[71] This synergistic effect of the two intracellular pathways on CT secretion could not be confirmed in rMTC 6–23[48,73] and the human TT cell line.[54] Activation of protein kinase C by phorbol ester (TPA) or of PKA by forskolin leads to a comparable level of CT secretion, while a combined treatment of TT cells with the two compounds simply produces an additive effect on CT secretion. This suggests that TPA and cAMP may act independently to bring about increased CT secretion by TT cells. TPA has no effect on intracellular cAMP. in rMTC 6–23 cells, glucagon-stimulated intracellular cAMP was not influenced by $[Ca^{2+}]_e$ and glucagon failed to increase $[Ca^{2+}]_i$.[44] A prolongation of the ionomycin-stimulated secretion of CT by TPA in human TT cells could be shown.[63] These results can be interpreted in terms of a model of cell activation in which $[Ca^{2+}]_i$ is involved in initiating secretion of stored CT and protein kinase C in the continuing of the response, perhaps by secretion of newly synthesized CT.

4.3. OSCILLATION OF INTRACELLULAR CALCIUM

More recent studies of $[Ca^{2+}]_i$ have used the technology of single-cell recordings to demonstrate that single neuroendocrine cells show oscillations in $[Ca^{2+}]_i$, either spontaneously or after stimulation by hormones or neurotransmitters. Such oscillations depend on the entry of extracellular Ca^{2+} and/or Ca^{2+} release from intracellular stores. In many cell types, both mechanisms are involved in the generation of Ca^{2+} oscillations.[74,75] $[Ca^{2+}]_i$ oscillation appears to sustain hormone secretion from neuroendocrine cells.[76]

4.3.1. Extracellular Ca^{2+}-Activated Ca^{2+} Oscillation

A rise of $[Ca^{2+}]_e$ induces either a steady increase of $[Ca^{2+}]_i$ or oscillations of $[Ca^{2+}]_i$.[40,44,45] Responses of individual cells are quite heterogeneous, with some cells exhibiting complex oscillatory patterns and others demonstrating a biphasic response. The latter consists of a rapid rise, reaching a peak within 5 s and then declining to a lower plateau. At higher $[Ca^{2+}]_e$, significantly more cells exhibit oscillations of $[Ca^{2+}]_i$. The oscillatory frequency varies from <1/min to 6/min.[45] When rMTC 44–2 cells are superfused with a solution containing high K^+

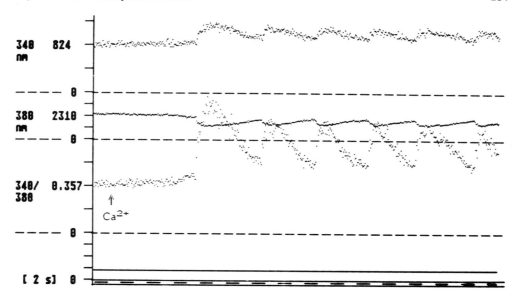

FIGURE 5. Effect of high $[Ca^{2+}]_e$ on $[Ca^{2+}]_i$ in a single rMTC cell pretreated with glucagon. The rMTC cell was pretreated with 1 µM glucagon for 45 min. As indicated by the arrow, $[Ca^{2+}]_e$ was raised from 1.0 to 3.0 mM. Changes in $[Ca^{2+}]_i$ are shown as relative changes in the fluorescence ratio 340 nM/380 nM. (From Eckert, R. W. et al., *Mol. Cell. Endocrinol.*, 64, 267–270, 1989. With permission.)

concentrations (7.5–20 mM), the pattern of $[Ca^{2+}]_i$ oscillations resembles that seen with stimulation by $[Ca^{2+}]_e$. The $[Ca^{2+}]_e$- and K^+-evoked $[Ca^{2+}]_i$ response is virtually eliminated by Cd^{2+} and partly by nifedipine. These results suggest an involvement of DHP-sensitive, voltage-dependent Ca^{2+} channels in the oscillations of $[Ca^{2+}]_i$ in C cells.

4.3.2. Extracellular Ca^{2+} and cAMP-Activated Ca^{2+} Oscillation

A more complex response of $[Ca^{2+}]_i$ to elevation of $[Ca^{2+}]_e$ is seen in C cells that have been preactivated either by glucagon or 8-bromo cAMP: $[Ca^{2+}]_i$ starts to oscillate at a frequency of 0.1–0.2 Hz, and the oscillations can be observed for about 5 min (Figure 5).[44] These fluctuations of $[Ca^{2+}]_i$ can be stopped by chelating the $[Ca^{2+}]_e$ with EGTA or by adding Ca^{2+} channel blockers. This points again to an important role for voltage-dependent Ca^{2+} channels in maintaining $[Ca^{2+}]_i$ oscillations in C cells. An involvement of intracellular Ca^{2+} stores in these periodic changes of $[Ca^{2+}]_i$ appears likely, but has not yet been studied.

4.3.3. Mechanism of Ca^{2+} Oscillation

$[Ca^{2+}]_i$ can be mobilized in many cell types via the generation of IP_3. In C cells, no increase of IP_1, IP_2, or IP_3 is found when $[Ca^{2+}]_e$ is increased above 1.0 mM.[51] Thus, the increase in $[Ca^{2+}]_i$ due to an increase in $[Ca^{2+}]_e$ is not likely to be mediated by IP_3, but is due to a Ca^{2+} influx through voltage-dependent Ca^{2+} channels.

The oscillation of $[Ca^{2+}]_i$ may be mediated by cAMP-and/or Ca^{2+}-dependent mechanisms occurring at the inner mouth of Ca^{2+} channels. The activity of voltage-dependent Ca^{2+} channels and thereby Ca^{2+} influx can be regulated by phosphorylation/dephosphorylation of the Ca^{2+} channel protein or by regulatory components linked to protein kinases. Activators of protein kinase C have been found to modulate Ca^{2+} conductance through voltage-dependent Ca^{2+} channels.[77] cAMP-dependent phosphorylation of Ca^{2+} channels as a prerequisite for channel opening has been reported for pituitary cells. Besides Ca^{2+} channel modulation by phosphorylation, another direct mechanism via G proteins has become apparent. For example, SST- and adenosine A_1 receptor agonists cause an inhibition of Ca^{2+} currents via pertussis toxin-sensitive G proteins[78] (see Section 4.4.1).

The mechanisms underlying the complex oscillatory pattern in $[Ca^{2+}]_i$ still require further studies. Coincident measurements of $[Ca^{2+}]_i$ and action potentials in excitable pituitary cells have suggested that Ca^{2+} entry during action potentials may be responsible for oscillation of $[Ca^{2+}]_i$ in these cells.[79]

Although it has been suggested that the spontaneous electrical activity is responsible for oscillation of $[Ca^{2+}]_i$ and thereby for basal Ca^{2+}-dependent hormone release, some caution appears warranted. Thus, in C cells spontaneous action potentials are not required for a $[Ca^{2+}]_e$-induced increase in $[Ca^{2+}]_i$ or CT release.[30] Suppression of the spontaneous electrical activity by tetrodotoxin, a Na^+ channel blocker, fails to affect Ca^{2+}-dependent CT release. Thus, action potentials are not essential for basal CT release due to Ca^{2+} influx.[30,33]

4.4. G PROTEIN-MEDIATED MODULATION OF VOLTAGE-DEPENDENT Ca^{2+} CHANNELS

4.4.1.   Inhibitory Effect of Somatostatin

Voltage-dependent Ca^{2+} channels can be modulated by $[Ca^{2+}]_i$, by protein kinases A and C and by G proteins.[70,79] Therefore, neuropeptides and hormones involved in the control of CT secretion in C cells may operate through modulation of voltage-dependent Ca^{2+} channels.

SST is known to inhibit hormone secretion from CT-secreting cells and has recently been used to lower serum CT levels in patients with MTC.[80] Similar to reports on pituitary cells,[81] both cAMP-dependent and cAMP-independent mechanisms appear to underlie SST inhibitory action on CT secretion. SST-induced inhibition of GRF- or glucagon-stimulated CT release as well as SST-induced inhibition of GRF- or glucagon-stimulated cAMP accumulation occur via pertussis toxin-sensitive G_i proteins.[62,69] However, the mechanisms by which SST inhibits Ca^{2+}-stimulated CT release appear to be independent of an effect on the intracellular cAMP concentration.[78] Thus, the intracellular cAMP concentration is unchanged during Ca^{2+}-induced CT secretion, and intracellularly administered cAMP rails to affect the control Ca^{2+} channel current or its reduction by SST. SST reduces Ca^{2+} channel currents, the $[Ca^{2+}]_e$-induced rise in $[Ca^{2+}]_i$, and Ca^{2+}-induced CT secretion in C cells. Consistent with these findings in C cells, SST inhibits voltage-dependent Ca^{2+} channels in various cell types, including the β cells of the pancreas, pituitary, and neuronal cells.[70,82–84] There seems to be a causal relationship between the inhibition of Ca^{2+} channel currents and inhibition of secretion of neurotransmitters or hormones. Similar to the effect of pertussis toxin on cAMP-dependent CT secretion, pretreatment of rMTC cells with pertussis toxin attenuates the inhibitory effect of SST on the $[Ca^{2+}]_e$-induced release of CT, on the rise in $[Ca^{2+}]_i$, and on Ca^{2+} channel currents.[78] This indicates that SST receptors inhibit Ca^{2+} channel currents in C cells via pertussis toxin-sensitive G proteins. The regulation of voltage-dependent Ca^{2+} channels by specific G proteins activated by SST receptors could be demonstrated in GH_3 pituitary cells, in which the expression of α_o subunits of G proteins had been suppressed by specific antisense oligonucleotides. The α_{o2} protein was found to specifically mediate the inhibition of Ca^{2+} channels in response to SST receptor activation.[83] In several neuronal and neuroendocrine cells, the inhibitory action of SST on Ca^{2+} entry and hormone release is additionally mediated by a G_i protein-activated K^+ conductance, leading to membrane hyperpolarization. This concept, however, does not apply to C cells. SST neither hyperpolarized C cells nor activated a K^+ conductance. It is possible that G proteins regulate membrane-associated, not yet identified enzymes, e.g., protein kinases or phosphatases, which control channel activity by phosphorylation or dephosphorylation. Alternatively, G proteins themselves may modify the channel protein and, as a consequence, its function by an inherent enzymatic activity.

4.4.2.   Stimulatory Effect of Norepinephrine

NE is an example of an agonist that activates both Ca^{2+}- and cAMP-dependent signaling systems. It stimulates cAMP accumulation and $[Ca^{2+}]_i$ via β_2 receptors, probably linked to G_s proteins.[67,85] NE induces a virtually identical change in $[Ca^{2+}]_i$, in terms of duration and shape,

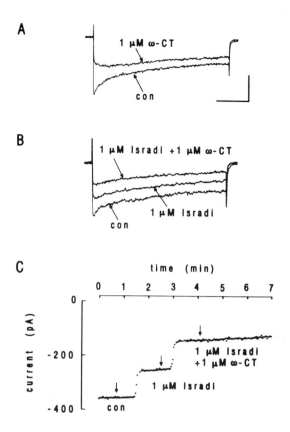

FIGURE 6. Effects of the Ca^{2+} channel blockers isradipine and ω-conotoxin (ω-CT) on voltage-dependent Ca^{2+} channels in rMTC cells. Stimulation frequency was 0.5 Hz. (A and B) Original current traces as recorded during voltage steps from –80 to 0 mV. Calibration marks correspond to 50 ms and 200 pA. A: effect of ω-CT (1 μ*M*). CON, control. B: first isradipine (Isradi, 1 μ*M*) was applied, followed second by the combination of isradipine (1 μ*M*) with ω-CT (1 μ*M*). (C) Time course of peak current amplitudes. Vertical arrows indicate time points at which current traces in B were obtained. (From Scherübl, H. et al., *Am. J. Physiol.,* 264, E354–E360, 1993. With permission.)

as does an increase in $[Ca^{2+}]_e$.[67] Furthermore, chelating $[Ca^{2+}]_e$ with EGTA or blocking Ca^{2+} influx through voltage-dependent Ca^{2+} channels with verapamil reverses the NE-induced rise in $[Ca^{2+}]_i$. These results strongly suggest that the NE-induced rise in $[Ca^{2+}]_i$ is due to an influx of Ca^{2+} from the extracellular space, most probably through voltage-dependent Ca^{2+} channels. The link between β receptors and activation of Ca^{2+} channels may be direct by G proteins, or indirect by NE-induced accumulation of cAMP and activation of PKA. In pituitary cells, cAMP-dependent phosphorylation of Ca^{2+} channels may be a prerequisite for channel opening.[86] Both intracellular signals, cAMP and Ca^{2+}, are necessary for NE-stimulated CT secretion; by blocking calcium channels with verapamil or β receptors with propranolol, CT secretion drops. Apparently, synergistic interactions between signaling systems activated in parallel are necessary to guarantee NE-induced CT secretion.

5. PHARMACOLOGICAL STUDIES

C cells express multiple types of Ca^{2+} channels.[20,28–30,34,87–89] Electrophysiological, molecular biological, and pharmacological studies demonstrate not only DHP-sensitive L-type, but also DHP-insensitive, fast-inactivating T-type and DHP-insensitive, ω-conotoxin (ω-CT)-sensitive N-type Ca^{2+} channels in C cells. Moreover, a component of the Ca^{2+} inward current in rMTC cells can be blocked by funnel web spider toxin, suggesting the existence of so-called

P-type Ca²⁺ channels.[89] The combined administration of isradipine, a DHP Ca²⁺ channel blocker, and ω-CT fails to suppress the depolarization-induced Ca²⁺ current completely (Figure 6).[20] The remaining component of the Ca²⁺ current is completely blocked by 0.1 mM Cd²⁺ or Ni²⁺; this remaining component of the Ca²⁺ current still has to be characterized in detail.

6. SCHEME

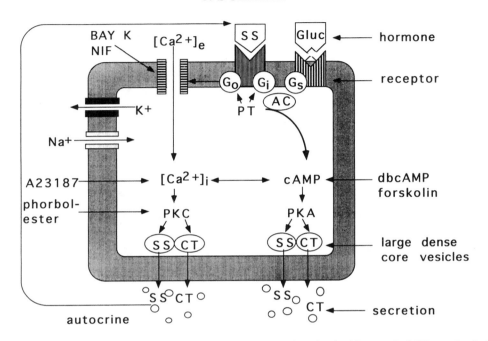

FIGURE 7. Schematic diagram of extra- and intracellular mechanisms involved in control of CT secretion in C cells by extracellular Ca²⁺ and other factors.

ACKNOWLEDGMENT

We would like to thank the Deutsche Forschungsgemeinschaft for continuing support of our research projects.

REFERENCES

1. **Deftos, L. J., Weisman, M. H., Williams, G. W., Karpf, D. B., Frumar, A. M., Davidson, B. J., Parthemore, J. G., and Judd, H. L.,** Influence of age and sex on plasma calcitonin in human beings, *N. Engl. J. Med.,* 302, 1351, 1980.
2. **Austin, L. A. and Heath, H., III,** Calcitonin — physiology and pathophysiology, *N. Engl. J. Med.,* 304, 269, 1981.
3. **Austin, L. A., Heath, H., III and Go, V. L. W.,** Regulation of calcitonin secretion in normal man by changes of serum calcium within the physiologic range, *J. Clin. Invest.,* 64, 1721, 1979.
4. **Barlet, J. P.,** in *Endocrinology of Calcium Metabolism,* Parsons, J. A., Ed., Raven Press, New York, 1982, 235.
5. **Parthmore, J. G., Bronzert, D., Roberts, G., and Deftos, L. J.,** A short calcium infusion in the diagnosis of medullary thyroid carcinoma, *J. Clin. Endocrinol. Metab.,* 39, 108, 1974.

6. **Lambert, P. W., Heath, H., III, and Sizemore, G. W.,** Pre- and postoperative studies of plasma calcitonin in primary hyperparathyroidism, *J. Clin. Invest.,* 63, 602, 1979.

7. **Parthemore, J. G. and Deftos, L. J.,** Calcitonin secretion in primary hyperparathyroidism, *J. Clin. Endocrinol. Metab.,* 49, 223, 1979.

8. **Adachi, I., Abe, K., Tanaka, M., Yamaguchi, K., Miyakawa, S., Hirakawa, H., and Tanaka, N.,** Plasma human calcitonin (hCT) levels in normal and pathologic conditions, and their responses to short calcium or tetragastrin infusion, *Endocrinol. Jpn.,* 23, 517, 1976.

9. **Raue, F., Deutschle, I., Küntzel, C., and Zigler, R.,** Reversible diminished calcitonin secretion in the rat during chronic hypercalcemia, *Endocrinology,* 115, 2362, 1984.

10. **Deftos, L. J., Powell, D., Parthemore, J. G., and Potts, J. T.,** Secretion of calcitonin in hypocalcemic states in man, *J. Clin. Invest.,* 52, 3109, 1973.

11. **Raue, F., Wieland, U., Weiler, C., and Ziegler, R.,** Enhanced calcitonin secretion in the rat after parathyroidectomy and during chronic calcium deprivation, *Eur. J. Clin. Invest.,* 18, 284, 1988.

12. **Chambers, T. J. and Moore, A.,** The sensitivity of isolated osteoclasts to morphological transformation by calcitonin, *J. Clin. Endocrinol. Metab.,* 57, 819, 1983.

13. **Jaeger, P., Jones, W., Clemens, T. L., and Haystett, J. P.,** Evidence that calcitonin stimulates 1,25-dihydroxyvitamin D production and intestinal absorption of calcium in vivo, *J. Clin. Invest.,* 78, 456, 1986.

14. **Naveh-Many, T. and Silver, J.,** Regulation of calcitonin gene transcription by vitamin D metabolites in vivo in the rat, *J. Clin. Invest.,* 81, 270, 1988.

15. **Gagel, R. F., Zeytinoglu, F., Voelkel, E. F., and Tashijan, A. H.,** Establishment of a calcitonin-producing rat medullary carcinoma cell line. II. Secretory studies of the tumor and cells in culture, *Endocrinology,* 107, 516, 1980.

16. **Leong, S. S., Horoszewicz, J. S., Shimoaka, K., Friedmann, M., Kawinski, E., Song, M. J., Zeigel, R., Chu, T. M., Baylin, S., and Miraud, E. A.,** in *Advances in Thyroid Neoplasia,* Andreoli, M., Monaco, F., and Robbins, J., Eds., Field Educational Italia, Rome, 1981, 95.

17. **Tischler, A. S., Dichter, M. A., Biales, B., De-Lellis, R. A., and Wolfe, H.,** Neural properties of cultured human endocrine tumor cells of proposed neural crest origin, *Science,* 192, 902, 1976.

18. **Sand, O., Ozswa, S., and Gautvik, K. M.,** Sodium and calcium action potentials in cells derived from a rat medullary thyroid carcinoma, *Acute Physiol. Scand.,* 112, 287, 1981.

19. **Sand, O., Jonsson, L., Nielsen, M., Holm, R., and Gautvik, K. M.,** Electrophysiological properties of calcitonin-secreting cells derived from human medullary thyroid carcinoma, *Acta Physiol. Scand.,* 126, 173, 1986.

20. **Scherübl, H., Schultz, G., and Hescheler, J.,** A slowly inactivating calcium current works as a calcium sensor in calcitonin-secreting cells, *FEBS,* 273, 51, 1990. (Errata: 278, 289, 1991.)

21. **Hamill, O. P., Marty, A., Neher, E., Sakmann, B., and Sigworth, F. J.,** Improved patch-clamp techniques for high-resolution current recording from cells and cell-free membrane patches, *Pflügers Arch.,* 391, 85, 1981.

22. **Kleppisch, T., Scherübl, H., and Hescheler, J.,** Dihydropyridine-sensitive steady-state Ca^{2+} channel currents in calcitonin-secreting cells: multiple conductance levels, submitted.

23. **Belles, B., Malecot, C. O., Hescheler, J., and Trautwein, W.,** "Run-down" of the Ca current during long whole-cell recordings in guinea pig heart cells: role of phosphorylation and intracellular calcium, *Pflügers Arch.,* 411, 353, 1988.

24. **Scherübl, H. and Hescheler, J.,** Steady-state currents through voltage-dependent, dihydropyridine-sensitive calcium channels in GH_3 pituitary cells, *Proc. R. Soc. London,* 245, 127, 1991.

25. **Scherübl, H., Brandi, M. L., and Hescheler, J.,** Extracellular Ca^{2+} sensing in C-cells and parathyroid cells, *Henry Ford Hosp. Med. J.,* 40, 303, 1992.

26. **Haller-Brem, S., Muff, R., Petermann, J. B., Born, W., Roos, B. A., and Fischer, J. A.,** Role of cytosolic free calcium concentration in the secretion of calcitonin gene-related peptide and calcitonin from medullary thyroid carcinoma cells, *Endocrinology,* 121, 1272, 1987.

27. **Tsien, R. W., Lipscombe, D., Madison, D. V., Bley, K. R., and Fox, A. P.,** Multiple types of neuronal calcium channels and their selective modulation, *TINS,* 11, 431, 1988.

28. **Biagi, B. A., Mlinar, B., and Enyeart, J. J.,** Membrane currents in a calcitonin-secreting human C-cell line, *Am. J. Physiol.,* 263, C986, 1992.

29. **Yamashita, N. and Hagiwara, S.,** Membrane depolarization and intracellular Ca^{2+} increase caused by high external Ca^{2+} in a rat calcitonin-secreting cell line, *J. Physiol.,* 431, 243, 1990.

30. **Scherübl, H., Kleppisch, T., Zink, A., Raue, F., Krautwurst, D., and Hescheler, J.,** Major role of dihydropyridine-sensitive Ca^{2+} channels in Ca^{2+}-induced calcitonin secretion, *Am. J. Physiol.,* 264, E354, 1993.

31. **Tomiko, S. A., Taraskevich, P. S. and Douglas, W. W.,** Effects of veratridine, tetrodotoxin and other drugs that alter electrical behavior on secretion of melanocyte-stimulating hormone from melanotrophs of the pituitary pars intermedia, *Neuroscience,* 12, 1223, 1984.

32. **Ozawa, S. and Sand O.,** Electrophysiology of excitable endocrine cells, *Physiol. Rev.,* 66, 887, 1986.

33. **Scherübl, H. and Hescheler, J.,** Steady-state Ca^{2+} influx and electrical activity in endocrine cells, *Trends Neurosci.,* 15, 126, 1992.

34. **Krautwurst, D., Scherübl, H., Kleppisch, T., Hescheler, J., and Schultz, G.,** Dihydropyridine binding and Ca^{2+} channel characterization in clonal calcitonin-secreting cells, *Biochem. J.,* 289, 659, 1993.

35. **Scherübl, H., Raue, F., Zopf, G., Hoffmann, J., and Ziegler, R.,** Reversible desensitization of calcitonin secretion by repetitive stimulation with calcium, *Mol. Cell. Endocrinol.,* 63, 263, 1993.

35a. **Scherübl, H. and Hescheler, J.,** Unpublished observations.

36. **Eckert, R. and Chad, J. E.,** Inactivation of Ca channels, *Prog. Biophys. Mol. Biol.,* 44, 215, 1984.

37. **Augustine, G. J. and Neher, E.,** Calcium requirements for secretion in bovine chromaffin cells, *J. Physiol. (London),* 450, 247, 1992.

38. **Cooper, C. W., Ramp, W. K., Becker, D. I., and Ontjes, D. A.,** In vitro secretion of immunoreactive rat thyrocalcitonin, *Endocrinology,* 101, 304, 1977.

39. **Raue, F., Serve, H., Grauer, A., Rix, E., Scherübl, H., Schneider, H. G., and Ziegler, R.,** Role of voltage dependent calcium channels in secretion of calcitonin from human medullary thyroid carcinoma cells, *Klin. Wochenschr.,* 67, 635, 1989.

40. **Fried, R. M. and Tashjian, A. H.,** Unusual sensitivity of cytosolic free Ca^{2+} to changes in extracellular Ca^{2+} in rat C-cells, *J. Biol. Chem.,* 261, 7669, 1986.

41. **Muff, R., Nemeth, E. F., Haller-Brem, S., and Fischer, J. A.,** Regulation of hormone secretion and cytosolic Ca^{2+} by extracellular Ca^{2+} in parathyroid cells and C-cells: role of voltage-sensitive Ca^{2+} channels, *Arch. Biochem. Biophys.,* 265, 128, 1988.

42. **Seitz, P. K. and Cooper, C. W.,** Cosecretion of calcitonin and calcitonin gene-related peptide from cultured rat medullary thyroid C-cells, *J. Bone Miner. Res.,* 4, 129, 1989.

43. **Zeytinoglu, F. N., Gagel, R. F., Tashjian, A. H., Hammer, R. A. and Leeman, S. E.,** Characterization of neurotensin production by a line of rat medullary thyroid carcinoma cells, *Proc. Natl. Acad. Sci.,* 77, 3741, 1980.

44. **Eckert, R. W., Scherübl, H., Petzelt, C., Raue, F., and Ziegler, R.,** Rhythmic oscillations of cytosolic free calcium in rat C-cells, *Mol. Cell. Endocrinol.,* 64, 267, 1989.

45. **Fajtova, V. T., Quinn, S. J., and Brown, E. M.,** Cytosolic calcium responses of single rMTC 44–2 cells to stimulation with external calcium and potassium, *Am. J. Physiol.,* 261, E151, 1991.

46. **Cooper, C. W., Borosky, S. A., Farrell, P. E., and Steinsland, O. S.,** Effects of the calcium channel activator BAY K 8644 on in vitro secretion of calcitonin and parathyroid hormone, *Endocrinology,* 118, 545, 1986.

47. **Hishikawa, R., Fukase, M., Takenaka, M., and Fujita, T.,** Effect of calcium channel agonist BAY K 8644 on calcitonin secretion from a rat C-cell line, *Biochem. Biophys. Res. Commun.,* 130, 454, 1985.

48. **Hishikawa, R., Fukase, M., Yamatani, T., Kadowaki, S., and Fujita, T.,** Phorbol ester stimulates calcitonin secretion synergistically with A 23187, and additively with dibutyryl cyclic AMP in a rat C-cell line, *Biochem. Biophys. Res. Commun.,* 132, 424, 1985.

49. **Berridge, M. J. and Irvine, R. F.,** Inositol phosphates and cell signalling, *Nature,* 341, 197, 1989.

50. **Prentki, M., Biden, T. J., Janjic, D., Irvine, R. F., Berridge, M. J., and Wollheim, C. B.,** Rapid mobilization of Ca^{2+} from rat insulinoma microsomes by inositol-1,4,5-triphosphate, *Nature,* 309, 562, 1984.

51. **Fried, R. M. and Tashjian, A. H.,** Action of rat growth hormone-releasing factor and norepinephrine on cytosolic free calcium and inositol trisphosphate in rat C-cells, *J. Bone Miner. Res.,* 2, 579, 1987.

52. **Cooper, C. W., Yi, S. J., and Seitz, P. K.,** Inhibition by phenytoin of in vitro secretion of calcitonin from rat thyroid glands and cultured rat C-cells, *J. Bone Miner. Res.,* 3, 219, 1988.

53. **Cooper, C. W. and Borosky, S. A.,** Inhibition of secretion of rat calcitonin by calmodulin inhibitors, *Calcif. Tissue Int.,* 38, 103, 1986.

54. **de Bustros, A., Baylin, S. B., Levine, M. A., and Nelkin, B. D.,** Cyclic AMP and phorbol esters separately induce growth inhibition, calcitonin secretion, and calcitonin gene transcription in cultured human medullary thyroid carcinoma, *J. Biol. Chem.,* 261, 8036, 1986.

55. **Mekonnen, Y., Raue, F., and Ziegler, R.,** In vitro secretion of calcitonin from a rat C-cell line: effect of repetitive stimulation with the calcium channel agonist BAY K 8644, *Horm. Metab. Res.,* 24, 272, 1992.

56. **Brown, E. M., Gamba, G., Riccardi, D., Lombardi, M., Butters, R., Kifor, O., Sun, A., Hediger, M. A., Lytton, J., and Hebert, S. C.,** Cloning and characterization of an extracellular Ca^{2+}-sensing receptor from bovine parathyroids, *Nature,* 366, 575, 1993.

57. **Ridefelt, P., Liu, Y. J., Rastad, J., Åkerstrom, G., and Gylfe, E.,** Calcium sensing by human medullary thyroid carcinoma cells, *FEBS Lett.,* 337, 243, 1994.

58. **Raue, F., Zink, A., and Scherübl, H.,** *Recent Research in Cancer Research: 125 Medullary Thyroid Carcinoma,* Raue, F., Ed., Springer-Verlag, Heidelberg, 1992, 1.

59. **Raue, F., Zink, A., and Scherübl, H.,** Regulation of calcitonin secretion in vitro, *Horm. Metab. Res.,* 25, 473, 1993.

60. **Davis, J. R. E., Bidey, S. P., and Tomlinson, S.,** Signal transduction in endocrine tissues, *Clin. Endocrinol.,* 36, 437, 1992.

61. **Rasmussen, H.,** The calcium messenger system, *N. Engl. J. Med.,* 314, 1094, 1986.

62. **Zink, A., Scherübl, H., Raue, F., and Ziegler, R.,** Somatostatin acts via a pertussis toxin-sensitive mechanism on calcitonin secretion in C-cells, *Henry Ford Hosp. Med. J.,* 40, 289, 1992.

63. **Haller-Brem, S., Muff, R., and Fischer, J. A.,** Calcitonin gene-related peptide and calcitonin secretion from a human medullary thyroid carcinoma cell line: effects of ionomycin, phorbol ester and forskolin, *J. Endocrinol.,* 119, 147, 1988.

64. **Murray, S. S., Burton, D. W., and Deftos, L. J.,** The effects of forskolin and calcium ionophore A23187 on secretion and cytoplasmic RNA levels of chromogranin-A and calcitonin, *J. Bone Miner. Res.,* 3, 447, 1988.

65. **Zeytin, F. N. and DeLellis, R.,** The neuropeptide-synthesizing rat 44–2 C-cell line: regulation of peptide synthesis, secretion, 3′,5′-cyclic adenosine monophosphate efflux, and adenylate cyclase activation, *Endocrinology,* 121, 352, 1987.

66. **Zeytinoglu, F. N., Gagel, R. F., Tashijian, A. H., Hammer, R. A., and Leeman, S. E.,** Regulation of neurotensin release by a continuous line of mammalian C-cells: the role of biogenic amines, *Endocrinology,* 112, 1240, 1983.

67. **Zink, A. and Raue, F.,** Somatostatin inhibits the norepinephrine-activated calcium channels in rMTC 6–23 cells: possible involvement of pertussis toxin-sensitive G-protein, *Acta Endocrinol.,* 127, 378, 1992.

68. **Zeytin, F. N. and Brazeau, P.,** GRF (somatocrinin) stimulates release of neurotensin, calcitonin and cAMP by a rat C-cell line, *Biochem. Biophys. Res. Commun.,* 123, 497, 1984.

69. **Zink, A., Scherübl, H., Kliemann, D., Höflich, M., Ziegler, R., and Raue, F.,** Inhibitory effect of somatostatin on cAMP accumulation and calcitonin secretion in C-cells: involvement of pertussis toxin-sensitive G-proteins, *Mol. Cell. Endocrinol.,* 86, 213, 1992.

70. **Birnbaumer, L., Abramowitz, J., and Brown, A. M.,** Receptor-effector coupling by G-proteins, *Biochim. Biophys. Acta,* 1031, 163, 1990.

71. **Aron, C. C., Muszynski, M., R. S., B., Sabo, S. W., and Roos, B. A.,** Somatostatin elaboration by monolayer cell cultures derived from transplantable rat medullary thyroid carcinoma: synergistic stimulatory effects of glucagon and calcium, *Endocrinology,* 109, 1830, 1981.

72. **Gagel, R. F., Palmer, W. N., Leonhardt, K., Chan, L., and Leong, S. S.,** Somatostatin production by a human medullary thyroid carcinoma cell line, *Endocrinology,* 118, 1643, 1986.

73. **Scherübl, H., Raue, F., Zopf, G., and Ziegler, R.,** Calcitonin secretion and cAMP efflux from C-cells stimulated by glucagon and either calcium or Bay K 8644, *Horm. Metab. Res.,* 21, Suppl., 18, 1989.

74. **Berridge, M. J.,** Calcium oscillation, *J. Biol. Chem.,* 265, 9583, 1990.

75. **Tsien, R. W. and Tsien, R. Y.,** Calcium channels, stores and oscillation, *Annu. Rev. Cell. Biol.,* 6, 715, 1990.

76. **Malgaroli, A. and Meldolesi, J.,** $[Ca^{2+}]_i$ oscillations from internal stores sustain exocytic secretion from the chromaffin cells of the rat, *FEBS Lett.,* 283, 169, 1991.

77. **Haymes, A. A., Kwan, Y. W., Arena, J. P., Kass, R. S., and Hinkle, P. M.,** Activation of protein kinase C reduces L-type calcium channel activity of GH_3 pituitary cells, *Am. J. Physiol.,* 262, C1211, 1992.

78. **Scherübl, H., Hescheler, J., Schultz, G., Kliemann, D., Zink, A., Ziegler, R., and Raue, F.,** Inhibition of Ca^{++}-induced calcitonin secretion by somatostatin: roles of voltage dependent Ca^{++} channels and G-proteins, *Cell Sign.,* 4, 77, 1992.

79. **Stojilkovic, S. S. and Catt, K. J.,** Calcium oscillation in anterior pituitary cells, *Endocrinol. Rev.,* 13, 256, 1992.

80. **Mahler, C., Verhelst, J., De Longueville, M., and Harris, A.,** Long-term treatment of metastatic medullary thyroid carcinoma with the somatostatin analogue octreotide, *Clin. Endocrinol.,* 33, 261, 1990.

81. **Koch, B. D., Blalock, J. B., and Schonbrunn, A.,** Characterization of the cyclic AMP-independent actions of somatostatin in GH cells. I. An increase in potassium conductance is responsible for both the hyperpolarisation and the decrease in intracellular free calcium produced by somatostatin, *J. Biol. Chem.,* 263, 216, 1988.

82. **Dolphin, A. C.,** G Protein modulation of calcium currents in neurons, *Annu. Rev. Physiol.,* 52, 243, 1990.

83. **Kleuss, C., Hescheler, J., Ewel, C., Rosenthal, W., Schultz, G., and Wittig, B.,** Assignment of G-protein subtypes to specific receptors inducing inhibition of calcium currents, *Nature,* 353, 43, 1991.

84. **Schultz, G., Rosenthal, W., Hescheler, J., and Trautwein, W.,** Role of G proteins in calcium channel modulation, *Annu. Rev. Physiol.,* 52, 243, 1990.

85. **Komitowska, J., Zink, A., and Raue, F.,** Norepinephrine acts via β-receptor on calcitonin secretion in a neoplastic C-cell line, *Exp. Clin. Endocrinol.,* 101 (Abstr.), 112, 1993.

86. **Holl, B. W., Thorner, M. O., and Leong, D. A.,** Intracellular calcium concentration and growth hormone secretion in individual somatotropes: effects of growth hormone-releasing factor and somatostatin, *Endocrinology,* 122, 2927, 1988.

87. **Biagi, B. A. and Enyeart, J. J.,** Multiple calcium currents in a thyroid C-cell line: biophysical properties and pharmacology, *Am. J. Physiol.,* 260, C1253, 1991.

88. **Scherübl, H., Schultz, G., and Hescheler, J.,** Electrophysiological properties of rat calcitonin-secreting cells, *Mol. Cell. Endocrinol.,* 82, 293, 1991.

89. **Snutch, T. P. and Reiner, P. B.,** Ca^{2+} channels: diversity of form and function, *Curr. Opin. Neurobiol.,* 2, 247, 1992.

89a. **Scherübl, H.,** Unpublished observations.

Chapter 8

EXTRACELLULAR CALCIUM SENSING IN PARATHYROID CELLS

A. Tanini, H. Scherübl, and M. L. Brandi

CONTENTS

1. SUMMARY

The extracellular ionized Ca^{2+} concentration ($[Ca^{2+}]_e$) is the main regulator of parathyroid hormone (PTH) release. Increases in $[Ca^{2+}]_e$ inhibit and decreases in $[Ca^{2+}]_e$ stimulate PTH secretion. Available data indicate that the dose-response relationship for PTH secretion as a function of $[Ca^{2+}]_i$ concentration is biphasic, with a peak at about 200 nM $[Ca^{2+}]_i$. A dose-response relationship for the membrane potential of parathyroid cells as a function of extracellular Ca^{2+} has been demonstrated. Furthermore, the calcium dependence of the open probability for calcium-activated potassium channel is also biphasic, with a peak at 160 mM $[Ca^{2+}]_i$ concentration. It has been suggested that opening of the potassium channels is required for secretion. Several evidences exclude the presence of voltage-dependent calcium channels in the parathyroid cells. Parathyroid cells possess a cell-surface Ca^{2+}-sensing mechanism that also recognizes trivalent and polyvalent cations. Recently, the cloning of complementary DNA, encoding an extracellular Ca^{2+}-sensing receptor from bovine parathyroid cells, has been reported.

0-8493-2477-7/95/$0.00+$.50
© 1995 by CRC Press Inc.

2. INTRODUCTION

Mammalian species are characterized by a complex homeostatic system which has the function of maintaining constant the extracellular ionized calcium concentration.[1-3] This system mainly acts through two essential components: (a) the effector system, made up by specialized cells in the kidney, bone, and intestine, which responds to calciotropic hormones with changes in the transport of mineral ions; (b) specialized endocrine cells (parathyroid and thyroid C cells) sensitive to changes in the extracellular Ca^{2+} concentration, which respond by altering the secretion of their products (PTH and calcitonin).The parathyroid glands play a fundamental role in maintaining the serum calcium concentrations through their secretory product, PTH.[3]

As a general rule, adult human beings possess two pairs of parathyroid glands, each weighing 30–50 mg, placed posteriorly to the four poles of the thyroid gland. A low percentage (2–5%) of individuals show supernumerary parathyroids. The parathyroids originate as paired entodermal structures from the third (parathyroids III) and fourth (parathyroids IV) branchial pouches. Parathyroids III, which develop in close association with the thymus, migrate to become the inferior glands: usually they pinch off just inferolateral to the lower pole of the thyroid. Parathyroids IV (superior glands) form in close association with the ultimobranchial body from which they usually pinch off as it is incorporated into the thyroid. Superior glands are placed dorsolaterally to the thyroid at the level of the thyroid isthmus. Each gland is surrounded by a not clearly defined fibrous capsule and is endowed with a generous capillary plexus derived from hilar vessels. Parathyroid glands are composed of two kinds of cells: the main parenchymal parathyroid cell, also known as chief cell, including water-clear cells, and the oxyphil cells. It is through the chief cells that PTH is synthesized and secreted. They have the typical appearance of active and secretory cells, with a prominent Golgi apparatus and rough endoplasmic reticulum, together with many secretory granules. Chief cells may be arranged in a follicular way, but usually show a cord or nest arrangement. On the contrary, the oxyphil cells have prominent mitochondria and a scarcely developed rough endoplasmic reticulum and Golgi apparatus. They are also not so abundant as the chief cells and usually appear as single cells among the cords of chief cells. Their precise function is unknown. There is also a third type of cell, the so-called water-clear cells (or clear cells), considered large polygonal nonstaining chief cells. The chief cells of the parathyroid gland are rich in electron-dense secretory granules[4] in which PTH is contained.[5] Newly synthesized proteins, including PTH, are transferred from the endoplasmic reticulum to the Golgi apparatus and, finally, to mature secretory vesicles. This process takes about 30 min and involves several covalent modifications of the molecule, independently of alterations in the extracellular Ca^{2+} concentration.[6]

Conversely, PTH secretion is regulated by the concentration of extracellular Ca^{2+}, which represents the primary physiological stimulus. "Ca^{2+} sensing" indicates the capacity of the cell to either recognize or respond to a physiologically relevant change in the extracellular Ca^{2+} concentration.

The relationship between extracellular Ca^{2+} concentration and PTH secretion is inversely proportional. Increases in the extracellular Ca^{2+} concentration depress secretion of PTH, while PTH levels increase in response to a decrease in the extracellular Ca^{2+} concentration. Studies carried out either *in vivo*[7,8] or *in vitro*[9,10] have clearly demonstrated the relationship existing between the extracellular Ca^{2+} concentration and the parathyroid secretory activity. For maintaining the extracellular Ca^{2+} concentration within the very narrow range observed *in vivo*, even with the wide variations in short-term mineral ion availability, PTH secretion must show the following properties: (1) quick and sensitive responsiveness to variations in the extracellular Ca^{2+} concentration, making it possible therefore to amplify the cellular signal into a large secretory response capable of modifying the cellular functions of distant target tissue; (2) adaptive responsiveness, which enables it to adjust the PTH secretion to severe and/ or long-term disorders in extracellular Ca^{2+} concentration.

An important aspect of the secretory control of the parathyroid cell is the inverse relationship between Ca^{2+} and hormonal secretion, which is different than that observed in the majority of secretory cells.

Although in recent years there has been a renewed interest in the mechanism underlying the long-term control of secretion, such as the level of PTH gene expression and the control of cellular proliferation, for a number of years the primary focus of research on parathyroid cell was the acute control of its secretory function.

According to the classical stimulus-secretion coupling theory, each cell ought to respond to a particular stimulus by releasing a secretory product. As a consequence, on the ground of the assumption according to which parathyroid cells are a homogeneous population, an "all or none" response to a decrease in calcium concentrations was expected. Recently, a heterogeneous secretory response to a given agonist has been recognized to occur in several types of secretory cells. Sun and co-workers,[11] using the reverse hemolytic plaque assay (RHPA) for determining the amount of PTH per cell released in response to a given stimulus,[12] have demonstrated that parathyroid cells are homogeneous for what concerns PTH content and capability of synthesis, but do not respond in a comparable homogeneous way to changes in extracellular calcium.

3. PHYSIOLOGY OF CALCIUM-REGULATED PARATHYROID HORMONE SECRETION

Earlier *in vitro* studies[13] suggested that the relationship between the extracellular Ca^{2+} concentration and PTH release was of an inverse linear nature. In later studies, Habener and Potts[14] demonstrated that the relationship between these two parameters is of an inverse sigmoidal nature. These data were confirmed by Brown and co-workers in dispersed bovine[15] and human[16] parathyroid cells. Brown,[17] employing a computer program to curve fit data relating PTH release *in vitro* to the extracellular calcium concentration in normal bovine and human as well as in pathological human parathyroid tissue, provided a mathematical model based on the following equation:

$$PTH\ release = \{(A - D)/[1+ (Ca^{2+}/C)^B]\}+ D$$

where A is the maximal secretory rate, B is the slope of the curve at its midpoint, C is the Ca^{2+} concentration producing half of the maximal inhibition of secretion (the set point), and D is the maximum suppression of PTH release (Figure 1). The steepness of the slope ensures that there will be large changes in the rate of PTH secretion in response to small changes in Ca^{2+}, which contributes to maintaining the extracellular Ca^{2+} concentration within a narrow range. The set point of normal human parathyroid cells is around 1 mM Ca^{2+},[16] which is close to the extracellular Ca^{2+} concentrations (1.1–1.3 mM) in humans.[2,3] The initial response of the parathyroid gland to any alteration in the extracellular Ca^{2+} concentration is immediate, taking place within seconds, either *in vivo* or *in vitro*.[18] In fact, the kinetics of the secretory response to a rapid decrease in the extracellular Ca^{2+} concentration cannot be readily separated from the time course for the alteration in the extracellular Ca^{2+} concentration (Figure 2).

The inversely proportional relationship between PTH secretion and any variation in the extracellular calcium concentration in the physiological range has been described as paradoxical, being the majority of secretory systems stimulated to secrete their products by increases in intracellular calcium concentrations.[19] However, when the extracellular calcium concentration was taken to levels far below the physiological range, PTH secretion decreased.[20,21] Using intracellular quin-2 as indicator for $[Ca^{2+}]_i$, it was possible to demonstrate that PTH release

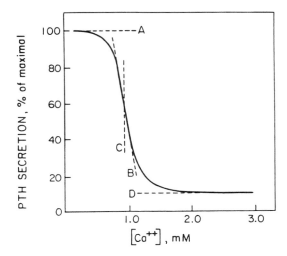

FIGURE 1. Four-parameter model of inverse sigmoidal relationship between extracellular Ca^{2+} and PTH release. (A), maximal secretory rate; (B), slope of curve at its midpoint; (C), midpoint or set point (Ca^{2+} concentration producing half-maximal change in PTH release); (D), minimal secretory rate. (From Brown, E. M., *J. Clin. Endocrinol. Metab.*, 56, 572, 1983. ©The Endocrine Society. With pemission.)

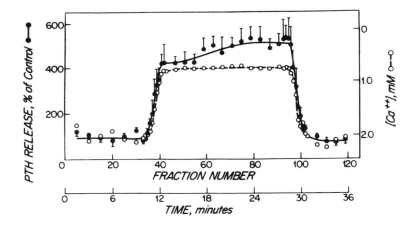

FIGURE 2. Relationship between time course for changes in PTH release and extracellular Ca^{2+} concentration in perifused bovine parathyroid cells. (From Brown, E. M., et al., *Endocrinology*, 116, 1123, 1985. ©The Endocrine Society. With permission.)

increased with $[Ca^{2+}]_i$ up to 200 nM, but gradually decreased above this concentration[22] (Figure 3). The existence of a biphasical dose-response relationship with a peak at an intracellular concentration of about 200 nM and the ascending part of the curve corresponding to subphysiological concentrations, can therefore be proposed. These results were confirmed by another method which involved the control of intracellular calcium concentration through cell permeabilization with electric pulses.[23] The shape of the calcium dose-response curve for PTH secretion was also biphasic, with a peak at 10^{-7} M calcium. In conclusion, the dose-response curves produced by intracellular calcium for PTH secretion in parathyroid cells differ from those typical of other secretory cells. PTH secretion takes place at unusually low calcium concentration and has a clear peak at $1–2 \times 10^{-7}$ M $[Ca^{2+}]_i$. What is unique about parathyroid cells is the inhibitory effect occurring at low calcium concentrations, which becomes the

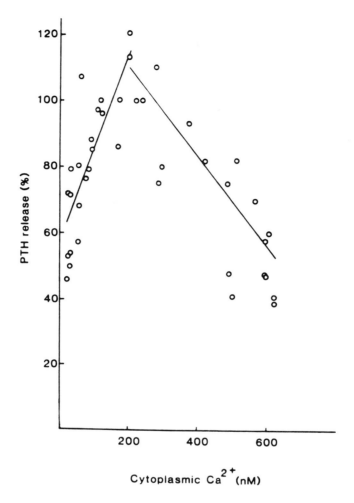

FIGURE 3. Dose-response relationship for PTH secretion as a function of intracellular Ca^{2+} concentration, in quin-2-loaded intact bovine cells. (From Nygren, P. et al., *FEBS Lett.*, 213, 195, 1987. With permission.)

dominant striking feature in the physiological range. The results we have reported demonstrate that low calcium concentrations produce a dose-response curve for PTH secretion which is qualitatively similar to that of other hormones. Furthermore, PTH secretion occurs at an intracellular calcium concentration of about 200 n*M*, both in permeabilized cells (where extracellular calcium concentration is in the submicromolar range) and in intact cells (where extracellular calcium concentration is in the millimolar range). We can conclude that extracellular calcium regulates the intracellular calcium concentration which, in turn, modulates PTH secretion.

4. CYTOSOLIC CALCIUM AND CALCIUM-REGULATED PARATHYROID HORMONE RELEASE

As stated above, within the physiological range there is a well-established negative sigmoid relationship between extracellular Ca^{2+} concentration and PTH secretion.[6,17] The development of intracellular Ca^{2+}-sensitive fluorescent dyes made clear that $[Ca^{2+}]_i$ was related in a positive sigmoidal fashion to extracellular Ca^{2+} (and Mg^{2+}), but inversely to PTH release[24] (Figure 4).

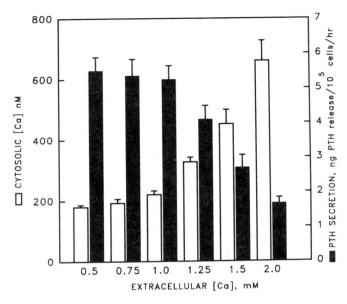

FIGURE 4. Cytosolic Ca^{2+} determination and PTH release at varying extracellular Ca^{2+} concentration in quin-2 loaded cells. White bars represent cytosolic Ca^{2+}_i concentration, while black bars represent PTH secretion. (Data from Reference 24.)

It was suggested, therefore, that cytosolic Ca^{2+} might act as a second messenger mediating the effects of these divalent cations on PTH release. Moreover, adding the divalent cation ionophore ionomycin to bovine parathyroid cells at 1.0 mM extracellular calcium, PTH release decreased, while at the same time $[Ca^{2+}]_i$ increased, reproducing the effect due to an increase in extracellular Ca^{2+} concentration.[25] Several other studies confirmed the mediator role $[Ca^{2+}]_i$ plays in Ca^{2+}-regulated PTH release. In bovine parathyroid cells, high extracellular K^+, while stimulating hormonal secretion, lowered $[Ca^{2+}]_i$.[26,27] Furthermore, either in cells from neonatal calves or in cells from adult cows cultured for several days, there was a gradual decline in the sensitivity of PTH release to extracellular calcium, associated with a decrease in the cytosolic calcium concentration at high calcium.[28] A similar reduction in the sensitivity of both $[Ca^{2+}]_i$ and hormone release to changes in extracellular Ca^{2+} was observed in pathologic parathyroid tissue from many human subjects with primary and secondary hyperparathyroidism.[28] Direct measurements of $[Ca^{2+}]_i$ in response to other modulators of PTH secretion (such as tumor-promoting phorbol ester,[27] Mg^{2+},[25] D-600,[21] and WR2721[29]) have all supported the concept of a causal relationship between $[Ca^{2+}]_i$ and PTH release under these conditions.

In order to confirm the role of $[Ca^{2+}]_i$ in regulating PTH secretion, the relationship existing between extracellular and intracellular calcium concentrations has been evaluated through use of calcium-sensitive fluorescent dyes, quin-2 and fura-2.[22,25,27,30] Both dyes, although showing quantitatively different results, yield similar qualitative results regarding the proportional changes in extracellular and intracellular calcium concentrations. In fact, a 1 mM change in $[Ca^{2+}]_e$ corresponds to a change in the range of a few hundred nanomoles in $[Ca^{2+}]_i$. However, unlike what happens using quin-2, when parathyroid cells were loaded with the Ca^{2+} indicator fura-2, the measurements showed that an increase in the extracellular Ca^{2+} concentration elicited two different responses of bovine parathyroid cells: a rapid and transient increase in $[Ca^{2+}]_i$ followed by a lower, yet sustained increase in $[Ca^{2+}]_i$.[30] (Figure 5).

In cells loaded with quin-2, which has the ability of buffering rapid cytosolic calcium changes, Nemeth et al.[27] and Nemeth and Scarpa[30] demonstrated that high concentrations of extracellular Ca^{2+} or Mg^{2+} inhibited PTH release, suggesting that the high Ca^{2+}-evoked cytosolic

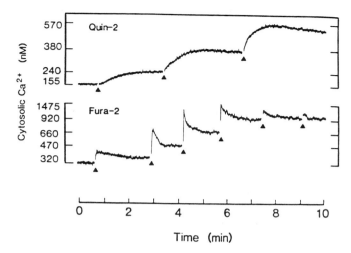

FIGURE 5. Effects of increases in concentrations of extracellular Ca^{2+} on Ca^{2+}_i in parathyroid cells. Cells were loaded with quin-2 (top trace) or fura-2 (bottom trace). Indicator-loaded cells were initially equilibrated at 0.5 mM $CaCl_2$. At each arrowhead, the concentration of Ca^{2+} was increased by 0.5 mM. (From Nemeth, E. F. and Scarpa, A., *J. Biol. Chem.*, 262, 5188, 1987. With permission of ©The American Society for Biochemistry and Molecular Biology.)

Ca^{2+} transients were not required for the accompanying inhibition of PTH release. However, other studies in the kinetics of the secretory response to changes in extracellular Ca^{2+} have suggested that the modulation of $[Ca^{2+}]_i$ dynamics due to the presence of Ca^{2+} buffer does modify the secretory response in parathyroid cells. Therefore, additional studies are necessary to establish whether or not PTH secretion can be altered independently of transient increases in $[Ca^{2+}]_i$.

Despite the many evidences confirming the mediator role played by $[Ca^{2+}]_i$ for Ca^{2+}-regulated PTH release, the causal relationship between $[Ca^{2+}]_i$ and PTH release has been seriously debated. In fact, at 0.5 mM Ca^{2+}, the initial transient, but not the sustained increase, could be elicited by several divalent cations, which inhibit PTH release.[30] Although it was argued that PTH secretion may be regulated by factors other than $[Ca^{2+}]_i$, it is possible that the measurement of changes in $[Ca^{2+}]_i$ in parathyroid cells may not reflect the contribution of $[Ca^{2+}]_i$ in subcompartments within the cytosol that are more intimately related to the control of hormonal secretion. Moreover, the interactions between Mn^{2+} and intracellular fura-2 offer new ways of interpreting the effects of membrane-permeant ions other than Ca^{2+} on steady-state values for $[Ca^{2+}]_i$. In addition, La^{3+}, an ion which is restricted to the extracellular space, inhibits PTH release, mimicking both the initial transient and the sustained increase in Ca^{2+}.[29]

Despite the close correlation between Ca^{2+}-mediated changes in $[Ca^{2+}]_i$ and PTH release, there are a number of instances in which these two parameters may be dissociated. For example, the addition of 20 mM LiCl to bovine parathyroid cells suspended in 2 mM $CaCl_2$ results in a twofold increase in the PTH secretion rate, without any change in $[Ca^{2+}]_i$.[27] Furthermore, at a low concentration of extracellular calcium, Li^+ increases $[Ca^{2+}]_i$, enhancing secretion.[31] Other examples of dissociation between PTH secretion and changes in $[Ca^{2+}]_i$ can be found. Thus, the addition of ionomycin, a divalent cation ionophore, at a low $[Ca^{2+}]_e$ results in a twofold rise in $[Ca^{2+}]_i$, despite the little inhibition of PTH secretion.[27] Although these observations seem to demonstrate that PTH secretion can be readily dissociated from the steady-state level of $[Ca^{2+}]_i$ made artificially high with ionomycin, it is possible that the ionomycin-evoked changes in $[Ca^{2+}]_i$ may have different temporal and spatial characteristics compared with those elicited by an increase in extracellular calcium.

Moreover, various agents (e.g., dopamine and isoproterenol) can modulate PTH secretion in the absence of changes in extracellular calcium and without detectable changes in $[Ca^{2+}]_i$.

A possible explanation for the lack of correlation between PTH secretion and the measured $[Ca^{2+}]_i$ concentrations is that secretion may be correlated with local changes in calcium concentrations at discrete sites of vesicle fusion, and fluorescent dye measurements may not always detect these local changes.

5. MECHANISM BY WHICH $[Ca^{2+}]_E$ REGULATES $[Ca^{2+}]_I$ IN PARATHYROID CELLS

Many secretory cells respond to their appropriate stimuli with the receptor activation that often results in rapid cytosolic Ca^{2+} transients.[32] Therefore, the existence of similar transients in the parathyroid cell responding to its physiological stimulus, together with other evidences we will discuss in the next section, suggested the hypothesis that the parathyroid plasma membrane contained a receptor that coupled extracellular calcium to a second-messenger system that stimulated the release of calcium from intracellular stores. On the ground of other experimental data, it was postulated that the receptor might not be selective for calcium while it could be considered as a polyvalent cation "receptor". Indeed, other divalent and trivalent cations (Mg^{2+}, Sr^{2+}, Ba^{2+}, La^{3+}) also evoked transient increases in $[Ca^{2+}]_i$, depressing PTH secretion and competitively interacting with extracellular calcium. A consistent body of evidence supports the concept that elevated extracellular calcium as well as of other di- and trivalent cations modulate parathyroid function by a receptor-like mechanism.[27,33] This putative receptor seems to be coupled with intracellular effector systems[27,33] by one or more G proteins.[34,35] The early hypothesis concerning the existence of this putative receptor was due to the electrophysiological studies by Lòpez-Barneo and Armstrong,[36] which will be fully discussed later. Recently, the cloning of complementary DNA encoding an extracellular Ca^{2+}-sensing receptor from bovine parathyroids with pharmacological and functional properties nearly identical to those of the native receptor has been described.[37,38] The receptor contains a cluster of acidic amino acid residues, possibly involved in calcium binding, coupled to a seven-membrane spanning domain like those in the G protein-coupled receptor superfamily.[37] Furthermore, it has been demonstrated that mutations in human Ca^{2+}-sensing receptor gene cause familial hypocalciuric hypercalcemia and neonatal severe hyperparathyroidism, two inherited conditions characterized by altered calcium homeostasis.[38]

Usually, Ca^{2+}-mobilizing hormones bind to receptors known to either release intracellular Ca^{2+}, thus evoking the initial spike in $[Ca^{2+}]_i$, or to promote uptake of $[Ca^{2+}]_e$. Likewise, in parathyroid cells the initial spike in $[Ca^{2+}]_i$ is due to the release of a calcium ion from the endoplasmic reticulum,[27,30] probably induced by inositol 1,4,5-triphosphate (IP_3).

The other variable, involved in the coupling between extracellular and intracellular Ca^{2+} concentrations, is the influx across the plasma membrane that contributes to the steady-state levels of $[Ca^{2+}]_i$ in parathyroid cells.[27] There is a direct movement of calcium ions across the plasma membrane through calcium-selective channels that allow controlled entry of extracellular calcium into cells. Next, we describe the role of these channels and the effects of agonists and antagonists in the mechanism regulating secretory events.

It is not yet clear whether the two mechanisms function separately or if they are coupled to each other.

6. ELECTRICAL PROPERTIES OF PARATHYROID CELLS

The first attempts to measure membrane potentials from parathyroid cells were by Bruce and Anderson,[40] who determined the magnitude of the resting membrane potential (RMP) in the mouse as -20.1 ± 1.0 mV in 2.5 mmol/l Ca^{2+} solution.[40] As in many other secretory cells,[41] the main factor in determining the magnitude of the membrane potential was shown to be the transmembrane gradient of K^+.[40] In later studies, the RMP was measured in the parathyroid of different species in 2.5 mmol/l Ca^{2+} solution[42,43,36] (reviewed by Green).[44] The results of several studies indicate that the RMP of the parathyroid cells is clearly of a similar magnitude to that found in the liver, the lacrimal gland, and the pancreatic acinar cells, while higher potentials were recorded in the thyroid follicular, in the parotid acinar, and in the adrenocortical cells (reviewed by Green).[44]

Indeed, membrane electrical activity plays a pivotal role in stimulus-secretion coupling in many secretory cells.[45] The first electrophysiological study on parathyroid cells[40] indicated that the magnitude of RMP of mouse parathyroid cells is exquisitely sensitive to the concentration of extracellular calcium. Other secretory cells, including the mouse thyroid[40] and the exocrine pancreatic,[46] failed to demonstrated this calcium responsiveness.

Another characteristic of parathyroid cells is that the transition between low and high potentials occurs over a narrow range of the calcium concentration (2.25–1.5 mM), with low potentials in high calcium solution and high potentials in low calcium solution. The RMP of mouse parathyroid cells in 1.5 mM calcium solution strongly depended on the extracellular potassium concentration, while in 2.5 mM calcium the membrane potential was independent of extracellular potassium. This led to the hypothesis, which was later confirmed by Jia et al.,[47] that parathyroid cells had a calcium-dependent potassium permeability different from the one observed in other cells. In fact, the ability of calcium to modulate membrane potassium permeability has been clearly demonstrated in other cells, where an increase in $[Ca^{2+}]_i$ levels increased membrane potassium permeability.[48]

Although it has been reported that goat parathyroid cells maintained in culture depolarized at low Ca^{2+},[42] several studies confirmed that a reduction in external Ca^{2+} concentration induced parathyroid hyperpolarization.[36,43] The dose-response relationship for the membrane potential of rat parathyroid cells as a function of extracellular calcium has a biphasic nature, with peak hyperpolarization at a calcium concentration of about 1 mM (Figure 6). On the other hand, while depolarization by calcium of parathyroid cell is associated with inhibition of hormone secretion (Figure 6), depolarization by potassium in a medium at low calcium ion concentration brings an increase in hormone secretion.[43] These data suggest that depolarization per se cannot account for the decrease in hormone secretion at high extracellular $[Ca^{2+}]$ concentrations.

The depolarizing action of high Ca^{2+} is mimicked by several divalent cations (Sr^{2+}, Mg^{2+}, Ba^{2+}, Cd^{2+}, Mn^{2+}, Co^{2+}, Zn^{2+}).[36] Indeed, Co^{2+}, Cd^{2+}, and Mn^{2+}, which are Ca^{2+} channel blockers, mimic the effects of Ca^{2+} rather than block its response.[36] Furthermore, the Ca^{2+} channel blocker D-600 has little, if any, effect on the response produced by high Ca^{2+}.[36] The order of effectiveness of divalent cations in depolarizing parathyroid cells is very similar to the affinity of Ca-binding proteins and calmodulin-like receptors for several divalent cations. The interaction of divalent cations with parathyroid cells is not based on their different permeability through the membrane and points towards the possible existence of a divalent cation receptor in the external surface of these cells, with a different affinity for each of the divalent cations that affect membrane voltage.[36]

Direct electrical communication between cells via gap junctions ("cell-to-cell coupling") is a recognized property of most secretory tissues.[49] The measure of input resistance in whole rat parathyroid gland, indicating a very low value for input resistance in the gland (i.e., ~24 MΩ), suggests that parathyroids, as other endocrine glands,[45] are coupled by junctions of lower resistance. In fact, the value for input resistance predicted for rat parathyroid cells that are ~8 µm in diameter,[50] using 1000 Ω cm^2,[51] would be 498 MΩ if the cells were not coupled to each other by junctions of lower resistance. Indeed, for cultured goat parathyroid cells, the reported figure is ~500 MΩ in 2.4 mM Ca^{2+}. The existence of numerous gap junctions in rat parathyroid cells was confirmed by freeze-fracture studies.[52]

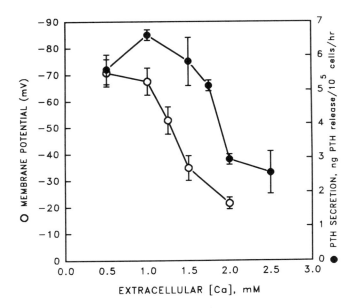

FIGURE 6.   Dose-response relationship for membrane potential and PTH secretion as functions of extracellular Ca^{2+} concentration. Membrane potential data for rat parathyroid cells are taken from Reference 36 and PTH secretion data for bovine parathyroid cells, from Reference 24.

In the range in which increasing calcium concentrations induce a large membrane depolarization, the same calcium concentrations cause a large increase in the input resistance of the membrane.[36] In a tissue that is electrically interconnected by gap junctions, the change observed in the input resistance on raising external Ca^{2+} could arise from altering either junctional or membrane resistance. Although the mechanism by which extracellular Mg^{2+} or transition elements could raise junctional resistance is unknown, it is probable that divalent cations alter membrane resistance. Results obtained by Lòpez-Barneo and Armstrong[36] suggested that the depolarization seen in high Ca^{2+} was due to a suppression of K^+ permeability.

In different species, the relationship between membrane potential and external calcium,[36,40,43] is very similar to the one existing between PTH release and external Ca^{2+} observed in the cow, either *in vitro* [9,24] or *in vivo*.[53] In Figure 6, PTH release and membrane potentials are illustrated as a function of external Ca^{2+}. The data suggest that in parathyroid cells external Ca^{2+}-induced depolarization is associated with a decrease in secretion. However, as already noted earlier, the effects of increased K^+ ion concentration on increase in hormone secretion and decrease in membrane potential,[21,26,43] indicate that depolarization per se cannot account for the decrease in hormone secretion.

The presence of a Ca^{2+}- and voltage-activated K^+ conductance in bovine and rat parathyroid cells was demonstrated by Castellano et al.[54] using the "whole-cell" and "inside-out excised patch" methods, which are variants of the patch clamp technique.[55] They demonstrated that parathyroid cells had K^+ channels of large unitary conductance which resemble the Ca^{2+}-activated K^+ channels found in several electrically excitable cells.[56,57] An important contribution for explaining the unusual responses produced by parathyroid cells to an increase in the external calcium concentration (i.e., a reduction in secretion and a depolarization) was derived from voltage-clamped inside-out patches studies in bovine parathyroid cells.[47] Differently from other calcium-activated potassium channels, where open probability increases with increasing intracellular Ca^{2+} concentration,[48] in parathyroid cells the calcium dependence of the open probability for calcium-activated potassium channels is biphasic, with a peak at an intracellular calcium concentration of about 160 nM [47] (Figure 7). In the absence of calcium,

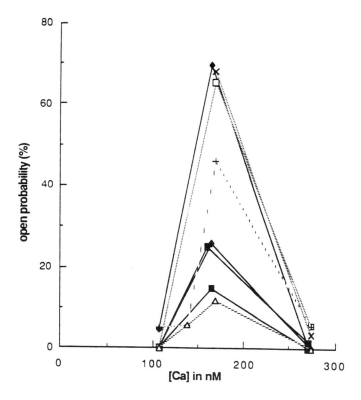

FIGURE 7.   Dose-response relationship of calcium-activated potassium channel open probability as a function of Ca^{2+} concentrations, in the bath at membrane potentials of –30mV (four solid lines, ♦, ■), 0 mV (three lines with horizontal dashes, □, **X**, △), and –50 mV (one line with vertical dashes, +). (From Jia, M. et al., *Proc. Natl. Acad. Sci. U.S.A.,* 85, 7236, 1988. With permission.)

no channel opening was observed, confirming the data obtained by Castellano et al.[54] Parathyroid cells show a K+ channel activity directly dependent on increase in calcium concentration, but only in a very low and narrow range. As shown in Figure 7 the maximum open probability for most channels corresponds to an internal Ca^{2+} concentration of about 160 nM.

When the internal Ca^{2+} concentration is increased above 160 nM, calcium-activated potassium channels tend to close, thus explaining the membrane depolarization that occurs when the external calcium concentration of parathyroid cells is increased. Indeed, an increase in external calcium concentrations has been shown to cause an increase in $[Ca^{2+}]_i$ levels in bovine parathyroid cells.[25] Furthermore, at submicromolar intracellular calcium concentrations, channel opening was relatively independent of membrane potential between –50 and 0 mV.

The proposal that in parathyroid cells an increase in extracellular calcium induces membrane depolarization through closing of calcium-activated potassium channels requires that they represent a large fraction of potassium conductance of the membrane. Accordingly, the depolarization is accompanied by a significant increase in membrane resistance.[36] Other experimental observations support the hypothesis that the block of potassium channels depolarizes the parathyroid cells. In fact, it has been demonstrated that the sensitivity of the membrane potential to changes in potassium concentration strongly depends on the extracellular calcium concentration.[36,40] The above reported data suggest the possibility that in parathyroid cells secretion requires the opening of calcium-activated potassium channels (reviewed by Pocotte et al.).[58] Therefore, secretion could require the opening of calcium-activated potassium channels either in plasma membrane or in secretory vesicles. A possible mechanism

PT−r

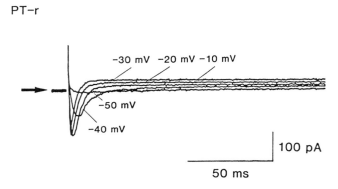

FIGURE 8.   Whole-cell recordings of Ca^{2+} channel currents in PT-r, rat parathyroid cells. Ca^{2+} channel currents, elicited by test pulses to the various indicated potentials from a holding potential of −80mV, are shown. The horizontal arrow marks the zero current level. (From Scherübl, H. et al., *Henry Ford Hosp. Med. J.,* 40, 303–306, 1992. Copyright 1990 *Henry Ford Hospital Medical Journal.* With permission.)

based on the opening of calcium-activated channels in secretory vesicles has been recently proposed.[59] According to this hypothesis, vesicle membrane contains a calcium-activated cation channel and an anion channel, while the plasma membrane contains a calcium channel. When a calcium channel in the plasma membranes opens, a calcium-activated potassium channel in the vesicle membrane opens too, and this, in turn, leads to a flow of anions and water into the vesicles. As a result of water loss from the intermembrane space, the two membranes come in close contact and fuse. On the other hand, Brown et al.[60] demonstrated the existence of the anion transport system which could play a role in exocytosis of PTH in bovine parathyroid cells. In fact, their results showed that PTH release from dispersed parathyroid cells was inhibited by drugs known to block anion transport.

Finally, the patch clamp technique was applied in a rat parathyroid cell line, PT-r cells,[61] in order to investigate whether or not Ca^{2+} influx through voltage-dependent Ca^{2+} channels coupled the extracellular with the intracellular Ca^{2+}.[62] As for parathyroid cells from other species, PT-r parathyroid cells RMP was around −42 mV at 1.2 mM $[Ca^{2+}]_e$; spontaneous action potentials were never noticed. When Ba^{2+} was used as divalent charge carrier and its currents through Ca^{2+} channels were elicited by test pulse to various potentials from a holding potential of −80 mV, Pt-r cells showed fast inactivating, low-threshold Ca^{2+} channel currents with no steady-state conductivity for Ca^{2+} at the resting potential of around −40 mV (Figure 8). The depolarization-induced Ca^{2+} channel currents of Pt-r cells were neither significantly affected by the dihydropyridine Ca^{2+} channel stimulator BAY K 8644 nor by the Ca^{2+} channel blocker israpidine. The absence of any effects of BAY K 8644 and israpidine on voltage-dependent Ca^{2+} channels was demonstrated under steady-state conditions, too. When PT-r cells were voltage clamped close to their resting potential (−40 mV), they exhibited no dihydropyridine-sensitive steady-state conductance (Figure 9). Therefore, on the ground of electrophysiological studies in PT-r cells, it was inferred that voltage dependent dihydropyridine-sensitive Ca^{2+} channels did not play any role in the Ca^{2+} sensitivity of parathyroid cells.

7. PARATHYROID CELL CALCIUM CHANNELS

In other secretory systems, Ca^{2+} influx in response to primary stimuli often results from the opening of voltage-dependent Ca^{2+} channels[63] and calcium channel agonists increase cytosolic calcium[64] and stimulate secretion, while calcium channel antagonists act as inhibitory agents.[65] In the case of parathyroid cells, with the inverse relationship existing between intra- and extracellular Ca^{2+} concentrations and PTH secretion, the dihydropyridine calcium channel

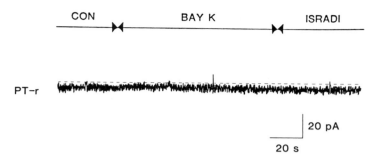

FIGURE 9. No effects of BAY K 8644 and israpidine on steady-state Ca^{2+} channel current in parathyroid PT-r cells. PT-r cells were voltage clamped at a potential of –40mV for 2 min before recording was begun. After the control phase (CON), 500 nM BAY K 8644 followed by 10^{-6} M israpidine was added as indicated. The zero current level is indicated by the broken line. (From Scherübl, H. et al., *Henry Ford Hosp. Med. J.*, 40, 303–306, 1992. Copyright 1990 *Henry Ford Hospital Medical Journal*. With permission.)

agonist (+)202-791 inhibited PTH secretion, while the antagonist (–)202-791 stimulated it.[66] Moreover, the incubation with pertussis toxin, which ADP-ribosylated and inactivated a guanine nucleotide regulatory protein,[66] released the inhibition by the calcium channel agonist, suggesting the existence of a G protein coupling to the calcium channel. Parathyroid cells have a 150,000 mol wt protein analogous to the α_1 subunit of the dihydropyridine-sensitive calcium channel.[66] Antibodies directed against the α_1 subunits of dihydropyridine-sensitive Ca^{2+} channels inhibited PTH release and produced a marked uptake of ^{45}Ca into bovine parathyroid cells, providing further evidence for calcium channel existence.[67] The demonstration of specific binding sites for $[^{125}I]$iodipine, a dihydropyridine-sensitive calcium ligand, suggested the existence of dihydropyridine-sensitive calcium channels in parathyroid cells.[68]

Several investigations on the role played by agonists and antagonists for these channels in the regulation of secretory events have been carried out. Some of these agents, including BAY K 8644, a dihydropyridine, and maitotoxin, a potent activator of voltage-sensitive calcium channels, have proved to have Ca^{2+} channel agonist properties.[69] Pertussis toxin did not block the maitotoxin-induced inhibition, thus suggesting that the agonist does not inhibit PTH release through the same second messenger (pertussis toxin-sensitive G protein) as the 1,4-dihydropiridine derivative (+)202-791.[70] Diltiazem, a calcium channel antagonist of cardiac and vascular muscle cells, produced both a dose-dependent inhibition of PTH release and an increase in the cytosolic Ca^{2+} concentrations in dispersed neonatal bovine parathyroid cells, thus behaving in the parathyroid cells as a Ca^{2+} agonist.[70] Its agonist action was confirmed in bovine parathyroid cells by TA-309, an even more potent derivative of diltiazem.[71] On the other hand, D-600, a verapamil derivative, behaves as an agonist or an antagonist, depending on the extracellular Ca^{2+} concentrations.[21] This Ca^{2+} agonistic behavior of a Ca^{2+} "antagonist" is not a totally unusual phenomenon, because also nifedipine and nitrendipine can shift from a Ca^{2+} agonistic to a Ca^{2+} antagonist role. In fact, although being voltage-dependent Ca^{2+} channel blockers, they can stimulate cellular efflux of Ca^{2+} while, at the same time, their effects can be competitively inhibited by other blockers.[72] Despite the indication of the existence in parathyroid cell membrane of dihydropyridine-sensitive calcium channels, the lack of effects of K^+ depolarization on ^{45}Ca uptake and efflux,[73] together with the observation that such depolarization depresses $[Ca^{2+}]_i$ while stimulating PTH secretion,[26] excluded Ca^{2+} influx through voltage-dependent Ca^{2+} channels in parathyroid cells. Furthermore, as noted earlier, the electrophysiological studies by Scherübl et al.[62] in PT-r cells brought to conclude that voltage-dependent Ca^{2+} channels could not be considered as the mechanism by which extracellular calcium could produce increases in Ca^{2+} influxes in parathyroid cells. On the other hand, effects of dihydropyridines on current flow through voltage-insensitive calcium

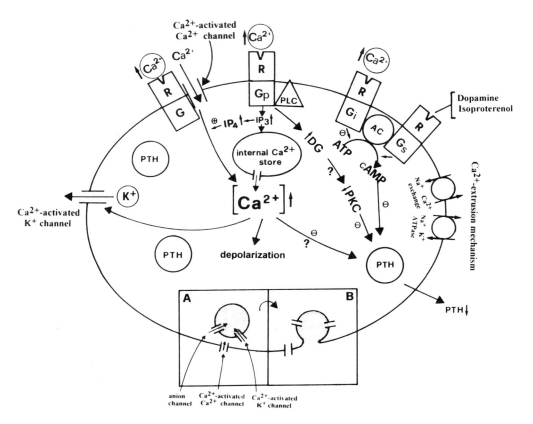

FIGURE 10. Schematic diagram of extra- and intracellular mechanisms involved in control of PTH secretion in bovine parathyroid cells by extracellular Ca^{2+} and other factors. High extracellular Ca^{2+} binds to one or more types of Ca^{2+} receptors, and every one of them may interact with separate intracellular effector systems. In the adenylate cyclase (AC)-cAMP system, the receptor is coupled via an inhibitory G protein (G_i). High extracellular Ca^{2+} also activates phospholipase C (PLC), perhaps via a G protein (G_p). Resultant increase in inositol trisphosphate (IP_3) may mediate the Ca^{2+}-induced spike in cytosolic Ca^{2+}; inositol tetrakisphosphate (IP_4), a product of the phosphorylation of IP_3, may act as a stimulator of extracellular Ca^{2+} influx. Ca^{2+} inhibition of PTH release is associated with parathyroid cell depolarization and Ca^{2+} influx. The parathyroid cell plasma membrane appears to possess voltage-independent Ca^{2+} channels activated by Ca^{2+} binding to an external receptor. Moreover, a Na^+-Ca^{2+} exchange system in the parathyroid cell plasma membrane may contribute significantly to the maintenance of cytosolic calcium homeostasis. The energetics of this Na^+-Ca^{2+} exchange mechanism rely upon the activity of the plasma membrane Na^+-K^+-ATPase. Parathyroid cells possess Ca^{2+}-dependent K^+ channels; their role, possibly involved in PTH secretion (inset), is fully discussed in the text.

channels have been described.[74] Many nonexcitable cells show a sustained phase of elevated $[Ca^{2+}]_i$ due to influx of extracellular calcium.[75] On the ground of electrophysiological and pharmacological evidences, an influx mediated by second messengers, rather than by classic voltage-dependent channels, can be inferred. The role of a potential calcium influx mediator has been attributed to $[Ca^{2+}]_i$[76] and to IP_3,[77] but inositol 1,3,4,5-tetrakisphosphate (IP_4) could also be involved in this process.[78] A number of evidences indicated that IP_4, a product of the phosphorylation of IP_3 by a IP_3-3-kinase, may act as a stimulator of influx of extracellular Ca^{2+}, perhaps through IP_4-dependent Ca^{2+} channels.[78] As we will fully discuss later, the concentration of IP_3 in the parathyroid cell increases in response to the increase in extracellular calcium concentration.[79] Furthermore, both Ca^{2+} and Mg^{2+} produce dose-dependent increases in IP_4.[80] The concept of a Ca^{2+}-activated Ca^{2+} permeability was tested further by using the trivalent ion La^{3+}, which, despite being extracellular, was found to be able to stimulate ^{45}Ca uptake and to raise $[Ca^{2+}]_i$.[29]

8. ROLE OF SECOND MESSENGERS

8.1. cAMP

cAMP does seem to be an important second messenger in mediating PTH secretion in response to agonists, such as dopamine or isoproterenol, which stimulate PTH release (reviewed by Brown)[81] in the absence of any measurable change in $[Ca^{2+}]_i$.[25] Such secretagogues activate adenylate cyclase and cAMP accumulation through the stimulatory guanine nucleotide-binding protein, G_s.[82] Because cAMP rises in parathyroid cells at low extracellular Ca^{2+},[83] this cyclic nucleotide could potentially contribute to low-Ca^{2+}-stimulated PTH release. Conversely, α-adrenergic agents and prostaglanolin $F_{2\alpha}$ ($PGF_{2\alpha}$) inhibit cAMP production and hormone release from parathyroid cells (reviewed by Brown).[81] The effect of pertussis toxin on the inhibitory actions of α-adrenergic agents and $PGF_{2\alpha}$ on cAMP accumulation and PTH secretion indicates that parathyroid cells contain "inhibitory" G proteins (G_i) that couple the actions of those agents to inhibition of cAMP production and PTH secretion.[84] Elevation in intracellular cAMP are not associated with any changes in either $[Ca^{2+}]_i$[81] or in the levels of inositol phosphate (IP).[81] Thus, available evidences suggest that cAMP can regulate PTH secretion independent of other messengers.

Ca^{2+}, as well as several other divalent cations, reduces cAMP content in parathyroid cells and potently inhibits dopamine-stimulated cAMP accumulation.[85] The inhibitory effects of Ca^{2+}, as well as of the other divalent cations, are totally prevented by preincubation with pertussis toxin.[85] Thus, polyvalent cations may reduce cAMP accumulation in parathyroid cells by a G_i-coupled receptor analogous to the mechanisms observed for hormone receptors. As an alternative mechanism by which Ca^{2+}, as well as the other divalent cations, could modulate cAMP levels in the parathyroid cells, it has been suggested that this action might result from divalent-induced increases in cytosolic Ca^{2+}, which could inhibit adenylate cyclase and/or activate phosphodiesterase activities.[28] However, later results obtained from the same authors did not confirm the previous data.[85] The pharmacological comparison of the effects of divalent and trivalent cations on cAMP and PTH release to their actions on IP metabolism raises the possibility that there might be more than one form of the putative Ca^{2+} receptor.[86]

The role of cAMP in Ca^{2+}-regulated secretion is considerably less clear. It is generally agreed that cAMP decreases and $[Ca^{2+}]_i$ increases with elevation in extracellular Ca^{2+}, but these changes can be dissociated from the accompanying changes in the secretion.[81] Currently available data therefore do not support an important mediatory role for cAMP in Ca^{2+}-regulated PTH secretion.

8.2. THE PI-PROTEIN KINASE C SYSTEM AND PTH SECRETION

In other cell types the turnover of polyphosphoinositides (PI) is an important signal transduction pathway that generates the second messengers IP_3 and diacylglycerol (DAG). IP_3 mobilizes calcium from intracellular stores, and DAG activates protein kinase C (PKC), a calcium-activated, phospholipid-dependent enzyme that plays an important role in controlling secretion and cell proliferation.[87]

In parathyroid cells, high extracellular Ca^{2+},[79] divalent cations,[79] and fluoride[88, 89] all produce marked increases in IP_3 and related products. This observation supports the idea that increases in IP_3 may mediate intracellular Ca^{2+} mobilization and that second messengers, generated through polyphosphoinositide turnover, may be involved in inhibiting PTH release. Recent evidences have confirmed that high extracellular Ca^{2+} rapidly reduces the levels of [^{3}H]inositol-labeled phosphoinositides and increases the level of 1,4,5-trisphosphate, thus, like more classic Ca^{2+}-mobilizing hormones, activating phospholipase C (PLC).[90] The changes in the metabolism of IP may be related to a guanine nucleotide-binding protein (G protein)-dependent process, since exposure to F^-,[89] a known activator of G protein-mediated phosphatidylinositol hydrolysis, also enhances accumulation of IP. In addition, high extracellular Ca^{2+} rapidly increases intracellular levels of DAG as well as IP_3 in bovine parathyroid cells.[91]

Unlike other secretory cells, in which phosphoinositide breakdown is accompanied by enhanced hormonal release, in the parathyroid cell this process is coupled to inhibition of PTH secretion. It seems unlikely that increases in PLC activity per se inhibit PTH release, since addition of exogenous PLC stimulates hormone secretion from parathyroid cells.[92] Furthermore, two studies have demonstrated that acute exposure of bovine parathyroid cells to low extracellular calcium (0.5 mM) caused an increase in PKC activity associated with the particulate fraction, while high extracellular Ca^{2+} reduced the translocation of PKC from cytosol to particulate fractions.[93, 94] The reported data suggest that in parathyroid cells increased PKC activity produces a concomitant stimulation of hormonal secretion. To furthermore support this hypothesis, it has been demonstrated that phorbol ester and synthetic DAG enhance the rate of PTH secretion at high extracellular Ca^{2+} concentration.[93] Moreover, in the RHPL, PKC inhibitors such as H_7, tamoxifen, and sphinganine inhibited PTH release at low extracellular Ca^{2+} and decreased recruitment over the physiological range of extracellular Ca^{2+}.[95] In addition, the same study has indicated that PKC stimulation by 12-*O*-tetradecanoylphorbol 13 acetate (TPA) increases steady-state PTH mRNA levels.

However, it may appear paradoxical that in parathyroid cells, PKC activity will increase at low extracellular calcium concentration, when cytosolic calcium is reduced, since the enzyme requires calcium for its activity.[96] Moreover, DAG, which appears to serve as an intracellular mediator by activating PKC, is low at low extracellular calcium concentration. The possible ways in which this paradox can be resolved is that in the presence of adequate levels of DAG, PKC can be activated even at "resting" levels of intracellular calcium (i.e., 100 mM)[96] and therefore it is possible that in parathyroid cells the resting levels of DAG at low extracellular calcium are sufficiently high to activate the enzyme.[81] The other question that arises is how can the observed increase in DAG and $[Ca^{2+}]_i$ at high Ca^{2+} in parathyroid cells lead to a decrease, rather than to the expected increase, in PKC activity? Perhaps additional mechanisms would have to exist that inhibit PKC in the face of increasing levels of factors (DAG and $[Ca^{2+}]_i$) known to stimulate the enzyme. For example, Oishi et al.[97] have found that hormones activating phosphoinositide hydrolysis may concurrently activate phospholipase A_2 (PLA_2) and sphingomyelinase, yielding lysophosphatityl choline and sphingosine, which can both inhibit PKC. Accordingly, recent evidences indicate that increasing extracellular calcium concentrations increases free [^{3}H] arachidonic acid (AA) release and decreases PTH secretion from labeled parathyroid cells.[98] Exogenous PLA_2 and AA inhibited PTH secretion in a dose-dependent manner. The effects of two inhibitors of the lipoxygenase pathway (LO), phenidone and baicalein, a relatively selective 12-LO inhibitor, on high Ca^{2+}-induced inhibition of PTH secretion, indicated that the AA may act via the 12-LO pathway, suggesting that LO pathway products may act as second messengers in parathyroid cells. Further studies are necessary to clarify the role of AA metabolites in the action of Ca^{2+} on parathyroid cells, in particular whether or not they might alter calcium influx and/or affect PKC.

Finally, it is possible that additional mediators known to be regulated by extracellular Ca^{2+} also contribute to the regulation of PTH secretion by extracellular Ca^{2+} in parathyroid cells. The Ca^{2+}-mediated changes in intracellular pH,[99] hexose monophosphate shunt activity,[100] and cellular respiration[101] could play some role — whose nature is not clear — in the control of PTH secretion. Moreover, it has been demonstrated that plasma membrane of bovine parathyroid cells contains a Na^+-Ca^{2+} exchange mechanism capable of regulating cytosolic calcium concentration, which, in turn, regulates PTH secretion.[102] The energetics of this mechanism rely upon the activity of Na^+-K^+-ATPase of the plasma membrane.

9. OTHER PARATHYROID SECRETORY PRODUCTS

9.1. PARATHYROID HORMONE-LIKE PROTEIN

PTH-related protein (PTH-rP) is a recently discovered protein which shares N-terminal homology with PTH and displays many of its biological properties.[103]

Recently, biological activity and secretion of PTH-rP were described in bovine parathyroid glands.[104] Moreover, it has been demonstrated that in normal bovine parathyroid cells, PTH-rP release did not respond to various stimuli which are known to modulate PTH secretion (e.g., medium calcium concentration, cAMP, or isoproterenol). The available data seem to indicate that in bovine parathyroid cells, PTH-rP is not secreted via the regulated pathway utilizing PTH secretory granules.[104] Most of the cellular PTH-rP occurs in the soluble portion of cell lysate, while PTH occurs in the crude granule fraction. The role of PTH-rP in the parathyroid cell is unknown. It is possible that PTH-rP works as a paracrine/autocrine factor in parathyroid cells.

9.2. CHROMOGRANIN A

Chromogranin A (CgA) is a member of a family of acidic glycoproteins that are found in endocrine and neuroendocrine tissues and which are contained in secretory granules (large, dense-cored granules).[105] In the parathyroid glands, CgA that was initially described under the name secretory protein-I,[106] is cosecreted with PTH in response to low calcium and other stimuli, such as β agonists.[107] Recently, a 26-kDa N-terminal fragment of CgA has been described that inhibits low calcium-stimulated secretion of PTH and CgA to the same extent as high calcium concentrations.[108] Thus, in normal endocrine tissues, intact CgA could serve as a precursor for an N-terminal fragment that might be capable of inhibiting cellular secretion. Likewise, an inhibitory effect of PTH (1-34) on PTH secretion has been reported,[109] thus suggesting the likely existence of a cooperative mechanism between calcium and other factors to inhibit secretion. Moreover, bovine parathyroid cells secrete a 44-kDa plasminogen activator, whose secretion is regulated by extracellular calcium.[110] Finally, recent researches have shown the parathyroid to be a source of endothelin-1[111] and a yet unidentified vasoconstrictor humoral factor.[112]

REFERENCES

1. Brown, E. M., Regulation of the synthesis, secretion and actions of parathyroid hormone, in *Contemporary Issues in Nephrology. Divalent Cation Homeostasis,* Vol. 11, Brenner, B. M. and Stein, J. H., Eds., Churchill-Livingstone, New York, 1983, 151.
2. Aurbach, G. D., Marx, S. J., and Spiegel, A. M., Parathyroid hormone, calcitonin, and the calciferols, in *Textbook of Endocrinology,* 7th ed., Wilson, J. D. and Foster, D. W., Eds., W. B. Saunders, Philadelphia, 1985, 1137.
3. Stewart, A. F. and Broadus, A. E., Mineral metabolism, in *Endocrinology and Metabolism,* 2nd ed., Felig, P., Baxter, J. D., Broadus, A. E., and Frohman, L. A., Eds., McGraw-Hill, New York, 1987, 1317.
4. Roth, S. I., and Shiller, A. L., Comparative anatomy of the parathyroid glands, in *Handbook of Physiology. Endocrinology. Parathyroid Gland,* Sect. 7, Vol. VII, American Physiological Society, Washington, D.C., 1976, 281.
5. Ravazzola, M., Orci, L., Habener, J. F., and Potts, J. T., Jr., Parathyroid secretory protein, immunocytochemical localization within cells that contain parathyroid hormone, *Lancet,* 2, 371, 1978.
6. Habener, J. F., Rosenblatt, M., and Potts, J. T., Parathyroid hormone: biochemical aspects of biosynthesis, secretion, action, and metabolism, *Physiol. Rev.,* 64, 985, 1984.
7. Ramberg, C. F., Mayer, G. P., Kronfeld, D. S., Aurbach, G. D., Sherwood, L. M., and Potts, J. T., Plasma calcium and parathyroid hormone responses to EDTA infusion in the cow, *Am. J. Physiol.,* 213, 878, 1967.
8. Mayer, G. P. and Hurts, J. G., Sigmoidal relationship between parathyroid hormone secretion rate and plasma calcium concentration in calves, *Endocrinology,* 102, 1036, 1978.
9. Brown, E. M., Hurwitz, S., and Aurbach, G. D., Preparation of viable isolated bovine parathyroid cells, *Endocrinology,* 99, 1582, 1976.
10. Wallace, J. and Scarpa, A., Regulation of parathyroid hormone secretion in vitro by divalent cations and cellular metabolism, *J. Biol. Chem.,* 257, 10613, 1982.
11. Sun, F., Ritchie, C. K., Hasseger, C., and Fitzpatrick, L. A., Heterogeneous response to calcium by individual parathyroid cells, *J. Clin. Invest.,* 91, 595, 1993.

12. Fitzpatrick, L. A. and Leong, D. A., Individual parathyroid cells are more sensitive to calcium than a parathyroid cell population, *Endocrinology,* 126, 720, 1990.
13. Sherwood, L. M., Mayer, L. M., Ramberg, C. J., Kronfeld, D. S., Aurbach, G. D., and Potts, J. J., Regulation of parathyroid hormone secretion: proportional control by calcium, lack of effect of phosphate, *Endocrinology,* 83, 1043, 1968.
14. Habener, J. F. and Potts, J. T., Jr., Relative effectiveness of magnesium and calcium on the secretion and biosynthesis of parathyroid hormone in vitro, *Endocrinology,* 98, 197, 1976.
15. Brown, E. M., Brennan, M. F, Hurwitz, S., Windeck, R. A., Marx, S. J., Spiegel, A. M., Koheler, S., Gardner, J. D., and Aurbach, G. D., Dispersed cells prepared from human parathyroid glands: distinct calcium sensitivity of adenomas *vs.* primary hyperplasia, *J. Clin. Endocrinol. Metab.,* 46, 267, 1978.
16. Brown, E. M., Gardner, D. G., Brennan, M. F., Marx, S. J., Spiegel, A. M., Attie, M. F., Downs, R. W., Doppman, J. L., and Aurbach, G. D., Calcium-regulated parathyroid hormone release in primary hyperparathyroidism, *Am. J. Med.,* 66, 923, 1979.
17. Brown, E. M., Four-parameter model of the sigmoidal relationship between parathyroid hormone release and extracellular calcium concentration in normal and abnormal parathyroid tissue, *J. Clin. Endocrinol. Metab.,* 56, 572, 1983.
18. Brown, E. M., Leombruno, R., Thatcher, J., and Burrowes, M., The acute secretory response to alterations in the extracellular calcium concentration and dopamine in perifused bovine parathyroid cells, *Endocrinology,* 116, 1123, 1985.
19. Rubin, R. P., The role of calcium in the release of neurotransmitter substances and hormones, *Pharmacol. Rev.,* 22, 389, 1970.
20. Ramp, W. K., Cooper, C. W., Ross, A. J., III, and Wells, S. A. Jr., Effects of calcium and cyclic nucleotides on rat calcitonin and parathyroid hormone secretion, *Mol. Cell. Endocrinol.,* 14, 205, 1979.
21. Larsson, R., Akerstrom, G., Gylfe, E., Johansson, H., Ljiunghall, S., Rastad, J., and Wallfelt, C., Paradoxical effects of K^+ and D-600 on parathyroid hormone secretion and cytoplasmatic Ca^{2+} in normal bovine and pathological human parathyroid cells, *Biochim. Biophys. Acta,* 847, 263, 1985.
22. Nygren, P., Larsson, R., Lindh, E., Ljunghall, S., Rastad, J., Akestrom, G., and Gylfe, E., Bimodal regulation of secretion by cytoplasmatic Ca^{2+} as demonstrated by the parathyroid, *FEBS Lett.,* 213, 195, 1987.
23. Pocotte, S. L. and Ehrestein, G., The biphasic calcium dose-response curve for parathyroid hormone secretion in electropermeabilized adult bovine parathyroid cells, *Endocrinology,* 125, 1587, 1989.
24. Shoback, D. M., Thatcher, J., Leombruno, R., and Brown, E. M., Effects of extracellular Ca^{2+} and Mg^{2+} on cytosolic Ca^{2+} and PTH release in dispersed bovine parathyroid cells, *Endocrinology,* 113, 424, 1983.
25. Shoback, D. M., Thatcher, J., Leombruno, R., and Brown, E. M., Relationship between parathyroid hormone secretion and cytosolic calcium concentration in dispersed bovine parathyroid cells, *Proc. Natl Acad. Sci. U.S.A.,* 81, 3113, 1984.
26. Shoback, D. and Brown, E. M., PTH release stimulated by high extracellular potassium is associated with a decrease in cytosolic calcium in bovine parathyroid cells, *Biochem. Biophys. Res. Commun.,* 123, 684, 1984.
27. Nemeth, E. F., Wallace, J., and Scarpa, A., Stimulus-secretion coupling in bovine parathyroid cells, *J. Biol. Chem.,* 261, 2668, 1986.
28. LeBoff, M. S., Shoback, D., Brown, E. M., Thatcher, J., Leombruno, R., Beaudoin, D., Henry, M., Wilson, R., Pallotta, J., Marynick, S., Stock, J., and Leight, G., Regulation of parathyroid release and cytosolic calcium by extracellular calcium in dispersed and cultured bovine and pathological human parathyroid cells, *J. Clin. Invest.,* 75, 49, 1985.
29. Gylfe, E., Larsson, R., Johansson, H., Nygren, P., Rastad, J., Wallfelt, C., and Akerstrom, G., Calcium-activated permeability in parathyroid cells, *FEBS Lett.,* 205, 132, 1986.
30. Nemeth, E. F. and Scarpa, A., Rapid mobilization of cellular Ca^{2+} in bovine parathyroid cells evoked by extracellular divalent cations, *J. Biol. Chem,* 262, 5188, 1987.
31. Wallace, J. and Scarpa, A., Similarities of Li^+ and low Ca^{2+} in the modulation of secretion by parathyroid cells in vitro, *J. Biol. Chem.,* 258, 6288, 1983.
32. Gill, D. L., Receptors coupled to calcium mobilization, in *Advances in Cyclic Nucleotide and Protein Phosphorylation Research,* Cooper, D. M. F. and Searmon, K. B., Eds., Raven Press, New York, 1985, 307.
33. Brown, E. M., Chen, C. J., LeBoff, M. S., Kifor, O., and El-Hajj, G., Mechanism underlying the inverse control of parathyroid hormone secretion by calcium, in *Secretion and Its Control,* Oxford, G. and Armstrong, C. M., Eds., Rockfeller University Press, New York, 1989, 252.
34. Fitzpatrick, L. A. and Aurbach, G. D., Calcium inhibition of parathyroid hormone secretion is mediated via a guanine nucleotide regulatory protein, *Endocrinology,* 119, 2700, 1986.
35. Chen, C. J., Barnet, J. V., Congo, D. A., and Brown, E. M., Divalent cations suppress 3′,5′-adenosine monophosphate accumulation by stimulating a pertussis toxin-sensitive guanine nucleotide-binding protein in cultured bovine parathyroid cells, *Endocrinology,* 124, 233, 1989.
36. Lòpez-Barneo, J. and Amstrong, C. M., Depolarizing response of rat parathyroid cells to divalent cations, *J. Gen. Physiol.,* 82, 269, 1983.

37. Brown, E. M., Gamba, G., Riccardi, D., Lombardi, M., Butters, R., Kifor, O., Sun, A., Hediger, M. A., Lytton, J., and Hebert, S. C., Cloning and characterization of an extracellular Ca^{2+} receptor from bovine parathyroid, *Nature*, 366, 575, 1993.
38. Pollak, M. R., Brown, E. M., Wu Chou, Y. H., Hebert, S. C., Marx, C. E., Seidman, C. E., and Seidman, J. G., Mutations in the human Ca^{2+}-sensing receptor gene cause familial hypocalciuric hypercalcemia and neonatal severe hyperparathyroidism, *Cell*, 75, 1297, 1993.
39. Epstein, P. A., Prentky, M., and Attie, M. F., Modulation of intracellular Ca^{2+} in parathyroid cells. Release of Ca^{2+} from non-mitochondrial pools by inositol trisphosphate, *FEBS Lett.*, 188, 141, 1985.
40. Bruce, B. R. and Anderson, N. C., Jr., Hyperpolarization in mouse parathyroid cells by low calcium, *J. Physiol.*, 236, C15, 1979.
41. Williams, J. A., Origin of transmembrane potentials in non-excitable cells, *J. Theor. Biol.*, 28, 287, 1970.
42. Sand, O., Ozawa, S., and Hove, K., Electrophysiology of cultured parathyroid cells from the goat, *Acta Physiol. Scand.*, 113, 45, 1981.
43. Morrissey, J. J. and Klahr, S., Dissociation of membrane potential and hormone secretion in bovine parathyroid cells, *Am. J. Physiol.*, 245, E102, 1983.
44. Green, S. T., The electrophysiological properties of the parathyroid cell: results of a study employing Sprague-Dawley rats and a review of the literature, *Biomed. Pharmacother.*, 42, 61, 1988.
45. Petersen, O. H., *The Electrophysiology of Gland Cells,* Academic Press, New York, 1980.
46. Nishiyama, A. and Petersen, O. H., Pancreatic acinar cell: ionic dependence of acetylcholine-induced membrane potential and resistance change, *J. Physiol. London*, 244, 431, 1975.
47. Jia, M., Ehrestein, G., and Iwasa, K., Unusual calcium-activated potassium channel of bovine parathyroid cells, *Proc. Natl. Acad. Sci. U.S.A.*, 85, 7236, 1988.
48. Latorre, R., Oberhauser, A., Labarca, P., and Alvarez, O., Varieties of calcium activated potassium channels, *Annu. Rev. Physiol.*, 51, 385, 1989.
49. Green, S. T., Direct intercellular communication in health and disease: an overview, *Scott. Med. J.*, 26, 315, 1981.
50. Roth, S. I. and Raisz, L. G., Effect of calcium concentration on the ultrastructure of rat parathyroid in organ culture, *Lab. Invest.*, 13, 331, 1968.
51. Cole, K. S., *Membranes, Ions and Impulses*, University of California Press, Berkeley, 1968, 569.
52. Ravazzola, M. and Orci, L., Intercellular junctions in the rat parathyroid gland: a freeze-fracture study, *Biol. Cell.*, 28, 137, 1977.
53. Mayer, G. P. and Hurst, J. G., Sigmoidal relationship between parathyroid hormone secretion rate and plasma calcium concentration in calves, *Endocrinology*, 102, 1036, 1978.
54. Castellano, A., Pintado, E., and López-Barneo, J., Ca^{2+}- and voltage-dependent K^+ conductance in dispersed parathyroid cells, *Cell Calcium,* 8, 377, 1987.
55. Hamil, O. P., Marty, A., Neher, E., Sakmann, B., and Sigworth, F., Improved patch-clamp techniques for high resolution current recording from cells and cell-free membrane patches, *Pfluegers Arch.*, 381, 85–100, 1981.
56. Marty, A. and Neher, E., Potassium channels in cultured bovine adrenal chromaffin cells, *J. Physiol.*, 367, 117, 1985.
57. Blatz, A. and Magleby, K. L., Ion conductance and selectivity of single calcium-activated potassium channel in cultured rat muscle, *J. Gen. Physiol.*, 84, 1, 1984.
58. Pocotte, S. L., Ehrenstein, G., and Fitzpatrick, L. A., Regulation of parathyroid hormone secretion, *Endocr. Rev.,* 12, 291, 1991.
59. Ehrenstein, G., Stanley, E. F., Pocotte, S. L., Jia, M., Iwasa, K., and Krebs, K., E., Evidence for a model of exocytosis that involves calcium-acytivated channels, *Ann. N.Y. Acad. Sci.,* 635P, 297, 1991.
60. Brown, E. M., Pazoles, C. J., Creutz, C. E., Aurbach, G. D., and Pollard, H. B., Role of anions in parathyroid hormone release from dispersed bovine parathyroid cells, *Proc. Natl. Acad. Sci. U.S.A.*, 75, 876, 1978.
61. Sakaguchi, K., Santora, A., Zimering, M., Curcio, F., Aurbach, G. D., and Brandi, M. L., Functional epithelial cell line cloned from rat parathyroid glands, *Proc. Natl. Acad. Sci. U.S.A.*, 84, 3269, 1987.
62. Scherübl, H., Brandi, M. L., and Hescheler, J., Extracellular Ca^{2+} sensing in C-cells and parathyroid cells, *Henry Ford Hosp. Med. J.,* 40, 303, 1992.
63. Rasmussen, H. and Barret, P. Q., Calcium messenger system: an integrated view, *Physiol. Rev.*, 64, 938, 1984.
64. Hishikawa, R., Fukase, M., Takenaka, M., and Fujita, T., Effect of calcium channel agonist Bay K 8644 on calcitonin secretion from a rat C-cell line, *Biochem. Biophys. Res. Commun.*, 130, 454, 1985.
65. Aguilera, G. and Catt, K. J., Participation of voltage-dependent calcium channels in the regulation of adrenal glomerulosa function by angiotensin II and potassium, *Endocrinology,* 118, 112, 1986.
66. Fitzpatrick, L. A., Brandi, M. L., and Aurbach, G. D., Control of PTH secretion is mediated through calcium channels and is blocked by pertussis toxin treatment of parathyroid cells, *Biochem. Biophys. Res. Commun.*, 138, 960, 1986.
67. Fitzpatrick, L. A., Chin, H., Nirenberg, M., and Aurbach, G. D., Antibodies to an α subunit of skeletal muscle calcium channels regulated parathyroid cell secretion, *Proc. Natl. Acad. Sci. U.S.A.*, 85, 2115, 1988.

68. Jones, J. I. and Fitzpatrick, L. A., Binding of [125]Iiodipine to parathyroid cell membranes: evidence of a dihydropyridine-sensitive calcium channel, *Endocrinology,* 126, 2015, 1990.

69. Fitzpatrick, L. A., Yasumoto, T., and Aurbach, G. D., Inhibition of parathyroid hormone release by maitotoxin, a calcium channel activator, *Endocrinology,* 124, 97, 1989.

70. Seely, E. W., LeBoff, M. S., Brown, E. M., Chen, C., Posillico, J. T., Hollenberg, N. K., and Williams, G. H., The calcium channel blocker diltiazem lowers serum parathyroid hormone levels in vivo and in vitro, *J. Clin. Endocrinol. Metab.,* 68, 1007, 1989.

71. Chen, C. J. and Brown, E. M., The diltiazem analog TA-3090 mimics the actions of high extracellular Ca^{2+} on parathyroid function in dispersed bovine parathyroid cells, *J. Bone Miner. Res.,* 5, 581, 1990.

72. Towart, R. and Schramm, M., Recent advances in the pharmacology of the calcium channel, *Trends Pharmacol. Sci.,* 5, 111, 1984.

73. Wallfelt, C., Åkerström, G., Ljunghall, S., and Gylfe, E., Stimulus-secretion coupling of parathyroid hormone release: studies of ^{45}Ca and ^{86}Rb fluxes, *Acta Physiol. Scand.,* 124, 239, 1985.

74. Young, W., Chen, J., Jung, F., and Gardner, P., Dihydropyridine Bay K 8644 activates T lymphocyte calcium-permeable channels, *Mol. Pharmacol.,* 34, 239, 1988.

75. Merrit, J. and Rink, T., Regulation of cytosolic free calcium in fura-2-loaded rat parotid acinar cells, *J. Biol. Chem.,* 262, 17362, 1987.

76. von Tsharner, V., Prod'hom, B., Baggiolini, M., and Reuter, H., Ion channels in human neutrophils activated by a rise in free cytosolic calcium concentration, *Nature,* 324, 369, 1986.

77. Kuno, M. and Gardner, Ph., Ion channels activated by inositol 1,4,5-triphosphate in plasma membrane of human T-lymphocytes, *Nature,* 326, 301, 1987.

78. Irvine, R. F. and Moor, R. M., Micro-injection of inositol 1,3,4,5-tetrakisphosphate activates sea urchin eggs by a mechanism on external Ca^{2+}, *Biochem. J.,* 240, 917, 1986.

79. Brown, E. M., Enyedi, P., LeBoff, M., Rotberg, J., Preston, J., and Chen, C., High extracellular Ca^{2+} and Mg^{2+} stimulate accumulation of inositol phosphates in bovine parathyroid cells, *FEBS Lett.,* 218, 113, 1987.

80. Hawkins, D., Enyedi, P., and Brown, E. M., The effects of high extracellular Ca^{2+} and Mg^{2+} concentrations on the levels of inositol 1,3,4,5-tetrakisphosphate in bovine parathyroid cells, *Endocrinology,* 124, 838, 1989.

81. Brown, E. M., Extracellular Ca^{2+} sensing, regulation of parathyroid cell function, and role of Ca^{2+} and other ions as extracellular (first) messengers, *Physiol. Rev.,* 71, 371, 1991.

82. Attie, M. F., Brown, E. M., Gardner, D. G., Spiegel, A. M., and Aurbach, G. D., Characterization of the dopamine-response adenylate cyclase of bovine parathyroid cells and its relationship to parathyroid hormone secretion, *Endocrinology,* 107, 1776, 1980.

83. Brown, E. M., Gardner, D. G., Windeck, R. A., and Aurbach, D. G., Relationship of intracellular 3′,5′-monophosphate accumulation and parathyroid hormone release from dispersed bovine parathyroid cells, *Endocrinology,* 103, 2323, 1978.

84. Fitzpatrick, L. A., Brandi, M. L., and Aurbach, G. D., Prostaglandin $F_{2\alpha}$ and α-adrenergic agonists regulate parathyroid cell function via the inhibitory guanine nucleotide regulatory protein, *Endocrinology,* 118, 2115, 1986.

85. Chen, C. J., Barnet, J. V., Congo, D. A., and Brown, E. M., Divalent cations suppress 3′,5′-adenosine monophosphate accumulation by stimulating a pertussis toxin-sensitive guanine nucleotide-binding protein in cultured bovine parathyroid cells, *Endocrinology,* 124, 233, 1989.

86. Brown, E. M., El-Hajj Fuleihan, G., Chen, C. J., and Kifor, O., A comparison of the effects of divalent and trivalent cations on parathyroid hormone release, 3′,5′-cyclic-adenosine monophosphate accumulation, and the levels of inositol phosphates in bovine parathyroid cells, *Endocrinology,* 127, 1064, 1990.

87. Berridge, M. J., Inositol trisphosphate and diacylglycerol: two interacting second messengers, *Annu. Rev. Biochem.,* 56, 159, 1987.

88. Shoback, D. M. and McGhee, J. M., Fluoride stimulates the accumulation of inositol phosphates, increases intracellular free calcium, and inhibits parathyroid hormone release in dispersed bovine parathyroid cells, *Endocrinology,* 122, 2833, 1988.

89. Chen, C. J., Anast, C. S., and Brown, E. M., Effects of fluoride on parathyroid hormone secretion and intracellular second messengers in bovine parathyroid cells, *J. Bone Miner. Res.,* 3, 279, 1988.

90. Kifor, O., Kifor, I., and Brown, E. M., Effects of high extracellular calcium concentration on phosphoinositide turnover and inositol phosphate metabolism in dispersed bovine parathyroid cells, *J. Bone Miner. Res.,* 7, 1327, 1992.

91. Kifor, O. and Brown, E. M., Relationship between diacylglycerol levels and extracellular Ca^{2+} in dispersed bovine parathyroid cells, *Endocrinology,* 123, 2723, 1988.

92. Posillico, J. T., Burrowes, M., and Brown, E. M., Phospholipase C and Concanavalin A Stimulate PTH Secretion from Dispersed Parathyroid Cells by Different Mechanisms, 8th Annual Scientific Meeting of the American Society for Bone and Mineral Research, Anaheim, 1986, Ab. 311.

93. Morrissey, J. J., Effect of phorbol myristate acetate on secretion of parathyroid hormone, *Am. J. Physiol.,* 254, E63, 1988.

94. Kobayashi, N., Russel, J., Lettieri, D., and Sherwood, L. M., Regulation of protein kinase C by extra cellular calcium in bovine parathyroid cells, *Proc. Natl. Acad. Sci.U.S.A.*, 85, 4857, 1988.

95. Clarke, B. L., Hassager, C., and Fitzpatrick, L. A., Regulation of parathyroid hormone release by protein kinase-C is dependent on extracellular calcium in bovine parathyroid cells, *Endocrinology*, 132, 1168, 1993.

96. Nishizuka, Y., The role of protein kinase C in cell surface signal transduction and tumor promotion, *Nature, London*, 308, 693, 1984.

97. Oishi, K., Raynor, R. L., Charp, P. A., and Kuo, J. F., Regulation of protein kinase C by lysophospholipids, *J. Biol. Chem.*, 263, 6865, 1988.

98. Bordeau, A., Sorberbielle, J. C., Bonnet, P., Herviaux, P., Sachs, C., and Lieberherr, M., Phospholipase-A$_2$ action and arachidonic acid metabolism in calcium-mediated parathyroid hormone secretion, *Endocrinology*, 130, 1339, 1992.

99. Sugimoto, T., Civitelli, R., Ritter, C., Slatopolsky, E., and Morrissey, J., Regulation of cytosolic pH in bovine parathyroid cells, *Endocrinology*, 124, 149, 1989.

100. Morrissey, J. J. and Klahr, S., Role of hexose monophosphate shunt in parathyroid hormone secretion, *Am. J. Physiol.*, 245, (*Endocrinol. Metab.*, 8) E468, 1983.

101. Hansson, C. G. and Hamberger, L., Influence of calcium and magnesium on respiration of isolated parathyroid cells from the rat, *Endocrinology*, 92, 313, 1973.

102. Rothstein, M., Morrissey, J., Slatopolsky, E., and Klahr, S., The role of Na$^+$-Ca^{2+} exchange in parathyroid hormone secretion, *Endocrinology*, 111, 225, 1982.

103. Burtis, W. J., Wu, T., Bunch, C., Wysolmerrsky, J. J., Insogna, K., Weir, E., Broadus, A. E., and Stewart, A. F., Identification of a novel 17,000-Dalton parathyroid hormone-like adenilate cyclase-stimulating protein from a tumor associated with humoral hypercalcemia of malignancy, *J. Biol. Chem.*, 262, 7151, 1987.

104. Connor, C., Drees, B., and Hamilton, J., Parathyroid hormone-like peptide and parathyroid hormone are secreted from bovine parathyroid via different pathways, *Biochim. Biophys. Acta*, 1178, 81, 1993.

105. Iacangelo, A., Affolter, H.-U., Eiden, L. E., Hervert, E., and Grimes, M., Bovine chromogranin A sequence and distribution of its messanger RNA in endocrine tissue, *Nature*, 323, 82, 1986.

106. Cohn, D. V., Morrissey, J. J., Shofstall, R. E., and Chu, L. L. H., Cosecretion of secretory protein-I and parathormone by dispersed bovine parathyroid cells, *Endocrinology*, 110, 625, 1982.

107. Chu, L. L. H., McGregor, R. R., and Hamilton, J. W., Effects of isoproterenol and cycloheximide on parathyroid secretion, *Mol. Cell. Endocrinol.*, 33, 157, 1983.

108. Drees, B. M., Rouse, J., Johnson, J., and Hamilton, J. W., Bovine parathyrroid glands secrete a 26-kD N-terminal fragment of chromogranin-A which inhibits parathyroid cell secretion, *Endocrinology*, 129, 3381, 1991.

109. Fujimi, T., Baba, H., Fukase, M., and Fujita, T., Direct inhibitory effect of amino-terminal parathyroid hormone fragment [PTH(1-34)] on PTH secretion from bovine parathyroid primary cultured cells *in vitro*, *Biochem. Biophys. Res. Commun.*, 178, 953, 1991.

110. Bansal, D. D. and MacGregor, R. R., Calcium-regulated secretion of tissue plasminogen activator and parathyroid hormone from human parathyroid cells, *Endocrinology*, 74, 266, 1992.

111. Fujii, Y., Moreira, J. E., Orlando, C., Maggi, M., Aurbach, G. D., Brandi, M. L., and Sakaguchi, K., Endothelin as an autocrine factor in the regulation of parathyroid cells, *Proc. Natl. Acad. Sci. U.S.A.*, 88, 4235, 1991.

112. Schlüter, H., Quante, C., Buchholz, B., Dietl, K. H., Spieker, C., Karas, M., and Zidek, W., A vasopressor factor partially purified from human parathyroid glands, *Biochem. Biophys. Res. Commun.*, 188, 323, 1992.

Section V
Heart

Chapter 9

ELECTRICAL ACTIVITY AND SECRETION OF ATRIAL NATRIURETIC FACTOR FROM ATRIAL HEART CELLS

Geir Christensen

CONTENTS

1. SUMMARY

The cardiac atria secrete hormones belonging to a family termed natriuretic peptides: atrial natriuretic factor (ANF) and brain natriuretic peptide (BNP). ANF reduces systemic arterial pressure and cardiac output. These effects are primarily caused by reduction in circulating blood volume due to increased urinary fluid loss and shift of fluid to the interstitial space. BNP has similar effects. An important stimulus for secretion is increased atrial wall tension. A mechanochemical transducer is most likely located in the atrial myocytes. The cell membrane is one candidate for transduction of mechanical changes into secretion. Mechanosensitive ion channels exist in the membrane of atrial myocytes and appear to be cation selective. Although the depolarizations necessary for contractions produce great changes in Na^+, K^+, and Ca^{2+}, the mechanosensitive currents may cause the slow increase in intracellular concentration of Ca^{2+} observed during distension. Involvement of Ca^{2+} as a second messenger in stimulus-secretion coupling in atrial myocytes is supported by several studies. Increased phosphoinositide hydrolysis also occurs during distension, and the resulting activation of the Ca^{2+}-dependent enzyme protein kinase C increases ANF secretion.

0-8493-2477-7/95/$0.00+$.50
© 1995 by CRC Press Inc.

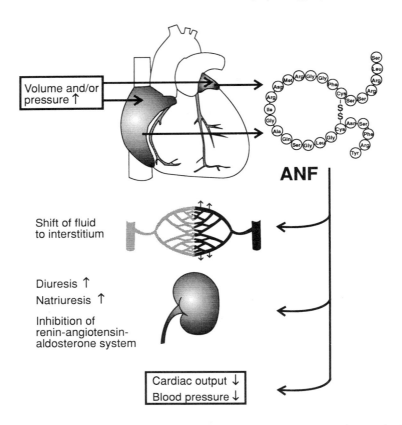

FIGURE 1.   Atrial natriuretic factor (ANF) is secreted from the cardiac atria in response to increased atrial pressure or volume. Circulating ANF acts on the microvasculature to induce a shift of fluid into the interstitial space. ANF also increases fluid loss from the kidneys and inhibits the renin-angiotensin-aldosterone system. The combined effects of ANF on various organs cause a reduction in cardiac output and systemic blood pressure.

2. THE ENDOCRINE FUNCTION OF THE HEART

The heart is not only a pump, but also an endocrine organ in the classical sense. In response to certain stimuli the cardiac atria produce and secrete hormones belonging to a family termed the natriuretic peptides. Natriuretic peptides have been classified as A-, B-, and C-type based on amino acid homology.[1] The A-type was discovered in the early 1980s[2] and was originally called atrial natriuretic factor (ANF). ANF consists of 28 amino acids.[3] BNP was initially identified in porcine brain and consists of 32 amino acids.[4,5] Both ANF and BNP are produced in the atrial myocytes,[3,6] whereas CNP is not.[7]

The effects of ANF have been well studied during the last decade. Infusion of ANF almost always results in reduced cardiac output and a small decline in systemic arterial pressure[8,9] (Figure 1). A substantial reduction in circulating blood volume has been shown to accompany infusion of ANF,[10,11] and is considered to be the main cause of the reduction in cardiac output.

The reduction in circulating blood volume is caused by both a shift of fluid from the intravascular to the interstitial space and an increase in urinary fluid loss. ANF acts on the microvasculature to increase capillary pressure[12] and capillary hydraulic conductivity,[13] which both enhance the transcapillary movement of water. The urinary fluid loss induced by infusing ANF is associated with an increase in glomerular filtration rate.[14] ANF-induced diuresis and natriuresis do not, however, result solely from increments in glomerular filtration rate, but are in large part due to inhibition of tubular sodium reabsorption.[15] ANF regulates blood volume and pressure also by inhibiting the renin-angiotensin-aldosterone system.[14] *In vitro* studies have shown that ANF possesses vasorelaxant properties,[16] but total peripheral resistance often remains unchanged during ANF infusion *in vivo*.[8] BNP has actions similar to those described for ANF,[4,5] but circulates in concentrations four to six times lower.[17]

Although it is certain that the natriuretic peptides will take their place as important regulatory hormones, further studies are needed to define their exact role in both normal and pathophysiological conditions. Great changes in plasma ANF levels have been reported secondary to diseases associated with alterations in atrial wall tension. One such disease is congestive heart failure. The measurement of plasma ANF during that condition may serve as a useful noninvasive marker for cardiac overload. Moreover, the natriuretic peptides are at present undergoing extensive evaluation as agents for treatment of congestive heart failure, hypertension, and states of fluid retention.

3. ANATOMY, ENDOCRINE FUNCTION, AND ELECTROPHYSIOLOGICAL PROPERTIES OF THE ATRIAL MYOCYTE

3.1. ULTRASTRUCTURAL ANATOMY

The ultrastructural anatomy of the atrial myocyte expresses the unique functions of these cells, namely excitability, contractility, and hormone secretion.[18,19] Atrial myocytes are approximately 20 μm in length, about five times smaller than the ventricular myocytes. The cells of the working atrial myocardium contain large numbers of myofibrils and mitochondria (Figure 2). The myofibrils are striated and exhibit a characteristic repeating pattern of light and dark transverse bands. This striated pattern reflects the disposition of interdigitating myofilaments made up of contractile proteins. Thin filaments composed of actin and the regulatory proteins tropomyosin and troponin are attached to dark transverse bands called the Z lines. The thin filaments interdigitate with an array of thicker filaments composed of myosin molecules. Interactions between the thick and thin filaments generate force and shortening of the myocardium. The mitochondria, which are largely responsible for aerobic production of chemical energy in the myocardium, are located immediately adjacent to the myofilaments.

The sarcolemma surrounds the cardiomyocytes and is the site of electrical polarization. This boundary between intra- and extracellular spaces consists of a bimolecular phospholipid layer and a basement membrane. The sarcolemma contains wide invaginations, the transverse tubular or t-system, which extends through the interior of the cells. Although this system is more rare in the atrial than in the ventricular myocytes, it has all the features typical of that in the ventricular cells. The t-system penetrates the myocytes not only in transverse direction, but the tubules can also turn and run longitudinally. The tubules are continuous with the extracellular space. One of the most important functions of this tubular system is, presumably, propagation of the action potential through the cells. The t-tubules are closely coupled to the sarcoplasmic reticulum by a narrow space that contains electron-dense material called "foot processes". The sarcoplasmic reticulum is a system of tubules formed by phospholipid membranes. They are not continuous with the extracellular space. The tubules of the sarcoplasmic reticulum are located around the bundles of contractile proteins and play an important role in the excitation-contraction coupling, since they store and release Ca^{2+}.

The atrial myocytes contain secretory granules which resemble granules found in other peptide-secreting cells. The granules are located near the nucleus of the atrial myocytes. In lesser number they are also found between the myofibrils. These granules were first observed by Kisch[20] in 1956.[20] Electron micrographs showed small bodies in the cytoplasm of guinea pig atria, but not in the ventricles of the same animal. Detailed descriptions by Jamieson and Palade[21] published in 1964 showed that these granules were intimately associated with the Golgi apparatus. This intimate association suggested that the granules form by condensation of a cell product in the Golgi complex.

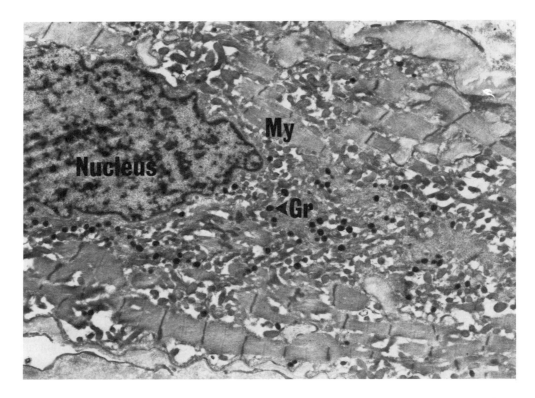

FIGURE 2.   Electron micrograph of a human atrial myocyte. My = myofibrils; Gr = secretory granules. Original magnification ×3600. (Courtesy of Torstein Lyberg, M.D., Ph.D.)

Atrial granules are larger and more numerous in smaller species, and generally are smaller and less numerous in larger mammals (dogs, humans).[21] In humans and other large mammals the amount of granules is similar in left and right atrial cells.[22,23] Ultrastructural evidence of exocytosis of the granules was published in 1986.[24] It was shown that extrusion of the secretion product occurs after fusion of the granular membrane and sarcolemma.

3.2.   ENDOCRINE PROPERTIES OF THE ATRIAL MYOCYTE

Based on their paper published in 1964, Jamieson and Palade[21] stated that the highly differentiated contractile atrial cell appears to possess a second specialized function in its ability to form and store a population of granules, presumably secretory in nature. Studies showing that changes in dietary sodium and water altered the granularity of rat atrial myocytes,[25] suggested that the atria may be involved in the control of extracellular fluid volume. It was not until 1981, however, that de Bold and associates[2] concluded that the granules store a hormone with natriuretic properties. In experiments on rats,[2] they found that infusion of atrial tissue extract induced an impressive sodium diuresis and a reduction in systemic arterial pressure, whereas ventricular extracts had no such effect. de Bold and colleagues[2] suspected that the active factor was derived from the atrial granules and referred to the substance as atrial natriuretic factor.

Within a short time after the experiments by de Bold et al.,[2] the complete chemical nature of ANF was identified. The active substance was shown to be a peptide consisting of 28 amino acids with a cysteine-cysteine disulfide cross link forming a 17-residue ring[3] (Figure 1). The disulfide cross link between cysteines 7 and 23 is essential for all known biological activity.[26]

The ANF precursor initially translated from atrial mRNA is a 151-amino acid peptide in humans, termed pre-pro-ANF.[27] The first 23- to 25-amino (N-) terminal amino acids of pre-pro-ANF contain a group of residues that is believed to facilitate transduction of the elongating polypeptide chain across the microsomal membrane into the rough endoplasmatic reticulum as the peptide is being synthesized. The remaining amino acid residues comprise a peptide that is termed pro-ANF$_{1-126}$. This peptide is the major molecular form of ANF stored in atrial granules.[28,29] The extrusion of the secretion product occurs by exocytosis.[24] When appropriate signals for hormone release are given, pro-ANF is cleaved into an NH$_2$-terminal fragment, pro-ANF$_{1-98}$, and the biological active form, ANF$_{99-126}$. Studies indicate that this maturation of ANF takes place in the atrial myocyte concurrently with secretion.[30] However, the enzyme responsible for pro-ANF processing has not yet been definitively identified.

BNP is synthesized by the atrial myocytes in a similar manner as ANF. BNP is derived from a precursor termed pre-pro-BNP and stored as pro-BNP.[6] In humans, BNP has been isolated from the atrial myocytes and brain and is a 32-amino acid peptide.[4,5] Like ANF, BNP contains a 17-member ring that is formed by a disulfide cross link between two cysteins. The greatest divergence from ANF is in the COOH extension of the molecule.

3.3 ELECTROPHYSIOLOGICAL PROPERTIES OF THE ATRIAL MYOCYTE

Electrophysiology of the cardiac myocytes has been well studied (for overview see References 31 and 32). The main focus has been on initiation of the heart beat and coupling between excitation and contraction. However, data on electrophysiological properties and secretion of natriuretic peptides are accumulating. Since secretion from the atrial myocytes is regulated by mechanical stimuli, data on how changes in wall tension affect the electrophysiology of the atrial myocytes are particularly important. These data will be discussed after describing the phases of the action potential.

The phospholipid bilayer of the sarcolemma has high resistance and selective permeability to ions achieved by various channel proteins. These properties are essential for the resting membrane potential and all other electrophysiological activity. The resting membrane potential across the sarcolemma is the basis for the action potential. The inside of the cell is 80 to 90 mV negative relative to the outside due to distribution of ions. Most important for the resting membrane potential is K$^+$, which has a large concentration gradient out of the cell. The high K$^+$ concentration within the cell is due to the Na$^+$-K$^+$ pump, which transports two K$^+$ ions into the cell in exchange with three Na$^+$ ions. The K$^+$ permeability during diastole ensures that the membrane potential reaches a value where the electrical driving force on K$^+$ balances the effect of its concentration gradient.

An action potential is initiated when the membrane potential becomes depolarized beyond a threshold. When exceeding this threshold, the depolarization continues independent of the initiating stimulus. This part of the action potential is called the upstroke or phase 0 (Figure 3) and is due to a sudden increase in membrane permeability to Na$^+$ ions. The voltage-dependent Na$^+$ channels open and Na$^+$ rushes down its electrochemical gradient into the cell.

Early rapid repolarization, termed phase 1, follows the upstroke of the action potential. This phase is due to a fall in sodium permeability and a transient outward K$^+$ current. During the following plateau of the action potential (phase 2) the membrane potential is near zero. During that period it is not possible to initiate another action potential. Sodium permeability is substantially reduced because of inactivation of the sodium channels. Similarly, the permeability to K$^+$ ions is low. Important for maintenance of the plateau is opening of Ca^{2+} channels, which leads to a slow inward current. The Ca^{2+} channels open when the membrane potential changes during phase 0. The inward Ca^{2+} current is due to a concentration gradient, since the concentration of Ca^{2+} within the cell is much lower than outside. Two types of voltage-dependent Ca^{2+} channels exist: the L- and the T-type. The L-type Ca^{2+} channels appear to be

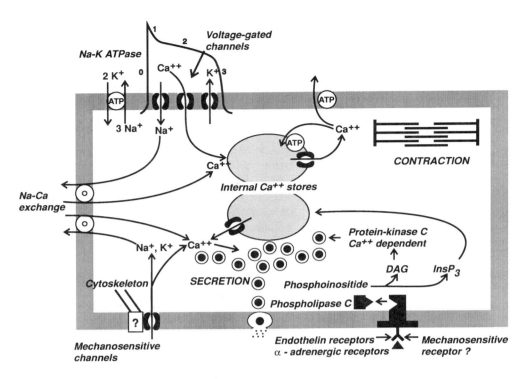

FIGURE 3. Scheme summarizing cellular mechanisms involved in secretion of atrial natriuretic factor (ANF) from an atrial myocyte. Ca^{2+} influx during the action potential determines release of Ca^{2+} from intracellular stores. Exocytosis of secretory granules is regulated by intracellular Ca^{2+}. Mechanosensitive channels may transduce increased tension into secretion by increasing intracellular Ca^{2+} concentration. Phosphoinositide hydrolysis to inositol trisphosphate (InsP$_3$) and diacylglycerol (DAG) participate in regulation of ANF secretion by activating the Ca^{2+}-dependent enzyme protein kinase C.

responsible for the majority of Ca^{2+} entry in myocytes. The Ca^{2+} current is inactivated at the end of the plateau phase and the final rapid repolarization begins (phase 3). During this phase an outward K^+ currents, often referred to as delayed rectifier, is activated, and K^+ ions move out of the cells, leaving behind a negative intracellular environment as described above.

The kinetics and amplitude of the Ca^{2+} entry during the action potential are critical factors in determining the amount of Ca^{2+} released by the sarcoplasmic reticulum (for overview see Reference 33). Entry of small amounts of Ca^{2+} across the plasma membrane induces release of Ca^{2+} by binding to a Ca^{2+} receptor at the surface of the sarcoplasmic reticulum. The Ca^{2+} concentration within the sarcoplasmic reticulum is high. Thus, release of Ca^{2+} through the sarcoplasmatic Ca^{2+} release channels is a passive process. Reuptake, however, requires energy and is effected by an ATP-dependent Ca^{2+} pump. Another important mechanism which brings Ca^{2+} ions out of the cytoplasm is the sodium-calcium exchange mechanism of the sarcolemma.

Several hormones participate in the regulation of cytosolic Ca^{2+} concentration. Stimulation of β-adrenergic receptors by adrenaline increases the number of open Ca^{2+} channels in the sarcolemma. This results in an increase in the transarcolemmal influx of Ca^{2+}. At the same time, sarcoplasmic reticulum release and reuptake of Ca^{2+} are enhanced. Many of these effects are mediated by cyclic AMP, which, acting via protein kinase A, increases phosphorylation of Ca^{2+}-regulating proteins. Other second messengers, as diacylglycerol (DAG) and inositol trisphosphate (InsP$_3$), are also involved in the regulation of intracellular Ca^{2+} homeostasis. These second messengers are activated as a result of the binding of α-adrenergic agonists, angiotensin II and endothelin to membrane receptors.

The transient rise in Ca^{2+} during the action potential triggers contraction by repressing troponin, the inhibitor of actin-myosin interaction. However, Ca^{2+} is most likely also involved in the regulation of hormone secretion from the atrial myocytes, as will be discussed below.

3.3.1. Effects of Mechanical Stimuli on Electrophysiological Properties

Increased tension in the atrial wall is an important stimulus for secretion of natriuretic peptides. There are a number of potential pathways by which such mechanical changes can induce exocytosis, and the mechanochemical transducer may be located at a number of different cell loci. However, the cell membrane and the cytoskeleton appear to be the best candidates for transduction of mechanical changes into intracellular chemical signals leading to exocytosis. At the membrane level activation of specific ion channels may be important.

Mechanosensitive channels may affect the action potential and particularly Ca^{2+} homeostasis. These channels have been shown to exist not only in specialized mechanoreceptors, but also in a great number of other cell types. In 1984 Guharay and Sachs[34] characterized mechanosensitive channels in skeletal muscle and named them stretch-activated channels (SAC).[34] To increase stress at the membrane, negative pressure was applied to the pipette electrode after formation of a cell-attached patch. In cardiac muscle this method has revealed a potassium-selective ion channel in molluscan ventricular cells[35] and a nonselective cation channel in neonatal rat,[36] chick embryo ventricular cells,[37] and rat atrial cells.[38] Using other techniques, a mechanosensitive Ca^{2+} channel has been shown to exist in the chick heart.[39] The ATP-sensitive K^+ channel has also been shown to be gated by mechanical stimuli.[40] Thus, mechanosensitive channels appear to be either cation selective with weak discrimination among the cations or K^+ selective.[41] There are indications that gadolinium blocks SAC in the heart.[42]

Mechanosensitive ion channels may be regulated either by mechanical changes affecting the membrane or possibly by the cytoskeleton (Figure 3). The role of the cytoskeleton in mechanotransduction is not well known, but it is possible that the cytoskeleton is linked to ionic channels and thereby regulates their gating.[43]

Channel density and conductance indicate that the channels contribute little to current relative to the voltage-gated Na^+, Ca^{2+}, or K^+ systems. However, the mechanosensitive channels may have a direct effect on the high impedance portion of the action potential, such as the plateau. Thus, it is possible that mechanosensitive currents may permit the cell to increase cell Ca^{2+} without producing excessive depolarization or changes in the action potential. Stretching has been reported to increase, decrease, or leave unchanged both resting membrane potential and action potential duration.[44] These findings are consistent with the hypothesis that mechanosensitive channels do not primarily affect the action potential, but act in concert with the action potential to regulate intracellular Ca^{2+}.

As will be discussed below, Ca^{2+} may play an important role in exocytosis, hence regulation of secretion of natriuretic peptides. A few studies have examined the effect of a length change on intracellular Ca^{2+}.[45,46] Using the chemiluminescent protein, aequorin, Allen and Kurihara[45] measured the effect of length change on cytoplasmic Ca^{2+} concentration. Their findings indicate that more Ca^{2+} is released into the cytoplasm at longer lengths. Enhancement of Ca^{2+} did not occur immediately, but developed slowly over several minutes. Mechanosensitive channels are likely to produce changes in Ca^{2+} and Na^+ influx, causing a change in the respective intracellular ion concentrations. An increase in intracellular Na^+ would activate sodium-calcium exchange, hence increase Ca^{2+} concentration.

4. STIMULUS-SECRETION COUPLING IN ATRIAL MYOCYTES

4.1. MECHANICAL FACTORS REGULATING SECRETION OF ANF

Dietz[47] first demonstrated, in an isolated heart-lung preparation, that a natriuretic factor was secreted by an increase in central venous pressure. In that study, secretion of the natriuretic factor was determined by bioassay. Lang et al.[48] also showed that a rise in atrial pressure increased ANF secretion. In that study ANF was assessed by radioimmunoassay of the fluid perfusing an isolated rat heart in a modified Langendorff preparation. That preparation has no extrinsic neural connections, and circulating hormones are absent. Thus, it was suggested that ANF secretion is regulated through a direct effect of mechanical changes in the atrial wall.

In vivo studies demonstrated a good correlation between mean atrial pressure and plasma concentration of ANF. An increase in atrial pressure was accomplished either by partial obstruction of the mitral orifice with a balloon,[49,50] or by blood volume expansion.[48,51,52] Since a rise in mean atrial pressure may reflect an increase in atrial dimensions, it was suggested that stretch or distension of the atrial wall is the stimulus which is sensed by the atrial myocytes.

However, mean atrial pressure does not always reflect the dimensions of the atrium, and, hence, it is important to consider phase-specific changes in pressure and dimensions. Atrial dimensions increase as the atrium is passively filled, and both pressure (v-wave) and dimensions increase in parallel. During atrial contraction (a-wave) atrial pressure increases, whereas the dimensions decrease as the blood is pumped into the ventricle. Hintze et al.[53] investigated the relationship between plasma ANF and phasic changes in atrial pressure and dimensions in the conscious dog. They found that changes in atrial dimensions are small and greatly underpredict plasma ANF during blood volume expansion. By calculating atrial wall stress during atrial filling, the v-wave became a better predictor of plasma ANF. Also, other studies[54,55] indicate that changes in atrial tension or stress determine ANF release and not only changes in atrial dimensions. Tension in a given atrial region is derived from the radius of that particular wall segment and the transmural pressure, and thus incorporates dimensional changes as well as pressure. Atrial wall stress also takes wall thickness into account.

Tachycardia has been shown to be an effective stimulus for ANF release and has been used for examining mechanical factors regulating ANF release. The increase in ANF release observed during an increase in heart rate does not require a change in atrial dimensions.[54-56] In a study on dogs,[55] an increase in heart more than doubled plasma ANF, whereas the atrial segment length, measured with sonomicrometer crystals inserted into the atrial wall, tended to decrease. King and Ledsome[54] and Riddervold et al.[56] also concluded from experiments on rabbits and dogs, respectively, that release of ANF during pacing tachycardia is not due to simple mechanical stretch of the atria.

During tachycardia, both the increase in wall stress during atrial systole and heart rate are determinants of ANF release. In support of this, Christensen et al.[55] reported that in barbiturate-anesthetized, closed-chest dogs the product of systolic right atrial pressure and the frequency of contractions were more closely related to ANF release than mean right atrial pressure or heart rate alone. Also, in other studies[54,57] release of ANF in response to tachycardia was most closely related to the minute index of systolic atrial wall stress. These results suggest that, in addition to atrial diastolic wall stress, atrial systolic wall stress represents an important stimulus for release of ANF.

Experiments on isolated atria are in accordance with the *in vivo* studies. Schiebinger and Linden[58] examined in isolated rat atria the effect of atrial distension on ANF release. They demonstrated that ANF release was stimulated by increasing resting atrial wall tension. In a subsequent study,[59] they found that ANF release from the isolated atria increased when frequency of contractions was raised from 2 to 8 Hz. In a similar preparation, Bilder et al.[60] demonstrated that diastolic tension, systolic tension, and frequency of stimulation all were determinants of ANF release.

4.1.1. Cellular Mechanisms for Tension-Induced ANF Release; Role of Calcium

Although several studies have examined the cellular mechanisms for ANF release (for overview see Reference 61), the precise signals linking the mechanical stimuli to secretion of ANF have not been fully established. As discussed above, mechanosensitive channels may act as mechanosensors and produce changes in intracellular Ca^{2+}. By analogy to other secretory systems, in which release of hormones is regulated, secretion of natriuretic peptides may be a Ca^{2+}-dependent process. However, the role of Ca^{2+} in stimulus-secretion coupling has been difficult to examine in atrial cells, since release and reuptake of intracellular Ca^{2+} are necessary for contractile function. Changes in intracellular Ca^{2+} and tension have therefore been difficult to separate experimentally.

In contracting atria an increase in cytosolic Ca^{2+} concentration has been shown to stimulate release of ANF. In anesthetized dogs, intracoronary infusion of Ca^{2+} increased plasma ANF.[23] In that study, atrial filling pressure remained constant, but left atrial systolic pressure increased. In the isolated beating rat heart, Saito et al.[62] and Ruskoaho et al.[63] have shown that ANF release is stimulated by increasing intracellular Ca^{2+} concentration. Similarly, in electrically paced isolated atria, BAY-K, which increases Ca^{2+} current by direct action on voltage-dependent Ca^{2+} channels, was reported to increase ANF secretion.[64] During such interventions developed tension increases. All these studies clearly indicate a role for Ca^{2+} as an intracellular second messenger regulating ANF secretion, but they do not provide final evidence for Ca^{2+} being the intracellular signal for stimulation of ANF release. The rise in intracellular Ca^{2+} may have stimulated release by increasing tension in the myocytes.

In experiments by Page et al.,[65] stretch-augmented ANF secretion was tested using quiescent rat atria. The experiments implicated transplasmalemmal Ca^{2+} influx, but not sarcoplasmic reticulum Ca^{2+} release in control of stretch-augmented ANF secretion. In a subsequent study,[66] they showed that increasing extracellular Ca^{2+} concentration increased stretch-augmented ANF secretion from isolated atria in the absence of contractions, and concluded that a component of ANF secretion is positively modulated by Ca^{2+}.

A role for Ca^{2+} as an intracellular signal for secretion of ANF is also indicated by studies performed on noncontracting isolated neonatal atrial myocytes. Ionomycin or BAY-K, which both increase intracellular Ca^{2+} concentration, were found to be potent secretagogues for ANF.[67,68] Moreover, Sei and Glembotski[69] showed that depolarization of neonatal atrial cells by KCl produced dose-dependent release of ANF. Nifedipine inhibited KCl-stimulated ANF secretion.[70] In a recent study by the same group,[71] the cellular mechanisms for pace-induced ANF secretion was examined in neonatal atrial cells. Pace-induced ANF secretion was inhibited by nifedipine and KN-62. The latter is a specific inhibitor of Ca^{2+}/calmodulin-dependent protein kinase II.

Presently, it may be concluded that Ca^{2+} most likely is involved as a second messenger in stimulus-secretion coupling in atrial myocytes, but the exact role is not yet clear. Although negative modulation has been described under some experimental conditions,[72] most studies indicate positive modulation of an increase in intracellular Ca^{2+} concentration.

4.1.2. Cellular Mechanisms for Tension-Induced ANF Release; Role of Phosphoinositide Hydrolysis and Protein Kinase C

Several studies indicate that the second messengers DAG and $InsP_3$ are involved in regulating ANF secretion. When activated, the membrane-bound enzyme phospholipase C generates the two products, DAG and $InsP_3$ (Figure 3). DAG activates the Ca^{2+} dependent enzyme protein kinase C. $InsP_3$ releases Ca^{2+} from intracellular stores by binding to specific receptors at the surface membrane of intracellular compartments. The combined effects of $InsP_3$ and DAG regulate the function of many endocrine organs.[73,74]

Several lines of evidence show that protein kinase C is importantly involved in regulating secretion of ANF. Intracellular activation of protein kinase C can be achieved by tumor-promoting phorbol esters such as 12-*O*-tetradecanoylphorbol 13-acetate (TPA), which probably binds to the same sites normally occupied by DAG. In the perfused, spontaneously beating rat heart, Ruskoaho et al.[63] showed that TPA stimulated ANF secretion. Similarly, TPA caused secretion of ANF from cultured neonatal atrial myocytes.[67,68,70] In these cells increasing intracellular Ca^{2+} had a synergistic effect with TPA,[67,68] giving support to a role for Ca^{2+} activated protein kinase C in ANF secretion.

A few studies have examined more directly the role of these second messengers in tension-induced ANF release. Increased formation of $InsP_3$ was noted when right atria were dilated.[75,76] Moreover, experiments by Page et al.[65] support a role for protein kinase C in mechanical stimulation of ANF release. They observed that H-7, a protein kinase C inhibitor, decreased the ANF secretory rate in isolated stretched atria.

In summary, stimulating phosphoinositide hydrolysis to $InsP_3$ and DAG increases secretion of ANF, presumably by involving protein kinase C. This second-messenger system may represent a cellular mechanism linking mechanical stimuli to release of ANF. However, more studies are required to examine the role of $InsP_3$-induced Ca^{2+} release from intracellular pools and the importance of intracellular Ca^{2+} for regulation of protein kinase C-induced ANF release during mechanical stimulation.

4.2. NEUROHUMORAL FACTORS REGULATING SECRETION OF ANF

In addition to mechanical factors, both humoral and neuronal factors have been suggested to contribute to regulation of ANF release. Several hormones have been shown to stimulate secretion of ANF *in vitro*. However, in the intact circulation the effect of these hormones appears to be much less important than the effect of mechanical factors.

Infusion of *angiotensin II in vivo* results in ANF release which is closely related to changes in right[77] and left[78-81] atrial pressure. In a study on dogs,[82] similar changes in left ventricular afterload induced by injection of angiotensin II, vasopressin, or mechanical constriction of the thoracic aorta caused parallel changes in left atrial pressure and plasma ANF. These *in vivo* studies did not indicate any direct stimulatory effect of these hormones.

However, a direct effect of angiotensin II *in vivo* has been suggested by some investigators.[83,84] Angiotensin II is coupled to phosphoinositide hydrolysis[85] and probably causes stimulation of protein kinase C and changes in cytosolic Ca^{2+}. However, angiotensin II induces much less $InsP_3$ accumulation than α-adrenergic stimulation,[74] and *in vitro* studies[80,86] have, with one exception,[87] demonstrated that angiotensin II does not exert any direct humoral effect on ANF release. Thus, it may be concluded that angiotensin II affects ANF release mainly by an effect on cardiac afterload and left atrial hemodynamics. However, presently, a small direct effect mediated by stimulation of angiotensin II receptors can not be excluded.

On the other hand, several studies show that *endothelin* directly stimulates ANF secretion from the atria. Endothelin was originally identified in *endothelial cells*,[88] and is the most potent vasoconstrictor yet known. Studies in dogs[89,90] have shown that infusion of endothelin increases both plasma ANF and atrial pressure. The rise in atrial pressure is partly responsible for ANF secretion under those experimental conditions.

Although circulating in plasma, endothelin more importantly functions in a paracrine or autocrine mode. Thus, it is possible that a direct effect of endothelin locally produced by the endocardium or the vascular endothelium is important in the regulation of ANF release. When binding to receptors on isolated atrial myocytes, endothelin is a potent stimulator of ANF.[69,91,92] Endothelin stimulates phosphoinositide hydrolysis to $InsP_3$ and DAG,[93] and ANF secretion stimulated by endothelin seems to involve these second messengers and Ca^{2+}. Irons et al.[68] showed that endothelin activated protein kinase C in neonatal atrial myocytes, and inhibition with H-7 reduced ANF release. Reducing Ca^{2+} in the medium decreased ANF secretion.[69]

The effects of *cardiac sympathetic stimulation* and infusion of *catecholamines* on ANF release have been examined both *in vivo* and *in vitro*. Stimulation of sympathetic fibers to the heart did not increase plasma ANF,[50,94] whereas infusion of noradrenaline stimulated ANF release.[94,95] In those studies, the effect of noradrenaline was attributed to changes in atrial hemodynamics.

Direct effects of both α- and β-adrenergic stimulation have been observed, but these are small compared to the effect of hemodynamic changes.[23] By infusing the α-adrenergic agonist phenylephrine into the circumflex coronary artery, Christensen et al.[23] found increased ANF release in the absence of changes in atrial pressure. However, the increase was much smaller than a moderate rise in filling pressure induced by constriction of the aorta. That α-adrenergic stimulation increases release of ANF through a direct effect is supported by studies on isolated myocytes,[70] isolated atria,[96-98] and isolated heart preparations.[99] The α-adrenergic receptors are coupled to phosphoinositide hydrolysis,[100] and the protein kinase C inhibitor H-7 has been reported to inhibit phenylephrine-stimulated ANF release.[101] The α-stimulated secretion of ANF appears to be dependent on Ca^{2+} entry [69].

By infusing the β-adrenergic agonist isoproterenol into the circumflex coronary artery, a small reduction in ANF release was observed in the absence of changes in atrial filling pressure.[23] On the other hand, intravenous infusion of isoproterenol has been shown to increase plasma ANF in some studies.[102-104] Similarly, studies carried out in isolated beating preparations[64,97,98] have shown a stimulating effect of β-adrenergic stimulation on ANF release. This stimulating effect may reflect increased development of tension. In support of this view, studies performed on isolated rat myocytes[70,105] have shown reduced ANF release during stimulation with isoproterenol or other agents that increase the intracellular concentration of cyclic AMP. Increasing intracellular concentration of cyclic AMP may suppress ANF by stimulation of protein kinase A.[70] If Ca^{2+} is involved in the response to β-adrenergic stimulation, the diastolic level of intracellular Ca^{2+}, which falls during β-adrenergic stimulation,[106] may reduce ANF release.

5. PHARMACOLOGY

ANF is at the present undergoing extensive evaluation as an agent for treatment of congestive heart failure, hypertension, and states of fluid retention. Since ANF increases sodium excretion and reduces systemic blood pressure, it is a potential drug for treating patients with such diseases. However, the susceptibility of ANF to enzyme attack in the gut has made oral administration very difficult. ANF can be administered intravenously, but has a half-life of only 3–4 min.[107]

Presently, it is not possible to selectively increase synthesis or secretion of ANF. However, more knowledge about mechanisms of ANF release might enable development of effective drugs which selectively control hormone secretion from the atria. An orally effective drug increasing ANF secretion would be a significant breakthrough.

It has been proposed to increase the plasma level of ANF by decreasing the catabolism of endogenous ANF. At least two mechanisms may operate in the inactivation of the peptide: a receptor-mediated clearance process (ANF-C receptors)[108] and an enzymatic process involving neutral metallopeptidase.[109] Both processes can be inhibited,[108,110] and it is possible that such inhibitors of ANF metabolism may become a new class of therapeutic agents for treatment of cardiovascular diseases.

REFERENCES

1. Sudoh, T., Minamino, N., Kangawa, K., and Matsuo, H., C-type natriuretic peptide (CNP): a new member of natriuretic peptide family identified in porcine brain, *Biochem.Biophys.Res.Commun.*, 168, 863, 1990.
2. de Bold, A. J., Borenstein, H. B., Veress, A. T., and Sonnenberg, H., A rapid and potent natriuretic response to intravenous injection of atrial myocardial extract in rats, *Life Sci.*, 28, 89, 1981.

3. Flynn, T. G., de Bold, M. L., and de Bold, A. J., The amino acid sequence of an atrial peptide with potent diuretic and natriuretic properties, *Biochem. Biophys. Res. Commun.*, 117, 859, 1983.
4. Sudoh, T., Kangawa, K., Minamino, N., and Matsuo, H., A new natriuretic peptide in porcine brain, *Nature*, 332, 78, 1988.
5. Sudoh, T., Minamino, N., Kangawa, K., and Matsuo, H., Brain natriuretic peptide-32: N-terminal six amino acid extended form of brain natriuretic peptide identified in porcine brain, *Biochem. Biophys. Res. Commun.*, 155, 726, 1988.
6. Maekawa, K., Sudoh, T., Furusawa, M., Minamino, N., Kangawa, K., Ohkubo, H., Nakanishi, S., and Matsuo, H., Cloning and sequence analysis of cDNA encoding a precursor for porcine brain natriuretic peptide, *Biochem. Biophys. Res. Commun.*, 157, 410, 1988.
7. Kojima, M., Minamino, N., Kangawa, K., and Matsuo, H., Cloning and sequence analysis of a cDNA encoding a precursor for rat C-type natriuretic peptide (CNP), *FEBS Lett.*, 276, 209, 1990.
8. Breuhaus, B. A., Saneii, H. H., Brandt, M. A., and Chimoskey, J. E., Atriopeptin II lowers cardiac output in conscious sheep, *Am. J. Physiol.*, 249, R776, 1985.
9. Lappe, R. W., Smits, J. F., Todt, J. A., Debets, J. J., and Wendt, R. L., Failure of atriopeptin II to cause arterial vasodilation in the conscious rat, *Circ. Res.*, 56, 606, 1985.
10. Almeida, F. A., Suzuki, M., and Maack, T., Atrial natriuretic factor increases hematocrit and decreases plasma volume in nephrectomized rats, *Life Sci.*, 39, 1193, 1986.
11. Flückiger, J. P., Waeber, B., Matsueda, G., Delaloye, B., Nussberger, J., and Brunner, H. R., Effect of atriopeptin III on hematocrit and volemia of nephrectomized rats. *Am. J. Physiol.*, 251, H880, 1986.
12. Faber, J. E., Gettes, D. R., and Gianturco, D. P., Microvascular effects of atrial natriuretic factor: interaction with α1- and α2-adrenoceptors, *Circ. Res.*, 63, 415, 1988.
13. Huxley, V. H., Tucker, V. L., Verburg, K. M., and Freeman, R. H., Increased capillary hydraulic conductivity induced by atrial natriuretic peptide, *Circ. Res.*, 60, 304, 1987.
14. Maack, T., Marion, D. N., Camargo, M. J., Kleinert, H. D., Laragh, J. H., Vaughan, E. D., Jr., and Atlas, S. A., Effects of auriculin (atrial natriuretic factor) on blood pressure, renal function, and the renin-aldosterone system in dogs. *Am. J. Med.*, 77, 1069, 1984.
15. Sonnenberg, H., Cupples, W. A., de Bold, A. J., and Veress, A. T., Intrarenal localization of the natriuretic effect of cardiac atrial extract, *Can. J. Physiol. Pharmacol.*, 60, 1149, 1982.
16. Currie, M. G., Geller, D. M., Cole, B. R., Boylan, J. G., YuSheng, W., Holmberg, S. W., and Needleman, P., Bioactive cardiac substances: potent vasorelaxant activity in mammalian atria, *Science*, 221, 71, 1983.
17. Mukoyama, M., Nakao, K., Saito, Y., Ogawa, Y., Hosoda, K., Suga, S., Shirakami, G., Jougasaki, M., and Imura, H., Human brain natriuretic peptide, a novel cardiac hormone [letter], *Lancet*, 335, 801, 1990.
18. McNutt, N. S. and Fawcett, D. W., The ultrastructure of the cat myocardium. II. Atrial muscle, *J. Cell Biol.*, 42, 46, 1969.
19. Legato, M. J., *The Myocardial Cell for the Clinical Cardiologist*, Futura Publishing, Mount Kisco, NY, 1973.
20. Kisch, B., Electron microscopy of the atrium of the heart. I. Guinea pig, *Exp. Med. Surg.*, 14, 99, 1956.
21. Jamieson, J. D. and Palade, G. E., Specific granules in atrial muscle cells, *J. Cell. Biol.*, 23, 151, 1964.
22. Rodeheffer, R. J., Tanaka, I., Imada, T., Hollister, A. S., Robertson, D., and Inagami, T., Atrial pressure and secretion of atrial natriuretic factor into the human central circulation, *J. Am. Coll. Cardiol.*, 8, 18, 1986.
23. Christensen, G., Aksnes, G., Ilebekk, A., and Kiil, F., Release of atrial natriuretic factor during selective cardiac α- and β-adrenergic stimulation, intracoronary Ca^{2+} infusion, and aortic constriction in pigs, *Circ. Res.*, 68, 638, 1991.
24. Page, E., Goings, G. E., Power, B., and Upshaw-Earley, J., Ultrastructural features of atrial peptide secretion, *Am. J. Physiol.*, 251, H340, 1986.
25. Marie, J. P., Guillemot, H., and Hatt, P. Y., Le degré de granulation des cardiocytes auriculaires. Étude planimétrique au cours de différents apports d'eau et de sodium chez le rat, *Pathol. Biol.*, 24, 549, 1976.
26. Atlas, S. A., Kleinert, H. D., Camargo, M. J., Januszewicz, A., Sealey, J. E., Laragh, J. H., Schilling, J. W., Lewicki, J. A., Johnson, L. K., and Maack, T., Purification, sequencing and synthesis of natriuretic and vasoactive rat atrial peptide, *Nature*, 309, 717, 1984.
27. Oikawa, S., Imai, M., Ueno, A., Tanaka, S., Noguchi, T., Nakazato, H., Kangawa, K., Fukuda, A., and Matsuo, H., Cloning and sequence analysis of cDNA encoding a precursor for human atrial natriuretic polypeptide, *Nature*, 309, 724, 1984.
28. Miyata, A., Kangawa, K., Toshimori, T., Hatoh, T., and Matsuo, H., Molecular forms of atrial natriuretic polypeptides in mammalian tissues and plasma, *Biochem. Biophys. Res. Commun.*, 129, 248, 1985.
29. Glembotski, C. C., Wildey, G. M., and Gibson, T. R., Molecular forms of immunoactive atrial natriuretic peptide in the rat hypothalamus and atrium. *Biochem. Biophys. Res. Commun.*, 129, 671, 1985.
30. Sei, C. A., Hand, G. L., Murray, S. F., and Glembotski, C. C., The cosecretional maturation of atrial natriuretic factor by primary atrial myocytes, *Mol. Endocrinol.*, 6, 309, 1992.
31. Katz, A. M., *Physiology of the Heart*, Raven Press, New York, 1992.

32. Zipes, D. P., Genesis of cardiac arrhythmias: electrophysiological considerations, in *Heart Disease*, Braunwald, E., Ed., W. B. Saunders, Philadelphia, 1992, 588.

33. Bers, D. M., *Excitation-Contraction Coupling and Cardiac Contractile Force*, Kluwer Academic Publishers, Dordrecht, The Netherlands, 1991.

34. Guharay, F. and Sachs, F., Stretch-activated single ion channel currents in tissue-cultured embryonic chick skeletal muscle, *J. Physiol. (Lond)*, 352, 685, 1984.

35. Sigurdson, W. J., Morris, C. E., Brezden, B. L., and Gardner, D. R., Stretch activation of a K^+ channel in molluscan heart cells, *J. Exp. Biol.*, 127, 191, 1987.

36. Craelius, W., Chen, V., and el-Sherif, N., Stretch activated ion channels in ventricular myocytes, *Biosci. Rep.*, 8, 407, 1988.

37. Bustamante, J. O., Ruknudin, A., and Sachs, F., Stretch-activated channels in heart cells: relevance to cardiac hypertrophy, *J. Cardiovasc. Pharmacol.*, 17 (Suppl. 2), S110, 1991.

38. Kim, D., Novel cation-selective mechanosensitive ion channel in the atrial cell membrane, *Circ. Res.*, 72, 225, 1993.

39. Sigurdson, W., Ruknudin, A., and Sachs, F., Calcium imaging of mechanically induced fluxes in tissue-cultured chick heart: role of stretch-activated ion channels. *Am. J. Physiol.*, 262, H1110, 1992.

40. Van Wagoner, D. R., Mechanosensitive gating of atrial ATP-sensitive potassium channels, *Circ. Res.*, 72, 973, 1993.

41. Morris, C. E., Mechanosensitive ion channels, *J. Membrane. Biol.*, 113, 93, 1990.

42. Hansen, D. E., Borganelli, M., Stacy, G. P., Jr., and Taylor, L. K., Dose-dependent inhibition of stretch-induced arrhythmias by gadolinium in isolated canine ventricles. Evidence for a unique mode of antiarrhythmic action, *Circ. Res.*, 69, 820, 1991.

43. Davies, P. F. and Tripathi, S. C., Mechanical stress mechanisms and the cell. An endothelial paradigm, *Circ. Res.*, 72, 239, 1993.

44. Lab, M. J., Contraction-excitation feedback in myocardium. Physiological basis and clinical relevance, *Circ. Res.*, 50, 757, 1982.

45. Allen, D. G. and Kurihara, S., The effects of muscle length on intracellular calcium transients in mammalian cardiac muscle, *J. Physiol. (London)*, 327, 79, 1982.

46. White, E., Le Guennec, J. Y., Nigretto, J. M., Gannier, F., Argibay, J. A., and Garnier, D., The effects of increasing cell length on auxotonic contractions; membrane potential and intracellular calcium transients in single guinea-pig ventricular myocytes, *Exp. Physiol.*, 78, 65, 1993.

47. Dietz, J. R., Release of natriuretic factor from rat heart-lung preparation by atrial distension, *Am. J. Physiol.*, 247, R1093, 1984.

48. Lang, R. E., Thölken, H., Ganten, D., Luft, F. C., Ruskoaho, H. and Unger, T., Atrial natriuretic factor — a circulating hormone stimulated by volume loading, *Nature*, 314, 264, 1985.

49. Goetz, K. L., Wang, B. C., Geer, P. G., Leadley, R. J., Jr., and Reinhardt, H. W., Atrial stretch increases sodium excretion independently of release of atrial peptides, *Am. J. Physiol.*, 250, R946, 1986.

50. Ledsome, J. R., Wilson, N., Rankin, A. J., and Courneya, C. A., Time course of release of atrial natriuretic peptide in the anaesthetized dog, *Can. J. Physiol. Pharmacol.*, 64, 1017, 1986.

51. Pettersson, A., Ricksten, S. E., Towle, A. C., Hedner, J., and Hedner, T., Effect of blood volume expansion and sympathetic denervation on plasma levels of atrial natriuretic factor (ANF) in the rat, *Acta Physiol. Scand.*, 124, 309, 1985.

52. Anderson, J. V., Christofides, N. D., and Bloom, S. R., Plasma release of atrial natriuretic peptide in response to blood volume expansion, *J. Endocrinol.*, 109, 9, 1986.

53. Hintze, T. H., McIntyre, J. J., Patel, M. B., Shapiro, J. T., DeLeonardis, M., Zeballos, G. A., and Loud, A. V., Atrial wall function and plasma atriopeptin during volume expansion in conscious dogs, *Am. J. Physiol.*, 256, H713, 1989.

54. King, K. A. and Ledsome, J. R., The effect of tachycardia on right atrial dynamics and plasma atrial natriuretic factor in anaesthetized rabbits, *J. Physiol. (London)*, 422, 289, 1990.

55. Christensen, G., Ilebekk, A., Aakeson, I., and Kiil, F., The release mechanism for atrial natriuretic factor during blood volume expansion and tachycardia in dogs. *Acta Physiol. Scand.*, 134, 263, 1988.

56. Riddervold, F., Smiseth, O. A., Hall, C., Groves, G., and Risøe, C., Rate-induced increase in plasma atrial natriuretic factor can occur independently of changes in atrial wall stretch, *Am. J. Physiol.*, 260, H1953, 1991.

57. King, K. A. and Ledsome, J. R., Atrial dynamics, atrial natriuretic factor, tachycardia, and blood volume in anesthetized rabbits, *Am. J. Physiol.*, 261, H22, 1991.

58. Schiebinger, R. J. and Linden, J., The influence of resting tension on immunoreactive atrial natriuretic peptide secretion by rat atria superfused in vitro, *Circ. Res.*, 59, 105, 1986.

59. Schiebinger, R. J. and Linden, J., Effect of atrial contraction frequency on atrial natriuretic peptide secretion, *Am. J. Physiol.*, 251, H1095, 1986.

60. Bilder, G. E., Siegl, P. K., Schofield, T. L., and Friedman, P. A., Chronotropic stimulation: a primary effector for release of atrial natriuretic factor, *Circ. Res.*, 64, 799, 1989.

61. Ruskoaho, H., Atrial natriuretic peptide: synthesis, release, and metabolism, *Pharmacol. Rev.*, 44, 479, 1992.

62. Saito, Y., Nakao, K., Morii, N., Sugawara, A., Shiono, S., Yamada, T., Itoh, H., Sakamoto, M., Kurahashi, K., and Fujiwara, M., Bay K 8644, a voltage-sensitive calcium channel agonist, facilitates secretion of atrial natriuretic polypeptide from isolated perfused rat hearts, *Biochem. Biophys. Res. Commun.*, 138, 1170, 1986.

63. Ruskoaho, H., Toth, M., and Lang, R. E., Atrial natriuretic peptide secretion: synergistic effect of phorbol ester and A23187, *Biochem. Biophys. Res. Commun.*, 133, 581, 1985.

64. Schiebinger, R. J., Calcium, its role in isoproterenol-stimulated atrial natriuretic peptide secretion by superfused rat atria, *Circ. Res.*, 65, 600, 1989.

65. Page, E., Goings, G. E., Power, B., and Upshaw-Earley, J., Basal and stretch-augmented natriuretic peptide secretion by quiescent rat atria, *Am. J. Physiol.*, 259, C801, 1990.

66. Page, E., Upshaw-Earley, J., Goings, G. E., and Hanck, D. A., Effect of external Ca^{2+} concentration on stretch-augmented natriuretic peptide secretion by rat atria, *Am. J. Physiol.*, 260, C756, 1991.

67. Matsubara, H., Hirata, Y., Yoshimi, H., Takata, S., Takagi, Y., Umeda, Y., Yamane, Y., and Inada, M., Role of calcium and protein kinase C in ANP secretion by cultured rat cardiocytes, *Am. J. Physiol.*, 255, H405, 1988.

68. Irons, C. E., Sei, C. A., Hidaka, H., and Glembotski, C. C., Protein kinase C and calmodulin kinase are required for endothelin-stimulated atrial natriuretic factor secretion from primary atrial myocytes, *J. Biol. Chem.*, 267, 5211, 1992.

69. Sei, C. A. and Glembotski, C. C., Calcium dependence of phenylephrine-, endothelin-, and potassium chloride-stimulated atrial natriuretic factor secretion from long term primary neonatal rat atrial cardiocytes, *J. Biol. Chem.*, 265, 7166, 1990.

70. Shields, P. P. and Glembotski, C. C., Regulation of atrial natriuretic factor-(99-126) secretion from neonatal rat primary atrial cultures by activators of protein kinases A and C, *J. Biol. Chem.*, 264, 9322, 1989.

71. McDonough, P. M., Stella, S. L., and Glembotski, C. C., Involvement of cytoplasmic calcium and protein kinases in the regulation of atrial natriuretic factor secretion by contraction rate and endothelin, *J. Biol. Chem.*, 269, 9466, 1994.

72. Greenwald, J. E., Apkon, M., Hruska, K. A., and Needleman, P., Stretch-induced atriopeptin secretion in the isolated rat myocyte and its negative modulation by calcium, *J. Clin. Invest.*, 83, 1061, 1989.

73. Nishizuka, Y., Studies and perspectives of protein kinase C, *Science*, 233, 305, 1986.

74. Berridge, M. J., Inositol trisphosphate and diacylglycerol: two interacting second messengers, *Annu. Rev. Biochem.*, 56, 159, 1987.

75. von Harsdorf, R., Lang, R. E., Fullerton, M., and Woodcock, E. A., Myocardial stretch stimulates phosphatidylinositol turnover, *Circ. Res.*, 65, 494, 1989.

76. von Harsdorf, R., Lang, R., Fullerton, M., Smith, A. I., and Woodcock, E. A., Right atrial dilatation increases inositol-(1,4,5)trisphosphate accumulation. Implications for the control of atrial natriuretic peptide release, *FEBS Lett.*, 233, 201, 1988.

77. Katsube, N., Schwartz, D., and Needleman, P., Release of atriopeptin in the rat by vasoconstrictors or water immersion correlates with changes in right atrial pressure, *Biochem. Biophys. Res. Commun.*, 133, 937, 1985.

78. Garcia, R., Gauquelin, G., and Lachance, D., Atrial natriuretic factor and experimental hypertension in the rat, *Int. J. Rad. Appl. Instrum.-Part B.*, 14, 333, 1987.

79. Lachance, D. and Garcia, R., Atrial natriuretic factor release during angiotensin II infusion in right and left atrial appendectomized rats, *J. Hypertens.*, 7, 293, 1989.

80. Lachance, D. and Garcia, R., Atrial natriuretic factor release by angiotensin II in the conscious rat, *Hypertension*, 11, 502, 1988.

81. Christensen, G., Ilebekk, A., and Kiil, F., Release of atrial natriuretic factor during infusion of isoproterenol and angiotensin II, *Am. J. Physiol.*, 257, R896, 1989.

82. Stewart, J. M., Wang, J., Singer, A., Zeballos, G. A., Ochoa, M., Patel, M. B., Gewitz, M. H., and Hintze, T. H., Regulation of plasma ANF after increases in afterload in conscious dogs, *Am. J. Physiol.*, 259, H1736, 1990.

83. Ruskoaho, H., Vakkuri, O., Arjamaa, O., Vuolteenaho, O., and Leppäluoto, J., Pressor hormones regulate atrial-stretch-induced release of atrial natriuretic peptide in the pithed rat, *Circ. Res.*, 64, 482, 1989.

84. Volpe, M., Atlas, S. A., Sosa, R. E., Marion, D. E., Mueller, F. B., Sealey, J. E., and Laragh, J. H., Angiotensin II-induced atrial natriuretic factor release in dogs is not related to hemodynamic responses, *Circ. Res.*, 67, 774, 1990.

85. Allen, I. S., Cohen, N. M., Dhallan, R. S., Gaa, S. T., Lederer, W. J. and Rogers, T. B., Angiotensin II increases spontaneous contractile frequency and stimulates calcium current in cultured neonatal rat heart myocytes: insights into the underlying biochemical mechanisms, *Circ. Res.*, 62, 524, 1988.

86. Dietz, J. R., The effect of angiotensin II and ADH on the secretion of atrial natriuretic factor, *Proc. Soc. Exp. Biol. Med.*, 187, 366, 1988.

87. Veress, A. T., Milojevic, S., Yip, C., Flynn, T. G., and Sonnenberg, H., In vitro secretion of atrial natriuretic factor: receptor-mediated release of prohormone, *Am. J. Physiol.*, 254, R809, 1988.

88. Yanagisawa, M., Kurihara, H., Kimura, S., Tomobe, Y., Kobayashi, M., Mitsui, Y., Yazaki, Y., Goto, K., and Masaki, T., A novel potent vasoconstrictor peptide produced by vascular endothelial cells, *Nature*, 332, 411, 1988.

89. Goetz, K. L., Wang, B. C., Madwed, J. B., Zhu, J. L., and Leadley, R. J., Jr., Cardiovascular, renal, and endocrine responses to intravenous endothelin in conscious dogs, *Am. J. Physiol.*, 255, R1064, 1988.

90. Miller, W. L., Redfield, M. M., and Burnett, J. C., Jr., Integrated cardiac, renal, and endocrine actions of endothelin, *J. Clin. Invest.*, 83, 317, 1989.

91. Fukuda, Y., Hirata, Y., Yoshimi, H., Kojima, T., Kobayashi, Y., Yanagisawa, M., and Masaki, T., Endothelin is a potent secretagogue for atrial natriuretic peptide in cultured rat atrial myocytes, *Biochem. Biophys. Res. Commun.*, 155, 167, 1988.

92. Gardner, D. G., Newman, E. D., Nakamura, K. K., and Nguyen, K. P., Endothelin increases the synthesis and secretion of atrial natriuretic peptide in neonatal rat cardiocytes, *Am. J. Physiol.*, 261, E177, 1991.

93. Shubeita, H. E., McDonough, P. M., Harris, A. N., Knowlton, K. U., Glembotski, C. C., Brown, J. H., and Chien, K. R., Endothelin induction of inositol phospholipid hydrolysis, sarcomere assembly, and cardiac gene expression in ventricular myocytes. A paracrine mechanism for myocardial cell hypertrophy, *J. Biol. Chem.*, 265, 20555, 1990.

94. Rankin, A. J., Wilson, N., and Ledsome, J. R., Effects of autonomic stimulation on plasma immunoreactive atrial natriuretic peptide in the anesthetized rabbit, *Can. J. Physiol. Pharmacol.*, 65, 532, 1987.

95. Uehlinger, D. E., Zaman, T., Weidmann, P., Shaw, S., and Gnadinger, M. P., Pressure dependence of atrial natriuretic peptide during norepinephrine infusion in humans, *Hypertension*, 10, 249, 1987.

96. Sonnenberg, H. and Veress, A. T., Cellular mechanism of release of atrial natriuretic factor, *Biochem. Biophys. Res. Commun.*, 124, 443, 1984.

97. Schiebinger, R. J., Baker, M. Z., and Linden, J., Effect of adrenergic and muscarinic cholinergic agonists on atrial natriuretic peptide secretion by isolated rat atria. Potential role of the autonomic nervous system in modulating atrial natriuretic peptide secretion, *J. Clin. Invest.*, 80, 1687, 1987.

98. Wong, N. L., Wong, E. F., Au, G. H., and Hu, D. C., Effect of α- and β-adrenergic stimulation on atrial natriuretic peptide release in vitro, *Am. J. Physiol.*, 255, E260, 1988.

99. Currie, M. G. and Newman, W. H., Evidence for α-1 adrenergic receptor regulation of atriopeptin release from the isolated rat heart, *Biochem. Biophys. Res. Commun.*, 137, 94, 1986.

100. Brown, J. H., Buxton, I. L., and Brunton, L. L., Alpha 1-adrenergic and muscarinic cholinergic stimulation of phosphoinositide hydrolysis in adult rat cardiomyocytes, *Circ. Res.*, 57, 532, 1985.

101. Ishida, A., Tanahashi, T., Okumura, K., Hashimoto, H., Ito, T., Ogawa, K., and Satake, T., A calmodulin antagonist (W-7) and a protein kinase C inhibitor (H-7) have no effect on atrial natriuretic peptide release induced by atrial stretch, *Life Sci.*, 42, 1659, 1988.

102. Rankin, A. J., Wilson, N., and Ledsome, J. R., Influence of isoproterenol on plasma immunoreactive atrial natriuretic peptide and plasma vasopressin in the anesthetized rabbit, *Pflügers Arch.*, 408, 124, 1987.

103. King, K. A., Wong, A., and Ledsome, J. R., The effect of isoproterenol on right and left atrial dynamics and plasma atrial natriuretic factor in anesthetized rabbits, *Can. J. Physiol. Pharmacol.*, 69, 464, 1991.

104. Lachance, D. and Garcia, R., Atrial natriuretic factor release during acute infusion of isoproterenol in the conscious rat, *Regul. Pept.*, 33, 31, 1991.

105. Iida, H. and Page, E., Inhibition of atrial natriuretic peptide secretion by forskolin in noncontracting cultured atrial myocytes, *Biochem. Biophys. Res. Commun.*, 157, 330, 1988.

106. Auffermann, W., Stefenelli, T., Wu, S. T., Parmley, W. W., Wikman-Coffelt, J., and Mason, D. T., Influence of positive inotropic agents on intracellular calcium transients. Part I. Normal rat heart, *Am. Heart J.*, 118, 1219, 1989.

107. Gnadinger, M. P., Lang, R. E., Hasler, L., Uehlinger, D. E., Shaw, S., and Weidmann, P., Plasma kinetics of synthetic alpha-human atrial natriuretic peptide in man, *Miner. Electrolyte Metab.*, 12, 371, 1986.

108. Maack, T., Suzuki, M., Almeida, F. A., Nussenzveig, D., Scarborough, R. M., McEnroe, G. A. and Lewicki, J. A., Physiological role of silent receptors of atrial natriuretic factor, *Science*, 238, 675, 1987.

109. Stephenson, S. L. and Kenny, A. J., The hydrolysis of α-human atrial natriuretic peptide by pig kidney microvillar membranes is initiated by endopeptidase-24.11, *Biochem. J.*, 243, 183, 1987.

110. Northridge, D. B., Jardine, A. G., Alabaster, C. T., Barclay, P. L., Connell, J. M., Dargie, H. J., Dilly, S. G., Findlay, I. N., Lever, A. F., and Samuels, G. M., Effects of UK 69 578: a novel atriopeptidase inhibitor, *Lancet*, 2, 591, 1989.

Section VI
Pancreas and Gastrointestinal Tract

Chapter 10

ELECTROPHYSIOLOGY OF PANCREATIC ISLET CELLS

Frances M. Ashcroft and Patrik Rorsman

CONTENTS

1. SUMMARY

The pancreatic islet hormones insulin, secreted by B cells, and glucagon, secreted by A cells, are essential for the control of the blood glucose concentration. These hormones act reciprocally to reduce and elevate plasma glucose levels, respectively. The A cell is electrically active in the absence of glucose, whereas glucose concentrations above 6 mM are required to elicit B cell electrical activity. The release of both hormones is triggered by the rise in intracellular Ca^{2+} concentration which results from Ca^{2+} influx through voltage-gated Ca^{2+} channels associated with this electrical activity. The primary insulin secretagogue is glucose, which depolarizes the B cell by closing ATP-sensitive K^+ channels as a consequence of its metabolism. Hormones and neurotransmitters act to inhibit or potentiate glucose-stimulated release via effects on electrical activity, intracellular Ca^{2+}, and the exocytotic machinery itself. Glucagon release is stimulated by amino acids, such as arginine, which produce membrane depolarization by their electrogenic transport into the A cell. The mechanism by which glucose inhibits glucagon secretion remains obscure, but may involve paracrine effects. How elevation of intracellular Ca^{2+} leads to the release of the secretory granules is not fully understood for either cell type, but it appears that in addition to Ca^{2+}, phosphorylation plays an important role.

In this review, we first focus on the properties of the various types of ion channels found in A and B cells as determined by both electrophysiological and molecular biological studies. We then provide a model for how these ion channels contribute to the electrical activity of their respective cell types. Finally, we discuss the mechanisms by which a rise in intracellular Ca^{2+} may give rise to exocytosis.

2. INTRODUCTION

The function of the islets of Langerhans, which comprise the endocrine pancreas, is to control the blood glucose concentration. They do so by secreting two major hormones with opposing actions on blood glucose: insulin, secreted from the B cells (which has a hypoglycemic action) and glucagon, released by A cells (which is hyperglycemic).

This year is the 125th anniversary of the discovery of the pancreatic islets by Paul Langerhans.[1] He described the islet cells as "Kleine Zellen von meist ganz homogenem Inhalt und polygonaler Form mit rundem Kern ohne Kernkörperchen, meist zu zweien oder zu kleinen Gruppen beisammen liegend"*. There are roughly 1 million islets in man which lie dispersed through the exocrine pancreas and comprise less than 1% of the total pancreatic mass. It was not until the beginning of the 20th century that it was recognized that the islet functioned as an endocrine gland. Of central importance for subsequent studies was the development, in 1963, of methods for isolating islets of Langerhans.[2] The pancreatic islet cells were among the first endocrine cells to be studied electrophysiologically. Intracellular micro-electrode recordings were reported by Dean and Matthews[3] as early as 1968,[3] but progress was limited because of technical difficulties and the inability to voltage clamp the islet cells. The properties of the B cell electrical activity as determined using intracellular electrodes have been excellently summarized by Henquin and Meissner.[4] With the application of patch clamp techniques to isolated islet cells there has been an explosion of information about both the electrophysiological and secretory properties of pancreatic B cells. Information about A cells is still rudimentary because of the difficulty in isolating and identifying these cells, and for the same reasons, nothing is known of the electrophysiological properties of D cells and PP cells.

The electrophysiology of the pancreatic B cells has been extensively reviewed[4-6] and interested readers are referred to these papers for more detailed accounts. Here we focus on the highlights of the first 10 years of patch clamp recordings from B cells and only briefly consider the A cell.

2.1. STRUCTURE OF THE ISLET

The islet is a highly organized structure[7] (Figure 1). It consists of a central core of B cells, which make up between 60–80% of the total cell number, surrounded by an outer mantle of A cells (10–20%). Somatostatin-secreting D cells (5%) and pancreatic polypeptide-secreting cells (PP cells; <1%) are scattered throughout the organ. Most of the B cells are juxtaposed to other B cells and are electrically connected to one another, as well as to neighboring A cells, by gap junctions. The pancreatic blood supply is provided by the celiac and anterior mesenteric arteries so that the islet monitors the systemic glucose level and not that draining the gut. The islets are extensively vascularized with the flow of the blood being from the core to the periphery: consequently, the peripheral cells (mainly A cells) are exposed to the secreted products of the B cells. The islet, as first noted by Langerhans,[1] receives a rich innervation with both parasympathetic and sympathetic nerve endings terminating close to islet cells and exerting effects on secretion.

The organization of the islet, with several different cell types located in close proximity to each other, has led to the proposal that local release of hormones within the islet influences secretion from adjacent islet cells (paracrine regulation). Evidence in favor of this idea that the microanatomy of the islet is important for B cell secretion is provided by the observation that isolated single pancreatic B cells release little insulin in response to glucose,[8] but that the presence of glucagon-secreting A cells, or agents which elevate intracellular cAMP (the intracellular mediator of glucagon action), markedly enhance the secretory response to the sugar.[8] A difficulty with this attractive hypothesis, however, is that the blood flows from the B cells to the A cells so that intracapillary glucagon cannot reach the B cells, and thus any effect of glucagon must occur by diffusion of the hormone through the islet interstitium. The islet anatomy is such that B cell secretory products may affect glucagon secretion both through the interstitium and via the circulation. As discussed below, glucose may mediate its inhibitory action on A cell secretion by such paracrine mechanisms.

---

* "Small cells, most of them with entirely homogeneous contents and a polygonal shape, having a round nucleus without nucleoli and usually lying side by side in pairs or small groups".

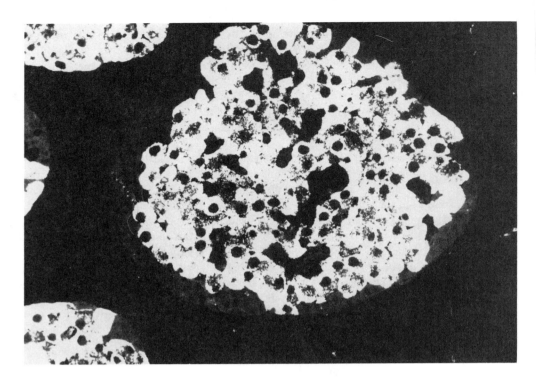

FIGURE 1.   Semi-thin section of an isolated rat islet of Langerhans after immunostaining for insulin. The insulin-containing cells are more numerous, largely confined to the center of the islets, and surrounded by one or two layers of insulin-negative cells. Original magnification ×400. (From Pipeleers et al., *Insulin,* Ashcroft, F. M. and Ashcroft, S. J., Eds., IRL Press, Oxford, 1992, 5. With permission of Oxford University Press.)

2.2.   STIMULATORS AND INHIBITORS OF ISLET SECRETION

Insulin secretagogues may be classified into two groups, the *initiators* and the *potentiators* (Figure 2).[9] The former, of which glucose is the most important, stimulate insulin release on their own. Other initiators include those carbohydrates and amino acids which are metabolized by the B cell, and drugs such as the sulfonylureas tolbutamide and glibenclamide. Potentiators of insulin release comprise neurotransmitters such as acetylcholine (ACh), the islet hormone glucagon, and circulating hormones such as glucagon-like peptide 1, gastrointestinal peptide, and vasoactive intestinal peptide. These substances are ineffective on their own, but are able to potentiate secretion induced by an initiator, and may even stimulate release in the presence of substimulatory concentrations of an initiator. Finally, insulin release is under inhibitory control by intra-islet hormones such as somatostatin, pancreastatin, and amylin, and by circulating hormones such as adrenaline and intra-islet neurotransmitters such as galanin.

The most important physiological stimulators of glucagon release are catecholamines, such as adrenaline, and amino acids, as exemplified by arginine.[10] Inhibitors of glucagon secretion include glucose and the islet hormone somatostatin. Insulin was formerly believed to exert an inhibitory effect on the A cell,[11] but with the demonstration that these cells have few receptors for insulin this is no longer thought to be the case.

3. ELECTROPHYSIOLOGICAL PROPERTIES OF ISLET CELLS

3.1.   A CELLS

Patch clamp experiments on isolated guinea pig A cells have shown that, unlike B cells, they produce spontaneous action potentials in the absence of glucose.[12] Activation of voltage-gated Na^+ and Ca^{2+} currents underlies the upstroke of these action potentials while K^+ currents

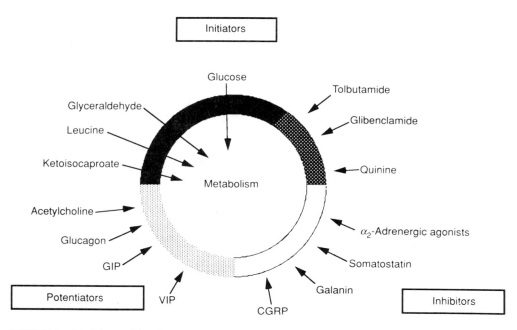

FIGURE 2. Modulators of insulin secretion. Initiators of secretion are able to release insulin in the absence of glucose and comprise metabolic substrates (solid segment) and drugs (hatched segment). Potentiators of secretion (stippled segment) act via classical second messenger systems to augment the secretory response to an initiator. Inhibitors (open segment) may act at various stages in the release process. VIP (vasoactive intestinal peptide), GIP (gastrointestinal peptide), CGRP (calcitonin-gene-related-peptide). (From Ashcroft, F. M. and Ashcroft, S. J. H., *Insulin,* IRL Press, Oxford, 1992, 97. With permission of Oxford University Press.)

are responsible for spike repolarization. In patch clamp experiments on intact mouse islets we find that the peripheral cells, which are likely to be mainly A cells (cf. Figure 1), have similar properties (unpublished observations).

3.1.1. Inward Currents

The Na^+ currents are tetrodotoxin (TTX) sensitive and are activated at potentials positive to -50 mV.[12] Steady-state inactivation is half maximal at about -60 mV and is steeply voltage-dependent. The time constant of inactivation is 2.5 ms at -30 mV and decreases with depolarization. At -70 mV, the most negative potential encountered in a spontaneously spiking A cell, the Na^+ current is already inactivated by about 20%.

There are a number of different families of voltage-dependent Ca^{2+} channels which have been classified into T, L, N, P, and Q types.[13] Both T- and L-types of Ca^{2+} channel are present in A cells.[14] Whereas openings of the T-type channel occur at potentials as negative as -65 mV, depolarization above -50 mV is required to elicit L-type channel openings. Inactivation of the T-type current is voltage dependent, with a time constant of 100 ms at -40 mV and half maximal activation at -60 mV. When $[Ca^{2+}]_i$ is buffered to nanomolar concentrations, the L-type channels do not show inactivation, but by analogy with other L-type Ca^{2+} channels they may be expected to undergo Ca^{2+}-dependent inactivation. The differences in the voltage dependence of activation suggest that the two types of Ca^{2+} current may have different functional roles. The T-type channels open around the action potential threshold and, as in other tissues,[15] may therefore play a role in action potential initiation and the control of pacemaker activity. The L-type channels activate at membrane potentials corresponding to the rapidly rising phase of the action potential, are of larger amplitude (~ 1 pA at 0 mV with 100 mM Ba^{2+} as the charge carrier), and are likely to be responsible for most of the Ca^{2+} entry required for secretion.

3.1.2. Outward currents

The whole-cell K^+ currents underlying action potential repolarization activate with a short delay at potentials > -10 mV and increase in amplitude with depolarization. Single-channel studies suggest a unitary current amplitude of 1.5–2 pA at -70 mV in symmetrical 140 mM $[K^+]$,[12] corresponding to a conductance of 20–30 pS. These data suggest the outward current in the A cell is carried by delayed-rectifying K^+ channels (K_{DR} channels) similar to those in the B cell (see below).

Little is known about the resting conductance of A cells. ATP-sensitive K^+ channels (K_{ATP} channels) appear to be absent in guinea pig A cells, since no time-dependent changes in whole-cell K^+ currents were observed when A cells were dialyzed with ATP-free solutions.[12] By contrast, K_{ATP} channels are present in the glucagon-secreting mouse cell line, α–TC.[16] Preliminary data, only presented in abstract form, indicate that the latter finding may also be extended to normal rat A cells.[17] The ATP sensitivity of the A cell K_{ATP} channel has not been reported, but if it resembles that of the B cell it is hard to explain how the A cell can be maximally active under hypoglycemic conditions, where the K_{ATP} current should be large. Although the functional role of the K_{ATP} channel in glucose-mediated inhibition of A cell secretion is thus far from clear, regulation of this channel by neurotransmitters and hormones may be important, as it is in smooth muscle.[18]

A γ-aminobutyric acid (GABA$_A$) receptor chloride channel has also been identified in guinea pig A cells.[19] This channel is activated by GABA binding to an extracellular site, which opens a chloride ionophore, and thereby repolarizes the cell and inhibits electrical activity. The current is elicited by GABA concentrations >1 μM and saturates at concentrations >30 μM. In addition to GABA, the channel is activated by muscimol, inhibited by picrotoxin and bicuculline, and modulated by diazepam. This pharmacology corresponds to that of neuronal GABA$_A$ receptors.[20] It has been proposed that activation of the A cell GABA$_A$-receptor channel by GABA released from adjacent B cells in response to glucose may provide a mechanism for paracrine control of glucagon secretion, at least in the guinea pig.[10,19] The reason for implicating such an indirect action of glucose, is that the sugar does not itself affect the electrical activity of the single A cell,[12] although it inhibits glucagon secretion from the islet. In favor of this idea is that GABA$_A$-receptor antagonists, such as bicuculline, reduce the inhibitory action of glucose on glucagon secretion in guinea pig islets,[19] although this is not the case in mouse or rat islets.[21]

3.2. B CELLS

Microelectrode recordings established the pattern of electrical activity in the B cell and its modulation by glucose.[4] At subphysiological glucose concentrations (<5 mM) the B cell is electrically silent with a resting membrane potential of about -70 mV. When glucose is increased above 6 mM, the B cell slowly depolarizes and electrical activity is initiated if this depolarization exceeds threshold (-50 mV). At concentrations between 7 and 16 mM, this electrical activity consists of slow oscillations in membrane potential (slow waves) between a depolarized plateau level, on which action potentials are superimposed, and a repolarized silent interval. The pattern of electrical activity is graded with glucose concentration, the slow waves increasing in duration and the intervals between them decreasing as glucose is elevated, until at concentrations above 16–18 mM electrical activity is continuous (Figure 3).

The advent of the patch clamp technique enabled the study of the ionic currents underlying B cell electrical activity. These studies demonstrated that a particular type of K^+ channel, the *ATP-sensitive* K^+ *channel* (K_{ATP} channel), plays a central role in stimulus-secretion coupling in the B cell (Figure 4). At the resting potential of the unstimulated B cell the K_{ATP} channel is spontaneously active. Glucose metabolism produces a concentration-dependent inhibition of the channel (IC$_{50}$ ~2 mM), which is probably mediated by changes in the intracellular

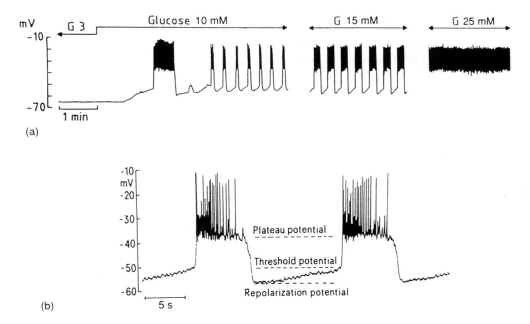

(a)

(b)

FIGURE 3. B cell electrical activity measured by intracellular microelectrode recordings. (a) Effect of increasing glucose from 3 to 10 m*M*, 15 and 25 m*M*. All traces come from the same cell, but an interval separates the recordings at higher glucose concentrations (From Henquin, J. C., *Arch. Int. Physiol. Biochim.*, 98, A61, 1990. With permission.) (b) Expanded section of two successive bursts recorded in 10 m*M* glucose. (From Henquin, J. C. and Meissner, H. P., *Experientia*, 40, 1043, 1984. With permission.)

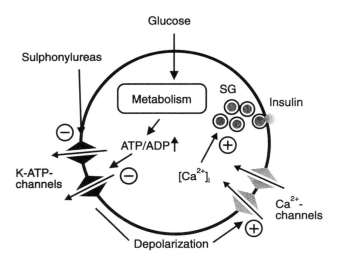

FIGURE 4. Model of stimulus-secretion coupling in the pancreatic B cell. The resting potential of the B cell is determined by the activity of K_{ATP} channels. When the plasma glucose concentration is increased, the rate of metabolism is stimulated and consequently the cytoplasmic ATP/ADP ratio increases. This leads to the closure of the K_{ATP} channels, membrane depolarization, and the opening of voltage-gated Ca^{2+} channels. The increased Ca^{2+} influx produces a rise in the cytoplasmic Ca^{2+} concentration ($[Ca^{2+}]_i$) which, via a series of undefined reactions, triggers the exocytosis of the insulin-containing secretory granules (SG). Sulfonylureas "short-circuit" this pathway by inducing the closure of the K_{ATP} channels without any changes in B cell metabolism or ATP/ADP ratio.

concentrations of ATP and ADP. The activity of the K_{ATP} channel is responsible for the maintenance of a negative membrane potential so that channel closure results in membrane depolarization. Activation of voltage-gated Ca^{2+} channels, followed by Ca^{2+} influx and an increase in the intracellular Ca^{2+} concentration ($[Ca^{2+}]_i$), produces exocytosis of the insulin-containing secretory granules. Here we summarize the properties of the different types of ion channel in the B cell.

3.2.1. K_{ATP} Channel

The K_{ATP} channel has a single-channel conductance of 50–80 pS in symmetrical 150–mM K^+ concentrations and of 10–30 pS (depending on the intracellular concentrations of Na^+ and Mg^{2+}; see below) when measured under physiological ionic gradients.[6] Outward currents are blocked in a voltage-dependent manner by intracellular Na^+ and Mg^{2+}, giving rise to a current-voltage relation that shows inward rectification.[22,23] The channel open probability is unaffected by voltage, but is dramatically reduced by a rise in the intracellular ATP concentration.[6] In excised membrane patches, the K_{ATP} channel is inhibited by intracellular ATP with half maximal inhibition of activity being produced by 10–15 μM ATP and almost complete block by 100 μM.[24-26] The inhibitory action of ATP does not involve phosphorylation, since nonhydrolyzable ATP analogs, and ATP in the absence of Mg^{2+}, are also effective; thus ATP is considered to interact directly with an inhibitory site on the channel protein. The active species appears to be the free ion, ATP^{4-}.[26,27] Other nucleotides are also inhibitory, but are less effective than ATP; the order of potency being ATP>ADP>AMP. The inhibition of channel activity by ADP, GTP, and GDP is seen more clearly in the absence of intracellular Mg^{2+} since in the presence of this cation the dinucleotides have an additional stimulatory action[23,28] which is discussed more extensively below (Section 3.2.1.1.1).

In intact cells, the K_{ATP} channel is inhibited by nutrient secretagogues including glucose, glyceraldehyde, leucine, and ketoisocaproate and is activated by substances which decrease B cell metabolism such as mannoheptulose, iodoacetate, rotenone, and azide.[29-32] In this way B cell electrical activity (and thereby insulin secretion) is modulated by B cell metabolism.

It was originally suggested that ATP serves as the second messenger that couples B cell metabolism to channel closure because this nucleotide directly blocks the K_{ATP} channel in both inside-out patches (see above) and whole-cell recordings.[31] The main argument against this idea is that the channel is almost fully blocked by 0.1 mM ATP in excised patches, whereas the ATP concentration measured biochemically in the B cell is 2–5 mM and changes little with glucose.[33,34] However, it is clear that in the intact B cell other cytosolic constituents must modulate the ATP sensitivity of the channel, because K_{ATP} channel activity can be recorded from intact cells despite the presence of millimolar concentrations of ATP, and estimates of the ATP sensitivity in the intact cell suggest a K_i of ~1 mM.[35,36] One factor which should be considered in this context is the concentration of ADP. This nucleotide both increases channel activity in the absence of ATP and relieves the blocking effect of ATP.[32,37,38] Furthermore, ADP varies concomitantly, but reciprocally, with the ATP concentration. Thus changes in ADP may be as important in coupling metabolism to channel activity as changes in ATP. The same evidence that has been put forward to support a coupling role for ATP[6,39] may also be used to argue that ADP is important. The evidence may be summarized as follows: (1) agents that elevate cellular ATP levels (and lower ADP) inhibit channel activity, while those that depress ATP production, and increase ADP levels, activate K_{ATP} channels; (2) there is a good correlation between hyperpolarization of the mitochondrial membrane potential (which is linked to ATP production and ADP utilization) and the block of K_{ATP} channels;[40] and (3) changes in K_{ATP} channel activity[41,42] and the intracellular ATP and ADP concentrations[43,44] occur over the same range of glucose concentrations and with a similar time course. Although ATP and ADP both vary with metabolism, the changes in ADP levels are much greater than those of ATP. Thus a decrease in ADP, at a relatively constant ATP, may be the link between metabolism and channel closure. This hypothesis is reminiscent of that put forward for the K_{ATP} channel of skeletal muscle, where changes in pH_i, rather than in ATP, have been postulated to serve as a coupling factor.[45]

In whole-cell recordings the K_{ATP} current activates following washout of ATP from the cell into the pipette and constitutes a time- and voltage-independent current which is highly K^+ selective.[46] With physiological ionic gradients the single-channel conductance is 15–20 pS,[46,47] and the whole-cell conductance in the absence of ATP is ~10 nS.[31] Since the whole-cell conductance is simply the product of the number of channels, the single-channel conductance, and the channel open probability, this indicates that there are at least 500 K_{ATP} channels per B cell (assuming an open probability of 1.0). However, this value must be considered an underestimate as the K_{ATP} channel is unlikely to be open continuously: with an open probability of 0.3, for example, the channel density is ~1500 per B cell.

3.2.1.1. Modulatory Influences

A variety of substances have been shown to modulate the K_{ATP} channel. It is important to understand the mechanisms underlying this modulation in order to fully comprehend the regulation of K_{ATP} channel activity in the intact B cell. Channel modulators include nucleotides such as ADP and GDP (see above), anions such as chloride,[48] divalent cations,[26,49] proteolytic enzymes[50] and pharmacological agents (review: Reference 6). It should also be noted that the properties of the channel change spontaneously after patch excision or formation of the whole-cell configuration. For example, channel activity usually declines with time (rundown) and the efficacy of various activators and inhibitors of channel activity also decreases.[28,51] One important exception is that the ATP sensitivity of the channel remains constant. This would be consistent with the ATP binding site being an integral part of the channel, whereas modulation may involve associated control elements which are only loosely attached to the membrane and are lost with time. A detailed account of modulation of the K_{ATP} channel is presented elsewhere;[6] here, we simply mention the more important findings and summarize recent data.

3.2.1.1.1. Nucleotides — Dinucleotides such as ADP and GDP are able to stimulate K_{ATP} channel activity in both the absence and presence of ATP. There is evidence that this effect involves a different regulatory site from that which mediates the inhibitory action of ADP or ATP.[28] First, activation, but not inhibition, requires Mg^{2+}. Secondly, the stimulatory effects of MgADP are not mimicked by non-hydrolyzable analogs such as α,β-methylene ADP, or by ADP^{3-}: indeed, these compounds actually block channel activity.[23,51,52] Thirdly, channel activation is produced by much lower nucleotide concentrations than those required to produce inhibition. Finally, the stimulatory actions of MgADP and MgGDP, in both the presence and absence of ATP, decline after patch excision.[28] The latter effect led to the proposal that dinucleotides might act on a separate regulatory protein which is washed out of the patch with time. Alternatively, the K_{ATP} channel protein itself is modified (for example, by endogenous proteases; see below) in a time-dependent way after patch excision. The latter idea seems more consistent with the recent observation that the cloned cardiac K_{ATP} channel is activated by ADP.[53]

3.2.1.1.2. Proteolysis — Application of the protease trypsin to the intracellular face of the membrane has several effects on the K_{ATP} channel.[50] It both increases channel activity and removes the rundown of channel activity observed in excised patches. The sensitivity of channel activity to inhibition by Mg^{2+} and ATP is also somewhat reduced. More importantly, proteolysis completely removes the ability of ADP to activate the channel or to modulate its ATP sensitivity and it abolishes the inhibitory effect of tolbutamide. These data suggest that sites involved in rundown, and in mediating the effects of ADP and sulfonylureas, are accessible from the inner side of the membrane. Trypsin had no effect on the single-channel conductance or the voltage-dependent block produced by intracellular Mg^{2+}, indicating that the pore is inaccessible to the protease.

3.2.1.1.3. Drugs — The activating effect of diazoxide disappears in excised patches, with approximately the same time course as that of MgADP. Furthermore, the effects of the K^+ channel opener and MgADP interact with each other, as low concentrations of MgADP potentiate the effect of diazoxide.[51] This suggests that they may act on the same accessory protein. The inhibitory effect of tolbutamide is also modulated by Mg^{2+} and adenine nucleotides with the drug being much less potent in the absence of both of these agents.[49]

3.2.1.1.4. Rundown — In inside-out patches, channel activity can be transiently restored following partial rundown by brief exposure to intracellular MgATP ("refreshment")[25,32,54] nonhydrolyzable analogs are ineffective. It has therefore been argued that the active K_{ATP} channel exists in a phosphorylated form and that the rundown of channel activity that is observed in excised patches and the whole-cell configuration results from channel dephosphorylation. Rundown of K_{ATP} channel activity in excised patches is markedly accelerated by micromolar concentrations of Mg^{2+} which would be consistent with the involvement of a Mg^{2+}-dependent phosphatase.[55] However, known Mg^{2+}-dependent phosphatases require more than 10 mM Mg^{2+} for activation,[56] which does not support the idea. In our view, the conclusion that rundown results from dephosphorylation is still premature. Evidence that the "refreshing" action of ATP results from phosphorylation by endogenous kinases is also limited, and it remains possible that MgATP acts by direct binding or via hydrolysis to MgADP.

3.2.1.2. Molecular Biology of the K_{ATP} Channel

A most exciting development in the past few years has been the recent cloning of the cardiac K_{ATP} channel.[53] The structure is quite different from that of the 140-kDa sulfonylurea receptor which has been purified from RINm5F cells,[57] indicating that the K_{ATP} channel and sulfonylurea receptor are different proteins. The primary structure suggests that the channel belongs to a family of inwardly rectifying K^+ channels which have a common molecular architecture. This consists of two hydrophobic (transmembrane) domains linked by a pore region which is homologous to that of other K^+ channels.[58] By analogy with the voltage-gated K^+ channels it seems likely that four of these subunits will come together to form the pore. The ATP binding site may involve the interaction of several parts of the protein, since there is no obvious consensus sequence for ATP binding. It seems likely that the B cell K_{ATP} channel will turn out to possess a similar structure.

3.2.2. Other Types of K^+ Channels

The B cell is equipped with several types of voltage-gated K^+ channels. The two most important are a *delayed-rectifying K^+ channel* (K_{DR}) with a single-channel conductance of ~10 pS (when measured with physiological ionic gradients) and a *Ca^{2+} activated K^+ channel* (K_{Ca}) with a single-channel conductance of ~100 pS under the same conditions.[59-62] The K_{DR} channels carry as much as 80–90% of the total voltage-dependent outward current and the K_{Ca} channels contribute only 10–20% of the delayed outward current.

The delayed rectifier opens gradually during depolarizations to membrane potentials more positive than –30 mV and is unaffected by changes in the cytoplasmic Ca^{2+} concentration. Both the steady-state activation and a slow time-dependent inactivation of the channel increase with depolarization.[59] As K_{DR} currents often run down gradually during whole-cell recordings with low intracellular ATP, channel activity may also be influenced by phosphorylation. Many types of voltage-gated K^+ channels have been cloned (review: Reference 58). Of these, $K_V 1.1$, $K_V 1.5$, and possibly also $K_V 1.4$ and $K_V 1.6$ are expressed in the B cell.[63,64] The properties of the native K_{DR} current in the B cell most closely resemble those of $K_V1.1$, in having a single-channel conductance of 9 pS, being half blocked at 0 mV by 1.4 mM tetraethylammonium (TEA), activating at potentials positive to −40 mV and showing both voltage-dependent activation and slow inactivation.[59,60,65]

As in other cells, the B cell K_{Ca} channel is activated by both Ca^{2+} and depolarization.[66] Its Ca^{2+} sensitivity is too low to permit activation at voltages around the resting potential, and thus the K_{Ca} channel does not contribute to the resting conductance of the B cell. During an action potential both the voltage and $[Ca^{2+}]_i$ change in a direction favoring channel opening, and it has therefore been argued that these channels participate in action potential repolarization. The Ca^{2+} responsible for K_{Ca} channel activation normally enters the B cell through voltage-gated Ca^{2+} channels, and thus the whole-cell K_{Ca} current has a voltage dependence which reflects that of the voltage-gated Ca^{2+} current and is blocked by inhibitors of Ca^{2+} influx.[62]

Pharmacological studies have defined the relative contribution of the K_{DR} channel and K_{Ca} channels to action potential repolarization. The K_{Ca} channels are almost completely blocked by 1 mM external TEA (K_d = 0.14 mM), whereas the K_{DR} channels are inhibited by only 30% (K_d = 1.4 mM). Since action potential repolarization is little affected by this concentration of TEA[60,61,68]) and is unaffected by charybdotoxin (a selective blocker of K_{Ca} channels)[69] it appears that spike repolarization results principally from activation of K_{DR} channels. In fact, in cell-attached patch recordings, B cell action potentials are often found to be associated with openings of K_{DR}-channels.[62]

In addition to the large-conductance K_{Ca} channel, a second *low-conductance Ca^{2+}-activated K^+ channel* ($K_{L,Ca}$) has been described in B cells.[70,71] Estimates from noise analysis suggest a single-channel conductance of less than 1 pS in 140 mM K^+. This channel is distinguished from other Ca^{2+}-activated K^+ channels by its insensitivity to both charybdotoxin and apamin. As discussed below, this channel may be important for the generation of the membrane potential oscillations characterizing B cell electrical activity in the presence of intermediary glucose concentrations.

A K^+ channel (K_I channel) activated by hormones and neurotransmitters that inhibit insulin secretion, such as galanin, somatostatin, and adrenaline, has also been reported in mouse B cells.[72] Again, the single-channel conductance is too small to be measured directly in patch clamp recordings, but it has been estimated as ~1 pS when measured with physiological ionic gradients by noise analysis. Activation of the channel is blocked by pretreatment with pertussis toxin, suggesting a G protein mediates the effects of inhibitors of secretion on the channel. It is likely that channel activation is responsible for the repolarization and inhibition of electrical activity produced by these agents, and thereby partly accounts for their ability to suppress insulin release in mouse B cells. In the rat insulinoma cell line RINm5F cells, by contrast, activation of K_{ATP} channels may be more important.[73-75]

3.2.3. Ca^{2+} Channels

To date, two types of voltage-dependent Ca^{2+} channels have been described in single-channel studies on mouse B cells (review: Reference 6), with properties resembling those of L-type and T-type channels in other tissues (terminology as proposed by Nowycky et al.[76]) The *T-type* (low voltage-activated or slowly deactivating) *channel* is activated at potentials around −50 mV, exhibits voltage-dependent inactivation, and has a single-channel conductance of 8 pS with 110 mM external Ba^{2+} as the charge carrier.[77] The *L-type channel* (high voltage-activated or fast-deactivating channel) is distinguished by its sensitivity to the dihydropyridines (DHP), being inhibited by DHP antagonists like nifedipine and potentiated by DHP agonists like BAY K 8644.[59,78] It is activated at potentials around −40 mV, inactivation of the current is Ca^{2+} dependent, and the single-channel conductance is 20–25 pS with 110 mM Ba^{2+}.[78,79] Mouse B cells appear to possess only the L-type Ca^{2+} channel,[78,79] whereas T-type channels have been identified at the single-channel level in rat,[77,80] human,[81] and

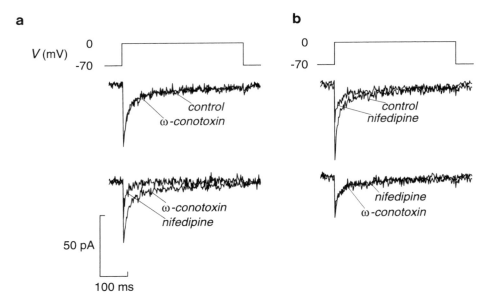

FIGURE 5. Pharmacological properties of B cell Ca^{2+} currents. Voltage-gated Ca^{2+} currents were evoked by 500-ms depolarizations from -70 to 0 mV applied at 2-min intervals. Currents were recorded from metabolically intact mouse B cells using the perforated-patch whole-cell configuration. (a) Top: Ca^{2+} currents were unaffected by $2 \mu M$ ω-conotoxin GVIA, a blocker of N-type Ca^{2+} channels. The integrated Ca^{2+} currents were 3.7 and 3.9 pC in the absence and presence of the neurotoxin, respectively; below: subsequent inclusion of nifedipine ($10 \mu M$), a blocker of L-type Ca^{2+} channels, in the continued presence of ω-conotoxin reduced the integrated current to 0.6 pC. (b) Top: the same type of experiment but nifedipine was applied first. The integrated current was reduced from 6.3 pC under control conditions to 2.2 pC in the presence of $10 \mu M$ nifedipine (below). Subsequent inclusion of ω-conotoxin ($2 \mu M$) produced no further block and the integrated current amounted to 2.5 pC.

RINm5F[82,83] B cells. It has also been argued, on the basis of whole-cell recordings, that RINm5F cells possess additional types of Ca^{2+} channel, which are insensitive to DHP.[84,85] As much as 45% of the total Ca^{2+} current may be carried by these channels, and since this fraction of the current is partially blocked by ω-conotoxin it may, in part, flow through N-type Ca^{2+} channels. In mouse B cells, we have observed no ω-conotoxin-sensitive Ca^{2+} component (Figure 5).

3.2.3.1. Molecular Biology

Molecular biological studies have revealed an even greater complexity than that indicated above by demonstrating the presence of multiple isoforms of the L-type Ca^{2+} channel. In skeletal muscle the L-type Ca^{2+} channel is composed of five distinct subunits (α_1, α_2, β, γ, δ. The α_1 subunit functions both as the Ca^{2+} channel pore and as the DHP receptor[86] and comprises four repeated domains, each of which contains six putative transmembrane segments and a pore-forming region. Two isoforms of the α_1 subunit of the L-type Ca^{2+} channel have been identified in rat B cells.[87,88] One of these is 98% homologous to the cardiac-type α_1 subunit and the other isoform shares 98% identity with the neuroendocrine-type α_1 subunit. A number of splice variants of the latter have been reported to be expressed in HIT cells.[87]

3.2.3.2. Inactivation

There is evidence that inactivation of the L-type Ca^{2+} channel is Ca^{2+} dependent.[89,90] This is suggested by two-pulse experiments which demonstrate that inactivation shows a voltage-dependence similar to that of the Ca^{2+} current itself, and by the fact that inactivation is markedly reduced when Ba^{2+} carries the inward current. Further support for this idea is provided by the finding that the decline of the peak Ca^{2+} current during a train of depolarizing pulses parallels the rise in intracellular Ca^{2+} [91] (cf. Reference 92). Surprisingly, high intracellular concentrations of the Ca^{2+} buffer EGTA do not reduce or slow inactivation in standard whole-cell recordings,[93] and this is also the case for the more rapid Ca^{2+} chelator BAPTA.[89] These results suggest that inactivation is determined by the Ca^{2+} concentration in the immediate vicinity of the Ca^{2+} channel pore, which the Ca^{2+} buffers cannot access (see below and Figure 11).

Whole-cell Ca^{2+} currents inactivate with a biphasic time course, a rapid inactivation to about one third of the initial amplitude of the current ($\tau \sim 20$ ms) being followed by a much slower phase of inactivation ($\tau > 0.1$–1 s). Evidence suggests that the initial phase of inactivation is Ca^{2+} dependent and that the slow component is voltage dependent:[94] this may be the reason why some inactivation persists when Ba^{2+} is the charge carrier.[79,95]

3.2.3.3. *Ca^{2+} Channels and Intracellular Ca^{2+} Transients*

A single action potential is sufficient to induce a substantial $[Ca^{2+}]_i$ transient in the B cell and elevates the average $[Ca^{2+}]_i$ from a resting concentration of 100 nM to a peak concentration of ~500 nM.[91] The duration of the Ca^{2+} transient (5–10 s) is much longer than the action potential (100 ms), and $[Ca^{2+}]_i$ recovers exponentially with a time constant of 1–2 s. This means that during a burst of action potentials, which has a spike frequency of >4 Hz, the interval between two successive spikes is insufficient for $[Ca^{2+}]_i$ to return to basal levels. Comparison of the amount of Ca^{2+} entering the cell during a depolarizing pulse with the increase in $[Ca^{2+}]_i$ measured using a fluorescent Ca^{2+} indicator reveals that only a fraction (<1%) of the Ca^{2+} entering the cell actually appears as free Ca^{2+} in the cytoplasm.[91,92] Thus most of the Ca^{2+} is rapidly buffered by the B cell. The identity of the Ca^{2+} buffers is not known, but they are not easily diffusible as they remain within the cell during standard whole-cell recordings.

Early recordings of single Ca^{2+} channel activity demonstrated that the density of Ca^{2+} channels varies considerably between different patches, suggesting that the channels are concentrated in certain regions of the cell.[78] Confirmation of this idea has been provided by digital imaging of $[Ca^{2+}]_i$ which showed that Ca^{2+} entry is confined to one side of the cell (Figure 6). Areas of high density of Ca^{2+} channels co-localize with the secretory granules and thus may represent "hot spots" of secretion.[96]

3.2.3.4. *Function*

The function of the different Ca^{2+} channels is twofold: to shape B cell electrical activity and to mediate the Ca^{2+} influx which triggers exocytosis. The L-type channel appears to be responsible for most of the Ca^{2+} influx required for secretion, at least in mouse B cells. Furthermore, since L-type Ca^{2+} channel openings occur at the B cell resting potential,[79] Ca^{2+} influx through these channels may contribute to the background Ca^{2+} influx into the B cell. Indeed, when nifedipine is applied to a voltage-clamped B cell, a slight reduction of the resting $[Ca^{2+}]_i$ is observed.[91] This would explain why DHPs can modulate basal insulin secretion.[97,98] The T-type Ca^{2+} channel may play a role in burst initiation in rat B cells, since it activates at voltages close to the threshold for electrical activity, but its contribution to Ca^{2+} influx is likely to be very small, since it will be largely inactivated at the plateau potential. T-type Ca^{2+} channels are clearly not essential for the characteristic bursting pattern of electrical activity observed in mouse B cells, which lack this type of Ca^{2+} channel.

3.2.3.5. *Modulation*

3.2.3.5.1. Protein Kinase A (PKA) — It is well established that L-type Ca^{2+} channels in a variety of tissues are potentiated by activation of cAMP-dependent protein kinase (PKA; review: Reference 99). In B cells, agents that elevate cAMP reduce the rate of Ca^{2+} channel inactivation without substantially altering the peak amplitude of the Ca^{2+} current.[100] As

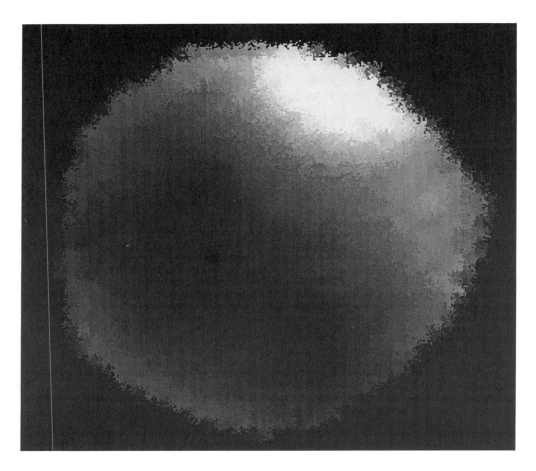

FIGURE 6. Ca^{2+} entry in mouse pancreatic B cell is localized to specific regions of the cell. The cell was stimulated by a 500-ms step depolarization from −70 to 0 mV. The B cell had been loaded with 0.5 μM of the fluorescent $[Ca^{2+}]_i$ indicator fura-2/AM for 25 min before the experiment. The image was obtained 250 ms after the onset of the depolarization and the increase in $[Ca^{2+}]_i$ is clearly localized to the upper part of the cell. This depolarization elicited an exocytotic response of 20 fF, which corresponds to the exocytosis of ten secretory granules. The black-shaded area corresponds to a Ca^{2+} concentration of 0.1 μM and the white area, to a concentration larger than 1 μM. Before the onset of the depolarization, the entire cell had a resting $[Ca^{2+}]_i$ of 0.1 μM.

discussed below, this effect may contribute to the lengthening of the slow waves found in the presence of agents that increase cAMP.[101,102] Since the effect of cAMP is reversed by the specific PKA inhibitor Rp-cAMPs, it appears that PKA activation mediates the effect of the nucleotide.[100] It seems likely that glucagon and other secretagogues, which elevate cAMP in B cells, will also be found to modulate the B cell Ca^{2+} current in the same way.

3.2.3.5.2 Protein Kinase C (PKC) — PKC also regulates Ca^{2+} channels in B cells. In perforated-patch recordings from mouse B cells, the phorbol ester PMA, which directly activates PKC, produced a 30% increase in the peak amplitude of the Ca^{2+} current, but did not affect inactivation.[103] Downregulation of PKC by long-term exposure to a high concentration of PMA, however, resulted in a 50% reduction of the Ca^{2+} current amplitude in mouse B cells.[104] A number of membrane-permeable analogs of diacylglycerol, which activate PKC, have also been reported to influence whole-cell or single-channel Ca^{2+} currents. For example, DiC_{10} increases the activity of single L-type Ca^{2+} channels in RINm5F cells,[105] whereas DiC_8 inhibits whole-cell Ca^{2+} currents in mouse B cells.[106] It seems possible that the synthetic diacylglycerols mediate their effects via a mechanism which is not associated with activation of PKC.[107]

3.2.3.5.3. GTP-Binding Proteins — Many inhibitors of insulin secretion are known to mediate their actions by GTP-binding proteins. In HIT cells, adrenaline reversibly blocks whole-cell Ca^{2+} currents[108] and similar results are found with somatostatin[109] and galanin.[110] Inhibition by the first two agents appears to involve an inhibitory G protein, and it is abolished by preincubation of the cells with pertussis toxin. In addition, adrenergic inhibition is irreversible in the presence of intracellular $GTP\gamma S$ and is abolished by inclusion of $GDP\beta S$ in the intracellular solution.[108]

Although catecholamines and galanin appear to be without effect on L-type Ca^{2+} channels in mouse B cells,[111,112] inhibition of the whole-cell Ca^{2+} current is observed following intracellular application of the nonhydrolyzable GTP analog, $GTP\gamma S$,[113] suggesting that the Ca^{2+} currents are under the control of G proteins. It should be emphasized that the inability of catecholamines and galanin to inhibit the Ca^{2+} current is not due to a loss of the receptors for these agents, since activation of the K_I current can be demonstrated in primary cultured B cells prepared the same way as those used for the Ca^{2+} current recordings.[72] Although it is possible that the biochemical coupling machinery which links the receptor to the Ca^{2+} channel is disturbed, the observation that the upstroke velocity of the action potential in the intact mouse islet is unaffected by adrenergic agonists argues that adrenaline, at least, is without effect on the Ca^{2+} current in mouse B cells.[114]

3.2.3.5.4. Diabetic Serum — Serum isolated from newly diagnosed insulin-dependent (IDDM) diabetics produces a marked upregulation of L-type Ca^{2+} channel activity in mouse B cells and RINm5F cells.[83] This effect is restricted to the L-type Ca^{2+} channels and requires overnight culture in the presence of serum, indicating that the antibody does not act as a direct activator of the channel. The mechanism of this effect is unclear, but there is some evidence suggesting the involvement of IgM. Exposure of the B cells to the serum resulted in apoptosis. This effect was prevented by L-type Ca^{2+} channel blockers, suggesting that B cell destruction in IDDM[115] may involve increased Ca^{2+} influx.

3.2.4. Na⁺ Channels

B cells possess several different types of Na⁺-conducting channels. *Voltage-gated, TTX-sensitive Na⁺ channels* are present in all species thus far investigated, but the voltage dependence of inactivation varies considerably. In mouse B cells, the Na⁺ current is completely inactivated at the resting potential[6,116] and it is therefore unlikely to serve a physiological function. By contrast, in rat B cells, the Na⁺ current is only partially inactivated at rest and contributes to B cell electrical activity. The Na⁺ current may also be important for stimulus-secretion coupling in rat B cells, as inhibition of Na⁺ channels by tetrodotoxin decreases insulin release at high glucose concentrations.[117] The dog B cell constitutes an extreme case which generates Na⁺-dependent action potentials.[118]

In addition to the voltage-gated Na⁺ current there must be a *background Na⁺ current* carried by Na⁺ ions, because removal of external Na⁺ is associated with hyperpolarization and a decrease in the resting inward current.[91,119] The presence of this background Na⁺ current accounts for, or at least contributes to (cf. Section 3.2.3.4), the depolarization of the B cell which occurs when K_{ATP} channels close: in its absence K_{ATP} channel closure would have no effect on the membrane potential. The background Na⁺ current has not been identified at the single-channel level. It may be produced by the resting activity of voltage-gated Na⁺-channels, by a non-selective cation channel (which would produce a Na⁺ current at the resting potential), or even result from the activity of an electrogenic Na⁺/Ca^{2+} exchanger.

Finally, there is evidence that activation of an *ACh-activated, Na⁺-permeable channel*[120] underlies the B cell depolarization produced by this neurotransmitter, since depolarization is abolished by Na⁺-free solutions.[120]

3.2.5. Pumps and Transporters

A number of amino acids which are poorly metabolized by the B cell stimulate electrical activity and thus insulin release because their transport into the cell is electrogenic. These include the positively charged amino acids arginine, ornithine, and lysine, probably carried by a y^+ transporter,[121] and neutral amino acids such as alanine which are cotransported with Na⁺.[122] Uptake of these amino acids thus generates an inward current which depolarizes the B cell and mediates their stimulatory effect on insulin release.

Inhibition of the Na⁺/K⁺-ATPase results in B cell depolarization.[123] There are two explanations for this result. First, as the pump is electrogenic, transporting three Na⁺ out for every two K⁺ in, it provides a small constant outward current which normally hyperpolarizes the B cell by a few millivolts. A second possibility is that because the Na⁺ pump is likely to consume a substantial part of cellular ATP, its inhibition will lead to a local rise in ATP (and decrease in ADP). Consequently K_{ATP} channels will be blocked, producing depolarization.[124] Modulation of the K_{ATP} channel by the activity of the Na⁺/K⁺-ATPase is established in epithelial cells.[125]

In muscle cells, the Na⁺/Ca²⁺ exchanger is electrogenic and transports three Na⁺ into the cell for each Ca²⁺ out, thus giving rise to an inward Na⁺ current which produces depolarization. Although B cells possess a Na⁺/Ca²⁺ exchanger,[91,126,127] its electrogenicity has not yet been explored in detail. If it were similar to that of muscle cells, however, it would give rise to an inward current at the resting potential. Clamping the B cell membrane potential to +80 mV slows the rate at which $[Ca^{2+}]_i$ returns to the resting concentration following a voltage clamp depolarization relative to that observed when the cell was held at −80 mV;[92] this finding is consistent with a voltage-dependent operation of the Na⁺/Ca²⁺ exchanger.

3.3. THE MECHANISM OF BURSTING ELECTRICAL ACTIVITY

One question which has intrigued electrophysiologists considerably is the mechanism which underlies the bursting pattern of B cell electrical activity. It was originally believed that the slow waves were terminated by activation of K_{Ca} channels resulting from Ca²⁺ influx entering the B cell during the burst of action potentials, and that the intervals reflected the time necessary for $[Ca^{2+}]_i$ to return to resting levels and K_{Ca} channels to close (review: Reference 128). However, the observation that the bursting pattern is unaffected by charybdotoxin[69] suggested that this idea is not correct.

In this section we consider the ionic basis of bursting (see also Reference 129 for a more detailed account). Although it is clear that this process requires the co-ordinated interaction of a number of different types of B cell channel, there is no general consensus and the mechanism underlying bursting remains controversial. The membrane potential of the B cell is governed by the balance between outward background K⁺ conductances, which tend to repolarize the B cell, and inward currents, which lead to depolarization. Models of B cell electrical activity thus variously attribute the termination of the burst to a gradual decrease in an inward current (for example, Ca²⁺ channel inactivation) or to the activation of an outward current (Ca²⁺-activated K⁺ channels or K_{ATP} channels), with the converse occurring during the silent intervals. It is worth remembering that small variations in the amplitude of these currents can markedly affect the membrane potential both because of the high input resistance of the B cell and because the currents involved are voltage activated. We now consider in turn each of the main models that have been proposed to explain B cell oscillatory electrical activity.

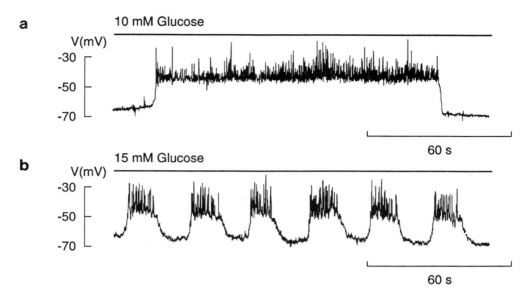

a 10 mM Glucose

V(mV)

-30

-50

-70

60 s

b 15 mM Glucose

V(mV)

-30

-50

-70

60 s

FIGURE 7. Glucose-stimulated electrical activity recorded from a cluster of mouse pancreatic B cells maintained in tissue culture (a) and from an intact, freshly isolated mouse islet (b). The recordings were obtained using the perforated patch whole-cell configuration. Note that the burst of electrical activity in the cultured cell lasts >2 min whereas the bursts in the intact islet last only 15 s.

3.3.1. Ca²⁺ Current Inactivation

As discussed above, inactivation of whole-cell Ca^{2+} currents is biphasic and consists of both a rapid and a slow component. The slow (voltage-dependent) inactivation takes place on a time scale comparable to that of the burst duration. It has therefore been proposed that the plateau depolarization results from a sustained Ca^{2+} current and that the rate of slow inactivation may determine the duration of the burst.[6,130] This hypothesis predicts that the B cell repolarizes, thus terminating the burst, when the persistent Ca^{2+} current becomes too small to balance the background hyperpolarizing K^+ conductance (the delayed-rectifying K^+ conductance is not active at the plateau potential). Once the cell starts to repolarize, the regenerative closure of Ca^{2+} channels will accelerate repolarization. The fact that the periodicity of the burst can be reset by current injected into the islet[131] is consistent with this idea. Furthermore, the burst duration is increased by agents that slow inactivation, such as those which elevate cAMP.[132] A problem with this model, however, is that while single primary cultured B cells (i.e., the cells that were used for the characterization of the inactivation properties) show slow inactivation of the Ca^{2+} current, they usually do not exhibit the characteristic pattern of electrical activity observed in intact islets at intermediary glucose concentrations. Rather, the bursts are markedly prolonged, and electrical activity may even be continuous (Figure 7a).[133] The difference cannot be attributed to differences in experimental details (such as temperature, pH buffer, recording method, etc.), since electrical activity recorded from freshly isolated islets, using the perforated patch whole-cell patch clamp method and conditions similar to those used in studies on single B cells, is virtually identical with that reported for microelectrode recordings from intact islets (Figure 7b).

3.3.2. Cyclic Variations in the Activity of the K_ATP Channel

A second hypothesis is that cyclic oscillations in K_{ATP} channel activity might generate the bursts. Such oscillations might arise from variations in the B cell metabolic rate, or as a consequence of cyclic activation of ATP-consuming reactions in the vicinity of the plasma membrane.[134,135] For example, it has been postulated that activation of the plasma membrane Ca-ATPase by Ca^{2+} entering the cell during a burst of action potentials could result in a localized increase in ADP (and an associated decrease in ATP), thus activating K_{ATP} channels for as long as the Ca^{2+} is elevated. It has been estimated that a local increase in ADP to a concentration as low as 10 μM would be sufficient to repolarize the B cell and terminate the burst.[136] However, this idea is difficult to reconcile with the observation that low concentrations of tolbutamide, applied in the absence of glucose, can produce oscillations in membrane potential similar to those evoked by glucose,[137] because the drug is unlikely to produce an oscillatory block of the K_{ATP} channel. It is also hard to account for the dependence of the slow waves on extracellular Ca^{2+}.[138] Indeed, it has recently been demonstrated that even in the presence of high concentrations of tolbutamide, oscillations in membrane potential can be induced by elevation of the extracellular Ca^{2+} concentration.[139]

3.3.3. Association of B Cells and Changes in Interstitial Ion Concentrations

It has been argued, on the basis of mathematical models, that the characteristic pattern of electrical activity in the intact islet might arise from the consequence of "channel sharing" between adjacent B cells[140] or changes in the extracellular K^+ and Ca^{2+} concentration in the islet.[141] The observation that the electrical activity of large clusters of B cells[70] more closely resembles that of a single cell than of an intact islet, however, mitigates against this suggestion.

3.3.4. Activation of Low-Conductance, Ca^{2+}-Activated K Channels

This possibility can be regarded as a revival of the original hypothesis that periodic activation of K_{Ca} channels produces the bursts.[128] Although the lack of effect of inhibitors of the high-conductance K_{Ca} channel led to the rejection of this idea, the discovery of another type of Ca^{2+}-activated K^+ channel with a different pharmacology (the $K_{L,Ca}$ channel) allows this attractive hypothesis to be resurrected. All arguments in favor of the idea that K_{Ca} channels account for the burst repolarization are equally valid in the case of $K_{L,Ca}$ channels. Further support for this idea is provided by the observation that conditions which produce oscillations in $[Ca^{2+}]_i$ result in associated oscillations in membrane conductance and potential as a consequence of periodic activation of $K_{L,Ca}$ channels.[70,71] Similar oscillations in $[Ca^{2+}]_i$ and membrane potential (or conductance) can also be evoked by IP_3-dependent mobilization of intracellular Ca^{2+} stores. Moreover, when IP_3-gated Ca^{2+} stores are prevented from refilling by thapsigargin, an inhibitor of the endoplasmic reticulum Ca^{2+}-ATPase, the bursting pattern is abolished and continuous electrical activity results.[142]

3.4. A PROPOSAL FOR B CELL ELECTRICAL ACTIVITY

In this section we attempt to provide a simple model of how the various ion channels contribute to the electrical activity of the B cell. There is no doubt that the K_{ATP} channel provides the resting K^+ conductance of the B cell and thereby maintains the negative resting membrane potential. The gradual closure of these channels in response to glucose metabolism results in a slow depolarization. This is because the relative contribution of a resting, as yet unidentified (cf. Sections 3.2.3.4 and 3.2.4) inward current to the membrane potential becomes more significant as the K_{ATP} current is reduced. Once the depolarization reaches ~ –50 mV, the threshold for the initiation of regenerative B cell electrical activity is reached. The depolarizing phase of the action potential results from increased activation of Ca^{2+} channels (and in some B cells, Na^+ channels also). The repolarization of the spike is largely attributable to the activation of K_{DR} channels with some possible contribution of large-conductance K_{Ca} channels. As we have previously suggested,[6] the plateau potential may arise as a consequence of the voltage dependence of the K_{DR} channels. These channels are almost completely closed at –40 mV (the plateau potential). The "afterhyperpolarization" following each spike results from the K_{DR} channels not closing instantaneously on repolarization, which thus produces a small transient hyperpolarizing K^+ conductance. At the plateau potential, the time constant of deactivation of this K_{DR} tail current is >30 ms and thus the pacemaker potential between successive action potentials may be explained by the decay of the K_{DR} current. The amplitude of the plateau potential, and the frequency of spikes superimposed on the plateau, will thus be influenced by both the inward Ca^{2+} current(s) and the K_{DR} current.

We favor the idea that the slow waves result from the interaction of several mechanisms, the most important being Ca^{2+} current inactivation and periodic activation of low-conductance $K_{L,Ca}$ channels. During a burst of action potentials Ca^{2+} enters the B cell and accumulates in the cytoplasm. Ultimately, this increase in $[Ca^{2+}]_i$ is sufficient to activate the $K_{L,Ca}$ channels and so produce repolarization and termination of the burst. Voltage-dependent inactivation of the Ca^{2+} current is also important. Repolarization of the burst will be accelerated by the voltage-dependent closing of Ca^{2+} channels, as discussed above. During the interval between successive bursts, the cytosolic $[Ca^{2+}]_i$ gradually returns to resting levels. Thus the biphasic decay of $[Ca^{2+}]_i$ seen in microfluorimetric measurements from unclamped preparations[143] results from both the rapid cessation of Ca^{2+} entry on repolarization and the slow buffering of $[Ca^{2+}]_i$. As $[Ca^{2+}]_i$ is reduced, $K_{L,Ca}$ channels gradually close, giving rise to the slow interburst depolarization, and Ca^{2+} channels recover from inactivation. These two processes produce increasing activation of Ca^{2+} channels and consequently a new burst of action potentials. To explain why bursting activity in cultured cells differs from that of intact islets, we hypothesize that the colocalization of Ca^{2+} channels and $K_{L,Ca}$ channels is lost in culture. An alternative idea is that intracellular Ca^{2+} handling is different in dissociated B cells.

In our model the primary source of the Ca^{2+} which activates the K_{Ca} channel during glucose-stimulated electrical activity is of extracellular origin. Ca^{2+} from internal stores may also contribute, but is of greater importance for the rapid oscillations in membrane potential induced by secretagogues which activate the phosphoinositide pathway. It is possible that glucose indirectly triggers release from IP_3-gated stores, both by increasing Ca^{2+}-influx, and thus IP_3-formation,[144] and by elevating $[ATP]_i$ and $[Ca^{2+}]_i$ which potentiate the response of the $InsP_3$ receptor to the agonist. Ca^{2+}-induced Ca^{2+} release may also contribute.[71]

Glucose, tolbutamide, and diazoxide modulate the frequency and duration of the bursts, suggesting that K_{ATP} channels are somehow involved in this regulation.[145] The K_{ATP} current constitutes a background K^+ conductance which tends to repolarize the membrane. In our view, the evidence favors the idea that glucose and sulfonylureas modulate the slow waves by varying the amplitude of this background K_{ATP} current. Thus we contend that the K_{ATP} channel is the primary glucose sensor in the B cell. We should emphasise that the K_{ATP} current is not directly involved in the *generation* of the bursts, but merely contributes a background K current which is superimposed on the oscillations produced by other current(s). When the K_{ATP} current is decreased by glucose metabolism or pharmacological blockers, the initiation of the burst is facilitated, as less inward current is now required to depolarize the B cell. Thus the duration of the interval is shortened. At the same time, the length of the burst is increased as the contribution of the K_{ATP} current to the repolarizing current at the end of the burst is reduced. This means that a larger outward current (or smaller inward current) is required for repolarization. In this way, an increase in glucose produces a lengthening of the burst duration. The fact that low concentrations of tolbutamide and diazoxide modulate burst frequency and duration in a manner similar to glucose supports this idea, as any change in the K_{ATP} current amplitude would be expected to affect burst activity.

Glucose has also been proposed to modulate bursting by enhancing the Ca^{2+} buffering of the B cell.[146,147] If this were the case, then a longer period of electrical activity (and Ca^{2+} influx) would be needed to elevate $[Ca^{2+}]_i$ sufficiently to activate the $K_{L,Ca}$ channels, thus accounting for the lengthening of the burst with glucose elevation. Such an effect would contribute to the effect exerted via inhibition of the background K_{ATP} conductance to convert oscillatory electrical activity into continuous spiking.

4. SECRETION

The functional significance of islet cell electrical activity is to elevate intracellular Ca^{2+} and thereby stimulate exocytosis. In this section we consider this process in detail.

Studies of secretion from A and B cells were initially confined to measurements of hormone release from perfused pancreas or isolated intact islets. An important development was the ability to permeabilize the plasma membrane and access the intracellular milieu, thus permitting the effects of Ca^{2+} and intracellular second messengers to be investigated under controlled conditions. Permeabilization is usually effected by high-voltage discharge or detergent treatment and has been used on both whole islets and dispersed-cell preparations. Recently, several methods have been introduced to permit the study of secretion from an individual cell. The first of these is the reverse hemolytic plaque assay, in which secretion is detected by the lysis of a surrounding lawn of red blood cells, which are coated with an anti-insulin antibody. Considerable heterogeneity in the extent of glucose-stimulated secretion from individual B cells was detected with this method.[117,148]

More recently, the capacitance method of monitoring exocytosis[149] has been applied to single B cells[93,100,103,150,151] This electrophysiological method is based on the fact that secretion of insulin involves the fusion of the secretory granules with the plasma membrane and consequently produces an increase in the cell surface area (Figure 8a). Because the cell capacitance is proportional to the cell surface area, there is an associated increase in cell capacitance which can be measured electrically using a patch electrode. This technique has several advantages: (1) there is a high temporal resolution; (2) the cell can be voltage clamped so that secretion can be initiated by voltage-dependent activation of Ca^{2+} influx in a controlled way; (3) using the patch pipette it is possible to manipulate the intracellular solution; (4) recovery of the secreted membrane (endocytosis) can also be investigated. However, the drawback is that the technique measures changes in an electrical property of the B cell and not insulin secretion per se. Confirmation that increases in cell capacitance are indeed linked to the release of secretory granule contents has been provided by simultaneous measurements of cell capacitance and the loss of quinacrine fluorescence (Figure 8b). This dye accumulates within the acidic environment of the secretory granule, and the release of the granule contents is observed as a decrease in fluorescence.[152,153] An alternative way to measure release from a single cell is to use amperometry, a technique which has been used to detect the release of electroactive species from neuroendocrine cells.[154] Since insulin is not (or at least much less) electroactive, its release cannot be measured by this method. However, when B cells are incubated in the presence of 5-hydroxytryptamine (5-HT) the amine is taken up and stored in the secretory granule and co-released with insulin.[155] This effect may be exploited to detect the release of secretory granule contents by measuring 5-HT with amperometry.[156]

4.1. REGULATION OF SECRETION FROM A CELLS

Our knowledge of the mechanisms regulating secretion from A cells is very limited. There is now a consensus that glucagon secretion is dependent on an increase in cytoplasmic Ca^{2+} and that activation of the kinases PKA and PKC may stimulate secretion by sensitizing the secretory machinery to $[Ca^{2+}]_i$ (review: Reference 10).

Arginine stimulates glucagon release by its electrogenic entry, which leads to depolarization, increased action potential frequency, and thereby to an increase in $[Ca^{2+}]_i$, as it does in B cells (see above). β-Adrenergic stimulation leads to elevation of cyclic AMP,[157] which in turn increases Ca^{2+} influx into the A cell[14] and mobilizes intracellular Ca^{2+}.[158]

The mechanism by which glucose inhibits glucagon release has not been clearly established, but a number of hypotheses have been put forward. First, glucose might act directly on the A cell to lower $[Ca^{2+}]_i$, either by inhibition of electrical activity or by promoting transport into stores and across the plasma membrane.[158] However, no effect of glucose on the

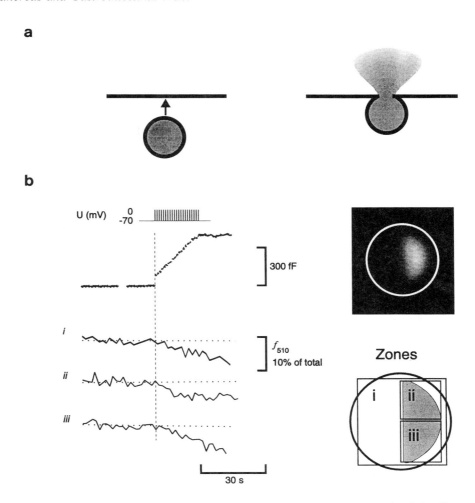

FIGURE 8. (a) Scheme explaining the principles of capacitance measurements of exocytosis. The insulin-containing secretory granules are enclosed by a lipid membrane of the same type as the plasma membrane. Exocytosis involves the fusion of the secretory granule(s) with the plasma membrane. The consequent increase in the surface area of the cell can be monitored as an increase in cell capacitance. (b) Parallel recordings of cell capacitance and quinacrine fluorescence in the same cell. Quinacrine accumulates within acidic compartments of which the secretory granules are the most abundant in the pancreatic B cell. A train of 21 200-ms depolarizations from –70 to 0 mV applied at 1-second intervals produces a capacitance increase of >300 fF. The associated loss of quinacrine fluorescence is analyzed in three different regions (i to iii) as indicated to the right; region i corresponds to the entire cell. Note that quinacrine is unevenly distributed within the cell and is concentrated on the right side of the cell.

electrical activity of single A cells has been found.[12] A second suggestion is that glucose instead acts by a paracrine mechanism by promoting the release of an inhibitory substance from adjacent B cells. Insulin was originally proposed to serve such a paracrine role, but this idea seems unlikely because there are very few insulin receptors on the A cell.[159] As discussed above, GABA release from B cells has also been postulated to inhibit glucagon release, since this transmitter has a marked inhibitory effect of A cell electrical activity.[19] As yet, a glucose-stimulated release of GABA from B cells has not been demonstrated. It remains possible, however, that GABA is continuously secreted from the B cell and that glucose metabolism in the A cell modulates the $GABA_A$ receptor Cl^- channel so that the current becomes larger in the presence of the sugar. This idea is suggested by analogy with neurones where the amplitude of the $GABA_A$ receptor Cl^- current is dependent on the metabolic state of the cell[160] or the internal ATP concentration.[161]

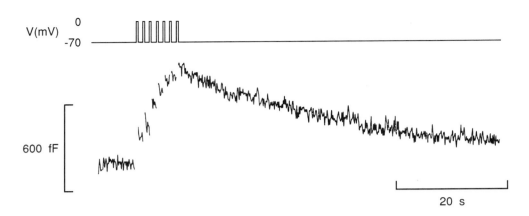

FIGURE 9. Exocytosis and endocytosis in a single B cell. The B cell was stimulated by a train of seven depolarizations from −70 to 0 mV applied at 1-second intervals (upper trace). Each depolarization produces a step increase in cell capacitance indicative of exocytosis. The capacitance increases become progressively smaller during the train (cf. Figure 10). Following stimulation, the cell capacitance gradually returns toward the prestimulatory level, probably reflecting the retrieval of secreted membranes.

4.2. REGULATION OF SECRETION FROM B CELLS

Electron microscopy has revealed that the B cell is polarized, with the secretory granules being confined to the apical part of the cell.[7,162] Further evidence for this polarization comes from imaging of quinacrine fluorescence (Figure 8). The B cell contains around 13,000 secretory granules with a diameter of between 200–300 nm.[163] This diameter predicts that the fusion of a single secretory granule with the plasma membrane will increase the capacitance by 2 fF and unitary events of this size are in fact observed.[93] In addition to insulin, the secretory granules contain many other constituents, such as C peptide, proteases, amylin, chromogranin, and ATP.[164] The B cell also possesses a second class of small synaptic vesicle-like granules which contain GABA (diameter 50 nm)[165] and which have been implicated in the paracine action of glucose on glucagon secretion (see above).

The bursts of action potentials characteristic of B cell electrical activity give rise to oscillations in intracellular $[Ca^{2+}]_i$ (review: Reference 147). As a consequence of electrical coupling between B cells, the average $[Ca^{2+}]_i$ measured from a whole islet also oscillates in synchrony with electrical activity.[143,166] These oscillations result in synchronous oscillations in secretion, as demonstrated by simultaneous measurements of $[Ca^{2+}]_i$ and insulin release.[167-169]

The maximum rate of insulin release varies according to the method of analysis. In a single intact islet, the maximum rate of secretion, measured at the peak of an oscillation, is around 0.2% of the granule population per second when a sample period of 3 s is used.[167] In capacitance studies, the maximum rate of release is much higher, being as much as 5% of the granule population per second when measured over a 50-ms period.[93] The major reason for this difference is that the exocytotic capacity is rapidly exhausted during sustained stimulation. This is clearly demonstrated in capacitance measurements when exocytosis is elicited in response to a train of voltage clamp depolarizations which mimic a burst of action potentials. As shown in Figure 9, the size of the secretory response diminishes throughout the train. Using the nomenclature that has been developed for neurones, we refer to this phenomenon as *depression*.

Only a small part of the depression of secretion that occurs during repetitive stimulation can be accounted for by Ca^{2+} current inactivation. Depression is therefore more likely to result from the depletion of a readily releasable pool of secretory granules (Figure 10) and once exhausted, secretion can only be reinitiated following its replenishment.[150] We estimate that this pool comprises 5–10% of the total granule population. The reason that electrical activity

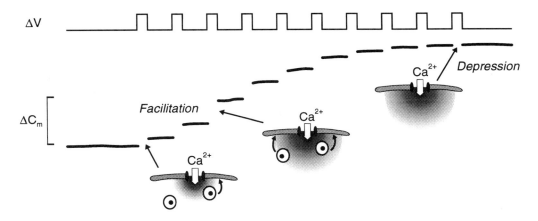

FIGURE 10. Schematic representation of facilitation and depression. The cell is stimulated by a train of depolarizing pulses. The first pulse produces a small exocytotic response reflected as an increased cell capacitance (C_m). Subsequent depolarizations produce larger exocytotic responses. This is because the $[Ca^{2+}]_i$ transients summate and thus extend deeper into the B cell. As a consequence, a greater number of secretory granules are recruited for release, thereby producing a larger exocytotic response. The pool of releasable granules is gradually depleted, resulting in the secretory responses becoming progressively smaller (depression), until eventually depolarization fails to produce a capacitance increase.

and insulin release from the B cell are pulsatile may be to provide sufficient time for the replenishment of the readily releasable pool.

Another feature of exocytosis commonly observed in capacitance measurements is *facilitation*, a term which describes the larger exocytotic response produced by a second depolarization applied shortly after the first (Figure 10). We attribute this behavior to the summation of the individual $[Ca^{2+}]_i$ transients which, because exocytosis is steeply $[Ca^{2+}]_i$-dependent,[93] results in a larger exocytotic response to the second stimulus.

It is worth pointing out that the latency between the opening of the Ca^{2+} channel and exocytosis is short, <50 ms.[93] Most of the delay between the onset of glucose stimulation and secretion seen in the intact B cell arises from the time required for metabolism to generate sufficient ATP (and/or reduce ADP) to close the K-ATP channels.[40]

4.2.1. Ca^{2+} Dependence

The central role of Ca^{2+} in the process of glucose-stimulated insulin secretion has been known for more than 25 years.[170] As discussed above, this Ca^{2+} is chiefly of extracellular origin and enters through the voltage-dependent Ca^{2+} channels. Consequently, the voltage dependence of exocytosis parallels that of the Ca^{2+} current, being maximal at potentials around 0 mV, and inhibition of Ca^{2+} influx suppresses release.[93]

Studies on both permeabilized islets and single B cells have suggested that secretion is regulated by Ca^{2+} concentrations between 0.1 and 10 μM.[93,171,172] Changes in the average cytosolic $[Ca^{2+}]_i$ evoked by a variety of secretagogues measured using microfluorimetric methods are generally smaller (<1 μM; cf. Reference 147). It is difficult, however, to estimate the true $[Ca^{2+}]_i$ dependence of exocytosis using microfluorimetry. This is because this technique reports the *average* cytosolic Ca^{2+} concentration and not that below the membrane close to the release sites.

Digital imaging of voltage-dependent $[Ca^{2+}]_i$ transients reveals the existence of steep Ca^{2+} gradients within the B cell (Figure 6). The importance of such localized Ca^{2+} increases is emphasized by the fact that exocytosis ceases immediately on membrane repolarization, although the average $[Ca^{2+}]_i$ remains high for several seconds.[93] This suggests that it is the $[Ca^{2+}]_i$ concentration close to the membrane which determines the rate of exocytosis, as is the

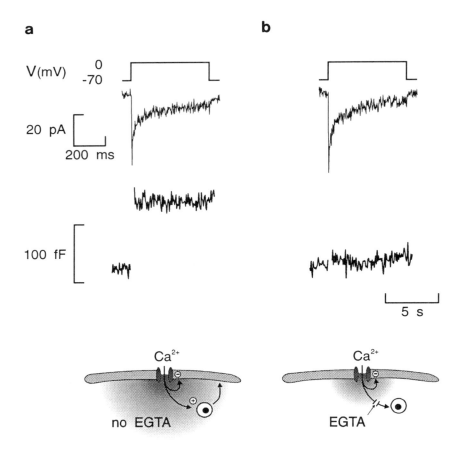

FIGURE 11. Effects of increasing the cytoplasmic Ca^{2+} buffering on exocytosis and Ca^{2+} current inactivation. Ca^{2+} currents were evoked by 500-ms depolarizations. The recordings were made using the standard whole-cell configuration with pipette solutions containing either 10 μM (a) or 2 mM EGTA (b). Increasing Ca^{2+} buffering abolishes exocytosis, but does not affect the inactivation of the Ca^{2+} current. As indicated schematically below, this may be because Ca^{2+} has to diffuse further to activate exocytosis than to inactivate the Ca^{2+} channel, and thus is more readily chelated by EGTA.

case at the nerve terminal. In the latter system, injection of Ca^{2+} buffers up to a concentration of 80 mM has no effect on neurotransmitter release.[173] By contrast, 2 mM EGTA (or a few hundred micromolar of Ca^{2+} indicators such as fura-2) blocks release in the B cell.[93,96] This is consistent with a longer diffusion path for Ca^{2+} to the secretory granule in the B cell (Figure 11). Interestingly, EGTA has no effect on Ca^{2+} current inactivation, a process which must therefore be governed by a more local $[Ca^{2+}]_i$ transient.

4.2.2. Ca^{2+}-Independent Exocytosis

There is accumulating evidence that $[Ca^{2+}]_i$-independent exocytosis also occurs in B cells. In electrically permeabilized RINm5F cells[174] and islet cells,[175] Ca^{2+}-independent insulin secretion can be stimulated by the nonhydrolyzable GTP analogs GTPγS and Gpp(NH)p. Similar results have been found in capacitance measurements.[175a] The effects of GTP analogs cannot be attributed to activation of protein kinases, since no concomitant changes in phosphoprotein pattern were observed.[176] Moreover, GTPγS stimulates secretion as effectively in the absence of cyclic AMP as in its presence, suggesting the nucleotide interacts directly with the exocytotic machinery. One possible target for GTPγS is G_e, a putative GTP-binding protein involved in exocytosis.[177] A GTP-dependent process also appears to mediate inhibition of Ca^{2+}-stimulated insulin secretion at a late stage in stimulus-secretion coupling, following activation of adrenergic, somatostatin or galanin receptors,[178-181] which may suggest the involvement of more than one G protein in the exocytotic process.

In other cell types a variety of monomeric small GTP-binding proteins have been implicated in exocytosis. A synthetic peptide corresponding to the effector domain of one of these proteins, rab3A, has been shown to promote secretion in mast cells[182] and to reconstitute exocytosis in permeabilized chromaffin cells[183] and pancreatic acinar cells.[184] Rab3A has also been detected in both pancreatic islets[185] and the insulin-secreting cell lines RINm5F and HIT T15.[186]

4.2.3. ATP Dependence

Exocytosis requires the presence of intracellular ATP[181,186a] which seems likely to be involved in both early and late steps in the release process. Apart from providing the energy required for secretory processes, ATP is also needed for phosphorylation reactions. We now discuss three kinases which play important roles in B cell exocytosis.

4.2.3.1 Role of CaM-Kinase

Calcium/calmodulin-dependent protein kinase II (CaM-kinase) plays a central role in neurotransmitter release by phosphorylating a synaptic vesicle protein, synapsin I.[187] This causes dissociation of vesicles from the cytoskeleton and so facilitates their docking with the plasma membrane. CaM-kinase is present in B cells, where it phosphorylates an endogenous protein with a molecular mass of 53 kDa[188] and there is both biochemical and electrophysiological evidence that CaM-kinase plays a key role in insulin secretion.[189-193] It is suggested that the B cell CaM-kinase may mediate the action of Ca^{2+} on exocytosis. Synapsin I has not been identified in B cells, and the substrate(s) for CaM-kinase is unknown. One candidate, however, is a subunit of tubulin,[189] which would be consistent with a role in regulating the interaction between the secretory vesicles and the cytoskeleton.

4.2.3.2. Role of Protein Kinase A

Agents which increase cytosolic cAMP markedly enhance insulin release, both via increased Ca^{2+} influx (and consequently $[Ca^{2+}]_i$) and by a direct effect on the secretory machinery itself. Both these effects are mediated by activation of PKA as they can be reversed by Rp-cAMPs, a specific inhibitor of this enzyme.[100,194] The increase in average cytosolic Ca^{2+} evoked by cAMP is small (10–20%) and it has been estimated that only about 20% of the effects of the nucelotide on exocytosis result from this effect.[100] There is good evidence that the more important mechanism by which cAMP enhances insulin secretion involves more distal steps in the exocytotic process. First, cAMP stimulates insulin from permeabilized B cells in which the Ca^{2+} concentration is buffered at a constant level.[195] Secondly, in single B cells cAMP stimulates exocytosis at Ca^{2+} concentrations which do not initiate release and also accelerates the rate of exocytosis at higher $[Ca^{2+}]_i$.[100] It is important to emphasize, however, that cAMP only acts to potentiate secretion and that it has no effect in the complete absence of Ca^{2+}.

Elevation of cAMP reduces the depression of exocytosis seen during a train of depolarizing pulses[151] and stimulates exocytosis following exhaustion of the response to such a train.[100] One explanation for these findings is that by increasing the Ca^{2+} sensitivity of exocytosis, cAMP extends the distance from the Ca^{2+} channels over which the secretory granules can be recruited. In addition, cAMP might directly affect granule translocation.

Although phosphorylation of several islet proteins by PKA has been reported,[196,197] the identity of these substrates, including that regulating exocytosis, is unknown.

4.2.3.3. Role of Protein Kinase C

PKC is activated by diacylglycerol which is generated in intact B cells via a receptor-linked phospholipase C by a number of potentiators of insulin secretion (reviews: References 198 and 199). Diacylglycerol is also produced following increases in $[Ca^{2+}]_i$, due to Ca^{2+}-dependent activation of phospholipase C, and thus PKC will be activated under all conditions in which intracellular Ca^{2+} is elevated.[144] For example, glucose and other nutrient secretagogues activate PKC in this way.

PKC activation by the phorbol ester TPA markedly potentiates insulin secretion,[196,198,200] an effect which is blocked by inhibitors of PKC.[201,202] PKC activation is essential for the stimulation of insulin secretion by certain potentiators of release such as acetylcholine, but it is not required for the stimulatory effects of glucose.[104,203-208]

The effects of PKC on exocytosis are similar to those of PKA; that is, to sensitize the secretory machinery to Ca^{2+}.[103,206,209] There is little effect of PKC activation (+30%) on the B cell Ca^{2+} current, so elevation of $[Ca^{2+}]_i$ by increased Ca^{2+} influx is unlikely to account for the massive (+300%) stimulation of exocytosis.[103] Likewise, mobilization of Ca^{2+} from intracellular stores is not important because TPA is able to promote insulin secretion while *lowering* intracellular Ca^{2+}.[206,208] In perforated-patch recordings, the effects of activation of PKC and activation of PKA on exocytosis are additive.[103] Furthermore, whereas the phosphatase inhibitor okadaic acid strongly potentiates forskolin-stimulated release, it has no effect on exocytosis stimulated by PMA.[103] This suggests that although PKA and PKC stimulate exocytosis in a similar fashion, they may do so by phosphorylating different substrates or different amino acid residues of the same protein.

4.2.4. Endocytosis

Following exocytosis, the secreted membrane is retrieved. In capacitance recordings from B cells using the standard whole-cell configuration, endocytosis is only occasionally observed following exocytosis evoked by Ca^{2+} entry.[93] It can, however, be elicited in response to rapid elevation of $[Ca^{2+}]_i$ to high levels, for example, by IP_3-mediated Ca^{2+} mobilization[210] or photolytic release of caged Ca^{2+} from DM-nitrophen.[96] Surprisingly, endocytosis is independent of the availability of ATP.[210a]

In intact B cells (studied using the perforated-patch whole-cell technique) membrane retrieval is seen in almost all cells (Figure 9). This difference between perforated-patch and standard whole-cell recordings might suggest the requirement for some diffusible cytosolic constituent, but the observation that in the latter configuration endocytosis can be elicited by high $[Ca^{2+}]_i$ suggests that this idea may be too simplistic.

Although endocytosis is often continuous, large step decreases in cell capacitance are also frequently observed. The mean rate of endocytosis in intact B cells is around 20 fF/s, which corresponds to ten secretory granules per second or <0.1% of the granule population.[96] This latter figure is similar to the maximum rate of exocytosis seen in intact islets in biochemical experiments (see above) and suggests that endocytosis may be rate limiting for exocytosis. Both the rate, and usually also the extent, of endocytosis are correlated with the amount of exocytosis, indicating that, whatever the mechanism of endocytosis, the cell has some means of recognizing the amount of secreted membrane.

One possibility that occurs to us is that the B cell may have a limited number of docking sites for the secretory granules on the plasma membrane (Figure 12). If this is the case, endocytosis will be a prerequisite for exocytosis and the docking sites will need to be cleared before further granules can dock for subsequent release. In this way endocytosis and exocytosis could be matched. This idea also provides an alternative explanation for the depression of exocytosis that occurs during repetitive stimulation; that is, it results from saturation of the docking sites, rather than depletion of a readily releasable granule pool (as discussed above). In addition, it can account for the use-dependent rundown of exocytosis in standard whole-cell recordings, since endocytosis is usually absent in this configuration. If our hypothesis is

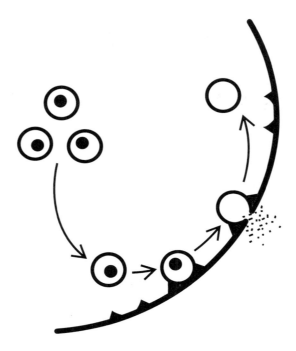

FIGURE 12. Model of exocytosis in pancreatic B cells. The secretory granules are hypothesized to exist in several pools, one of which is readily releasable (5–10%) and already docked with the plasma membrane. The docking sites are indicated by the hatched triangles associated with the plasma membrane. In addition, there is a much larger pool of granules that is not immediately available for release and that must be translocated or chemically modified in order to become releasable. Elevation of Ca^{2+} is believed to trigger the fusion of the secretory granules with the plasma membrane and subsequent release of the granule contents. Following exocytosis, empty secretory granules may be retrieved by endocytosis.

correct, then the number of docking sites may be calculated from the accumulated capacitance increase in standard whole-cell recordings: since the unitary event is 2 fF and the maximum capacitance increase we observe is 2–3 pF, we estimate around 1000 docking sites per B cell.

4.3. MODEL

Studies on purified B cell populations have revealed that the single B cell has a poor response to glucose.[8] We point out that this need not necessarily result from the failure of glucose to elicit B cell electrical activity. Studies of secretion using the capacitance method indicate that even if the Ca^{2+} channels are opened by a voltage clamp step, and there is substantial elevation of $[Ca^{2+}]_i$, exocytosis is small in the absence of PKA or PKC activation, but is increased up to 20-fold when both kinases are simultaneously activated, despite little change in the Ca^{2+} current.[103] These studies suggest that although elevation of cytosolic Ca^{2+} is certainly essential for exocytosis, it is not a sufficient stimulus. Instead, the magnitude of secretory response depends principally upon the extent of phosphorylation of certain proteins involved in the distal steps of exocytosis.

To account for the findings discussed above, we propose the following simple model (Figure 12). We assume that the secretory granules exist in several pools. One of these is readily releasable (5–10%) and comprises granules already docked with the plasma membrane. In addition, there is a much larger pool of undocked granules which are less (or not at all) available for release. In general, any condition which promotes phosphorylation of exocytosis-regulating proteins leads to the transfer of granules (either by chemical modification or, possibly, translocation) into the readily releasable (docked) pool. The observation that CaM-kinase inhibitors interfere with exocytosis suggests that even the fusion of the granule membrane with the plasma membrane involves phosphorylation. The potentiating effect of the phosphatase inhibitor okadaic acid on secretion indicates that dephosphorylation has the opposite effect. By analogy with what has been observed in chromaffin cells,[211] we further speculate that activation of a G protein-dependent phosphatase may be the way by which inhibitors, such as adrenaline, exert their distal effects on insulin release.[178-181]

5. PHARMACOLOGY

Clinically the most important pharmacological agents affecting insulin secretion are the hypoglycemic sulfonylureas, such as tolbutamide and glibenclamide, and the hyperglycemic K^+ channel opener diazoxide. The effects of these drugs on electrical activity and insulin secretion have been extensively reviewed elsewhere.[6,212] Here we focus on some of the more novel aspects of sulfonylurea action.

Several sulfonylurea binding proteins have been purified from B cells. Two low-affinity receptors for the sulfonylurea glyburide have been cloned and sequenced and found to be malate dehydrogenase (33 kDa) and calreticulin (55 kDa).[213] A 140-kDa high-affinity receptor has also been purified[214] and partially sequenced.[57] It has at least six putative transmembrane domains and shows no homology to any known K^+ channel, including the cloned cardiac K_{ATP} channel. Thus the sulfonylurea receptor appears to be a distinct protein, which may modulate the activity of the K_{ATP} channel, and may even be the putative regulatory protein (discussed above) that mediates the effects of ADP and diazoxide on K_{ATP} channel activity. Consistent with the hypothesis that the sulfonylurea receptor and the K_{ATP} channel are different proteins are the observations that the K_{ATP} channel density (as determined by patch clamp recordings) and the number of sulfonylurea receptors (as judged by sulfonylurea binding) can be dissociated in the insulin-secreting cell line CRI-D11.[215]

5.1. EFFECTS ON ION CHANNELS

Sulfonylureas selectively inhibit K_{ATP} channels in B cells with half maximal inhibition ocurring at ~10^{-5} M.[46,216,217] Mg^{2+} ions and a number of adenine nucleotides modulate the sensitivity to tolbutamide.[49,218] Furthermore, diazoxide activates the K_{ATP} channel in the presence of tolbutamide, indicating that the effect of the two drugs is interactive.[46]

Noninsulin-dependent diabetes (NIDDM or type II diabetes) is associated with impaired insulin release.[219-220] Sulfonylureas inhibit K_{ATP} channel activity in B cells at concentrations similar to those encountered therapeutically in the plasma of NIDDM patients,[216,217] This fact suggests that in NIDDM events distal to K_{ATP} channel closure are essentially normal and that the disturbance lies in the regulation of K_{ATP} channel activity. Indeed, mutations in both the glucokinase gene[221,222] and mitochondrial proteins (in particular, position 3234 of leucine tRNA)[223] have been associated with NIDDM, and the K_{ATP} channel itself is clearly also a candidate gene for the disease. The hypothesis that NIDDM is attributable to defective glucose metabolism, and thereby impaired K_{ATP} channel regulation, has been investigated in a diabetic animal model for NIDDM, the GK rat. In B cells prepared from these animals, the K_{ATP} channel is less sensitive to glucose, but responds normally to glyceraldehyde and ketoisocaproate.[224] The K_{ATP} channel also retains a normal sensitivity to ATP and sulfonylureas. These results suggest that the glucose intolerance observed in GK rats arises from a defect in glycolysis. Unfortunately, it has not yet been possible to do similar studies on B cells isolated from human type II diabetics.

5.2. EFFECTS ON EXOCYTOSIS

The stimulatory effect of sulfonylureas on insulin secretion results principally from their inhibitory effect on the K_{ATP} channel.[217] Surprisingly, these drugs also exert a direct stimulatory effect on exocytosis itself.[225,226] However, sulfonylureas are only able to *potentiate* Ca^{2+}-induced release and they are ineffective in the absence of Ca^{2+}. It is therefore clear that their therapeutic effect is largely dependent on K_{ATP} channel closure. Once these channels are closed, however, more insulin is likely to be released for a given electrophysiological stimulus. The mechanism by which sulfonylureas stimulate exocytosis remain unclear. A 145-kDa protein which is a substrate for PKC has been shown to be essential for reconstitution of Ca^{2+}-dependent exocytosis from neuroendocrine cells, including RINm5F cells.[227,228] It is tempting to speculate that this may be the same protein as the high-affinity sulfonylurea receptor which has a similar molecular weight. Circumstantial evidence in favor of this idea is that the stimulatory action of tolbutamide on exocytosis in single B cells is abolished by PKC inhibition, but is unaffected by inhibition of PKA. How this protein influences secretion is not known, but it is of interest that there are sulfonylurea receptors both in the plasma membrane and that of the secretory granules.[229]

6. CONCLUSION AND PERSPECTIVES

It is obvious that our knowledge of the pancreatic A cell is rudimentary, and thus considerable work is needed to understand the mechanisms underlying both electrical activity and secretion in these cells. As should be clear from this review, much more information is available about the B cell, particularly with regard to its ionic channels and the way in which they cooperate to produce electrical activity. There is still debate, however, about the way in which the cyclic oscillations in membrane potential (the slow waves) are produced, and progress in this area is to be expected. The recent cloning of the K_{ATP} channel will now allow clarification of the molecular mechanism of action of both physiological and pharmacological modulators of its activity. Although the basic properties of exocytosis have been defined, we are still ignorant of the molecular mechanisms involved, and clearly the identification and cloning of the proteins involved is also an exciting area for future research. Finally, it is of central importance to establish the possible involvement of defective ion channel regulation and exocytosis in diabetes.

ACKNOWLEDGMENTS

We thank our colleagues in Oxford and Göteborg for their stimulating discussions and hard work. Financial support was obtained from the Wellcome Trust, the British Diabetic Association, the British Medical Research Council (to FMA), the Juvenile Diabetes Foundation International, the Swedish Medical Research Council, the Swedish Diabetes Association, and Novo-Nordisk A/S (to PR).

REFERENCES

1. Langerhans, P., Beiträge zur mikroskopischen Anatomie der Bauchspeicheldrüse, Inaugural-dissertation zur Erlangung der Doktorwürde in der Medicin und Chirurgie vorgelegt der medicinischen Facultät der Friedrich-Wilhelms-Universität zu Berlin, 1869, 24.
2. Hellerström, C., Method for microdissection of intact pancreatic islets of mammals, *Acta Endocrinol.*, 45, 122, 1964.
3. Dean, P. M. and Matthews, E. K., Electrical activity in pancreatic islet cells, *Nature*, 219, 389, 1968.
4. Henquin, J. C. and Meissner, H. P., Significance of ionic fluxes and changes in membrane potential for stimulus-secretion coupling in pancreatic β-cells, *Experientia*, 40, 1043, 1984.
5. Dunne, M. J. and Petersen, O. H., Potassium selective ion channels in insulin-secreting cells: physiology, pharmacology and their role in stimulus-secretion coupling, *Biochim. Biophys. Acta*, 1071, 67, 1991.
6. Ashcroft, F. M. and Rorsman, P., Electrophysiology of the pancreatic β-cell, *Prog. Biophys. Mol. Biol.*, 54, 87, 1989.

7. Pipeleers, D. G., Kiekens, R., and In't Veld, P., Morphology of the pancreatic β-cell, in *Insulin*, Ashcroft, F. M. and Ashcroft, S. J. H., Eds., IRL Press, Oxford, 1992, 5.

8. Pipeleers, D., Islet cell interactions with pancreatic B-cells, *Experientia*, 40, 1114, 1984.

9. Ashcroft, F. M. and Ashcroft, S. J. H., Mechanism of insulin secretion, in *Insulin*, Ashcroft, F. M. and Ashcroft, S. J. H., Eds., IRL Press, Oxford, 1992, 97.

10. Rorsman, P., Ashcroft, F. M., and Berggren, P.-O., Regulation of glucagon release from pancreatic A-cells, *Biochem. Pharmacol.*, 41, 1783, 1990.

11. Gerich, J. E., Charles, M. A., and Grodsky, G. M., Regulation of pancreatic insulin and glucagon secretion, *Annu. Rev. Physiol.*, 38, 353, 1976.

12. Rorsman, P. and Hellman, B., Voltage-activated currents in guinea-pig α_2 cells, *J. Gen. Physiol.*, 91, 223, 1988.

13. Zhang, J. F., Randall, A. D., Ellinor, P. T., Horne, W. A., Sather, S. A., Tanabe, T., Schwarz, T. L., and Tisen, R. W., Distinctive pharmacology and kinetics of cloned neuronal Ca^{2+} channels and their possible counterparts in mammalian CNS neurons, *Neuropharmacology*, 32, 1075, 1993.

14. Rorsman, P., Two types of Ca^{2+} currents with different sensitivities to organic Ca^{2+} channel antagonists in guinea-pig pancreatic α_2 cells, *J. Gen. Physiol.*, 91, 243, 1988.

15. Bean, R. P., Classes of calcium channels in vertebrate cells, *Annu. Rev. Physiol.*, 51, 367, 1989.

16. Ronner, P., Matschinsky, F. M., Hang, T. L., Epstein, A. J., and Buettger, C., Sulfonylurea binding sites and ATP-sensitive K^+ channels in α-TC glucogonoma and β-TC insulinoma cells, *Diabetes,* 42, 1760, 1993.

17. Ronner, P., Higgins, T. J., Kraebber, M. J., and Najafi, H., α-Cells contain ATP-sensitive K^+ channels, *Diabetes*, 40, 155A, 1991.

18. Standen, N. B., Quayle, J. M., Davies, N. W., Brayden, J. E., Huang, Y., and Nelson, M. T., Hyperpolarising vasodilators activate ATP-sensitive K^+ channels in arterial smooth muscle, *Science*, 245, 177, 1989.

19. Rorsman, P., Berggren, P.-O., Bokvist, K., Ericson, H., Möhler, H., Östenson, C. G., and Smith, P. A., Glucose-inhbition of glucagon secretion involves activation of GABA$_A$-receptor chloride channels, *Nature*, 341, 233, 1989.

20. Bormann, J., Electrophysiology of GABA$_A$ and GABA$_B$ receptor subtypes, *Trends Neurosci.*, 11,112, 1988.

21. Gilon, P., Bertrand, G., Loubatieres-Mariani, M. M., Remacle, C., and Henquin, J. C., The influence of γ-aminobutyric acid on hormone release in the mouse and the rat endocrine pancreas, *Endocrinology*, 129, 2521, 1991.

22. Ciani, S. and Ribalet, B., Ion permeation and rectification in ATP-sensitive channels from insulin-secreting cells (RINm5F): effects of K^+, Na^+ and Mg^{2+}, *J. Membr. Biol.*, 103, 171, 1989.

23. Findlay, I., The effects of magnesium upon adenosine triphosphate-sensitive potassium channels in a rat-insulin-secreting cell line, *J. Physiol.*, 391, 611, 1987.

24. Cook, D. L. and Hales, C. N., Intracellular ATP directly blocks K^+ channels in pancreatic β-cells, *Nature*, 311, 269, 1984.

25. Ohno-Shosaku, T., Zünkler, B., and Trube, G., Dual effects of ATP on K^+- currents of mouse pancreatic β-cells, *Pflügers Arch.*, 408, 133, 1987.

26. Ashcroft, F. M. and Kakei, M., ATP-sensitive K-channels: modulation by ATP and Mg^{2+} ions, *J. Physiol.*, 416, 349, 1989.

27. Proks, P., Takano, M., and Ashcroft, F. M., Effects of intracellular pH on ATP-sensitive K-channels in mouse pancreatic β-cells, *J. Physiol.*, 475, 33, 1994.

28. Bokvist, K., Ämmälä, C., Ashcroft, F. M., Berggren, P.-O., Larsson, O., and Rorsman, P., Separate processes mediate nuclotide-induced inhibition and stimulation of the ATP-regulated K^+ channels in mouse pancreatic β-cells, *Proc. R. Soc. London B. Sci.*, 243, 139, 1991.

29. Ashcroft, F. M., Harrison, D. E., and Ashcroft, S. J. H., Glucose induces closure of single potassium channels in isolated rat pancreatic β-cells, *Nature*, 312, 446, 1984.

30. Ashcroft, F. M., Ashcroft, S. J. H., and Harrison, D. E., Effects of 2-ketoisocaproate on insulin release and single potassium channel activity in dispersed rat pancreatic β-cells, *J. Physiol.*, 385, 517, 1987.

31. Rorsman, P. and Trube, G., Glucose-dependent K^+-channels in pancreatic β-cells are regulated by intracellular ATP, *Pflügers Arch.*, 405, 305, 1985.

32. Misler, S., Falke, L. C., Gillis, K., and McDaniel, M. L., A metabolite-regulated potassium channel in rat pancreatic B cells, *Proc. Natl. Acad. Sci. U.S.A.*, 83, 7119, 1986.

33. Hellman, B., Idahl, L. Å., and Danielsson, Å., Adenosine triphosphate levels in mammalian β-cells after stimulation with glucose and hypoglycaemic sulfonylureas, *Diabetes*, 18, 509, 1969.

34. Ashcroft, S. J. H., Weerasinghe, L. C. C., and Randle, P. J., Interrelationships of islet metabolism, adenosine triphsophate content and insulin release, *Biochem. J.*, 132, 223, 1973.

35. Niki, I., Ashcroft, F. M., and Ashcroft, S. J. H., The dependence on intracellular ATP concentration of ATP-sensitive K-channels and of Na-K-ATPase in intact HIT-T15 β-cells, *FEBS Lett.*, 257, 361, 1989.

36. Schmid-Antomarchi, H., De Weille, K., Fosset, M., and Lazdunski, M., The receptor for antidiabetic sulfonylureas controls the activity of the ATP-modulated K^+-channel in insulin-secreting cells, *J. Biol. Chem.*, 262, 15840, 1987.

37. Kakei, M., Kelly, R. P., Ashcroft, S. J. H., and Ashcroft, F. M., The ATP-sensitivity of K+ channels in rat pancreatic β-cells is modulated by ADP, *FEBS Lett.*, 208, 63, 1986.

38. Dunne, M. J. and Petersen, O. H., Intracellular ADP activates K+ channels that are inhibited by ATP in an insulin-secreting cell line, *FEBS Lett.*, 208, 58–62, 1986.

39. Ashcroft, F. M., Williams, B., Smith, P. A., and Fewtrell, C. M. S., Ion channels involved in the regulation of nutrient-stimulated insulin secretion, in *Nutrient Regulation of Insulin Secretion*, Flatt, P. R., Ed., Portland Press, London, 1992, 193.

40. Duchen, M. R., Smith, P. A., and Ashcroft, F. M., Substrate-dependent changes in mitochondrial function, $[Ca^{2+}]_i$ and membrane channels in pancreatic β-cells, *Biochem. J.*, 294, 35, 1993.

41. Arkhammar, P., Nilsson, T., Rorsman, P., and Berggren, P.-O., Inhibition of ATP-regulated K+ channel precedes depolarization-induced increase in cytoplasmic free Ca^{2+}-concentration in pancreatic β-cells, *J. Biol. Chem.*, 262, 5448, 1987

42. Ashcroft, F. M., Ashcroft, S. J. H., and Harrison, D. E., Properties of single potassium channels modulated by glucose in rat pancreatic β-cells, *J. Physiol.*, 400, 501, 1988.

43. Malaisse, W. J., Hutton, J. C., Kawazu, S., Herchuelz, A., Valverde, I., and Sener, A., The stimulus-secretion coupling of glucose-induced insulin release. The links between metabolic and cationic events, *Diabetologia*, 16, 321, 1979.

44. Meglasson, M. D. and Matschinsky, F. M., Pancreatic islet glucose metabolism and regulation of insulin secretion, *Diabetes Metab. Rev.*, 2, 163, 1986.

45. Davies, N. W., Standen, N. B., and Stanfield, P. R., The effect of intracellular pH on ATP-dependent potassium channels of frog skeletal muscle, *J. Physiol.*, 445, 549, 1992.

46. Trube, G., Rorsman, P., and Ohno-Shosaku, T., Opposite effects of tolbutamide and diazoxide on the ATP-dependent K+ channel in mouse pancreatic β-cells, *Pflügers Arch.*, 407, 493, 1986.

47. Ashcroft, F. M., Kakei, M., and Kelly, R. P., Rubidium and sodium permeability of the ATP-sensitive K+ channel in single pancreatic β-cells, *J. Physiol.*, 408, 413, 1989.

48. Trube, G., Hescheler, J., and Schröter, K., Regulation and function of ATP-dependent K+-channels in pancreatic β-cells, in *Secretion and Its Control*, Oxford, G. S. and Armstrong, C. Eds., Society of General Physiologists, Series 44, Rockefeller University Press, New York, 1989, 84.

49. Lee, K., Ozanne, S. E., Hales, C. N., and Ashford, M. J. L., Mg-dependent inhibition of KATP by sulphonylureas in CRI-G1 insulin-secreting cells, *Br. J. Pharmacol.*, 111, 632, 1994.

50. Proks, P. and Ashcroft, F. M., Modification of K-ATP channels in pancreatic β-cells by trypsin, *Pflügers Arch.*, 424, 63, 1993.

51. Larsson, O., Ämmälä, C., Bokvist, K., Fredholm, B., and Rorsman, P., Stimulation of the K_{ATP} channel by ADP and diazoxide requires nucleotide hydrolysis in mouse pancreatic β-cells, *J. Physiol.*, 463, 349, 1993.

52. Dunne, M. J., West-Jordan, J. A., Abraham, R. J., Edwards, R. H. T., and Petersen, O. H., The gating of nucleotide-sensitive K+ channels in insulin-secreting cells can be modulated by changes in the ratio of ATP^{4-}/ADP^{3-} and by non-hydrolysable analogues of both ATP and ADP, *J. Membr. Biol.*, 104, 165, 1988.

53. Ashford, M. L. J., Bond, C. T., Blair, T. A., and Adelman, J. P., Cloning and functional expression of a rat heart K_{ATP} channel, *Nature*, 370, 456, 1994.

54. Findlay, I. and Dunne, M. J., ATP maintains ATP-inhibited K+-channels in an operational state, *Pflügers Arch.*, 407, 238, 1986.

55. Kozlowski, R. and Ashford, M. L. J., ATP-sensitive K+-channel rundown is Mg^{2+} dependent, *Proc. R. Soc. London, Ser. B.*, 240, 397, 1990.

56. Williams, B. A., Modulation of the ATP-sensitive K+ channel in pancreatic β-cells, D. Phil. thesis, Oxford University, Oxford, 1993.

57. Bryan, J., Aguilar-Bryan, L., and Nelson, D., Cloning of a sulfonylurea receptor (ATP-sensitive K+ channel?) from rodent α and β-cells, Abstracts of the Conference of ATP-sensitive K+ channels and Sulfonylurea Receptors, Houston, September 1993, 149, 1993.

58. Pongs, O., Molecular biology of voltage-dependent potassium channels, *Physiol. Rev.*, 72, S69, 1992.

59. Rorsman, P. and Trube, G., Calcium and delayed potassium currrents in mouse pancreatic β-cells under voltage-clamp conditions, *J. Physiol.*, 374, 531, 1986

60. Bokvist, K., Rorsman, P., and Smith, P. A., Block of ATP-regulated and Ca^{2+} activated K+ channels in mouse pancreatic β-cells by external tetraethylammonium and quinine, *J. Physiol.*, 423, 327, 1990.

61. Bokvist, K., Rorsman, P., and Smith, P. A., Effects of external tetraethylammonium ions and quinine on delayed rectifying K+ channels in mouse pancreatic β-cell, *J. Physiol.*, 423, 311, 1990.

62. Smith, P. A., Bokvist, K., Arkhammar, P., Bergrren, P.-O., and Rorsman, P., Delayed rectifying and calcium-activated K+-channels and their significance for action potential repolarisation in mouse pancreatic β-cells, *J. Gen. Physiol.*, 95, 1041, 1990.

63. Betsholtz, C., Baumann, A., Kenna, S., Ashcroft, F. M., Ashcroft, S. J. H., Berggren, P. O., Grupe, A., Pongs, O., Rorsman, P., Sandblom, J., and Welsh, M., Expression of voltage-gated K-channels in insulin-producing cells: analysis by polymerase chain reaction, *FEBS Lett.*, 263, 121, 1990.

64. Phillipson, L. H., Hice, R. E., Schaeffer, K., Lamndola, J., Bell, G., Nelson, D. J., and Steiner, D. F., Sequence and functional expression in *Xenopus* oocytes of a human insulinoma and islet potassium channel, *Proc. Natl. Acad. Sci. U.S.A.*, 88, 53, 1991.

65. Stühmer, W., Stocker, M., Sakmann, B., Seeburg, P., Baumann, A., Grupe, A., and Pongs, O., Potassium channels expressed from rat brain cDNA have delayed rectifier properties, *FEBS Lett.*, 242, 199, 1988.

66. Cook, D. L., Ikeuchi, M., and Fujimoto, W. Y., Lowering of pH$_i$ inhibits Ca^{2+}-activated K$^+$ channels in pancreatic β-cells, *Nature*, 311, 269, 1984.

67. Satin, L. S., Hopkins, W. F., Fatherazi, S., and Cook, D. L., Expression of a rapid, low-voltage threshold K$^+$ current in insulin-secreting cells is dependent on intracellular calcium buffering, *J. Membr. Biol.*, 112, 213, 1989.

68. Fatherazi, S. and Cook, D. L., Specificity of tetraethylammonium and quinine for three K channels in insulin-secreting cells, *J. Membr. Biol.*, 120, 105, 1991.

69. Kukuljan, M., Goncalves, A. A., and Atwater, I., Charybdotoxin-sensitive K(Ca) channel is not involved in glucose-induced electrical activity in pancreatic β-cells, *J. Membr. Biol.*, 119, 187, 1991.

70. Ämmälä, C., Larsson, O., Berggren, P.-O., Bokvist, K., Juntti-Berggren, L., Kindmark, H., and Rorsman, P., Inositol trisphosphate dependent periodic activation of a Ca^{2+}-activated K$^+$-conductance in glucose-stimulated pancreatic β-cells, *Nature,* 353, 849, 1991.

71. Ämmälä, C., Bokvist, K., Larsson, O., Berggren, P.-O., and Rorsman, P., Demonstration of a novel apamin-insensitive calcium-activated K$^+$ channel in mouse pancretic B cells, *Pflügers Arch.*, 422, 443, 1993.

72. Rorsman, P., Bokvist, K., Ämmälä, C., Arkhammar, P., Berggren, P.-O., Larsson, O., and Wåhlander, K., Activation by adrenaline of a low-conductance G-protein dependent K$^+$ channel in mouse pancreatic β-cells, *Nature*, 349, 77, 1991.

73. Dunne, M. J., Bullett, M. J., Li, G., Wollheim, C. B., and Petersen, O. H., Galanin activates nucelotide-dependent K$^+$ channels in insulin-secreting cells via a petussis toxin-sensitive G-protein, *EMBO J.*, 8, 413, 1989.

74. Fosset, M., Schmid-Antomarchi, H., De Weille, J. R., and Lazdunski, M., Somatostatin activates glibenclamide-sensitive ATP-regulated K$^+$ channels in insulinoma cells via a G-protein, *FEBS Lett.*, 242, 94, 1988.

75. De Weille, J. R., Schmid-Antomarchi, H., Fosset, M., and Lazdunski, M., Regulation of ATP-regulated K$^+$ channels in insulinoma cells: activation by somatostatin, protein kinase C and the role of cAMP, *Proc. Natl. Acad. Sci. U.S.A.*, 86, 2971, 1988.

76. Nowycky, M. C., Fox, A. P., and Tsien, R. W., Three types of neuronal calcium channel with different calcium agonist sensitivity, *Nature*, 316, 440, 1985.

77. Ashcroft, F. M., Smith, P. A., and Kelly, R. P., Two types of calcium channel in rat pancreatic β-cells, *Pflügers Arch.*, 415, 504, 1990.

78. Rorsman, P., Ashcroft, F. M., and Trube, G., Single Ca channel currents in mouse pancreatic B-cells, *Pflügers Arch.*, 412, 597, 1988.

79. Smith, P. A., Ashcroft, F. M., and Fewtrell, C. M. S., Permeation and gating properties of the L-type calcium channel in mouse pancreatic β cells, *J. Gen. Physiol.*, 101, 767, 1993.

80. Sala, S. and Matteson, D. R., Single channel recordings of two types of calcium channels in rat pancreactic β-cells, *Biophys. J.*, 58, 567, 1990.

81. Smith, P. A. and Quayle, J., Two types of Ca-channel in human pancreatic β-cells, *J. Physiol.*, 459, 238P, 1993.

82. Velasco, J. M., Calcium channels in rat insulin secreting cell line, *J. Physiol.*, 398, 15P, 1987.

83. Juntti-Berggren, L., Larsson. O., Rorsman, P., Ämmälä, C., Bokvist, K., Wåhlander, K., Nicotera, P., Dypbukt, J., Orrenius, S., Hallberg, A., and Berggren, P.-O., Increased activity of L-type Ca^{2+} channels exposed to serum from patients with Type I diabetes, *Science*, 261, 86, 1993.

84. Sher, E., Biancardi, E., Pollo, A., Carbone, E., Li, G., Wollheim, C. B., and Clementi, F., ω-Conotoxin-sensitive, voltage-operated Ca^{2+} channels in rat insulin-secreting cells, *Eur. J. Pharmacol.*, 216, 407, 1992.

85. Pollo, A., Lovallo, M., Biancardi, E., Sher, E., Socci, C., and Carbone, E., Sensitivity to dihydropyridines, ω-conotoxin and noradrenaline reveals multiple high voltage-activated Ca^{2+} channels in rat insulinoma and human pancreatic β-cells, *Pflügers Arch.*, 423, 462, 1993.

86. Campbell, K. P., Leung, A. T., and Sharp, A. H., The biochemistry and molecular biology of the dihydropyridine-sensitive calcium channel, *Trends Neurosci.*, 11, 425, 1988.

87. Perez-Reyes, E., Wei, X., Catellano, A., and Birnbaumer, L., Molecular diversity of L-type Ca^{2+} channels, *J. Biol. Chem.*, 265, 20430, 1990.

88. Seino, S., Chen, L., Seino, M., Blondel, O., Takeda, J., Johnson, J. H., and Bell, G. I., Cloning of the α$_1$ subunit of voltage-dependent calcium channels expressed in pancreatic β cells, *Proc. Natl. Acad. Sci. U.S.A.,* 89, 584, 1992.

89. Plant, T. D., Properties of calcium-dependent inactivation of calcium channels in cultured mouse pancreatic β-cells, *J. Physiol.*, 404, 731, 1988.

90. Kelly, R. P., Sutton, R., and Ashcroft, F. M., Voltage-activated calcium and potassium currents in human pancreatic β-cells, *J. Physiol.*, 443, 175, 1991.

91. Rorsman, P., Ämmälä, C., Berggren, P.-O., Bokvist, K., and Larsson, O., Cytoplasmic calcium transients due to single action potentials and voltage-clamp depolarizations in mouse pancreatic B-cells, *EMBO J.*, 11, 2877, 1992.

92. Dukes, I. D. and Cleeman, L., Calcium current regulation of depolarization-evoked calcium transients in β-cells (HIT T15), *Am. J. Physiol.*, 264, E348, 1993.

93. Ämmälä, C., Eliasson, L., Bokvist, K., Larsson, O., Ashcroft, F. M., and Rorsman, P., Exocytosis elicited by action potentials and voltage-clamp calcium currents in individual mouse pancreatic B-cells, *J. Physiol.*, 472, 665, 1993.

94. Satin, L. S. and Cook, D. L., Calcium current inactivation in insulin-secreting cells is mediated by calcium influx and membrane depolarization, *Pflügers Arch.*, 414, 1, 1989.

95. Hopkins, W., Satin, L., and Cook, D. L., Inactivation kinetics and pharmacology distinguish between two calcium currents in mouse pancreatic β-cells, *J. Membr. Biol.*, 119, 229, 1991.

96. Bokvist, K., Ämmälä, C., Eliasson, L., and Rorsman, P., Local intracellular Ca^{2+}-gradients associated with Ca^{2+} entry into and secretion from mouse pancreatic β-cells, *Diabetologia*, 36, A32, 1993.

97. Al-Mahmood, H. A., El-Katim, M. S., Gumaa, K. A., and Thulesius, O., The effects of calcium blockers nicardipine, darodipine, PN-200-110 and nifedipine on insulin release from isolated rat pancreatic islets, *Acta Physiol. Scand.*, 126, 295, 1986.

98. Boschero, A. C., Carrol, P. B., De Souza, C., and Atwater, I., Effects of Ca^{2+} channel agonist-antagonist enantiomers of dihyropyridine 202-791 on insulin release $^{45}Ca^{2+}$ uptake and electrical activity in isolated pancreatic islets, *Exp. Physiol.*, 75, 547, 1990.

99. Hille, B., Ionic Channels of Excitable Membranes, 2nd ed., Sinauer Associates, Sunderland, MA, 1992, 179.

100. Ämmälä, C., Ashcroft, F. M., and Rorsman, P., Calcium-independent potentiation of insulin release by cyclic AMP in single β-cells, *Nature*, 363, 356, 1993.

101. Henquin, J. C., The interplay betwen cyclic AMP and ions in the stimulus-secretion coupling in pancreatic B-cells, *Arch. Int. Physiol. Biochem.*, 93, 37, 1985.

102. Henquin, J. C., Bozem, M., Schmeer, W., and Nenquin, M., Distinct mechanisms for the two amplification systems of insulin release, *Biochem. J.*, 246, 393, 1987.

103. Ämmälä, C., Eliasson, L., Berggren, P.-O., Bokvist, K., and Rorsman, P., Permissive role of calcium in exocytosis in exocytosis promoted by protein phosphorylation in pancreatic β-cells, *Proc. Natl. Acad. Sci. U.S.A.*, 91, 4343, 1994.

104. Arkhammar, P., Juntti-Berggren, L., Larsson, O., Welsh, M., Nånberg, E., Sjöholm, Å., Köhler, M., and Berggren, P.-O., Protein kinase C modulates the insulin secretory process by maintaining a proper function of the β-cell voltage-activated Ca^{2+} channels, *J. Biol. Chem.*, 269, 2743, 1994.

105. Velasco, J. M. and Petersen, O. H., The effects of a cell-permeable diacylglycerol analogue on Ca^{2+} (Ba^{2+}) channel currents in an insulin-secreting cell line RINm5F, *Q. J. Exp. Physiol.*, 74, 367, 1989.

106. Plant, T. D., Effects of 1,2 dioctanoyl-sn-glycerol (DiC8) on ionic currents in cultured mouse pancreatic β-cells, *J. Physiol.*, 418, 21P, 1989.

107. Hockberger, P., Toselli, M., Swandulla, D., and Lux, H. D., A diacylglycerol analogue reduces neuronal calcium currents independently of protein kinase C activation, *Nature*, 338, 340, 1989.

108. Keahey, H. H., Boyd, A. E., and Kunze, D. L., Catecholamine modulation of calcium currents in clonal pancreatic β-cells, *Am. J. Physiol.*, 357, C1171, 1990.

109. Hsu, W. H., Xiang, H. D., Kunze, D. L., Rajan, A., and Boyd, A. E., Somatostatin decreases insulin secretion by inhibiting Ca^{2+} influx through voltage-dependent Ca^{2+}-channels in an insulin-secreting cell line (HIT cells), *J. Biol. Chem.*, 266, 837, 1991.

110. Homaidan, F. R., Sharp, G. W. G., and Nowak, L. M., Galanin inhibits a dihydropyridine-sensitive Ca^{2+} current in the RINm5F cell line, *Proc. Natl. Acad. Sci. U.S.A.*, 88, 8744, 1991.

111. Ahrén, B., Berggren, P.-O., Bokvist, K., and Rorsman, P., Does galanin inhibit insulin secretion by activation of the ATP-regulated K^+-channel in the pancreatic β-cell? *Peptides*, 10, 453, 1989.

112. Bokvist, K., Ämmälä, C., Berggren, P.-O., Rorsman, P., and Wåhlander, K., Alpha$_2$-adrenoreceptor stimulation does not inhibit L-type Ca^{2+} channels in mouse pancreatic β-cells, *Biosci. Rep.*, 11, 147, 1991.

113. Ämmälä, C., Berggren, P.-O., Bokvist, K., and Rorsman, P., Inhibition of L-type calcium channels by internal GTP$_\gamma$S in mouse pancreatic β-cells, *Pflügers Arch.*, 420, 72, 1992.

114. Cook, D. L. and Perara, E., Islet electrical pacemaker response to alpha adrenergic stimulation, *Diabetes*, 31, 985, 1982.

115. Christie, M. R., Aetiology of Type I diabetes: immunological aspects, in *Insulin*, Ashcroft, F. M. and Ashcroft, S. J. H., Eds., IRL Press, Oxford, 1992, 306.

116. Plant, T. D., Na^+ currents in cultured mouse pancreatic β-cells, *Pflügers Arch.*, 411, 429, 1988.

117. Hiriart, M. and Matteson, D. R., Na channels and two types of Ca channels in rat pancreatic β-cells identified with the reverse hemolytic plaque assay, *J. Gen. Physiol.*, 91, 617, 1988.

118. Pressel, D. and Misler, S., Sodium channels contribute to action potential generation in canine and human β-cells, *J. Membr. Biol.*, 116, 273, 1990.

119. Ribalet, B. and Beigelman, P. M., Effects of sodium on β-cell electrical activity, *Am. J. Physiol.*, 242, C296–303, 1982.

120. Henquin, J. C., Garcia, M. C., Bozem, M., Hermans, M. P., and Nenquin, M., Muscarinic control of pancreatic β-cell function involves sodium-dependent depolarisation and calcium influx, *Endocrinology*, 122, 2134, 1988.

121. Christensen, H. N., Role of amino acid transport and countertransport in nutrition and metabolism, *Physiol. Rev.*, 70, 43, 1990.

122. Henquin, J. C. and Meissner, H. P., Cyclic adenosine monophosphate differently affects the response of mouse pancreatic β-cells to various amino acids, *J. Physiol.*, 381, 77, 1986.

123. Henquin, J. C. and Meissner, H. P., The electrogenic sodium-potassium pump of mouse pancreatic B-cells, *J. Physiol.*, 332, 529, 1982.

124. Grapengiesser, E., Berts, A., Saha, S., Lund, P.-E., Gylfe, E., and Hellman, B., Dual effects of Na/K pump inhibition on cytoplasmic Ca^{2+} oscillations in pancreatic β-cells, *Arch. Biochem. Biophys.*, 300, 372, 1993.

125. Harvey, B. J., Cellular mechanisms of regulation of ion and water channels and pumps in high-resistance epithelia, in *Isotonic Transport in Leaky Epithelia*, Ussing, H. H., Fischbarg, J., Sten-Kudnsen, O., Larsen, E. H., and Willumsen, N. J., Eds., Munksgaard, Copenhagen, 1993, 312.

126. Hoenig, M., Culberson, L. H., Clement, J. M., and Ferguson, D. C., Na^+/Ca^{2+} exchange in plasma membrane veiscles from a glucose-responsive insulinoma, *Cell Calcium*, 13, 1, 1992.

127. Plasman, P. O. and Herchuelz, A., Regulation of the Na^+/Ca^{2+} exchange in the rat pancreatic B cell, *Biochem. J.*, 285, 123, 1992.

128. Atwater, I., Rosario, L., and Rojas, E., Properties of the Ca-activated K^+ channel in pancreatic β-cells, *Cell Calcium*, 4, 451, 1983.

129. Satin, L. S. and Smolen, P. D., Electrical bursting in pancreatic islets of Langerhans, *Endocr. J.*, in press.

130. Cook, D. L., Satin, L. S., and Hopkins, W. F., The B cell is bursting, but how? *Trends Neurosci.*, 14, 411, 1991.

131. Cook, D. L., Crill, W. E., and Porte, D., Plateau potentials in pancreatic islet cells are voltage-dependent action potentials, *Nature*, 286, 404, 1980.

132. Henquin, J. C. and Meissner, H. P., Dibutyrl cyclic AMP triggers Ca^{2+} influx and Ca^{2+}-dependent electrical activity in pancreatic β-cells, *Biochem. Biophys. Res. Commun.*, 112, 614, 1983.

133. Smith, P. A., Rorsman, P., and Ashcroft, F. M., Simultaneous recording of β-cell electrical activity and ATP-sensitive K-currents in mouse pancreatic β-cells. *FEBS Lett.*, 261, 187, 1990.

134. Corkey, B. E., Deeney, J. T., Glennon, M. C., Matschinsky, F. M., and Prentki, M., Regulation of steady-state Ca^{2+}-levels by the ATP/ADP ratio and ortophosphate in permeabilized RINm5F insulinoma cells, *J. Biol. Chem.*, 263, 4247, 1988.

135. Corkey, B. E., Tornheim, K., Deeney, J. T., Glennon, M. C., Parker, K. J., Matschinsky, F. M., Ruderman, N. B., and Prentki, M., Linked oscillations of free Ca^{2+} and the ATP/ADP ratio in permeabilized RINm5F insulinoma cells supplemented with a glycolyzing cell-free muscle extract, *J. Biol. Chem.*, 263, 5254, 1988.

136. Hopkins, W. F., Fatherazi, S., Peter-Riesch, B., Corkey, B. E., and Cook, D. L., Two sites for adenine-nucleotide regulation of ATP-sensitive potassium channels in mouse pancreatic β-cells and HIT cells, *J. Membr. Biol.*, 129, 287, 1992.

137. Henquin, J. C., Debuyser, A., Drews, G., and Plant, T. D., Regulation of K^+ permeability and membrane potential in insulin-secreting cells, in *Nutrient Regulation of Insulin Secretion*, Flatt, P. R., Ed., Portland Press, London, 1992, 173.

138. Meissner, H. P. and Schmelz, H., Membrane potential of pancreatic islets, *Pflügers Arch.*, 351, 195, 1974.

139. Santos, R. M., Barbosa, R. M., Silva, A. M., Antunes, C. M., and Rosario, L. M., High external Ca^{2+} levels trigger membrane potential oscillations in mouse pancreatic B-cells during blockade of K(ATP) channels, *Biochem. Biophys. Res. Commun.*, 187, 872, 1992.

140. Sherman, A., Rinzel, J., and Keizer, J., Emergence of organised bursting in clusters of pancreatic β-cells by channel sharing, *Biophys. J.*, 54, 411, 1988.

141. Chay, T. R. and Keizer, J., Theory of the effect of extracellular potassium on oscillations in the pancreatic β-cell, *Biophys. J.*, 48, 815, 1985.

142. Dukes, I. D., McIntyre, M. S., Roe, M. W., and Worley, J. F., Intracellular calcium store regulation of membrane potential in mouse islets; effects of thapsigargin, *Biophys. J.*, 66, A248, 1994.

143. Valdeolmillos, M., Santos, R. M., Contreras, D., Soria, B., and Rosario, L. M., Glucose-induced oscillations of intracellular Ca^{2+} concentration matching bursting electrical activity in single mouse islets of Langerhans, *FEBS Lett.*, 259, 19, 1989.

144. Biden, T. J., Peter-Riesch, B., Schegel, W., and Wolheim, C. B., Ca^{2+}-mediated generation of inositol 1,4,5-trisphosphate and inositol 1,3,4,5, tetrakisphosphate in pancreatic islets. Studies with K^+, glucose and carbomylcholine, *J. Biol. Chem.*, 262, 3567, 1987.

145. Henquin, J. C., ATP-sensitive K^+ channels may control glucose-induced electrical activity in pancreatic β-cells, *Biochem. Biophys. Res. Commun.*, 156, 568, 1988.

146. Hellman, B., Gylfe, E., Grapengiesser, E., Lund, P.-E, and Berts, A., Cytoplasmic Ca^{2+}-oscillations in pancreatic β-cells, *Biochim. Biophys. Acta*, 1113, 295, 1992.

147. Hellman, B., Gylfe, E., Grapengiesser, E., Lind, P. E., and Marcström, A., Cytoplasmic calcium and insulin secretion, in *Nutrient Regulation of Insulin Secretion*, Flatt, P. R., Ed., Portland Press, London, 1992, 213.

148. Salomon, D. and Meda, P., Heterogeneity and contact-dependent regulation of hormone secretion by individual B cells, *Exp. Cell Res.*, 162, 507, 1986.

149. Neher, E. and Marty, A., Discrete changes of cell capacitance observed under conditions of enhanced secretion in bovine adrenal chromaffin cells, *Proc. Natl. Acad. Sci. U.S.A.*, 79, 6712, 1982.

150. Gillis, K. D. and Misler, S., Single cell assay of exocytosis from pancreatic islet B-cells, *Pflügers Arch.*, 420, 121, 1992.

151. Gillis, K. D. and Misler, S., Enhancers of cytosolic cAMP augment depolarisation-induced exocytosis from pancreatic β-cells: evidence for effects distal to Ca^{2+} entry, *Pflügers Arch.*, 424, 195, 1993.

152. Pralong, W. F., Bartley, C., and Wollheim, C. B., Single islet β-cell stimulation by nutrients: relationship between pyridine nucleotides, cytosolic Ca^{2+} and secretion, *EMBO J.*, 9, 53, 1990.

153. Breckenridge, L. J. and Almers, W., Currents through the fusion pore that forms during exocytosis of a secretory vesicle, *Nature*, 328, 814, 1987.

154. Chow, R. H., von Rüden, L., and Neher, E., Delay in vesicle fusion revealed by electrochemical monitoring of single secretory events in adrenal chromaffin cells, *Nature*, 356, 60, 1992.

155. Gylfe, E., Association between 5-hydroxytryptamine release and insulin secretion, *J. Endocrinol.*, 78, 239, 1978.

156. Smith, P. A., Duchen, M., and Ashcroft, F. M., Voltammetric monitoring of 5-HT secretion from isolated mouse pancreatic β-cells, *J. Physiol.*, 475, 157P, 1994.

157. Schuit, F. C. and Pipeleers, D. G., Differences in adrenergic recognition by pancreatic A and B cells, *Science*, 232, 875, 1986.

158. Johansson, H., Gylfe, E., and Hellman, B., Cyclic AMP raises cytoplasmic calcium in pancreatic α_2-cells by mobilizing calcium incorporated in response to glucose, *Cell Calcium*, 10, 205, 1989.

159. Van Schravendijk, C. F. H., Foriers, A., Hooghe-Peters, E. L., Rogiers, V., De Meyts, P., Sodoyeez, J. C., and Pipeleers, D. G., Pancreatic hormone receptors on islet cells, *Endocrinology*, 117, 841, 1985.

160. Duchen, M. R., $GABA_A$ responses of dissociated mouse neurons are attenuated by metabolic blockade, *J. Physiol.*, 415, 48P, 1989.

161. Shirasaki, T., Aibara, K., and Akaike, N., Direct modulation of the $GABA_A$ receptor by intracellular ATP in dissociated nucleus tractus solitarii neurones of rat, *J. Physiol.*, 449, 551, 1992.

162. Bonner-Weir, S., Morphological evidence for pancreatic polarity of β-cells within islets of Langerhans, *Diabetes*, 37, 616, 1988.

163. Dean, P. M., Ultrastructural morphology of the pancreatic β-cell, *Diabetologia*, 9, 115, 1973.

164. Bailyes, E. M., Guest, P. C., and Hutton, J. C., Insulin synthesis, in *Insulin*, Ashcroft, F. M. and Ashcroft, S. J. H., Eds., IRL Press, London, 1992, 64.

165. Reetz, A., Solimena, M., Matteoli, M., Folli, F., Takai, K., and De Camilli, P., GABA and pancreatic β-cells: co-localization of glutamic acid decarboxylase (GAD) and GABA with synaptic-like microvesicles suggests their role in GABA storage and secretion, *EMBO J.*, 10, 1275, 1991.

166. Gilon, P. and Henquin, J. C., Influence of membrane potential changes on cytoplasmic Ca^{2+} concentration in an electrically excitable cell, the insulin- secreting pancreatic B-cell, *J. Biol. Chem.*, 267, 20713, 1992.

167. Bergsten, P. and Hellman, B., Glucose-induced amplitude regulation of pulsatile insulin secretion from individual pancreatic islets, *Diabetes*, 42, 670, 1993.

168. Bergsten, P., Grapengiesser, E., Gylfe, E., Tengholm, A., and Hellman, B., Synchronous oscillations of cytoplasmic Ca^{2+} and insulin release in glucose-stimulated pancreatic islets, *J. Biol. Chem.*, 269, 8749, 1994.

169. Gilon, P., Shepherd, R. M., and Henquin, J. C., Oscillations of secretion driven by oscillations of cytoplasmic Ca^{2+} as evidenced in single pancreatic islets, *J. Biol. Chem.*, 268, 22265, 1993.

170. Milner, R. D. G. and Hales, C. N., The role of calcium and magnesium in insulin secretion from the rabbit pancreas studied *in vivo*, *Diabetologia*, 3, 47, 1967.

171. Jones, P. M. and Howell, S. L., Insulin secretion studied in islets permeabilised by high voltage discharge, *Adv. Exp. Med. Biol.*, 211, 279, 1986.

172. Tamagawa, T., Niki, H., Niki, A., and Niki, I., Regulation of insulin release independent of changes of cytosolic Ca^{2+} concentration, *Adv. Exp. Med. Biol.*, 211, 293, 1986.

173. Adler, E. M., Augustine, G. J., Duffy, S. N., and Charlton, M. P., Alien intracellular calcium chelators attenuate neurotransmitter release at the squid giant synapse, *Neuroscience*, 11, 1496, 1991.

174. Vallar, L., Biden, T. J., and Wollheim, C. B., Guanine nucleotides induce Ca^{2+}-dependent insulin secretion from permeabilised RINm5F cells, *J. Biol. Chem.*, 262, 5049, 1987.

175. Wollheim, C. B., Ullrich, S., Meda, P., and Valler, L., Regulation of exocytosis in electrically permeabilised insulin-secreting cells — evidence for Ca^{2+}-dependent and independent secretion, *Biosci. Rep.*, 7, 443, 1987.

175a. Ashcroft, F. M. and Rorsman, P., Unpublished observations.

176. Regazzi, R., Li, G., Ullrich, S., Jaggi, C., and Wollheim, C. B., Different requirements for protein kinase C activation and Ca^{2+} independent insulin secretion in response to guanine nucleotides, *J. Biol. Chem.*, 264, 9939, 1989.

177. Lindau, M. and Gomperts, B. D., Techniques and concepts in exocytosis: focus on mast cells, *Biochim. Biophys. Acta*, 1071, 429, 1991.

178. Ullrich, S. and Wollheim, C. B., GTP-dependent inhibition of insulin secretion by epinephrine in permeabilized RINm5F cells. Lack of correlation between insulin secretion and cyclic AMP levels, *J. Biol. Chem.*, 263, 8615, 1988.

179. Ullrich, S. and Wollheim, C. B., Galanin inhibits insulin secretion by direct interaction with exocytosis, *FEBS Lett.*, 247, 401, 1989.

180. Nilsson, T., Arkhammar, P., Rorsman, P., and Berggren, P.-O., Inhibition of glucose-stimulated insulin release by α_2-adrenoceptor activation is paralleled by both a repolarization and a reduction in cytoplasmic free Ca^{2+} concentration, *J. Biol. Chem.*, 263, 1855, 1988.

181. Jones, P. M., Fyles, J. M., Persaud, S. J., and Howell, S. L., Catecholamine inhibition of Ca^{2+}-induced insulin secretion from electrically permeabilised islets of Langerhans, *FEBS Lett.*, 219, 139, 1987.

182. Oberhauser, A. F., Monck, J. R., Balch, W. E., and Fernandez, J. M., Exocytotic fusion is activated by Rab3a peptides, *Nature*, 360, 270, 1992.

183. Senyshyn, J., Balch, W. E., and Holz, R. W., Synthetic peptides of the effector binding domain of Rab enhance secretion from digitonin-permeabilised chromaffin cells, *FEBS Lett.*, 309, 41, 1992.

184. Padfield, P. J., Balch, W. E., and Jamieson, J. D., A synthetic peptide of the Rab3a effector domain stimulates amylase release from permeabilized pancreatic accini, *Proc. Natl. Acad. Sci. U.S.A.*, 89, 1656, 1992.

185. Jena, B. P., Gumkowski, F. D., Konieczko, E. M., von Mollard, G. F., Jahn, R., and Jamieson, J. D., Redistribution of a rab3-like GTP-binding protein from secretory granules to the Golgi complex in pancreatic acinar cells during regulated exocytosis, *J. Cell Biol.*, 124, 43, 1994.

186. Regazzi, R., Vallar, L., Ullrich, S., Ravazzola, M., Kikuchi, A., Takai, Y., and Wollheim, C. B., Characterisation of small-molecular mass guanine-nucleotide binding regulatory proteins in insulin-secreting cells and PC12 cells, *Eur. J. Biochem.*, 208, 729, 1992.

186a. Ashcroft, F. M. and Rorsman, P., Unpublished data.

187. Llinás, R., Gruner, J. A., Sugimori, M., McGuiness, T. L., and Greengard, P., Regulation by synapsin I and Ca-calmodulin dependent protein kinase II of transmitter release in squid giant synapse, *J. Physiol.*, 436, 257, 1991.

188. Harrison, D. E. and Ashcroft, S. J. H., Effects of Ca^{2+}, calmodulin and cyclic AMP on the phosphorylation of endogenous proteins by homogenates of rat islets of Langerhans, *Biochim. Biophys. Acta*, 714, 313, 1982.

189. Colca, J. R., Wolf, B. A., Comens, P. G., and McDaniel, M. L., Correlation of Ca^{2+} and calmodulin-dependent protein kinase activity with secretion from islets of Langerhans, *Biochem. J.*, 212, 819–827, 1983.

190. Colca, J. R., Kotagal, N., Brooks, C. L., Lacy, P. E., Landt, M., and McDaniels, M. L., Alloxan inhibition of a Ca^{2+} and calmodulin-dependent protein kinase activity in pancreatic islets, *J. Biol. Chem.*, 258, 7260, 1983.

191. Harrison, D. E., Poje, M., Rocic, B., and Ashcroft, S. J. H., Effects of dehydrouranil on protein phosphorylation and insulin secretion in rat islets of Langerhans, *Biochem. J.*, 237, 191, 1986.

192. Li, G., Hidaka, H., and Wollheim, C. B., Inhibition of voltage-gated Ca^{2+} channels and insulin secretion in HIT cells by the Ca^{2+}/calmodulin-dependent protein kinase II inhibitor KN-62: comparison with antagonists of calmodulin and L-type Ca^{2+} channels, *Mol. Pharmacol.*, 42, 489, 1992.

193. Niki, I., Okazaki, K., Saitoh, M., Niki, A., Niki, H., Tamagawa, T., Iguchi, A., and Hidaka, H., Presence and possible involvement of Ca/calmodulin-dependent protein kinases in insulin release from the rat pancreatic β cell, *Biochem. Biophys. Res. Commun.*, 191, 255, 1993.

194. Persaud, S. J., Jones, P. M., and Howell, S. L., Glucose-stimulated insulin secretion is not dependent on activation of protein kinase A, *Biochem. Biophys. Res. Commun.* 173, 833, 1990.

195. Jones, P. M., Fyles, J. M., and Howell, S. L., Regulation of insulin secretion by cAMP in rat islets of Langerhans permeabilised by high voltage discharge, *FEBS Lett.*, 205, 205, 1986.

196. Harrison, D. E., Ashcroft, S. J. H., Christie, M. R., and Lord, J. M., Protein phosphorylation in the pancreatic β-cell, *Experientia*, 40, 1057, 1984.

197. Hughes, S. J. and Ashcroft, S. J. H., Cyclic AMP, protein phosphorylation and insulin secretion, in *Nutrient Regulation of Insulin Secretion*, Flatt, P. R., Ed., Portland Press, London, 1992, 271.

198. Persaud, S. J., Jones, P. M., and Howell, S. L., The role of protein kinase C in insulin secretion, in *Nutrient Regulation of Insulin Secretion*, Flatt, P. R., Ed., Portland Press, London, 1992, 247.

199. Morgan, N. G. and Montague, W., Phospholipids and insulin secretion, in *Nutrient Regulation of Insulin Secretion*, Flatt, P. R., Ed., Portland Press, London, 1992, 125.

200. Howell, S. L., Jones, P. M., and Persaud, S. J., Protein kinase C and the regulation of insulin secretion, *Biochem. Soc. Trans.*, 18, 114, 1990.

201. Hughes, S. J. and Ashcroft, S. J. H., Effects of a phorbol ester and clomiphene on protein phosphorylation and insulin secretion in rat pancreatic islets, *Biochem. J.*, 249, 825, 1988.

202. Metz, S., Is protein kinase C required for physiologic insulin release?, *Diabetes*, 37, 3, 1988.

203. Peter-Riesch, B., Fathi, M., Schegel, W., and Wollheim, C. B., Glucose and carbachol generate 1,2 diacylglycerols by different mechanisms in pancreatic islet, *J. Clin. Invest.*, 81, 1154, 1988.

204. Regazzi, R. and Wollheim, C. B., Protein kinase C in insulin-releasing cells. Putative role in stimulus-secretion coupling, *FEBS Lett.*, 268, 376, 1990.

205. Persaud, S. J., Jones, P. M., Sugden, D., and Howell, S. L., The role of protein kinase C in cholinergic stimulation of insulin secretion from rat islets of Langerhans, *Biochem. J.*, 264, 753, 1989.
206. Hughes, S. J., Chalk, J. G., and Ashcroft, S. J. H., The role of cytosolic free Ca^{2+} and protein kinase C in acetylcholine-induced insulin release in the clonal β-cell line HIT T15, *Biochem. J.*, 267, 227, 1990.
207. Persaud, S. J., Jones, P. M., and Howell, S. L., Activation of protein kinase C is essential for sustained insulin secretion in response to cholinergic stimulation, *Biochim. Biophys. Acta*, 1091, 120, 1991.
208. Arkhammar, P., Nilsson, T., Welsh, M., Welsh, N., and Berggren, P.-O., Effects of protein kinase C activation on the regulation of the stimulus-secretion coupling in pancreatic β-cells, *Biochem. J.*, 264, 207, 1989.
209. Jones, P. M. and Howell, S. L., Role of protein kinase C in the regulation of insulin secretion, *Biochem. Soc. Trans.*, 17, 61, 1988.
210. Rorsman, P., Bokvist, K., Ämmälä, C, Eliasson, L., and Gäbel, J., Ion channels, electrical activity and insulin secretion, *Diabetes Metab.*, 20, 138, 1994.
210a. Ashcroft, F. M. and Rorsman, P., Unpublished observations.
211. Galindo, E., Zwiller, J., Bader, M.-F., and Aunis, D., Chromostatin inhibits catecholamine secretion in adrenal chromaffin cells by activating a protein phosphatase, *Proc. Natl. Acad. Sci. U.S.A.*, 89, 7398, 1992.
212. Ashcroft, F. M. and Ashcroft, S. J. H., Properties and functions of ATP-sensitive K-channels, *Cell. Signal.*, 2, 197, 1990.
213. Shaw, S. Y., Fenderson, W., Reiss, P., Bernatowicz, M., and Matseuda, G. R., Purification and identification of two low affinity glyburide binding proteins from HIT cell membranes, Abstracts of the Conference of ATP-sensitive K$^+$ Channels and Sulfonylurea Receptors, Houston, 1993, 159.
214. Aguilar-Bryan, L., Nelson, D. A., Vu, Q. A., Humphrey, M. B., and Boyd, A. E., III, Photoaffinity labelling and partial purification of the β-cell sulfonylurea receptor using a novel, biologically active glyburide analog, *J. Biol. Chem.*, 265, 8218, 1990.
215. Khan, R. N., Hales, C. N., Ozanne, S. E., Adogu, A. A., and Ashford, M. L., Dissociation of KATP channel and sulphonylurea receptor in the rat clonal insulin-secreting cell line, CRI-D11, *Proc. R. Soc. London Ser. B.*, 253, 225, 1993.
216. Zünkler, B. J., Lenzen, S., Männer, K., Panten, U., and Trube, G., Concentration-dependent effects of tolbutamide, meglitinide, glipizide, glibenclamide, and diazoxide on ATP-regulated K+ channels in pancreatic β-cells. *Naunyn-Schmiedebergis Arch. Pharmacol.*, 337, 225, 1988.
217. Panten, U., Burgfeld, J., Görke, F., Rennicke, M., Schwanstecher, M., Wallasch, A., Zünkler, B., and Lenzen, S., Control of insulin secretion by sulphonylureas, meglitinide and diazoxide in relation to their binding to the sulphonylurea receptor, *Biochem. Pharmacol.*, 38, 1217, 1989.
218. Zünkler, B. J., Lins, S., Ohno-Shosaku, T., Trube, G., and Panten, U., Cytosolic ADP enhances the sensitivity of tolbutmide of ATP-dependent K$^+$ channels from pancreatic β-cells, *FEBS Lett.*, 239, 241, 1988.
219. Vague, P. and Moulin, J. P., The defective glucose sensitivity of the B-cell in non-insulin dependent diabetes. Improvement after 24 hours of normoglycaemia, *Metabolism*, 31, 139, 1982.
220. Cerasi, E., Aetiology of Type II diabetes, in *Insulin*, Ashcroft, F. M. and Ashcroft, S. J. H., Eds., IRL Press, Oxford, 1992, 347.
221. Hattersley, A. T., Turner, R. C., Permutt, M. A., Patel, P., Tanizawa, Y., Chiu, K. C., O'Rahilly, S., Watkins, P. J., and Wainscoat, J. S., Linkage of type II diabetes to the glucokinase gene, *Lancet*, 339, 1307, 1992.
222. Vionnet, N., Stoffel, M., Takeda, J., Yasuda, K., Bell, G. I., Zouali, H., Lesage, S., Velho, G., Iris, F., Passa, Ph., Froguel, Ph., and Cohen, D., Nonsense mutation in the glucokinase gene causes early-onset non-insulin-dependent diabetes mellitus, *Nature*, 356, 721, 1992.
223. Reardon, W., Ross, R. J. M., Sweeney, M. G., Luxon, L. M., Pembrey, M. E., Harding, A. E., and Trembath, R. C., Diabetes mellitus associated with a pathogenic point mutation in mitochondrial DNA, *Lancet*, 340, 1376, 1992.
224. Tsuura, Y., Ishida, H., Okamoto, Y., Kato, S., Sakamoto, K., Horie, M., Ikeda, H., Okada, Y., and Seino, Y., Glucose sensitivity of ATP-sensitive K$^+$ channels is impaired in β-cells of the GK rat. A new genetic model of NIDDM, *Diabetes*, 42, 1446, 1993.
225. Shibier, O., Flatt, P. R., Efendic, S., and Berggren, P.-O., Intracellular action of sulfonylureas in the stimulation of insulin release, *Diabetologia*, 34, A91, 1991.
226. Ämmälä, C., Bokvist, K., Eliasson, P., Lindström, P., and Rorsman, P., Tolbutamide stimulates exocytosis by direct interaction with the secretory machinery in B-cells, Diabetologia, 36, A60, 1993.
227. Nishizaki, T., Walent, J. H., Kowalchyk, J. A., and Martin, T. F. J., A key role for a 145-kDa cytosolic protein in the stimulation of Ca^{2+}-dependent secretion by protein kinase C, *J. Biol. Chem.*, 267, 23972, 1992.
228. Walent, J. H., Porter, B. W., and Martin, T. F. J., A novel 145 kd brain cytosolic protein reconstitutes Ca^{2+}-regulated secretion in permeable neuroendocrine cells, *Cell*, 79, 765, 1992.
229. Carpentier, J. L., Sawano, F., Ravazzola, M. and Malaisse, W. J., Internalization of ^{3}H-glibenclamide in pancreatic islet cells, *Diabetologia*, 29, 259, 1986.
230. Henquin, J. C., Les mécanismes cellularires du contrÔle de la sécrétion d'insuline, *Arch. Int. Physiol. Biochim.*, 98, A61, 1990.

Chapter 11

ELECTRICAL ACTIVITY AND SEROTONIN SECRETION FROM ENTEROCHROMAFFIN CELLS

K. Racké, H. Schwörer, and A. Reimann

CONTENTS

1. INTRODUCTION

Mucosal cells which express neuroendocrine markers are distributed throughout the gastrointestinal tract and represent a diffuse neuroendocrine system. These cells, which represent a rather heterogen cell population, appear to differentiate from multipotent epithelial stem cells.[1,2] Classical **enterochromaffin cells** (EC) are probably the most important and currently the best characterized neuroepithelial cells in the intestinal tract. The present review will focus on this type of intestinal neuroendocrine cell and summarize available knowledge about the ionic mechanisms involved in the control of their secretions.

As with chromaffin cells of the adrenal medulla, EC display a pronounced argentaffin reaction (see Reference 3), which finally was responsible for their name, and this chromaffin reaction can be considered as one of the first indications that such mucosal cells might be a kind of **neuroendocrine cell**. ECs are further characterized by a very high content in serotonin (5-hydroxytryptamine, 5-HT). Actually, the intestinal mucosa is the tissue with the highest 5-HT concentration in the mammalian organism, and more than 95% of the intestinal 5-HT is present in the EC of the mucosa, whereas trace amounts of 5-HT are present in intestinal neurons. The intestinal mucosa was also the tissue from which 5-HT was first isolated in 1937 by Erspamer and Vialli (see Reference 4) who initially named it "enteramine".

In EC, 5-HT is stored predominantly in large, electron-dense secretory granules[5,6] which are concentrated at the base of the EC,[7-9] suggesting that 5-HT release at the basolateral membrane (i.e., the interstitial side) may be most important. EC can also contain, to variable degrees, several other secretory peptides, such as chromogranins, substance P or opioid peptides,[10-13] and the recently discovered peptide guanylin.[14]

As already mentioned, 5-HT in the intestine is almost exclusively confined to the EC and the 5-HT concentration in intestinal preparations with an intact mucosa exceeds about 100-fold that of mucosa-free preparations.[15,16] Therefore, 5-HT release from intestinal *in vitro* preparations with intact mucosa has been used to monitor 5-HT secretion from EC (e.g., References 15 to 20), and at present, most of the knowledge regarding mechanisms controlling secretion from EC is based on studies in which modulatory effects on 5-HT release were measured from *in vitro* preparations of intestinal tissue. These functional studies, which demonstrated that 5-HT release from the EC is controlled by a complex pattern of neural and paracrine regulatory mechanisms, have recently been reviewed.[20] Since direct electrophysiological studies on EC have not yet been published, the electrophysiological properties of EC have to be deduced from such pharmacological studies, and the present review will summarize present knowledge from this point of view.

2. 5-HT RELEASE FROM ENTEROCHROMAFFIN CELLS

2.1. ROLE OF VOLTAGE-REGULATED CHANNELS

2.1.1. Calcium Channels

Several studies carried out on isolated intestinal preparations of different species showed that 5-HT release from EC is largely calcium dependent.[15-19] Thus, like the release of catecholamines from chromaffin cells of the adrenal medulla[21] and all other neurosecretory processes, 5-HT secretion from EC may occur via calcium-dependent exocytosis. The most detailed characterization of the ionic mechanisms involved in 5-HT secretion from EC has been carried out in a study on a porcine intestinal preparation.[19] Figure 1 shows a series of experiments from this study, in which effects of the calcium channel blockers ω-conotoxin and gadolinium ions on the spontaneous and high potassium-evoked outflow of 5-HT are documented. In the presence of gadolinium, the spontaneous 5-HT outflow was reduced by about 75%, and a similar reduction was observed when calcium was removed from the extracellular medium (Table 1). This indicates that spontaneous 5-HT outflow may largely reflect calcium-dependent secretion and that calcium influx through specific channels is involved in triggering the secretion. ω-Conotoxin and the dihydropyridine nifedipine also inhibited the spontaneous outflow of 5-HT, although the effect was somewhat less pronounced (Table 1). Moreover, the effects of nifedipine and ω-conotoxin, although submaximal, were not additive, indicating that both drugs may affect the same channels. Thus, in EC calcium influx through several types of calcium channels may contribute to the spontaneous secretory activity of the cells, but dihydropyridine-sensitive channels (i.e., L-type channels) appear to be of particular importance. Most interestingly, these L-type channels appear also to be affected by ω-conotoxin, a toxin which is known to primarily block N-type channels, but which may also block a subtype of L-type channels.[22-24] Another particular property of the porcine EC appears to be that BAY K 8644, an opener of L-type calcium channels both on smooth muscle as well as on adrenal chromaffin cells,[25,26] did not evoke 5-HT secretion, but rather inhibited 5-HT release, i.e., displayed characteristics of a calcium channel blocker.

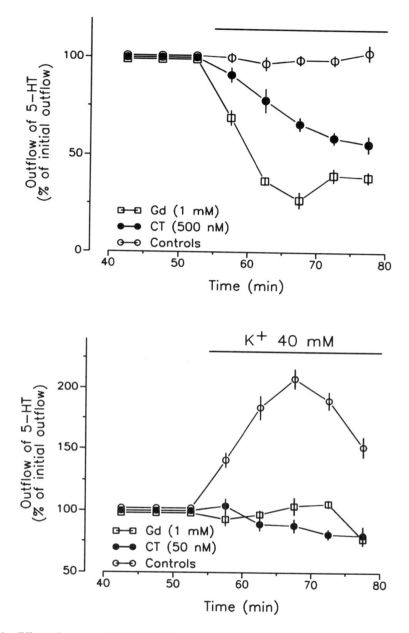

FIGURE 1. Effects of ω-conotoxin (CT, 500 n*M*), or gadolinium (Gd, 1 m*M*) on the spontaneous outflow of 5-HT (upper part) and of high potassium (K+ 40 m*M*, lower part) in the absence or presence of ω-conotoxin or gadolinium on the outflow of 5-HT from strips of the porcine small intestine incubated *in vitro*. The horizontal bar indicates the period of incubation with the respective test drug (upper part) or high potassium (lower part). The open circles show control experiments without drug application. *Ordinates:* outflow of 5-HT in the incubation medium expressed as percent of the mean outflow from 40–55 min of incubation in the individual experiments, means ± SEM of 4-12 experiments. *Abscissa:* time after onset of incubation. (Data from Reference 19.)

Finally, experiments in which 5-HT secretion was evoked by depolarizing concentrations of potassium indicate that the calcium channels of EC are voltage regulated. Thus, elevation of the extracellular potassium to 40 m*M* (Figure 1) caused a marked release of 5-HT, and this was effectively blocked by gadolinium and ω-conotoxin. Nifedipine also inhibited the high potassium-evoked 5-HT release, but was somewhat less effective than ω-conotoxin or gadolinium

TABLE 1
Maximal Effects of Omission of Extracellular Calcium and Different Calcium Channel Blockers on Spontaneous Outflow of 5-HT from Isolated Strips of the Porcine Small Intestine

	5-HT outflow (% inhibition)
Ca-free Gadolinium (1 mM)	70–75
ω-Conotoxin (100 nM) Nifedipine (10 μM) ω-Conotoxin (100 nM) + Nifedipine (10 μM)	45–50

Data from Reference 19.

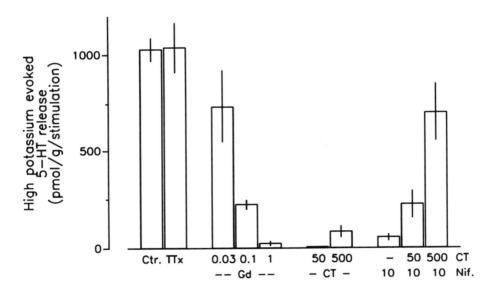

FIGURE 2. Effects of tetrodotoxin (TTX, 1 μM), gadolinium (Gd), ω-conotoxin (CT), and nifedipine (Nif.) on the release of 5-HT evoked by high potassium (40 mM). Isolated strips of the porcine small intestine were incubated as described in Figure 1. Test drugs were present already 20 min before the high potassium stimulus at the concentrations indicated below the columns (Gd, mM; CT, nM; Nif., μM). *Height of columns:* high potassium-evoked 5-HT release (above basal outflow) expressed as pmol/g/stimulation, means ± SEM of 3–12 experiments. (Data from Reference 19.)

(Figure 2). Similar to observations on spontaneous secretion, the "calcium channel opener" BAY K 8644 also inhibited, rather than facilitated, the high potassium-evoked 5-HT release. Basically, the pharmacological characteristics of the calcium channels activated during high potassium depolarization appear to be similar to those of the channels responsible for calcium influx-induced triggering of spontaneous 5-HT secretion. One interesting observation, however, was that ω-conotoxin in increasing concentrations attenuated the inhibitory effect of nifedipine (Figure 2) on high potassium-evoked 5-HT release. Since similar effects were not observed on spontaneous 5-HT secretion, it was concluded that ω-conotoxin may allosterically interact with the binding of dihydropyridines in a voltage-dependent manner. Moreover, these interactions could further be taken as evidence that both ligands act indeed at the same calcium channels.

The experiments summarized above indicate that EC are endowed with voltage-regulated calcium channels, and L-type channels appear to represent the major group. However, these L-type channels have some particular characteristics: they are blocked by ω-conotoxin and not activated by BAY K-8644. Furthermore, ω-conotoxin and dihydropyridines appear to interfere at these channels in a voltage-dependent manner.

Finally, species differences may exist, as recent experiments performed on rabbit isolated intestine showed that 5-HT secretion in this species was not affected by ω-conotoxin, whereas gadolinium and nifedipine were effective blockers of high potassium-evoked 5-HT release.[17]

2.1.2. Sodium Channels

Calcium-dependent 5-HT secretion from EC, and its modulation by various types of stimulatory and inhibitory receptor-mediated mechanisms, also operates in the presence of the neurotoxin tetrodotoxin (TTX).[27-32] Since TTX blocks fast, voltage-activated sodium channels,[33,34] these observations indicate that the fast, sodium-dependent, "neuronal-type" of action potential is not crucially involved in the stimulus secretion coupling of EC. In different intestinal *in vitro* preparations having an intact neuronal input to the EC, TTX consistently reduced 5-HT release, but this effect could always be attributed to a spontaneous cholinergic (nicotinic and/or muscarinic) stimulatory input to the EC[16,27,29] (see also below).

The lack of significance of fast sodium channels in EC is further indicated by experiments which showed that veratridine, a sodium channel activator,[34] had no effect on the release of 5-HT.[19]

2.2. ROLE OF LIGAND-GATED ION CHANNELS
2.2.1. Excitatory Receptors
2.2.1.1. Nicotine Receptors

Generally, nicotine receptors are the prototype of receptor-activated ion channels, and there is evidence that EC, at least in some species, may be endowed with stimulatory nicotine receptors. Thus, both in guinea pig as well as porcine intestinal preparations, nicotine and other nicotine receptor agonists, such as DMPP,[16,27] enhanced 5-HT release. Since the stimulatory effects of these nicotinic agonists was also observed in the presence of TTX, an indirect neuronal mediation could be excluded and an action direct on the EC appeared to be likely. The effect of nicotine was blocked by the nicotine receptor antagonist hexamethonium, confirming the involvement of specific nicotine receptors. In agreement with the general rapid desensitization of nicotine receptors,[35,36] the nicotinic effects on EC were only transient and preincubation of the tissue to a subthreshold concentration of nicotine abolished the response to an effective concentration of nicotine.[16] There is good evidence for heterogeneity of nicotine receptors,[37,38] and α-bungarotoxin is known to potently block the nicotine receptors at the neuromuscular junction, but not most of the neuronal nicotine receptors.[39-41] α-Bungarotoxin did not affect the nicotinic stimulation of 5-HT release,[16] indicating that a neuronal type of nicotine receptors may be present on EC.

It should be mentioned, however, that canine and rabbit EC do not appear to be endowed with stimulatory nicotine receptors. In these species nicotine also enhanced 5-HT release from intact intestinal preparations, but these effects were blocked by TTX, indicating an indirect neuronal mediation.[16,18] Further experiments indicated that in these species cholinergic neurons are activated via ganglionic nicotine receptors, and acetylcholine released in the vicinity of the EC stimulates 5-HT secretion via muscarine receptors[16] (see below). Similar species differences with regard to the role of stimulatory nicotine and muscarine receptors have also been observed at adrenal chromaffin cells.[42-47]

2.2.1.2. 5-HT₃ Receptors

A subtype of receptors for 5-HT, the 5-HT₃ receptor, is another important ligand-gated ion channel.[48] There is good pharmacological evidence that the EC are endowed with excitatory

5-HT$_3$ receptors. Thus, in the presence of TTX, the 5-HT$_3$ receptor agonist 2-methyl-5-HT enhanced 5-HT release from the guinea pig small intestine, and this effect was blocked by the 5-HT$_3$ receptor selective antagonist tropisetron.[49] Moreover, several 5-HT$_3$ receptor-selective antagonists, tropisetron, MDL 72222, and graniseteron, reduced 5-HT release in this preparation, indicating that the 5-HT$_3$ receptors may indeed be an autoreceptor, i.e., a target for endogenous 5-HT released from the EC. Interestingly, however, the pharmacological profile of the 5-HT$_3$ receptors on EC appears to differ from that of 5-HT$_3$ receptors on myenteric neurons.[49]

2.2.2. Inhibitory Receptors

2.2.2.1. *GABA$_A$-Benzodiazepine Receptors*

In the intestinal tract γ-aminobutyric acid (GABA) is an important mediator of neuronal and most likely also of paracrine origin,[50-53] and GABAergic mechanisms appear also to play a significant role in the control of 5-HT release from the EC. There are two major classes of GABA receptors (GABA$_A$ and GABA$_B$). The GABA$_A$ receptor is a chloride channel, and in most excitable cells the opening of this chloride channel results in a reduction of excitability.[54] 5-HT release from the guinea pig ileum in the presence of TTX is inhibited by the GABA$_A$ receptor agonist muscimol, and this effect is blocked by the GABA$_A$ receptor antagonist bicuculline,[31] indicating that GABA$_A$ receptors may mediate inhibition of 5-HT release by a direct action at the EC. In the presence of TTX, the GABA$_A$ receptor antagonist bicuculline alone enhanced 5-HT release, indicating that endogenous GABA, most likely of nonneuronal, paracrine origin, may exert a tonic inhibition to the EC even under *in vitro* conditions.[31]

Finally, the inhibitory tone by endogenous GABA could be augmented by benzodiazepine receptor activation. Thus, in the presence of TTX the benzodiazepine receptor agonist midazolam caused a reduction in 5-HT release, an effect blocked by the benzodiazepine receptor antagonist flumazenil,[32] indicating that the GABA$_A$ receptor at the EC, like neuronal GABA$_A$ receptors,[54] is modulated by benzodiazepine receptors.

2.2.2.2. *P$_{2y\alpha}$ Purinoceptors*

5-HT release from EC can also be modulated via purinergic mechanisms, since both adenosine, as well as ATP, have been shown to inhibit 5-HT release from porcine small intestine in the presence of TTX. Moreover, there is good pharmacological evidence that adenosine and ATP mediated their effects via different types of purinoceptors (A$_2$ and P$_{2y}$, respectively). Thus, the effect of ATP was blocked by the P$_{2y}$ receptor antagonist cibacron blue, mimicked by the P$_{2y}$ receptor agonist α-methyl-thio-ATP, and not affected by the A$_2$ receptor-selective antagonist CGS 15943A, whereas the effect of adenosine was effectively blocked by this antagonist.[55,56] The P$_{2y}$ receptors have recently been subdivided into P$_{2y\alpha}$ and P$_{2y\beta}$ subtypes, and the P$_{2y\alpha}$ subtype has been suggested to be a ligand-controlled cation channel[57] which can be blocked by *d*-tubocurarine. In the above experiments, the inhibitory effect of ATP could also be prevented by *d*-tubocurarine, indicating that the ATP receptors at the EC may belong to the P$_{2y\alpha}$ subtype. Moreover, the inhibitory effect of ATP was also abolished in the presence of apamin, a toxin known to block selectively calcium-activated potassium channels.[58] Therefore, it can be concluded that calcium influx through ATP-gated channels may activate potassium channels, resulting in hyperpolarization of the EC and inhibition of 5-HT secretion. It is interesting to note that the calcium influx via ATP-gated channels appears not to activate 5-HT secretion, indicating that a rise of intracellular calcium in different cellular compartments of the EC could result in opposing functional responses. The polarity of the EC with the secretory active zone at the basolateral side was already mentioned above.

2.3. ROLE OF G PROTEIN-LINKED RECEPTOR SYSTEMS

The EC appear to be endowed with several other receptors which belong to the G protein-linked receptor superfamily. At present there is little known about possible ionic mechanisms involved in the action of these different receptor systems, although it appears quite likely that at least some of them may act via effects on calcium and/or potassium channels of EC or affect the intracellular handling of calcium.

2.3.1. Muscarine Receptors and Role of Intracellular Calcium

In all species studied so far, 5-HT release from intestinal segments in the presence of TTX was markedly enhanced by muscarine receptor agonists,[16,27,59] indicating that EC are endowed with facilitatory muscarine receptors which, in the rabbit, belong to the M3 subtype.[59] In contrast to the above described observations on high potassium-evoked 5-HT release, the muscarine receptor-mediated stimulation of 5-HT release is not significantly inhibited by the calcium channel blocker nifedipine.[17] On the other hand, pretreatment with ryanodine, which can deplete intracellular calcium stores,[60] markedly attenuated the stimulatory effect of muscarine receptor activation without affecting 5-HT release evoked by high potassium,[59] indicating that 5-HT release evoked by muscarine receptor activation may be triggered by liberation of calcium from intracellular stores, most likely via activation of phospholipase C and generation of inositol 1,4,5-triphosphate.

2.3.2. Adrenoceptors and Role of cAMP

Activation of β-adrenoceptors increased the release of 5-HT from the guinea pig[28] as well as rabbit[61] small intestine, and the effect of β-adrenoceptor agonists was also observed in the presence of TTX, indicating that stimulatory β-adrenoceptors may be localized directly at the EC. Generally, β-adrenoceptors are linked positively to adenylyl cyclase,[62] and elevation of the intracellular cAMP, either by direct stimulation of the adenylyl cyclase with forskolin or by inhibition of the phosphodiesterase, resulted in an increased 5-HT release.[18,28] Likewise, the stable cAMP analog 8-bromo cAMP also stimulated 5-HT release,[18] as did cholera toxin,[63] which is known to cause a prolonged activation of adenylyl cyclase.[64,65] Finally, the stimulatory effect of isoprenaline on 5-HT release was markedly potentiated after inhibition of phosphodiesterase, indicating that the β-adrenoceptors on the EC may indeed be linked to adenylyl cyclase.[28] However, the mechanisms by which cAMP finally increases 5-HT secretion from the EC are unknown at present. Possibilities include cAMP-dependent phosphorylation of ion channels, as well as effects on the secretory machinery distal to the calcium signal.

The EC are, in addition, endowed with inhibitory α_2-adrenoceptors,[28] but it is unknown whether or not their effects are mediated via inhibition of cAMP generation, although α_2-adrenoceptors are often linked inhibitory to the adenylyl cyclase.[62]

2.3.3. Histamine H_3 Receptors

Histamine is an important paracrine mediator in the gastrointestinal tract[66] and has been shown to inhibit 5-HT secretion from porcine small intestine,[67] and this inhibition was also seen in the presence of TTX indicating a direct effect at the EC. There are three major classes of histamine receptors, H_1, H_2 and H_3 receptors,[68] and detailed pharmacological experiments indicated that the histamine receptors at EC belong to the H_3 subtype. Thus, the effect of histamine was mimicked by the H_3 receptor-selective agonists (R)-α-methyl-histamine and imetit,[67,69] and H_3 receptor-selective antagonists, such as thioperamide, burimamide, and dimaprit, antagonized the inhibitory effect of histamine and of H_3 receptor-selective agonists.[67,69,70] However, the cellular mechanisms underlying this histaminergic inhibition are currently unknown.

2.3.4. Peptide Receptors
2.3.4.1. Somatostatin

Somatostatin, and its stable analog octreotide, markedly inhibited 5-HT release from the rabbit small intestine evoked by different stimuli which directly activate secretion of 5-HT from EC, such as depolarizing concentrations of potassium, forskolin, or phosphodiesterase inhibitors (i.e., elevation of intracellular cAMP) as well as muscarine receptor agonists.[18,63,71] Currently, five different somatostatin receptors can be differentiated which may couple to different effector systems,[72,73] and inhibitory effects on adenylyl cyclase as well as on calcium channels are possible cellular responses following somatostatin receptor activation, and could be of significance in EC. Octreotide, in nanomolar concentrations, inhibited 5-HT release evoked by high potassium or forskolin, whereas micromolar concentrations of octreotide were required to inhibit 5-HT release evoked by the muscarine receptor agonist oxotremorine.[63,71] Thus, it appears that the EC are endowed with pharmacologically different somatostatin receptors which may affect multiple processes in the chain of events of stimulus-secretion coupling in EC. Since voltage-activated calcium channels are crucially involved in high potassium-evoked 5-HT release (see above), an inhibitory effect of somatostatin on the calcium channels of the EC is not unlikely. However, the observations that forskolin evoked 5-HT secretion, which is not affected by nifedipine,[71] is also potently inhibited by octreotide,[71] indicate that somatostatin receptors on EC may in addition inhibit adenylyl cyclase.

2.3.4.2. Vasoactive Intestinal Polypeptide (VIP)

In the intestinal tract, VIPergic neurons represent an important neuronal pathway and project directly into the mucosa.[74,75] 5-HT release from the guinea pig small intestine was inhibited by VIP in picomolar concentrations.[30] Moreover, the effects of VIP were also observed in the presence of TTX, indicating that the EC are endowed with *high affinity* VIP receptors. Although VIP has been shown to be a potent activator of adenylyl cyclase in mucosal cells of the intestine,[76,77] it is unlikely that such an effect may be responsible for the inhibition of 5-HT release from the EC, since activation of adenylyl cyclase by other stimuli resulted in enhanced release of 5-HT (see above). Thus, the cellular mechanisms of the action of VIP in EC remain unknown at present.

3. CONCLUDING REMARKS

Although, at present, direct electrophysiological recordings from EC are not available, there are a number of functional, pharmacological observations which indicate that EC are electrically excitable cells. Figure 3 summarizes the present ideas about the role of ionic mechanisms in the stimulus-secretion coupling of this type of neuroendocrine cells: secretion from EC can be triggered by calcium influx through voltage-regulated calcium channels or by liberation of calcium from intracellular stores (for example, by muscarine receptor activation). Furthermore, the EC appear to contain several receptor-linked channels, the activation of which can result in excitation (nicotine and 5-HT$_3$ receptors) or inhibition of excitation (GABA$_A$-benzodiazepine receptor or P$_{2y\alpha}$ purinoceptor). Finally, a number of other receptors (α- and β-adrenoceptors, histamine H$_3$ receptors, somatostatin and VIP receptors) appear to be involved in the control of secretory activity of EC, but at present very little is known about their cellular mechanisms.

ACKNOWLEDGMENT

The authors' own work was supported by the Deutsche Forschungsgemeinschaft, Forschungsrat Rauchen und Gesundheit, and the VERUM Foundation.

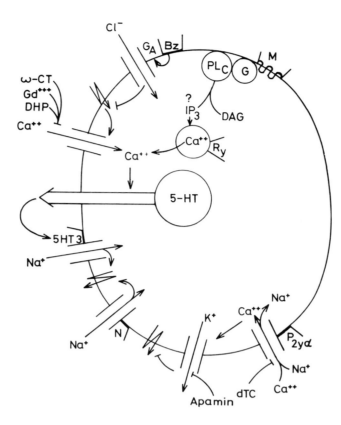

FIGURE 3. Schematic representation of the stimulatory and inhibitory ion channel-linked mechanisms involved in the control of stimulus-secretion coupling in enterochromaffin cells. Abbreviations: N, nicotine receptor; M, muscarine receptor; G_A, GABA$_A$ receptor; Bz, benzodiazepine receptor; 5-HT$_3$, 5-HT$_3$ receptor; $P_{2y\alpha}$, $P_{2y\alpha}$ receptor; ω-CT, ω-conotoxin; Gd^{+++}, gadolinium ions; dTC, d-tubocurarine; DHP, dihydropyridine; Ry, ryanodine; DAG, diacylglycerol; IP$_3$, inositol triphosphate; PL$_C$, phospholipase C; G, G protein.

REFERENCES

1. Cheng, H. and Leblond, C. P., Origin, differentiation and renewal of the four main epithelial cell types in the mouse small intestine. III. Entero-endocrine cells, *Am. J. Anat.*, 141, 503, 1973.
2. De Bruine, A. P., Dinjens, W. N. M., Zijlema, J. H. L., and Lenders, M. H., Renewal of enterochromaffin cells in the rat caecum, *Anat. Rec.*, 233, 75, 1992.
3. Vialli, M., Histology of enterochromaffin cell system, in *Handbuch der Experimentellen Pharmakologie*, Vol. 19, Eichler, O., and Farah, A., Eds., Springer-Verlag, Berlin, 1966, 1.
4. Erspamer, V., Occurrence of indolealkyl amines in nature, in *Handbuch der Experimentellen Pharmakologie*, Vol. 19, Eichler, O. and Farah, A., Eds., Springer-Verlag, Berlin, 1966, 132.
5. Solcia, E., Capella, C., Buffa, R., Usellini, L., and Tenti, P.. Morphological basis of gastrointestinal motility: ultrastructure and histochemistry of endocrine-paracrine cells in the gut, in *Handbook of Experimental Pharmacology*, Vol. 59,1, Bertaccini G., Ed. Springer-Verlag, Berlin, 1982, 55.
6. Dey, R. D. and Hoffpauir, J., Ultrastructural immunocytochemical localization of 5-hydroxytryptamine in gastric enterochromaffin cells, *J. Histochem. Cytochem.*, 32, 661, 1984.
7. Nilsson, O., Ericson, L. E., Dahlström, A., Ekholm, R., Steinbusch, H. W. M., and Ahlman, H., Subcellular localization of serotonin immunoreactivity in rat enterochromaffin cells, *Histochemistry*, 82, 351, 1985.
8. Nilsson, O., Ahlman, H., Geffard, M., Dahlström, A., and Ericson, L. E., Bipolarity of duodenal enterochromaffin cells in the rat, *Cell Tissue Res.*, 248, 49, 1987.
9. Nilsson, O., Dahlström, A., Geffard, M., Ahlman, H., and Ericson, L. E., An improved immunocytochemical method for subcellular localization of serotonin in rat enterochromaffin cells, *J. Histochem. Cytochem.*, 35, 319, 1987.

10. Nilsson, G., Larsson, L. I., Hakanson, R., Brodin, E., Pernow, B., and Sundler, F., Localization of substance P-like immunoreactivity in mouse gut, *Histochemistry,* 43, 97, 1975.
11. Pearse, A. G. E. and Polak, J. M., Immunocytochemical localization of substance P in mammalian intestine, *Histochemistry,* 41, 373, 1975.
12. Cetin, Y., Enterochromaffin (EC-) cells of the mammalian gastro-entero-pancreatic (GEP) endocrine system: cellular source of pro-dynorphin derived peptides, *Cell Tissue Res.,* 253, 173, 1988.
13. Buffa, R., Mare, P., Gini, A., and Saladore, M., Chromogranins A and B and secretogranin II in hormonally identified endocrine cells of the gut and the pancreas, *Basic Appl. Histochem.,* 32, 471, 1988.
14. Kuhn, M., Risse, K., Adermann, K., Forssmannm, W.-G., and Rechkemmer, G., Guanylin: stimulation of chloride secretion in human intestinal mucosa via cyclic GMP, *Naunyn-Schmiedeberg's Arch. Pharmacol.,* 349, R63, 1994.
15. Schwörer, H., Racké, K., and Kilbinger, H., Spontaneous release of endogenous 5-hydroxytryptamine and 5-hydroxyindoleacetic acid from the isolated vascularly perfused ileum of the guinea-pig, *Neuroscience,* 21, 297, 1987.
16. Racké, K. and Schwörer, H., Nicotinic and muscarinic modulation of the release of 5-hydroxytryptamine from the porcine and canine small intestine, *Clin. Invest.,* 70, 190, 1992.
17. Reimann, A., Bock, C., and Racké, K., The role of different calcium compartments in the muscarinic and high-potassium evoked release of 5-hydroxytryptamine (5-HT) from isolated rabbit ileum, *Br. J. Pharmacol.,* 112, 113P, 1994.
18. Forsberg, E. J., and Miller, R. J., Regulation of serotonin release from rabbit intestinal enterochromaffin cells, *J. Pharmacol. Exp. Ther.,* 227, 755, 1983.
19. Racké, K. and Schwörer, H., Characterization of the role of calcium and sodium channels in the stimulus secretion coupling of 5-hydroxytryptamine release from porcine enterochromaffin cells, *Naunyn-Schmiedeberg's Arch. Pharmacol.,* 347, 1, 1993.
20. Racké, K. and Schwörer, H., Regulation of serotonin release from the intestinal mucosa, *Pharmacol. Res.,* 21, 13, 1991.
21. Winkler, H., Occurrence and mechanism of exocytosis in adrenal medulla and sympathetic nerve, in *Handbook of Experimental Pharmacology,* Vol. 90, 1, *Catecholamines I,* Trendelenburg, U. and Weiner, N., Eds. Springer-Verlag, Heidelberg, 1989, 43.
22. Tsien, R. W., Lipscombe, D., Madison, D. V., Bley, K. R., and Fox, A. P., Multiple types of neuronal calcium channels and their selective modulation, *Trends Neurosci.,* 11, 431, 1988.
23. Bean, B. P., Classes of calcium channels in vertebrate cells, *Annu. Rev. Physiol.,* 51, 367, 1989.
24. Hess, P., Calcium channels in vertebrate cells, *Annu. Rev. Neurosci.,* 13, 337, 1990.
25. Brown, A. M., Kunze, D. L., and Yatani, A., The agonist effect of dihydropyridines on Ca channels, *Nature,* 311, 570, 1984.
26. Albus, U., Habermann, E., Ferry, D. R., and Glossmann, H., Novel 1,4-dihydropyridine (Bay K 8644) facilitates calcium-dependent [^{3}H]-noradrenaline release from PC 12 cells, *J. Neurochem.,* 42, 1186, 1984.
27. Schwörer, H., Racké, K., and Kilbinger, H., Cholinergic modulation of the release of 5-hydroxytryptamine from the guinea pig ileum, *Naunyn-Schmiedeberg's Arch. Pharmacol.,* 336, 127, 1987.
28. Racké, K., Schwörer, H., and Kilbinger, H., Adrenergic modulation of the release of 5-hydroxytryptamine from the vascularly perfused ileum of the guinea-pig, *Br. J. Pharmacol.,* 95, 923, 1988.
29. Schwörer, H., Racké, K., and Kilbinger, H., Characterization of the muscarine receptors involved in the modulation of serotonin release from the vascularly perfused small intestine of guinea pig, *Naunyn-Schmiedeberg's Arch. Pharmacol.,* 339, 263, 1989.
30. Schwörer, H., Racké, K., and Kilbinger, H., Effect of vasoactive intestinal polypeptide on the release of serotonin from the in vitro vascularly perfused small intestine of guinea pig, *Naunyn-Schmiedeberg's Arch. Pharmacol.,* 339, 540, 1989.
31. Schwörer, H., Racké, K., and Kilbinger, H., GABA receptors are involved in the modulation of the release of 5-hydroxytryptamine from the vascularly perfused small intestine of the guinea-pig, *Eur. J. Pharmacol.,* 165, 29, 1989.
32. Racké, K., Schwörer, H., and Kilbinger, H., Effects of the benzodiazepine receptor agonist midazolam and antagonist flumazenil on 5-hydroxytryptamine release from guinea-pig intestine *in vitro.* Indirect support for a "natural" benzodiazepine-like substance in the intestine, *Naunyn-Schmiedeberg's Arch. Pharmacol.,* 341, 1, 1990.
33. Narahashi, T., Moore, J. W., and Scott, W. R., Tetrodotoxin blockage of sodium conductance increase in lobster gigant axons, *J. Gen. Physiol.,* 47, 965, 1964.
34. Catterall, W. A., Neurotoxins that act on voltage sensitive sodium channels in excitable membranes, *Annu. Rev. Pharmacol. Toxicol.,* 20, 15, 1980.
35. Katz, B. and Thesleff, S., A study of the desensitisation produced by acetylcholine at the motor endplate, *J. Physiol. (London),* 138, 63, 1957.
36. Paton, W. D. M. and Perry, W. L. M., The relationship between depolarization and block in the cat's superior cervical ganglion, *J. Physiol. (London),* 119, 43, 1953.

37. Colquhoun, D., Ogden, D. C., and Mathie, A., Nicotinic acetylcholine receptors of nerve and muscle: functional aspects, *Trends Pharmacol. Sci.*, 8, 465, 1987.
38. Maelicke, A., Structure and function of the nicotinic acetylcholine receptor, in *Handbook of Experimental Pharmacology*, Vol. 86, Whittaker, V. P. Ed., Springer-Verlag, Berlin, 1988, 267.
39. Chiappinelli, V. A., Action of snake venom toxins on neuronal nicotinic receptors and other neuronal receptors, *Pharmacol. Ther.*, 31, 1, 1985.
40. Burstajan, S. and Gershon, D. D., Discrimination between nicotinic receptors in vertebrate ganglia and skeletal muscle by alpha-bungarotoxin and cobra venoms, *J. Physiol. (London)*, 269, 17, 1977.
41. Osswald, R. E. and Freeman, J. A., Alpha-bungarotoxin binding and central nervous system nicotinic acetylcholine receptors, *Neuroscience*, 6, 1, 1981.
42. Douglas, W. W. and Poisner, A., Preferential release of adrenaline from the adrenal medulla by muscarine and pilocarpine, *Nature*, 208, 1102, 1965.
43. Lee, F. L. and Trendelenburg, U., Muscarine transmission of preganglionic impulses to the adrenal medulla of the cat, *J. Pharmacol. Exp. Ther.*, 158, 73, 1967.
44. Kovacic, B. and Robinson, R., Drug-induced secretion of catecholamines by the perfused adrenal glands of the dog during nicotine blockade, *J. Pharmacol. Exp. Ther.*, 175, 178, 1970.
45. Yoshizaki, T., Participation of muscarinic receptors in splanchnic-adrenal transmission in the rat, *Jpn. J. Pharmacol.*, 23, 813, 1973.
46. Schneider, A. S., Herz, R., and Rosenbeck, K., Stimulus-secretion coupling in chromaffin cells isolated from bovine adrenal medulla, *Proc. Natl. Acad. Sci. U.S.A.*, 74, 5036, 1977.
47. Role, L. W. and Perlman, R. L., Both nicotinic and muscarinic receptors mediate catecholamine secretion by isolated guinea-pig chromaffin cells, *Neuroscience*, 10, 979, 1983.
48. Derkach, V., Surprenant, A., and North, R. A., 5-HT_3 receptors are membrane ion channels, *Nature*, 339, 706, 1989.
49. Gebauer, A., Merger, M., and Kilbinger, H., Modulation by 5-HT_3 and 5-HT_4 receptors of the release of 5-hydroxytryptamine from the guinea-pig small intestine, *Naunyn-Schmiedeberg's Arch. Pharmacol.*, 347, 137, 1993.
50. Jessen, K. R., Mirsky, R., and Hills, J. M., GABA as an autonomic neurotransmitter: studies on intrinsic GABAergic neurons in the myenteric plexus of the gut, *Trends Neurosci.*, 10, 255, 1987.
51. Tanaka, C., γ-Aminobutyric acid in peripheral tissues, *Life Sci.*, 37, 2221, 1985.
52. Erdö, S. L., Joo, F., Amenta, F., and Ezer, E., Characterization and function of a non-neuronal GABA system in rat stomach, *Neuroscience*, 22, S348, 1987.
53. Erdö, S. L. and Wolf, J. R., Releasable, non-neuronal GABA pool in rat stomach, *Eur. J. Pharmacol.*, 156, 165, 1988.
54. Olsen, R. W., GABA-benzodiazepine-barbiturate receptor interactions, *J. Neurochem.*, 27, 1, 1981.
55. Schwörer, H., Ramadori, G., and Racké, K., Modulation of 5-hydroxytryptamine (5-HT) release from porcine intestinal mucosa by purinoceptors, *Naunyn-Schmiedeberg's Arch. Pharmacol.*, 347, R123, 1993.
56. Schwörer, H., Ramadori, G., and Racké, K., Inhibitorische Modulation der Serotoninfreisetzung aus der Dünndarmmukosa durch Adenosin- und Purinrezeptoren, *Z. Gastroenterol.*, 31, 533, 1993.
57. Illes, P. and Nörrenberg, W., Neuronal ATP receptors and their mechanism of action, *Trends Pharmacol. Sci.*, 14, 50, 1993.
58. Cook, N. S., The pharmacology of potassium channels and their therapeutic potenial, *Trends Pharmacol. Sci.*, 9, 21, 1988.
59. Reimann, A., Bock, C., and Racké, K., Muscarinic M_3 receptor mediate stimulation of 5-hydroxytryptamine (5-HT) release from isolated segments of the rabbit small intestine incubated in vitro, *J. Physiol. (London)*, 367, 150P, 1993.
60. Kanmura, Y., Missiaen, L., Raeymaekers, L., and Casteels, R., Ryanodine reduces the amount of calcium in intracellular stores of smooth-muscle cells of the rabbit ear artery, *Pflügers Arch.*, 357, 327, 1988.
61. Kuemmerle, J. F., Kraus, H., and Kellum, J. M., Serotonin release is mediated by muscarine receptors on duodenal mucosal cells, *J. Surg. Res.*, 43, 139, 1987.
62. Lefkowitz, R. J., Stadel, J. M., and Caron, M. G., Adenylate cyclase-coupled beta-adrenergic receptors: structure and mechanisms of activation and desensitization, *Annu. Rev. Biochem.*, 52, 159, 1983.
63. Reimann, A., Bock, C., and Racké, K., Differential effects of octreotide on evoked 5-hydroxytryptamine outflow from isolated rabbit ileum, *Naunyn-Schmiedeberg's Arch. Pharmacol.*, 349, R63, 1994.
64. Sharp, G. W. G., and Hynie, S., Stimulation of intestinal adenylate cyclase by cholera toxin, *Nature*, 229, 266, 1971.
65. Kimberg, D. V., Field, M., Johnson, J., Henderson, A., and Gershon, E., Stimulation of intestinal mucosal adenyl cyclase by cholera enterotoxin and prostaglandins, *J. Clin. Invest.*, 50, 1218, 1971.
66. Rangachari, P. K., Histamine: mercurial messenger in the gut, *Am. J. Physiol.*, 262, G1, 1992.
67. Schwörer, H., Katsoulis, S., and Racké, K., Histamine inhibits 5-hydroxytryptamine release from the porcine small intestine. Involvement of H_3 receptors, *Gastroenterology*, 102, 1906, 1992.

68. Hill, S. J., Distribution, properties, and functional characteristics of three classes of histamine receptors, *Pharmacol. Rev.*, 42, 45, 1990.
69. Schwörer, H., Reimann, A., Ramadori, G., and Racké, K., Characterization of histamine H_3 receptors inhibiting 5-HT release from porcine enterochromaffin cells: further evidence for H_3 receptor heterogeneity, *Naunyn-Schmiedeberg's Arch. Pharmacol.*, 350, 375, 1994.
70. Racké, K. Schwörer, H., Novel histamine H_3 receptor subtype (H_{3C}) mediates inhibition of 5-hydroxytryptamine release from porcine enterochromaffin cells, *Naunyn-Schmiedeberg's Arch. Pharmacol.*, 346, R2, 1992.
71. Reimann, A. and Racké, K., Stimulus-dependent inhibition by octreotide of 5-HT release from isolated small intestine, *Z. Gastroenterol.*, 32, 82, 1994.
72. Bell, G. I. and Reisine T., Molecular biology of somatostatin receptors, *Trends Neurosci.*, 16, 34, 1993.
73. Raynor, K., O'Carroll, A-M., Kong, H., Yasuda, K., Mahan, L. C., Bell, G. I., and Reisine, T., Characterization of cloned somatostatin receptors SSTR4 and SSTR5, *Mol. Pharmacol.*, 44, 385, 1993.
74. Costa, M. and Furness, J. B., The origin, pathways and termination of neurons with VIP-like immunoreactivity in the guinea-pig small intestine, *Neuroscience*, 8, 665, 1983.
75. Schultzberg, M., Hökfelt, T., Nilsson, G., Terenius, L., Rehfeld, J. F., Brown, M., Elde, R., Goldstein, M., and Said, S., Distribution of peptide and catecholamine-containing neurons in the gastro-intestinal tract of rat and guinea-pig: immunohistochemical studies with antisera to substance P, vasoactive intestinal polypeptide, enkephalins, somatostatin, gastrin/cholecystokinin, neurotensin and dopamine β-hydroxylase, *Neuroscience*, 5, 689, 1980.
76. Schwartz, C. L., Kimberg, D. V., Sheerin, H. E., Field, M., and Said, S. J., Vasoactive intestinal peptide stimulation of adenylate cyclase and active electrolyte secretion in intestinal mucosa, *J. Clin. Invest.*, 54, 536, 1974.
77. Prieto, J. C., Laburthe, M., Hoa, D. H. B., and Rosselin, G., Quantitative studies of vasoactive intestinal peptide (VIP) binding sites and VIP-induced adenosine 3':5'-monophosphate production in epithelial cells from duodenum, jejunum, ileum, coecum, colon and rectum in the rat, *Acta Endocrinol.*, 96, 100, 1981.

Section VII
Adrenal and Juxtaglomerular Cells

Chapter 12

ELECTRICAL PROPERTIES OF ADRENAL CHROMAFFIN CELLS

Antonio R. Artalejo

CONTENTS

1. SUMMARY

Chromaffin cells from the adrenal medulla synthesize, store, and secrete catecholamines. Adrenomedullary catecholamine secretion contributes to the cardiovascular and metabolic adaptations of the body to stressful situations. Cytoplasmic Ca^{2+} ions play an essential role in coupling chromaffin cell stimulation to the exocytotic secretory response. As it is true for many other structures derived from the neural crest, chromaffin cells are electrically active; the electrical behavior influences Ca^{2+} entry and, hence, catecholamine release. In this article, the ionic conductances underlying the changes in the electrical activity of chromaffin cells are reviewed and their relation to secretion is discussed.

2. INTRODUCTION

The chromaffin cells, named for their ready staining with chromate-based fixatives, constitute the parenchymal cells of the adrenal medulla. A characteristic feature of chromaffin cells is their ability to synthesize and store large amounts of catecholamines, which are released in response to preganglionic sympathetic nerve stimulation. Ontogenetically, they derive from the neural crest and can be regarded, either morphologically or functionally, as modified sympathetic neurons.

Two main catecholamines, adrenaline (A) and noradrenaline (NA), are stored in the secretory granules ("chromaffin" granules) of distinct populations of chromaffin cells.[1] NA-storing cells are easily distinguished from A-storing cells by their strong chromaffinity at the light microscope, or the electron-dense content of their granules when fixed with glutaraldehyde for ultrastructural studies (Figure 1).[2] Apart from this, the morphology of the chromaffin cells is rather homogeneous. They have a polygonal appearance in the intact gland, but adopt a spherical shape (about 15 µm in diameter) after isolation. The nucleus is usually displaced to one pole of the cell, allowing up to 30,000 chromaffin granules to accumulate in the rest of the cytoplasm.[3] At the ultrastructural level, a chromaffin granule appears as a membrane-delimited organelle (300–400 nm diameter) containing a core of variable density with a translucent halo between the core and the membrane. Several electron microscope studies have shown that the chromaffin granules are discharged by exocytosis. However, the existence of specialized regions of the plasma membrane, similar to the active zones of the neuromuscular junction, where the secretory vesicles dock and fuse have not been demonstrated.[3-6]

Biochemical and immunocytochemical studies have revealed that chromaffin granules contain, in addition to A and NA, other amines like dopamine, histamine, and serotonin. These amines are stored in complexes with adenine nucleotides and acidic proteins named chromogranins.[7] Opioids and several other peptides, such as substance P, vasoactive intestinal peptide (VIP), somatostatin, neurotensin, calcitonin gene-related peptide, neuropeptide Y (NPY), and the atrial natriuretic factor (ANF), also coexist with catecholamines in the adrenal medulla, giving rise to multiple potential modulations of the function of chromaffin cells. Besides chromaffin granules, adrenomedullary chromaffin cells seem also to host synaptic-like microvesicles (60–90 nm diameter) which contain catecholamines.[8,9]

Adrenal chromaffin cells are densely innervated by preganglionic sympathetic fibers that arise in the intermediolateral column of the spinal cord and travel to the adrenal medulla, traversing the greater thoracic splanchnic nerve. The synaptic endings are usually found within indentations on the chromaffin cells, separated from the postsynaptic membrane by a cleft of about 30 nm. The nerve endings contain two types of vesicular structures. One is the small synaptic-type vesicle that is cholinergic; the other is the large, dense-core vesicle which has been associated with the presence of VIP, substance P, and opioids in these nerve terminals. These data suggest that peptides released from the presynaptic terminals may also modulate the cholinergic stimulation of chromaffin cells.

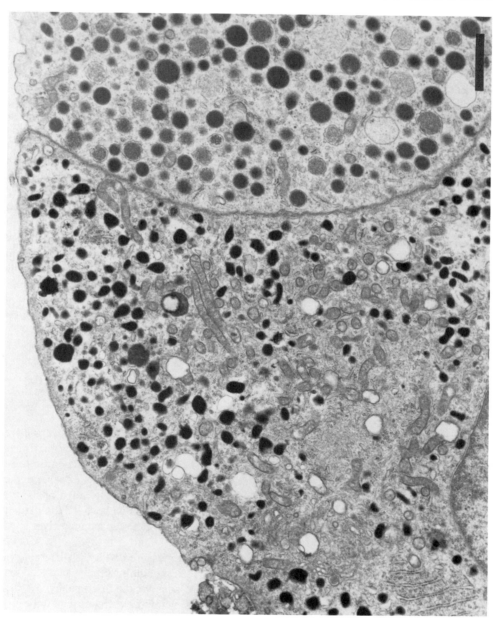

FIGURE 1. An electron micrograph of A- (right) and NA-storing (left) cells (according to Coupland[2]). The A cell contains granules that are spherical and have proteinaceous and low-density cores. NA-containing cells have granules that are irregularly shaped with highly dense core material that is often eccentrically positioned. Original magnification ×18,000. Bar = 1 μm. (Micrograph kindly provided by G. Q. Fox).

The chromaffin cells secrete catecholamines into the blood in response to acetylcholine (ACh) released from the terminals of the splanchnic nerve. Activation of nicotinic and/or muscarinic receptors of the plasma membrane triggers a chain of events leading ultimately to the appearance of secretory products in the extracellular environment. This notion of "stimulus-secretion coupling", as defined originally by Douglas and Rubin,[10] requires Ca^{2+} as an obligatory mediator, since the secretory response is triggered by an increase in the free cytosolic concentration of Ca^{2+} ions ($[Ca^{2+}]_i$). One way in which ACh can elevate $[Ca^{2+}]_i$ is by changing the membrane permeability to Ca^{2+} ions. In the mid 1960s, Douglas and colleagues[11,12] used microelectrodes to demonstrate that ACh depolarizes the chromaffin cells, and that this effect is mediated principally by the inward movement of Na^+ ions; however, raising the extracellular Ca^{2+} concentration also enhanced the ability of ACh to depolarize the cells.[11,12] The electrical excitability of chromaffin cells was demonstrated 10 years later, when two groups of investigators independently recorded action potentials from chromaffin cells depolarized through application of either high external K^+ or ACh.[13,14] Soon after, it was found that the rate of catecholamine release increases, dose dependently, with the frequency of action potential discharge induced by ACh;[15] this was the first attempt to correlate the electrical behavior of the chromaffin cell with secretory function.

Subsequently, a major advancement in our knowledge about the electrical properties of chromaffin cells has been possible due to the advent of the patch clamp technology.[16] These methods are particularly well suited to characterize ionic channels, whose activity underlie the overall cellular electrical activity. As a result of the patch clamp studies, over 15 different ionic channels have been described in chromaffin cells and their unitary conductances, ionic selectivity, voltage dependency, and modulation analyzed in great detail. Moreover, the whole-cell configuration of the patch clamp can also be used to determine the cell membrane capacitance, whose changes are a faithful and highly time-resolved indicator of exocytotic release.[17] The greatest strength of this technique is that it allows the study of stimulus-secretion coupling at the single-cell level and the correlation of the biophysics of the plasma membrane with the functional response. Finally, the recent application of electrochemical methods to monitor secretion of oxidizable substances from single cells provide the highest sensitivity and temporal resolution for the detection of single-vesicle secretion of catecholamines from chromaffin cells.[18,19] In short, we are just at the beginning of a period full of new prospects for a most refined analysis of the stimulus-secretion coupling in neuroendocrine cells and, as it has been the case for the last 40 years, chromaffin cells continue to serve as a privileged model for pioneering studies in this field.

3. ELECTRICAL PROPERTIES OF CHROMAFFIN CELLS

3.1. MEMBRANE PASSIVE ELECTRICAL PROPERTIES

By using the patch clamp method, Fenwick et al.[20] defined the passive electrical properties of the chromaffin cell membrane. The membrane input resistance of isolated chromaffin cells was estimated to be at least 5 GΩ, and the membrane time constant, calculated during the voltage responses to current pulses, was about 50 ms. These time constants correspond to cell membrane capacitances measured in the whole-cell configuration in the range of 5–10 pF. Considering that a chromaffin cell can be approximated by a sphere with 10–15 μm diameter, the ratio between the calculated surface area and the total membrane capacitance gives a value for the specific capacitance of 1.08 μF/cm^2.

Both membrane input resistance and time constant as measured with the different patch clamp configurations are many times larger than the corresponding figures calculated using conventional intracellular microelectrodes.[14] Possible reasons for the discrepancy are differences in the cell size, since successfully recorded cells using microelectrodes tend to be larger than the population average. However, the main factor accounting for these differences is believed to be the membrane damage caused by the intracellular electrodes, that would result in a current leak and, in consequence, the underestimation of the above mentioned parameters. Nevertheless, smaller input resistance values (25–200 MΩ), have also been observed in *in situ* measurements from perfused rat adrenal medulla,[21] intact mouse adrenal gland,[22] and bissected adrenal gland from guinea pig.[23] These results, in accordance with some morphological studies,[24] have been taken as evidence of electrical coupling between cells. However, this interpretation should again be balanced against the possibility that the membrane damage caused by penetration with a microelectrode results in a decrease in input resistance. To resolve this, application of the patch clamp technique to a slice preparation of the adrenal medulla would, undoubtedly, be the best way to address both the problem of electrical coupling of chromaffin cells *in situ* and the functional relevance of such a coupling.[25]

3.2. RESTING MEMBRANE POTENTIAL AND CURRENT-VOLTAGE RELATION

In the current clamp mode of the patch clamp, the cell resting potential can be measured as the zero-current voltage assuming that the pipette-bath conductance is small compared to the cell conductance. Typically, chromaffin cell resting potentials measured with this technique lie between –50 and –80 mV.[20,26-29] These values are slightly more negative than those reported using intracellular microelectrode recordings,[13,14,21-23] which are placed around –50 mV. The explanation for this good agreement, in spite of the likely shunt to ground induced by the microelectrodes, could be that Ca^{2+} entry caused by leak around the site of impalement results in the activation of Ca^{2+}-dependent K^+ conductance and membrane hyperpolarization.

The ionic mechanisms of the resting membrane potential were first studied by Douglas et al.[12] and, thereafter confirmed by other groups.[14,21,22] The resting potential was shown to vary linearly with the logarithm of the extracellular concentration of K^+, and followed the predictions of the Goldman-Hodgkin-Katz equation for a calculated ratio of membrane permeabilities to Na^+ and K^+ of 0.07. Substituting Cl^- ions by other anions had a negligible effect on the membrane potential, thus ruling out any significant contribution of Cl^- permeability to resting ionic cellular permeabilities.[21,22]

Patch clamp measurements have revealed that chromaffin cell "resting" membrane potential is not stable. The cells often display spontaneous action potentials,[20,22,27,30,31] which are interspersed among slow fluctuations in the membrane potential.[20] Taking into account the high input resistance of chromaffin cells, the random opening and closing of a small number of channels active under resting conditions would suffice to produce relatively large variations in membrane potential and, eventually, action potentials. What is the identity of these channels? This is an aspect of the electrophysiology of chromaffin cells which remains relatively obscure. Chromaffin cells depolarize upon bath application of Ba^{2+} and tetraethylammonium (TEA),[22,28,32] indicating the existence of K^+ channel activity at rest. Current-voltage relationships in rat, bovine, and mouse chromaffin cells show a normal rectification after injecting inward current (Figure 2).[14,20,22] This conductance increase is time dependent and is most likely attributable to the activation of delayed-rectifier K^+ channels.[33,34] Such an interpretation would also be consistent with the prolonged action potentials observed in rat and mouse chromaffin cells following TEA application.[15,22] However, with outward current pulses the current-voltage relation is linear, suggesting the lack of anomalous rectification both in bovine chromaffin cells using the patch clamp technique[20] and in rat chromaffin cells recorded intracellularly with microelectrodes.[14] Two recent reports have challenged the notion that chromaffin cells do not express inward rectifier K^+ channels. Inoue and Imanaga[35] demonstrated an inward rectifier in guinea pig chromaffin cells and discovered an inhibitory modulation of this channel by G proteins, much like what has been found for the inwardly rectifier K^+ channel of ventricular myocytes. Furthermore, a similar type of channel would be inhibited by muscarine and be responsible for a long-lasting depolarization in rat adrenal chromaffin cells.[27,36]

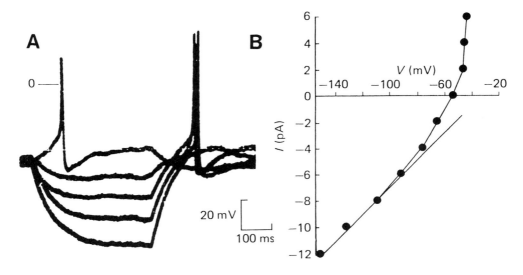

FIGURE 2. Current clamp records. (A) Responses to current pulses of +2, –2, –4, –6 and –8 pA. The responses to negative current stimulations of 4 pA or more display anodal-break action potentials. The responses to –2 and –4 pA stimulations are approximately exponential, with a time constant of 50 ms. Responses to larger stimulations clearly deviate from simple exponentials. (B) *I-V* relationship obtained from the data in A. A straight line through the most hyperpolarized points gives a slope conductance of 100 pS, corresponding to a resistance of 10 GΩ. Note that at the resting potential, the cell resistance is only about one half of that. Cell capacitance is 9.5 pF. (From Fenwick, E. M., Marty, A., and Neher, E., *J. Physiol.*, 331, 577, 1982. With permission.)

Other classes of channels potentially contributing to the resting membrane potential are those underlying the M current.[37] This current has not yet been described in chromaffin cells, but it has been reported in PC12 cells, the tumor counterpart of rat chromaffin cells.[38] As a rapid washout of cytoplasmic constituents may prevent observation of this current in the whole-cell configuration, the perforated-patch clamp method appears ideally suited to look for the M-current in chromaffin cells.[39] Also, under conditions of basal or slightly raised $[Ca^{2+}]_i$, small-conductance Ca^{2+}-dependent K^+ channels and Ca^{2+}-dependent cation currents could modulate the resting potential of chromaffin cells.[20,29,40]

3.3. THE ACTION POTENTIAL

Chromaffin cells are electrically excitable. This idea was already suggested at the end of the last century, based on the observation that adrenal glands which had been electrically stimulated produced an enhanced catecholamine output.[41] In 1976, 80 years later, direct demonstration of this was provided by Biales et al.[13] and Brandt et al.,[14] who independently recorded action potentials from gerbil and human chromaffin cells and rat adrenomedullary cells, respectively. An action potential of a chromaffin cell shows the typical features of Na^+-dependent action potentials from neuronal axons; they have large overshoots (30 mV), a rather steep rising phase (50 V/s maximal slope), and pronounced afterhyperpolarizations (Figure 2).[20] However, as has been shown in rat and mouse chromaffin cells, neither removal of Na^+ ions from the bath solution nor addition of tetrodotoxin (TTX) suppressed the generation of action potentials, and when Ca^{2+} ions in Na^+-free solution are replaced with Ba^{2+} prolonged action potentials are observed. Only by removing Na^+ and Ca^{2+} from the external medium is it possible to abort the firing of action potentials triggered by injecting current into the cell.[15,22] Therefore, it is widely held that action potentials in the chromaffin cells have both voltage-dependent Na^+ and Ca^{2+} components.

A distinctive property of action potential firing in chromaffin cells seems to be the existence of a relatively large refractory period for spike generation.[42] In experiments in which a voltage-sensitive dye was used to measure membrane potential changes in a slice preparation of the rat adrenal gland, the chromaffin cells failed to discharge action potentials when they were transsynaptically stimulated at a frequency of 100 Hz. These data indicate a refractory period longer than 10 ms which is considerably larger than that observed in the nerve terminals of the splanchnic nerve.

3.4. ION CHANNELS ACTIVATED DURING THE ACTION POTENTIAL
3.4.1. Na$^+$ Channels

Voltage-dependent Na$^+$ conductance of chromaffin cells is due to channels similar to those originally described by Hodgkin and Huxley[43] in the squid giant axon. By using a combined approach of single-channel recording, fluctuation analysis, and mean current kinetics, Fenwick et al.[20] provided the first description of Na$^+$ channels from chromaffin cells (Figure 3). Whole-cell recordings showed that Na$^+$ currents start to activate at around –30 mV and reach a maximum at +10 mV. The potential of half-inactivation was determined to be –35 mV. Single-channel measurements from outside-out membrane patches revealed an elementary current amplitude of 1 pA at –10 mV in 150 mM Na$_o^+$ and 10 mM Na$_i^+$. This current amplitude increased with hyperpolarization between –10 and –40 mV. The mean Na$^+$ channel open time was 1 ms at –30 mV and decreased both with depolarization and hyperpolarization. Ensemble noise analysis of the current fluctuations due to the stochastic openings and closings of Na$^+$ channels gave values consistent with those of single-channel measurements. By dividing values of maximum peak inward current (which ranges from 150–1,000 pA) by the unitary current amplitude, a Na$^+$ channel density of 1.5–10 channels/μm^2 was estimated.

3.4.2. Ca^{2+} Channels

The first characterization of mammalian Ca^{2+} channels underlying Ca^{2+} currents during action potentials was done in bovine chromaffin cells, and reported in a classical paper by the group of Neher in 1982.[44] Whole-cell Ca^{2+} (5 mM Ca$_o^{2+}$) currents were shown to activate at –30 mV and rise sharply between –20 and +10 mV, and then decline at more positive potentials to turn outward above +60 mV (Figure 4). These outward currents are most likely carried by Cs$^+$ ions, which were substituted for internal K$^+$, flowing through maximally activated Ca^{2+} channels. In isotonic BaCl$_2$, single-channel recordings in the cell-attached configuration revealed a unitary current of 0.9 pA at –5 mV, which became smaller for more positive potential values. Single Ba^{2+} currents appeared as bursts of 1.9 ms mean duration, with increasing lengths with larger depolarizations. Ensemble fluctuation analysis suggested that the opening probability of Ca^{2+} channels was at least 0.9 above +40 mV, and provided estimates of single-channel amplitudes (0.52 pA at –8 mV for isotonic Ba^{2+} and 0.086 pA in 5 mM Ca$^{2+}_o$ at –12 mV) consistent with the figures obtained in single-channel experiments in membrane patches. The calculated Ca^{2+} channel density ranged between 5 and 15 channels/μm^2 or 2500–7500 channels for a 12-μm-diameter cell, a value similar to that of Na$^+$ channels.[44]

Unlike Na$^+$ currents, Ca^{2+} currents tended to decay (run down) during prolonged episodes in the whole-cell recording configuration. At the single-channel level, it was noticed that the amplitude of the elementary current stayed constant during a large part of the run-down process; thus, it was suggested that it is the number of channels available for activation, rather than the unitary conductance or opening probability, which diminishes during run-down. This phenomenon becomes more pronounced when the [Ca^{2+}] of the internal solution is increased, suggesting that it may be due to intracellular Ca^{2+} accumulation induced by the channel activity leading to inhibition of Ca^{2+} current. However, the progressive loss of Ca^{2+} current does not depend only on [Ca^{2+}]$_i$, as a very fast run-down occurs after formation of isolated membrane patches. Taken together, all of these data indicate that run-down of Ca^{2+} channels is a multifactorial process involving the loss of intracellular constituents which diffuses out of the cells 10–20 min after establishing the whole-cell configuration.[44,45]

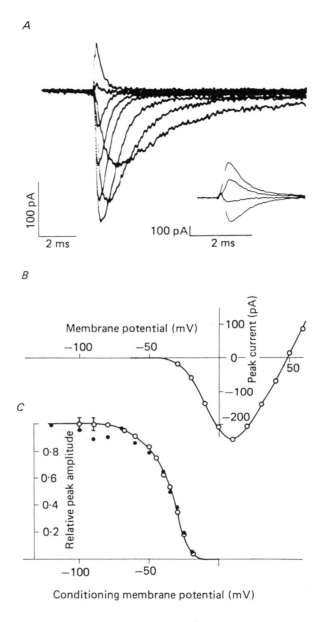

FIGURE 3. Na$^+$ currents in whole-cell recording mode. The bath contained normal saline with Ca^{2+} replaced by 1 mM Co^{2+}. The pipette was filled with a solution containing in mM: CsCl, 120; TEACl, 20; EGTA-NaOH, 11; CaCl$_2$, 1; Hepes-NaOH, 10; pH 7.2. (A) Superimposed traces for depolarizations starting from a holding potential of −80 mV to test levels of −10, 0, 10, 20, 30, 40, 50, and 60 mV. Each trace is the average of four records after subtraction of linear leak and capacitive currents. A period of 180 μs is blanked out at the time of the step depolarization. The inset shows four traces for voltages in the vicinity of the reversal potential at 10-mV increments from an experiment in which better time resolution was achieved. The clamp settled with a time constant of 30 μs. No blanking of residual capacitive artifact was employed in the inset. There is a clear indication of a nonlinear capacitive component. (B) Plot of peak current of traces in part A against test voltage. (C) Inactivation curve of Na$^+$ currents. A 25-ms conditioning pulse was given preceding a test pulse to 0 mV. Relative peak amplitude current during the test pulse is plotted against conditioning voltage; ●, holding potential −80 mV; O, −100 mV. Error bars give typical variations between successive experiments. (From Fenwick, E. M., Marty, A., and Neher, E., *J. Physiol.*, 331, 599, 1982. With permission.)

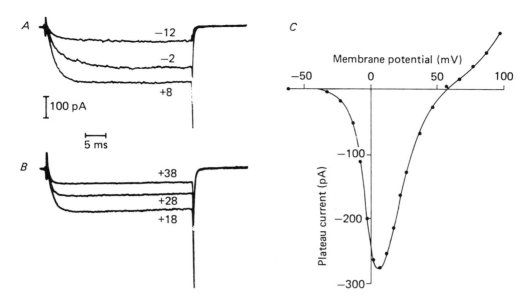

FIGURE 4. Voltage-clamp currents in 5 mM Ca$^{2+}_o$. (*A* and *B*) Test pulses were applied from a holding potential of –62 mV. Below +8 mV, single sweep records showed clear fluctuations (*A*). Above +8 mV, the fluctuations were much smaller and large tail currents were observed when the clamp voltage was returned to –62 mV (*B*). No subtraction of background current. Test potentials as indicated. (*C*) Plateau current amplitude as a function of test potential. The leakage current was subtracted in *C*. TTX (20 µg/ml) was present in the bath solution; internal solution contained (mM): CsCl, 120; TEACl, 20; EGTA-NaOH, 11; CaCl$_2$, 1; Mg Cl$_2$, 2; Hepes-NaOH, 10; pH 7.2. (From Fenwick, E. M., Marty, A., and Neher, E., *J. Physiol.*, 331, 599, 1982. With permission.)

Concomitantly with the irreversible reduction of activatable Ca^{2+} channels, Ca^{2+} currents also showed a decrease during a sustained depolarization.[44] Based on the different inactivating effects of Ca^{2+} and Ba^{2+} and the little inactivation observed when increasing the Ca^{2+}-buffering capacity of the internal solution,[44] this behavior was interpreted as a current-dependent inactivation.[46,47]

Ca^{2+} channel currents were also shown to have complex kinetics. The activation of Ca^{2+} (or Ba^{2+}) currents was well fitted by the sum of three exponentials The tail currents, which indicate relaxation of the conductance upon repolarization, were fitted by a biexponential function.[44] In spite of these observations, no inferences about the existence of multiple types of Ca^{2+} channels were made in this pioneer study. Since then, the diversity of Ca^{2+} channels in chromaffin cells (and the complexity of their classification) has become apparent.

Nowadays, the term voltage-dependent Ca^{2+} channels refers to a handful of entities classified according to a variety of biophysical and pharmacological criteria. From the biophysical point of view, Ca^{2+} channels are most clearly divided into low threshold or low voltage-activated (LVA) and high threshold or high voltage-activated (HVA) types. These two categories of channels are defined according to voltage dependence for activation. Low threshold, also referred to as T-type, Ca^{2+} channels require weak depolarizations to activate, and have a single-channel conductance of 8 pS in isotonic Ba^{2+}.[48] Holding potentials between –60 and –50 mV produce a marked steady-state inactivation of this current, which otherwise activates transiently. These channels are preferentially blocked by Ni^{2+} and amiloride. On the other hand, high-threshold Ca^{2+} channels activate following strong depolarizations (>–30 mV), are more sensitive to Cd^{2+} than to Ni^{2+}, and comprise several pharmacologically and kinetically different entities.[49] Among them, the so-called L-type Ca^{2+} channels are 1,4-dihydropyridine (DHP)-sensitive, show little inactivation during a depolarizing step, and possess a single-channel conductance in isotonic Ba^{2+} of 22–28 pS.

HVA Ca^{2+} channels of the N-type class are sensitive to ω-conotoxin GVIA (CgTx-GVIA), have a single-channel conductance in the 12- to 22-pS range (isotonic Ba^{2+}), and are insensitive to DHP. N-type Ca^{2+} channels inactivate faster than L-type channels, and negative holding potentials are usually required to reprime these channels completely.[49]

Finally, the P-type Ca^{2+} channels are 12-pS HVA channels (isotonic Ba^{2+}), first identified in Purkinje neurons.[50] These channels are sensitive to the toxin fraction (FTX) of the poison from the funnel web spider *Agenelopsis Aperta*, but are more selectively blocked by ω-agatoxin IVA (AGA-IVA). The P-type Ca^{2+} current seems to be relatively insensitive to changes in the holding potential.

Chromaffin cell Ca^{2+} channel subtypes, identified thus far, can be classified in some of the above-mentioned categories. With the exception of the work of Diverse-Pierluissi et al.,[51] evidence for the presence of T-type channels has not been provided in any of the reports on chromaffin cell Ca^{2+} channels. In a subpopulation of bovine chromaffin cells, Diverse-Pierluissi et al. found a transient current carried by either Ca^{2+} or Na^+ with a threshold for activation around –45 mV and a maximum near –20 mV. Moreover, steady-state inactivation curves allowed the calculation of a half-inactivation voltage of –73 mV.[51] Further studies will be required to better define the conditions under which this current is expressed, and to fully compare this LVA Ca^{2+} channel with the T-type currents characterized in other preparations.

Initial clues about the existence of a DHP-sensitive Ca^{2+} channel in bovine chromaffin cells were given by catecholamine secretion and $^{45}Ca^{2+}$ uptake studies in which potent inhibitory or facilitatory actions of DHP antagonists and agonists were demonstrated.[52,53] In electrophysiological studies, the effects of DHP agonists were first reported in 1987, when Hoshi and Smith[54] described that BAY K 8644 increased by at least a factor of 10 the open time duration of single Ca^{2+} channels recorded in the cell-attached configuration. Noticeably, the long-lasting openings induced by the DHP agonist were not accompanied by any significant change in the elementary current amplitude. DHP agonist effects on whole-cell Ca^{2+} currents were first explored by Ceña and co-workers.[55] They reported that in the presence of low concentrations of BAY K 8644 (0.1 μM) the midpoint of the activation curve of Ca^{2+} channels was shifted by 6 mV towards more negative membrane potentials, and a marked prolongation in the time constant of Ca^{2+} channel deactivation was induced.

The involvement of an L-type component in the HVA Ca^{2+} currents of chromaffin cells has been subsequently confirmed with the use of both DHP agonists and antagonists.[56-63] The effects of DHP antagonists on the whole-cell Ca^{2+} channel currents are generally not prominent, though variable degrees of inhibition among different cells and groups have been reported (Figure 5).[57,60,62,63] The blocking actions range from negligible effects[57] to up to 40% blockade.[62] Possible reasons for this discrepancy are (i) the different concentrations of DHP used (e.g., from 1 mM to 30 mM nisoldipine in different studies),[57,63] thus potentially causing more or less selective Ca^{2+} channel actions; (ii) the time-dependent block by DHP that may also be associated with a simultaneous time-dependent inactivation of other Ca^{2+} current components (N-type channels);[60] and (iii) the voltage-dependent effects of these drugs that arise from intrinsic features of the DHP mechanism of action, but also from distinct holding potential sensitivities of the Ca^{2+} channel types and slight differences in the activation threshold of the various components of the Ca^{2+} current.[64,65] As a consequence of this, the comparison between the reported data regarding the DHP effects on Ca^{2+} currents of chromaffin cells appears rather complicated.

At the single-channel level, the DHP-sensitive channel has been identified as a 27-pS (in 90 mM Ba^{2+}) or a 31-pS (in 110 mM Ba^{2+}) channel recorded in nonexcised membrane patches.[56,58,61] Typically, the activity of this channel is not inhibited at depolarized potentials, and DHP agonists increase its opening probability and/or induce long-lived openings (Figure 6).[54,58,61] Conversely, DHP antagonists suppress the unitary activity of 27-pS channels.[58]

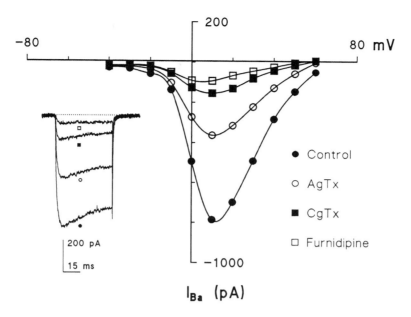

FIGURE 5. Effect of different types of Ca²⁺ channel blockers on chromaffin cell macroscopic Ba²⁺ currents. Depolarizing pulses were applied in the absence (control) and the presence of 1 μM Aga-IVA (AgTx), 1 μM CgTx-GVIA (CgTx), and 1 μM of the DHP, furnidipine. The inset shows capacitance and linear leak subtracted current traces elicited by a 50-ms pulse to 0 mV from a holding potential of –80 mV. The *I-V* relation was obtained by plotting peak inward currents vs. the different test potentials. Each concentration of drug was present at least 2 min before the depolarizing pulse. (From Albillos, A., García, A. G., and Gandía, L., *FEBS Lett.*, 336, 259, 1993. With permission.)

CgTx-GVIA was found to bind to bovine chromaffin cell membranes and inhibit ⁴⁵Ca²⁺ uptake induced by high K⁺ depolarization.[66] Consistent with these results, a component of the whole-cell Ca²⁺ current was shown to be sensitive to this toxin.[67] The effects of CgTx-GVIA were irreversible and highly variable among cells. In fact, a trimodal distribution of the chromaffin cells according to their sensitivity to the toxin has been proposed: some cells would be completely resistant to CgTx-GVIA, in others the toxin fully blocks the whole-cell currents, while the large majority of the cells (60%) showed an intermediate sensitivity manifested as a 40-50% inhibition of the Ca²⁺ channel currents.[59] The effects of the toxin have been described to be independent of the holding potential, and thus, indicative of the existence of a "nonclassical" N-type Ca²⁺ channel,[59,67] but other reports stated that CgTx-GVIA acts on a current component having a voltage for half-inactivation of –55 mV.[60] Differences in current kinetics are observed after toxin application. Faster activation and deactivation phases are associated with the CgTx-GVIA-resistant component, which suggests that CgTx-GVIA-resistant and CgTx-GVIA-sensitive channels possess different ranges for activation and rates of activation and deactivation, which are otherwise hardly separable by simple kinetic analysis.[59]

At the unitary level, CgTx-GVIA blocks a channel with a conductance of either 16 (110 mM Ba²⁺) or 14 pS (90 mM Ba²⁺).[58,59,61] This channel is insensitive to DHP, but CgTx-GVIA drastically reduces the number of channel openings. Moreover, bovine chromaffin cells appear to be endowed with two kinetically distinguishable 14-pS channels, having mean open times of 0.5 and 1.5 ms, respectively.[59] This finding led Artalejo and colleagues to hypothesize that the channel exhibiting brief open times would underlie a CgTx-GVIA-insensitive and DHP-insensitive component of the whole-cell current, and to suggest that it would be similar

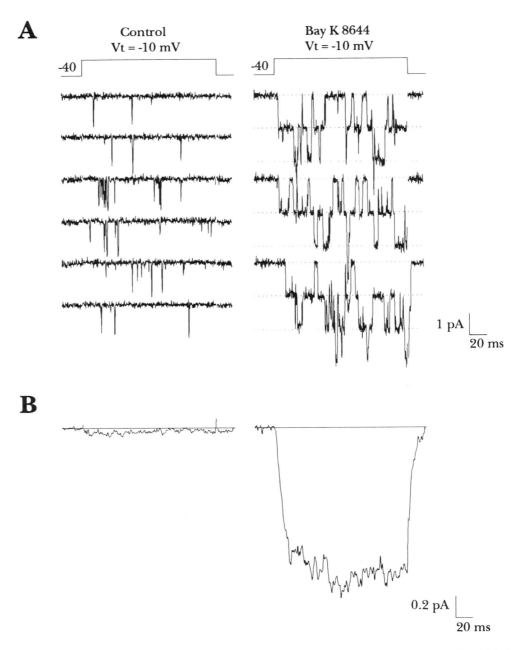

FIGURE 6. BAY K 8644 induces long-lived openings of a 27-pS Ca^{2+} channel in cell-attached patches. (*A*) Left panel, leak subtracted sweeps obtained in drug-free conditions. Right panel, current records obtained after addition of 1 μ*M* BAY K 8644. (*B*) Ensemble average currents are plotted immediately below the current traces. Pipette solution contained (in m*M*): $BaCl_2$, 90; TEACl, 15; CsCl, 15; HEPES-TEAOH, 10; pH 7.3. The bath solution contained (m*M*): potassium aspartate, 140; potassium EGTA, 10; $MgCl_2$, 1; HEPES-KOH, 10; pH 7.4. (From Artalejo, C. R., Mogul, D. J., Perlman, R. L., and Fox, A. P., *J. Physiol.*, 444, 213, 1991. With permission.)

to the P-type channel described in neurons. In fact, more recent experiments have revealed that most of the component of the whole-cell Ca^{2+} channel current that is resistant to CgTx-GVIA and DHP can be effectively blocked by FTX[68] and Aga-IVA.[69,70] Additionally, in many cells, a fraction of the current is preserved after application of maximally effective concentrations of L-type, N-type, and P-type blockers (Figure 5). At present, it is difficult to ascribe a definitive identity to this current component, but new types of Ca^{2+} channels are being defined on the basis of their sensitivity to novel toxins isolated from the marine snail *Conus magus*.[71] One of these toxins, termed ω-conotoxin MVIIC, has been shown to suppress currents mediated by all non-L-type Ca^{2+} channels and also to block a current fraction which is otherwise resistant to CgTx-GVIA and to low doses of Aga-IVA.[72] This current component, linked to the activation of the so-called Q-type Ca^{2+} channel, has been recently observed in bovine chromaffin cells, too.[73] Presumably, this class of Ca^{2+} channel was not previously detected due to the relatively high doses of Aga-IVA employed.[69]

All the electrophysiological studies reviewed above have been conducted in bovine chromaffin cells. However, marked differences have been documented in the sensitivity to Ca^{2+} channel blockers of the secretory response of intact adrenal glands from various species.[74] In particular, the high efficacy of DHP antagonists to inhibit the catecholamine output from cat adrenal glands may reflect a larger proportion of L-type Ca^{2+} channels in feline chromaffin cells than in the bovine ones. A recent study performed on isolated cat chromaffin cells has revealed that L-type, N-type, and P-type Ca^{2+} channels are also present in these cells, but their relative proportion is clearly different to that already described in bovine cells.[75] Thus, in feline cells the predominant subtype is the DHP-sensitive channel, which carries around 50% of the whole-cell Ca^{2+} current, the N-type accounts for about 35%, leaving a fraction of about 15% for the P-type Ca^{2+} channel.

Finally, one more property of chromaffin cell Ca^{2+} channel currents deserves to be mentioned. Double-pulse experiments, in which a test voltage pulse is preceded by a conditioning pulse to positive potentials, showed that Ca^{2+} currents can be potentiated by a prepulse.[44] This observation, henceforth referred to as facilitation, was corroborated by Hoshi et al.,[76] who demonstrated that it is a voltage-dependent phenomenon and not related to Ca^{2+} entry during the prepulse. In addition to large prepulses, short repetitive depolarizations were also found to progressively increase the amplitude of the current, suggesting that Ca^{2+} channel facilitation could enhance the release of catecholamines during a train of action potentials (Figure 7).[54,57] At the single-channel level, voltage- and frequency-dependent facilitation is associated with changes in the gating behavior of Ca^{2+} channels, resulting in the occurrence of long-lasting openings.[58,76]

Several molecular mechanisms have been invoked to explain Ca^{2+} current facilitation, from being solely a voltage-dependent conformational change of the Ca^{2+} channel itself[44,77] to reflect a more complex process requiring the participation of different regulatory proteins and enzymes.[56,63,78-80] Since these aspects are intimately related to the issue of Ca^{2+} channel regulation, they will be reviewed in the following section.

3.4.2.1. *Ca²⁺ Channel Regulation*

3.4.2.1.1. Phosphorylation-Dependent Regulation — Protein phosphorylation is a common mechanism of regulating enzymatic activity. Ionic channel proteins serving as regulators of membrane permeability to ions are also subjected to this mechanism of regulation. In particular, phosphorylation by cAMP-dependent protein kinase (protein kinase A) modulates voltage-dependent Ca^{2+} channels in several excitable tissues.[79] The functional consequences of protein kinase A-mediated phosphorylation have been most extensively studied in cardiac cells, where cAMP analogs and β-adrenergic stimulation of adenylate cyclase were found to increase the frequency by which L-type Ca^{2+} channels undergo long-lived openings.[81] In bovine chromaffin cells, Artalejo et al.[56] discovered that dopamine, acting through D_1 receptors, enhanced voltage-dependent Ca^{2+} currents by activating virtually silent Ca^{2+} channels. The effects were mediated by protein kinase A because they could be mimicked by membrane-permeant analogs of cAMP and prevented by specific protein kinase A inhibitors. Thus, any hormone, neurotransmitter, or drug that raises cAMP including dopamine, (which is co-released with the other catecholamines[82]), adenosine acting on P_1-purinergic receptors,[83] and VIP (which is believed to act as a cotransmitter at the splanchnic-adrenal synapse[84]) may

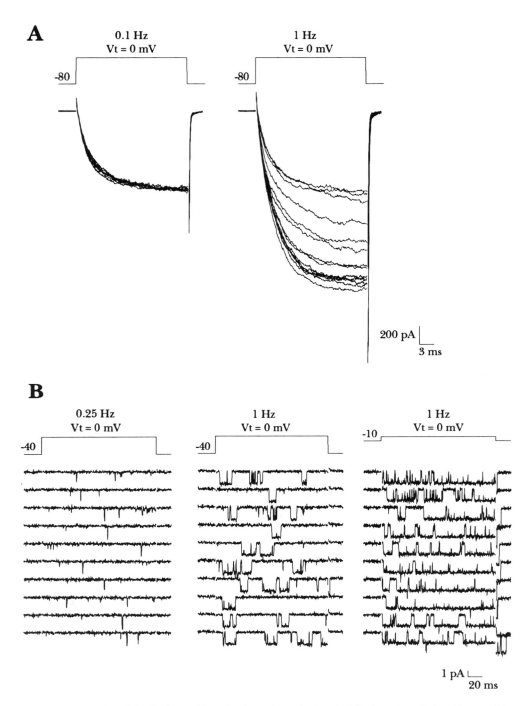

FIGURE 7. Whole-cell facilitation and long-lived openings of a 27-pS Ca^{2+} channel are induced by repetitive depolarizations to physiological potentials. (*A*) 13 Superimposed current traces recorded from a holding potential of -80 mV with a test depolarization to 0 mV, lasting 20 ms, are plotted. Currents shown in left panel were obtained when stimuli were delivered at 0.1 Hz. Right panel shows Ca^{2+} currents recruited by applying stimuli at 1 Hz. (*B*) Unitary current records obtained from a cell-attached patch from $V_h = -40$ mV to $V_t = 0$ mV stimulating the patch at 0.25 Hz (left panel). The stimulation frequency was then changed to 1 Hz (middle panel). With a stimulation frequency of 1 Hz, the holding potential of the patch was then changed to $V_h = -10$ mV (right panel). (From Artalejo, C. R., Mogual, D. J., Perlman, R. L., and Fox, A. P., *J. Physiol.*, 444, 213, 1991. With permission.)

recruit this type of current in chromaffin cells. The effects of protein kinase A were shown to be exerted at the 27-pS channel (the L-type channel), on which it induces the prolongation of the mean open time and increases the probability of opening.[56] At the macroscopic level, these changes also result in an augmented voltage sensitivity of Ca^{2+} channel gating.[85] Interestingly, the single-channel effects of cAMP are reminiscent of Ca^{2+} current facilitation by large predepolarizations; moreover, cAMP effects are not additive with those induced by the predepolarization itself or the treatment with BAY K 8644.[56-58] These results led Artalejo et al.[56] to conclude that those three pharmacological and biophysical manipulations would converge in the activation of a common Ca^{2+} channel, named the "facilitation" channel. More striking is the fact that recruitment of facilitation by large prepulses results from a voltage-dependent, cAMP-independent phosphorylation of the DHP-sensitive channel, as it can be suppressed by protein kinase inhibitors and enhanced in a sustained manner by antagonists of phosphatases, like okadaic acid and ATPγS. Conversely, the injection of phosphatase 2A prevented the recruitment of facilitation.[78] Therefore, it was suggested that depolarization may change the conformation of the facilitation Ca^{2+} channel, making it a better substrate for a kinase, thus producing a final effect similar to that of hormonal stimulation. Noticeably, a protein kinase A-mediated phosphorylation has been recently involved in the voltage-dependent potentiation of cardiac Ca^{2+} channels.[79] Both the voltage- and protein kinase A-dependent pathways of the phosphorylation-mediated facilitatory regulation of L-type Ca^{2+} channel activity would provide the adrenal gland with a positive feedback system to augment catecholamine release during the sympathetic stimulation that occurs in stress situations.

3.4.2.1.2. G Protein-Mediated Inhibition of Ca^{2+} Channels — A purported mechanism for the presynaptic inhibition by hormones and neurotransmitters of synaptic transmission involves G proteins. It is believed that G proteins directly couple the activation of a membrane receptor to a change in the activity of a Ca^{2+} channel serving as effector of the inhibition of transmitter release.[86] Evidence favoring the involvement of G proteins in the modulation of Ca^{2+} channels classically came from experiments in which nonhydrolyzable analogs of guanosine triphosphate (GTP) or the treatment with pertussis toxin (PTX) were shown to modify agonist-induced modulation of Ca^{2+} currents. More recently, intracellular application of G proteins or injection of oligonucleotides against certain G protein subunits have allowed the assignment of G protein subtypes to specific receptors inducing inhibition of Ca^{2+} currents.[87]

Several of the compounds stored in the chromaffin granules have been proposed to modulate catecholamine release. Likewise, Callewaert et al.,[88] found that the Ca^{2+} current measured in a cell located in a cluster of cultured bovine chromaffin cells, but not in a single isolated cell, was markedly suppressed when high K^+ was used to induce secretion, suggesting a possible autocrine/paracrine influence of chromaffin cell secretory products on Ca^{2+} channels. This mechanism for a negative feedback signal on Ca^{2+} currents has been reported for the catecholamines themselves, such that A acts on α_2-adrenoceptors to inhibit voltage-dependent Ca^{2+} channels through the activation of a PTX-sensitive G protein.[62] Additionally, dopamine, which is an intermediate in the synthesis of A and NA, reduces both Ca^{2+} currents and catecholamine release via a D_2-dopaminergic receptor in bovine and feline chromaffin cells.[89-91] Similar to catecholamines, other substances contained in the chromaffin granules, but also coming from the secretory vesicles of the splanchnic nerve terminals, are able to modulate Ca^{2+} channel activity. This is the case of the endogenous opioids, Leu-enkephalin and Met-enkephalin, which inhibit with different efficacies Ca^{2+} currents in chromaffin cells.[62,92] Although this effect is mediated by a PTX-sensitive G protein, the opioid receptor involved in Ca^{2+} channel modulation has not been characterized.[62] γ-Aminobutyric acid (GABA) also participates, though to a less extent, in this type of modulation through a PTX-sensitive G protein coupled to a (−)baclofen-sensitive $GABA_B$ receptor.[93]

Adenine nucleotides have been described to produce diverse electrophysiological effects on chromaffin cells. ATP exerts a dual action on Ca^{2+} currents that could partially account for both its enhancing and decreasing effects on depolarization-induced catecholamine release.[51,83,94] A potentiating effect of ATP on voltage-dependent Ca^{2+} currents was observed only in a subpopulation of chromaffin cells, and was consistently associated with the development of an inward holding current resulting from an increase in resting membrane conductance. The nature of the channel responsible for the increase of the holding current is not clear, but it may be similar to the ATP-gated, nonselective cation channel described in PC12 cells.[95] These two effects of ATP typically desensitize unveiling a sustained inhibition of Ca^{2+} currents which is observed in all cells assayed. These opposing actions on Ca^{2+} currents are mediated by different receptors and biochemical pathways, the inhibitory one involving a PTX-sensitive G protein stimulated via a P_{2y}-purinergic receptor.[51,63,96]

Despite the existence of G_{i1}, G_{i2}, $G_{o1,}$ and G_{o2} PTX-sensitive G proteins in chromaffin cells, at present there is no information available about their selective involvement in receptor-specific modulation of Ca^{2+} currents.[62] Moreover, the Ca^{2+} channel type under the inhibitory influence of extracellular signals has not been thoroughly investigated. For modulators like ATP, opioids, and A, the Ca^{2+} current suppressed exhibited a mixed L/N-type pharmacology, being sensitive to both DHP and CgTx-GVIA.[62,63] Certainly, in the near future more detailed information on this topic will provide important hints about the role of particular types of Ca^{2+} channels in controlling catecholamine secretion from chromaffin cells.

A characteristic feature of G protein-mediated modulation of neuronal Ca^{2+} channels is its voltage dependency. This property is manifested as a progressive reduction of neurotransmitter-induced inhibition of Ca^{2+} currents with an increase in depolarization, such that at large membrane potentials or after application of strong predepolarizations most of the inhibition is relieved.[97,98] This is the case for the ATP modulation of chromaffin cell Ca^{2+} channels, where conditioning pulses to potentials positive to +80 mV partially reverse the inhibitory effects on the Ca^{2+} current.[63]

This kind of observation has led to the proposal that Ca^{2+} current facilitation, as induced by a double-pulse protocol, could arise from the removal of a G protein-dependent tonic inhibition of Ca^{2+} channels, probably imposed by chromaffin cell secretory products that feed back and activate autoreceptors. The fact that local superfusion of chromaffin cells with normal bath solution produces a rapid and reversible increase in Ca^{2+} channel currents and also abolishes prepulse facilitation, is consistent with the idea that an uncontrolled secretion of hormones may have taken place at rest, causing a tonic inhibition of the currents.[80] G protein involvement in such a tonic inhibition has been suggested by experiments which show that maneuvers that inactivate G proteins, like guanosine-5′-*O*-[2-tiodiphosphate] (GDPβS) dialysis or PTX treatment of the cells, resulted in an increase of the size of the Ca^{2+} channel currents.[80] Moreover, both pharmacological manipulations prevented voltage-dependent current facilitation.[63,75,80] These results indicate that, as in neurons, large depolarizations in chromaffin cells may relieve voltage-dependent suppression of HVA Ca^{2+} channels induced by agonist activation of G protein-coupled membrane receptors. Interestingly, in cat chromaffin cells, prepulse facilitation of Ca^{2+} currents is abolished by CgTx-GVIA, suggesting that, in a manner analogous to what occurs in sympathetic neurons, large depolarizations primarily relieve G protein-suppressed N-type Ca^{2+} channels.[75,99]

3.4.2.1.3. Other Modulatory Mechanisms — A nicotinic inhibition of HVA Ca^{2+} channels of bovine chromaffin cells has also been described.[100] This type of modulation requires nicotinic receptor activation either by ACh or other nicotinic agonists, and does not seem to involve G proteins. Furthermore, the nicotinic agonist-mediated decrease in Ca^{2+} currents are neither voltage dependent nor related to intracellular accumulation of Na^+ or Ca^{2+}. Whatever the mechanism of this modulation might be, it would be interesting to determine whether or not a particular type of Ca^{2+} channel is targeted, as that would cause, in effect, a selective recruitment by nicotinic-induced depolarization of the unaffected voltage-dependent Ca^{2+} channels.

Several direct modulatory effects on Ca^{2+} channels have been reported, too. Both catecholamine secretion and Ca^{2+} currents are highly sensitive to changes in the external H^+ concentration ($[H^+]_o$), such that an increase of $[H^+]_o$ to 1 μM reduced the peak Ca^{2+} currents and the secretory response by 70%.[88] Since the pH of the secretory vesicles is about 5.6,[101] these results raise the possibility that the release of H^+ from chromaffin granules, through a local change in $[H^+]_o$, may regulate the activity of Ca^{2+} channels and, hence, the release of catecholamines.

From the standpoint of pathophysiology, it is worth mentioning the blocking effects of IgG antibodies from patients with Lambert-Eaton syndrome on chromaffin cell Ca^{2+} channels.[102] Lambert-Eaton syndrome is an autoimmune disorder characterized by impaired evoked release of ACh from motor nerve terminals. IgG from patients with the syndrome reduces Ca^{2+} currents in chromaffin cells by 40%, without any significant kinetic effect at the macroscopic level or changes in the elementary channel activity. The inhibitory effects of the antibodies on catecholamine secretion from chromaffin cells, and presumably on ACh release in the endplate, are related to a diminution of the number of channels available for activation and not to an intracellular action, because when Ca^{2+} was administered directly into the cytoplasm through the patch pipette, the IgG-treated cells exhibit normal exocytotic response as assayed by membrane capacitance measurements.

Other compounds directly interfering with the Ca^{2+} channel function in chromaffin cells include general anesthetics like halothane and isoflurane,[103,104] polications like ruthenium red,[105] and relatives of the Ca^{2+} antagonist flunarizine like R56865.[106]

3.4.3. K+ Channels
3.4.3.1. *Voltage-Dependent K+ Channels*

In chromaffin cells dialyzed with KCl-based solutions and bathed in normal Ringer medium containing TTX, voltage steps elicit outward currents as shown in Figure 8.[33] The underlying conductance is K^+ selective, since it is not observed when Cs^+ substitutes for K^+ in the intracellular solution, and because currents reverse at 0 mV upon replacing Na^+ by K^+ in the bath. K^+ currents start to activate at test potentials of -40 mV, and do so slowly and with some delay between -40 and -20 mV. At more positive potentials the amplitude of the outward current varies depending on whether or not Ca^{2+} ions are present in the extracellular medium. When Ca^{2+} is present, the outward current is enlarged and displays a "hump" in the intensity-voltage (*I-V*) relationship. This "hump" reaches its maximum at $+10$ mV and then starts to diminish at more positive potentials to approximate at $+50$ mV values similar to those seen in the absence of external Ca^{2+}. In Ca^{2+}-free medium, voltage-dependent K^+ currents exhibit normal rectification at potentials beyond -20 mV, which, together with their slow activation kinetics, identify these currents as conducted by the delayed rectifier K^+ channel.

In isolated outside-out patches, two classes of Ca^{2+}-independent K^+ channels were observed (Figure 9).[33] Under normal ionic conditions (140 mM K^+ inside, 140 mM Na^+ outside), these channels exhibited unitary conductances measured between -20 and $+40$ mV of 18 pS and 8 pS. Both channels displayed burst kinetics and both were inactivated by membrane depolarization. However, activation and inactivation time courses were faster for the 18-pS channel (FK channel in Figure 9). This channel has an open probability which saturates between $+20$ and $+40$ mV and shows a clear inactivation in tens of ms. On the other hand, the 8-pS channel possesses a less steep voltage dependence as compared with the fast transient channel and remains active longer at depolarized voltages (SK channel in Figure 9). Both channels contribute significantly to the whole-cell outward current with low internal Ca^{2+} concentration and no Ca^{2+} outside, being partly responsible for the observed normal rectification.[33]

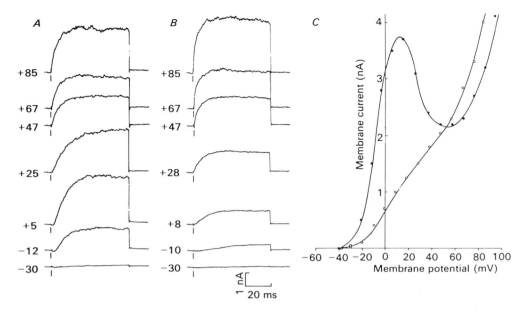

FIGURE 8.   Influence of external Ca^{2+} ions on outward currents. Responses to positive voltage steps to different test potentials, obtained in normal saline (*A*), or after replacement of Ca^{2+} (normal concentration 1 m*M*) with 0.5 m*M* EGTA (*B*). Cell capacitance 7.3 pF. Access resistance R_s = 3 MΩ; effective access resistance after compensation R_e = 1.2 MΩ. Holding potential –80 mV. Traces are corrected for leakage and capacitative current. The numbers next to each trace indicate the potential (in mV) which is reached at the plateau current level, taking into account the voltage drop across R_e. The onsets of the voltage pulses are marked by vertical bars below or above the individual traces. Na^+ currents were blocked with TTX (10 μg/ml). (*C*) *I-V* relationships of the plateau currents in the presence (filled symbols) and in the absence (open symbols) of external Ca^{2+}. The two curves cross over at +52 mV. (From Marty, A. and Neher, E., *J. Physiol.*, 367, 117, 1985. With permission.)

At the resting potential of –60 mV, about 20% of the delayed-rectifying K^+ channels would be inactivated.[34] At the macroscopic level, the inactivation of the delayed rectifier has been described as a slow process that develops in seconds and is enhanced by increasing the frequency or the duration of the depolarizations. Therefore, it has been proposed that inactivation of this repolarizing conductance may contribute to broadening the action potentials and thus, to enhance the entry of Ca^{2+} during the repetitive electrical activity superimposed to prolonged depolarizations.[34]

By modulating K^+ conductances, synaptic or hormonal stimulation may also change the width and height of action potentials, as G proteins can also act as signal transducers between membrane receptors and K^+ channels. It has been recently reported that G protein stimulation, through internal dialysis of guanosine 5'-*O*-[3-thiotriphosphate] (GTPγS) or external application of AlF^{-4}, caused an increase in voltage-gated outward K^+ currents accompanied by a negative shift in the half-maximal voltage required by its activation.[107] The K^+ current augmentation was reversibly inhibited by TEA, but not by the removal of external Ca^{2+}. Addition of fluoride to the incubation medium was found to reduce the catecholamine release stimulated by nicotine, suggesting that by increasing the rate of membrane repolarization G protein-mediated increase of K^+ channel activity may decrease nicotinic-evoked exocytosis in chromaffin cells.

Other K^+ channels modulated by G proteins include the inwardly rectifying K^+ channels in guinea pig adrenomedullary cells. In this animal species, a time- and voltage-dependent (between –80 and –140 mV) inactivating anomalous rectifier K^+ channel has been described.[35] Characteristic of this channel is the voltage-dependent block by Cs+ and Ba^{2+} ions, as well as, a PTX-insensitive G protein-mediated inhibition. This conductance could play a role in

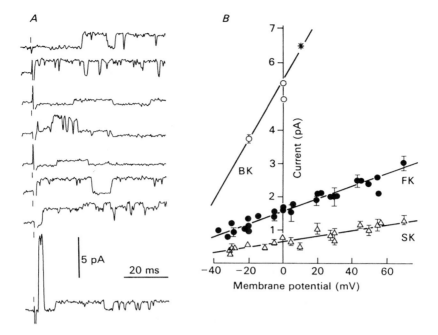

FIGURE 9. Single-channel amplitudes of K⁺ channels. Three types of outward currents can be distinguished by amplitudes. *A* shows representative examples from outside-out patches with a mixed population of channels. The upper seven traces are from a patch, the potential of which was stepped to 0 mV. FK and BK channels alternate or superimpose in consecutive traces. The last trace is from another patch that was jumped to 20 mV. Here one BK and one SK channel are active. Holding potential –80 mV. *B* gives pooled data from five outside-out patches. Error bars, where given, represent standard deviation within readings from one patch. The asterisk is a value taken from Marty.[108] The lines are linear regressions with the following values: SK: *x* intercept –82 mV; slope 8.4 pS. FK: *x* intercept –84 mV; slope 18.4 pS. BK: *x* intercept –60 mV; slope 96 pS (most of the data points lie outside the range of the figure). (From Marty, A. and Neher, E., *J. Physiol.,* 367, 117, 1985. With permission.)

maintaining the resting potential in this particular chromaffin cell, while in rat chromaffin cells a related channel has been proposed to close and depolarize the cell upon muscarinic stimulation.[27]

3.4.3.2. *Ca²⁺ Dependent K⁺ Channels*

As it is shown in Figure 8, external Ca²⁺-dependent K⁺ current is the major component of outward current at command potentials from –40 mV to about +50 mV. The voltage dependency of this Ca²⁺-dependent "hump" of the K⁺ current coincides with the voltage dependency of Ca²⁺ entry through Ca²⁺ channels in chromaffin cells. Thus, the typical "N" shape of K⁺ currents in cells bathed in physiological salines was proposed to result from the intracellular rise of Ca²⁺ and its action on K⁺ channels.[33] The Ca²⁺ sensitivity and conductance characteristics of these K⁺ channels were first studied in isolated membrane patches by Marty.[108] These channels have a typical large conductance, estimated to be of 180 and 275 pS in isotonic K⁺ solutions for bovine and rat chromaffin cells, respectively.[108,109] Under normal ionic conditions (140–160 mM K_i^+ and 140–154 mM Na_o^+) values from 96 to 210 pS have been reported for this channel in bovine chromaffin cells.[33,110] The probability of opening increases drastically above 0.2 μM Ca²⁺ on the cytoplasmic side, but it is not affected by changes in the external Ca²⁺ concentration. For low internal Ca²⁺ concentrations, this large-conductance Ca²⁺-activated K⁺ channel (often called the BK channel) is also voltage dependent. However, the sensitivity to the membrane potential is modulated by Ca²⁺, and at 1 mM inner Ca²⁺ no voltage dependence is observed.[108] The channel is strongly selective for K⁺ over Na⁺ and Cs⁺, but Rb⁺ also carries a significant current.[111] Millimolar concentrations of internal Na^+ reduce the average current through the channel by inducing short interruptions ("flickers") of the single-channel openings.[111,112] Pharmacologically, the BK channel is sensitive to low millimolar concentrations of TEA, particularly when is applied from the outside.[33,111] Similarly, quinine at micromolar concentrations blocks, although with low efficacy, the BK channels in bovine chromaffin cells,[33,110] while nanomolar concentrations of charybdotoxin fully suppress them when applied to outside-out patches.[92] Also, for potential therapeutic considerations, the marked inhibitory effects of several general anesthetics (isoflurane, enflurane, and others) on the BK channels of chromaffin cells should be pointed out.[103,104]

BK currents observed in rat chromaffin cells typically exhibit a pronounced inactivation, a feature which is not readily apparent in bovine cells.[40] Ensemble averages of single BK channels activated by voltage steps to +65 mV from a holding potential of –40 mV inactivate virtually completely within about 300 ms with 2 mM Ca^{2+} in the cytoplasmic side of an inside-out patch.[109] The inactivation can be removed by enzymatic treatment with trypsin and, strikingly, the channels can be blocked by internal application of a 20-amino acid peptide identical to the NH_2 terminal-inactivating domain of the Shaker K^+ channel. This finding indicates that rat chromaffin cell BK channels use a mechanism for inactivation analogous to that used by other voltage-gated K^+ channels, and demonstrates functional and structural homologies between voltage- and Ca^{2+}-dependent K^+ channels and members of the voltage-gated K^+ channel family.[109,113]

This observation adds a new level of complexity to the role that BK channels may play in regulating the electrical behavior of chromaffin cells. Ca^{2+}-dependent activation of BK channels has been observed in cell-attached recordings by exchanging the external solution with an isotonic KCl solution containing 1 mM Ca^{2+}.[33] Likewise, Fenwick et al.[20] showed that BK activation occurred during action potentials. These experiments prove that BK channels are activated by Ca^{2+} entry under conditions where the cells are kept in their normal intracellular medium, and suggest that, in accordance with their Ca^{2+} and voltage dependency, they participate in action potential repolarization. The inactivating properties of BK channels of rat chromaffin cells make it feasible to envisage that a cumulative inactivation of this current during prolonged cellular electrical activity may induce a progressive prolongation of action potential duration and of the associated Ca^{2+} influx. It should finally be mentioned that muscarinic stimulation has been reported to induce the inactivation of the BK current in rat chromaffin cells, thus providing a physiological scenario where this mechanism can operate to regulate the secretory response triggered not only by muscarinic agents, but also by ACh-mediated depolarization.[31]

BK channels are coupled to opioid receptors in bovine chromaffin cells.[92] Three endogenous opioids, Leu-enkephalin, Met-enkephalin, and dynorphin A, strongly potentiate the Ca^{2+}- and voltage-sensitive K^+ current, but they do not modulate Ca^{2+}-independent, voltage-activated K^+ currents. Moreover, the enkephalin potentiation of BK currents is specifically mediated by µ-opioid receptors. Application of opioids to the extracellular face of outside-out patches increases the opening probability of single BK channels, suggesting that coupling between the receptor and the channel does not require other cytoplasmic constituents than possibly the nucleotides usually included in the intracellular solutions. These results point to the fact that opioids exert a dual but complementary inhibitory effect on electrical excitability of chromaffin cells: potentiation of BK currents and inhibition of voltage-activated Ca^{2+} currents.[62,92] Both effects constitute plausible mechanisms for opioid inhibition of catecholamine release from the adrenal medulla, acting perhaps as negative feedback agents or as inhibitory cotransmitters released presynaptically with ACh.[114]

Next to the BK channel, a Ca^{2+}-dependent K^+-selective channel of small conductance, the so-called SK channel, has been described in chromaffin cells.[29,40] Distinctive features of SK channels are a very weak voltage-sensitivity, a unit conductance under physiological ionic

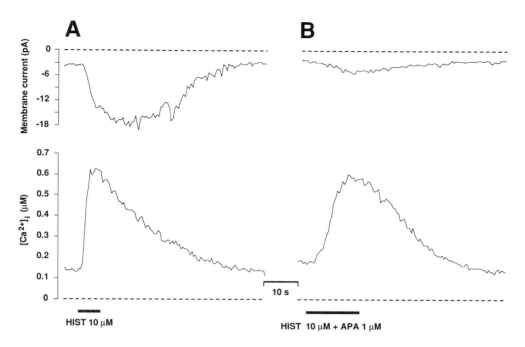

FIGURE 10. Block of Ca^{2+}-activated K^+ currents by apamin (APA). APA (1 μM) almost completely blocks the currents induced by histamine (HIST) (*B*), as compared with those observed in the same cell in the absence of the toxin (*A*). In both *A* and *B*, the upper trace represents the current through the membrane, while the lower trace shows the $[Ca^{2+}]_i$. The drug was added to the bath solution and also applied from a puffer pipette simultaneously with 10 μM HIST (continuous horizontal line). V_h: −78 mV. (From Artalejo, A. R., García, A. G., and Neher, E., *Pflügers Arch.,* 423, 97, 1993. Copyright Springer-Verlag. With permission.)

conditions of 3–5 pS, and block by apamin and (+)tubocurarine. Because of their ability to be activated at the resting membrane potential, these channels are very effective in hyperpolarizing the cell during periods of increased $[Ca^{2+}]_i$. So, the apamin-sensitive component of the Ca^{2+}-activated K^+ currents can be activated by inositol 1,4,5-trisphosphate (IP_3)-producing, Ca^{2+}-mobilizing agents such as muscarine in rat, methacoline in feline, and histamine in bovine adrenomedullary cells (Figure 10).[29,31,115] The immediate consequences of such activation are a transient hyperpolarization and a reduction in action potential firing rate.[29-31] SK channels are also involved in the afterhyperpolarization that follows an action potential in many excitable cells. By imposing hyperpolarizing gaps between consecutive action potentials, SK current is expected to act as a brake in chromaffin cell firing frequency, and thus, any mechanism by which SK channels were inhibited would lead to an increase in the rate of firing of action potentials and, consequently, the entry of Ca^{2+} and the ensuing catecholamine secretion. This hypothesis has been confirmed in cat chromaffin cells where muscarinic stimulation induces depolarization-mediated Ca^{2+} entry and catecholamine release, which were consistently enhanced by apamin.[115]

3.5. THE RESPONSE TO ACETYLCHOLINE
3.5.1. The Nicotinic Component

Upon application of ACh the adrenal chromaffin cells depolarize and fire action potentials (Figure 11).[11,13,14] This depolarization results from the opening of nicotinic ACh receptor (nAChR) channels, since a similar type of response can also be evoked by nicotine, and the effect of both drugs is suppressed by hexamethonium, a nicotinic receptor blocker, but not by the muscarinic antagonist atropine.[116] The nAChR from the adrenal medulla behave, both from the pharmacological and electrophysiological point of view, as neuronal nicotinic receptors.

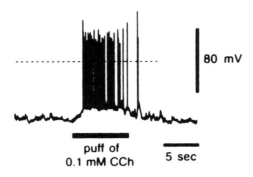

FIGURE 11. Perforated-patch current clamp recording of a rat chromaffin cell stimulated with the cholinergic agonist, carbamylcholine (CCh, 0.1 mM). Upon application of CCh from a puffer pipette, the cell displays a rapid depolarization and discharges a train of action potentials. The horizontal dashed line indicates the zero potential level. (This illustration was kindly provided by Drs. Z. Zhou and S. Misler.)

They possess a single-channel slope conductance in the voltage range from −40 to −120 mV of 40 pS in the cow[20,117-120] and about 30 pS in the rat and guinea pig.[121,122] At low micromolar concentrations (5 μM) the mean open time is characteristically long, of about 30 ms in the three mentioned species.[20,121,122] At the whole-cell level, ACh administration gives rise to inward currents at negative membrane potentials that desensitize in the continuous presence of the agonist. The observed time course of desensitization strongly depends on the speed of drug application, a factor which can possibly explain the different kinetic descriptions of this phenomenon reported in the literature. Thus, in early studies, desensitization was best described as a monoexponential process with a rate constant of 0.1 s^{-1},[117,123] while more recently two exponential components with rates of desensitization of about 10 and 25 s^{-1} has been postulated.[88,119] Interestingly, increasing the concentration of the agonist accelerates the rate of desensitization by changing the proportion of the two kinetic components, such that the faster one becomes more important.[88] At the single-channel level, desensitizing concentrations of ACh (>5 μM) induce the gating of the channel to proceed in bursts and clusters; this behavior has been interpreted as resulting from two distinct desensitized states: a short-lived state manifested by gaps between bursts of openings and a long-lived desensitized state accounting for the long closings between clusters of bursts.[117,123,124] The nicotinic receptor desensitization in chromaffin cells is a process essentially independent of membrane potential, which is not noticeably affected by maneuvers designed to modify the phosphorylation state of the cells.[88,117] Similarly, variations in the external Ca^{2+} from 10 mM to 10 mM did not influence the desensitization time course, hence not favoring an involvement of Ca^{2+} passing through the nAChR channels in inducing desensitization.[117] Recovery from desensitization is consistently described as a single exponential process with a time constant of about 7 s.[88,117,123]

Dose-response relationships performed at negative potentials (−40 to −50 mV) in the whole-cell configuration have provided estimates of the K_D of nAChR for ACh of 135 to 480 mM and of 53 mM for rat and cow chromaffin cells, respectively.[88,125] In both animal species, the Hill coefficient for this drug-receptor interaction was about 1.3, indicating that more than one agonist molecule must bind to the nAChR to open the channel. In fact, recent experiments suggest that the receptor is activated by the binding of either one or two agonist molecules in nonequivalent activation steps.[119] In the whole-cell configuration, nanomolar concentrations of ACh induce single-channel events with two distinct channel lifetimes of 0.6 and 11–15 ms. The relative excess of the very short openings at low agonist concentrations have led to the suggestion that the short lifetimes are openings of singly liganded receptors, while the longer lifetimes correspond to openings of doubly liganded receptors.

The ACh current-voltage relationship typically shows inward rectification, being linear at potentials between −100 and −10 mV, and displaying a region of very low conductance at more positive voltages.[88,120,122,125] In normal saline and CsCl- or KCl-based internal solutions, the reversal potential of the nicotinic current usually lies between −20 and 0 mV,[20,120,122,125] but values as positive as +20 mV have also been reported.[88] The proximity of E_{ACh} to 0 mV is consistent with the notion that the nAChR behaves as a nonselective cation channel. The ionic selectivity of this channel has been deduced from reversal potential measurements of the nAChR current under different isotonic solutions. The permeability values for monovalent cations relative to Na^+ have been established to be 1.32 for Cs^+, 1.03 for Li^+, and 0.18 for $Tris^+$.[120] The Ca^{2+} permeability of the nAChR was first proposed by Douglas et al.[12] when studying the ionic mechanisms of the ACh-induced depolarization. They observed that in Na^+-free saline the ACh response was reduced to about 30% of control, and that the amplitude of the residual response changed with the external Ca^{2+} concentration. Likewise, Kidokoro et al.[116] reported that by adding millimolar amounts of Co^{2+} to a Na^+-free saline, the depolarization induced by ACh was no longer observed. Therefore, it was concluded that the depolarizing effects of ACh are mediated by Na^+ and Ca^{2+} ions. In fact, Ca^{2+} ions experience a finite permeability through the nAChR which apparently increases at reduced Na^+ levels. Under physiological solutions, elevating external Ca^{2+} from 1 to 10 mM caused only a shift in the E_{ACh} of about 1.5 mV in the positive direction, while the change in the reversal potential was as large as 15 mV when reducing ten times the extracellular Na^+ concentration.[120] The estimates of the Ca^{2+} permeability of the nAChR from reversal potential measurements have yielded permeability ratios P_{Ca}/P_{Na} of 2.53 and P_{Ca}/P_{Cs} of 1.53 at E_{ACh}.[120,126] Recently, a method based on the simultaneous measurement of fura-2 fluorescence and Ca^{2+} current have allowed direct estimation (through the fluorescence of the Ca^{2+}-fura-2 complex) of the fractional contribution of Ca^{2+} to the net nAChR current over a wide range of membrane potentials.[127] At −70 mV and 2 mM external Ca^{2+}, it was found that Ca^{2+} contributes about 2.5% of the total current; under constant field assumptions and the Goldman-Hodgkin-Katz equation, this value corresponds to a permeability ratio P_{Ca}/P_{Cs} of 0.53, a number which is smaller than the ratios calculated based on reversal potential measurements. Furthermore, this fractional contribution is voltage dependent, increasing with the hyperpolarization by e-fold for 110 mV potential difference.

Ca^{2+} ions also modulate the activity of the nAChR channel. Removal of external Ca^{2+} causes a dramatic reduction of the size of the ACh-induced current,[126,127] while increasing the concentration potentiates the response to the agonist, reaching a maximum around 2 mM Ca^{2+}_o.[126,127] The biophysical and biochemical mechanisms underlying external Ca^{2+} modulation of nAChR are at present unknown.

Long single-channel open time, high Ca^{2+} permeability, and Ca^{2+} modulation are functional properties that differentiate neuronal-type nAChR channels from end-plate nAChR channels. The nAChR of chromaffin cells is similar to that of neurons also from the pharmacological point of view. Hence, α-bungarotoxin is not capable of suppressing chromaffin cell ACh-induced currents, but they are sensitive to κ-bungarotoxin.[20,128] The nicotinic agonists tetramethylammonium, nicotine, lobeline, and dimethyl-4-phenylpiperazinium activate nAChR in bovine chromaffin cells, whereas the depolarizing neuromuscular blocking drug decamethonium does not.[88,120,126,129] Among the nicotinic antagonists, mecamylamine, trimethaphan, (+)tubocurarine, and hexamethonium have been shown to cause dose-dependent reductions in the amplitude of ACh-evoked inward currents. With the exception of the ganglionic blocker trimethaphan, the antagonism of the other compounds is voltage dependent, increasing with membrane hyperpolarization. (+)Tubocurarine at concentrations higher than those necessary to produce blocking effects is also capable of activating nAChR-dependent macroscopic and elementary inward currents.[120]

It should be noted that all the studies referred to above have been performed on isolated chromaffin cells kept in culture. Therefore, the nAChR described most likely belong to the extrajunctional type. Chromaffin cells were denervated at the time of cell dissociation and their currents assayed after a few hours to several days in culture. During this time new receptors are presumably synthesized and inserted into the membrane; thus, the number and possibly the properties of these receptors may differ from those of synaptic receptors *in situ*. This may help to explain the disparate values estimated for the total number of nAChR on a cell. Based on the affinity of the nAChR for the agonist, the maximum peak current elicited by ACh, and the channel conductance the total number of activatable receptors on a cell has been calculated to be either 100–200 or 2600.[20,119] Certainly, differences in enzymatic treatment during cell dissociation, culture conditions, and time after isolation must account for this discrepancy.

A great variety of substances have been shown to modulate the nAChR channel-mediated currents. Potentially important from the physiological point of view are the effects of several peptides known to be present in chromaffin cells and being coreleased with catecholamines. These peptides include substance P, somatostatin, NPY, and the ANF. All of these substances, while capable of inhibiting nicotinic catecholamine release, do so by suppressing the nicotinic inward currents from bovine chromaffin cells. These effects, perhaps with the exception of NPY, seems to be due not to a conventional ligand-peptide receptor interaction, but to a direct action on nAChR channels. For the case of substance P, it has been suggested that this undecapeptide does not modulate the cholinergic response by interfering with channel activation (the peak current response is not affected) or with ion permeation (there is no change in single-channel amplitude), but by either increasing the rate of desensitization or by inducing a channel blockade which indirectly also leads to desensitization (Figure 12).[117] A similar interpretation has been given for the blocking effects of somatostatin.[130] On the other hand, a noncompetitive voltage-dependent block, much like that of local anesthetic drugs, has been proposed as the mechanism for the inhibitory effect of ANF.[131-133] Finally, NPY inhibition of nicotinic currents could be mediated by the activation of specific NPY receptors possibly coupled to a decrease of adenylate cyclase activity.[134] All of these effects suggest the existence within the adrenal medulla of a complex paracrine regulatory system involving endogenous bioactive peptides which control catecholamine release by affecting the nAChR channel.

Another type of possible paracrine interaction is exemplified by the inhibitory effects of several glucocorticoids (dexametasone, hydrocortisone, and prednisolone) on nAChR channel currents from guinea pig chromaffin cells.[135] These substances, simultaneously applied with ACh, reduce the size of the currents in a non-competitive manner without affecting its time course. Considering both the fast onset of the effect and its reversibility, it seems clear that the target for the glucocorticoids does not reside in its cytoplasmic receptor, but in the plasma membrane, namely, the nAChR-ion channel complex. It is known that secretion of glucocorticoids occurs in response to stress, which makes possible a short-time modulation of the secretory function of chromaffin cells by cortical steroids draining into the medullary vascular system.

nAChR channels are also targets for the nonspecific actions of several drugs. This is the case of the α_2-adrenergic agonist clonidine,[118] the D_2-dopaminergic agonist apomorphine,[91] the barbiturate pentobarbitone,[136] and the calmodulin inhibitor trifluoperazine.[123] Unlike the other drugs, trifluoperazine does not reduce the peak current upon a step application of ACh, but enhances the decay of the current following the initial peak. Generally, it is difficult to conclude that the inhibitory effects of these compounds on the catecholamine secretion are solely related to their actions on nicotinic currents. While this can be true for clonidine, it is well known that barbiturates interact with $GABA_A$ receptors, which are also present and functional in chromaffin cells; similarly, apomorphine via D_2-dopaminergic receptors has been shown to reduce voltage-gated Ca^{2+} currents, an effect which is shared by trifluoperazine. Moreover, this latter substance, in addition to its effects on electrical excitability, can independently abolish exocytosis acting at an intracellular site.[123]

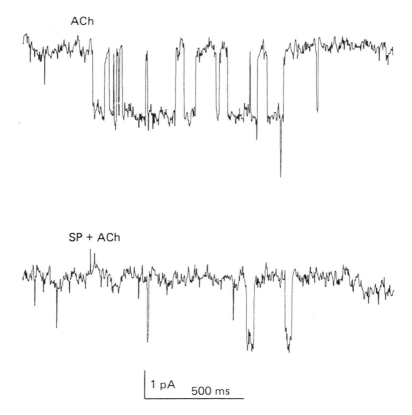

FIGURE 12. Bursting activity of single nAChR channels in 20 μM ACh compared with 20 μM ACh plus 10 μM substance P. Records are from a whole cell voltage clamped to -80 mV holding potential. Note the decrease in open time and openings per burst and increase in the time between openings. (From Clapham, D. E. and Neher, E., *J. Physiol.*, 347, 255, 1984. With permission.)

Finally, and because of its pharmacological importance, it is worth pointing out that Ca^{2+} channel blocking agents like DHP derivatives and methoxyverapamil have been found to block nAChR channels.[129,137] At the concentrations commonly used to block Ca^{2+} channels, these drugs also inhibit nicotinic currents. Furthermore, for the DHP compounds this mechanism of action is observed regardless of whether it is an agonist or antagonist of the Ca^{2+} channel. These results indicate that Ca^{2+} channel blocker-induced reduction of nicotinic catecholamine release can be attributed to the inhibition of nAChR currents rather than of voltage-dependent Ca^{2+} channels. This new mechanism of action may have clinical relevance, as drug-impaired cholinergic transmission, both at the adrenal medulla and the sympathetic ganglia, would contribute to the overall antihypertensive effects of Ca^{2+} antagonist drugs by limiting the vasoconstricting effects of excess circulating catecholamines.

3.5.2. The Muscarinic Component

ACh is the physiological transmitter at the splanchnic-chromaffin cell junction in the adrenal medulla. However, the contributions of nicotinic and muscarinic cholinergic receptors to the stimulated catecholamine release are highly variable among the different animal species. While in the chicken adrenal medulla the muscarinic receptors are protagonists in the ACh-triggered secretory response, bovine adrenomedullary cells secrete little if any upon application of muscarinic agonists. In other species like cat, rat, dog, and guinea pig, both muscarinic and nicotinic receptors mediate robust secretory responses. As could be expected, animal species differences have also been found when the electrical changes induced by muscarinic stimulation have been studied.

In rat chromaffin cells, muscarinic receptor activation produces a complex set of conductance changes. One effect of muscarine application is a transient increase in K^+ conductance.[30,31] Since this change in membrane conductance presumably occurs simultaneously with an IP_3-mediated transient elevation of cytosolic Ca^{2+}, activation of a Ca^{2+}-dependent K^+ conductance has been implicated in the initial steps of the electrical response to muscarinic stimulation. Furthermore, at potentials negative to −40 mV the muscarinic-activated current is largely blocked by apamin or (+)tubocurarine, while at more positive voltages a voltage-dependent TEA-sensitive and apamin-resistant component, most likely the BK channel, is recruited. However, this latter component can also be inactivated following its activation by muscarine.[31] The consequences of these actions on chromaffin cell excitability are quite complicated. A predictable effect of activation of apamin-sensitive SK channels would be membrane hyperpolarization, causing a transient reduction in action potential firing, and indeed such an effect has been observed while recording action currents in the cell-attached configuration.[30,31] On the other hand, the modulatory effects of muscarine on the BK channels are expected to affect the duration of the action potentials. Thus, action potential production appears as an obligatory condition for this type of modulation. Nevertheless, rat chromaffin cells are known to discharge action potentials at rest, the frequency of which is increased by nicotinic stimulation.[14,27,30,31,116]

Studies on freshly isolated rat chromaffin cells have revealed that, contrary to what is observed in cultured cells, muscarine stimulation produces a slow depolarization and, consequently, an increased rate of action potential firing.[27] At a holding voltage close to resting potential, muscarine induces a slow and long-lasting inward current associated with a reduction in membrane conductance. Because this current becomes almost undetectable by bringing the membrane potential to the reversal potential for K^+ channels, it is likely that it reflects a decrease in K^+ conductance. In this context, single-channel recordings have shown that rat chromaffin cells possess K^+ channels that open briefly at the resting potential.[36] In symmetrical K^+ (140 mM), the unitary conductance of these channels has been estimated in 37 pS. The frequency of channel openings decreases by shifting the holding voltage to more depolarized levels, indicative of inward rectification. Muscarine reduces the frequency of channel opening without producing detectable changes in the elementary conductance or the time constants of the open- and close-time histograms. These results rule out that muscarine effects are due to a direct blockade of K^+ channels, and suggests the involvement of a second-messenger system transducing the muscarinic activation. Recent data seem to implicate protein kinase C in this process, since protein kinase C activation by phorbol esters, like muscarinic agonists, caused a persistent depolarization by decreasing a K^+ conductance active at resting potential.[138] Therefore, it might well be that the signaling pathway following muscarinic activation of rat chromaffin cells could involve phospholipase C stimulation producing, on the one hand, IP_3 which, by releasing internal Ca^{2+} would activate Ca^{2+}-dependent K^+ channels to cause a transient inhibition of cell excitability and, on the other hand, diacylglycerol that activates protein kinase C to induce a longer-lasting excitation of chromaffin cells.

In guinea pig chromaffin cells, a different electrical mechanism seems to underlie the muscarinic-induced secretion. Muscarinic agonists induce an inward current at a holding potential of −40 mV.[139] The current-voltage relation has a negative slope below −30 or −20 mV, and the current reverses its polarity at approximately 0 mV in a physiological salt solution. Ion-substitution experiments has confirmed the nonspecific cationic permeability of the channel responsible for this current. As with other cation channels, the Ca^{2+} permeability of the muscarine-sensitive channel becomes electrically detectable in nominally Na^+-free solutions. Dialysis of GTPγS from the patch pipette mimics current activation by muscarinic agonists, while pretreatment with PTX prevents both the muscarinic- and the GTPγS-induced currents. These results suggest that activation of muscarinic receptors in guinea pig chromaffin cells activates a cation conductance through a PTX-sensitive G protein. Moreover, the receptor signal seems to be transmitted to the target channel via a phosphorylation process.[140] Different kinase inhibitors reversibly suppress the muscarinic activation of the nonselective cation current, whereas intracellular perfusion with vanadate, a nonselective inhibitor of protein phosphatases, or the removal of Mg^{2+} from the internal solution gradually leads to the irreversible generation of the current. Taken together, these data indicate that a kinase and a Mg^{2+}-dependent phosphatase are responsible, respectively, for the activation and the inactivation of the nonselective cation channels, and that these channels would be under the influence of both enzymes even in the absence of external signals.[140]

In guinea pig adrenal medulla, perfusion with isotonic K^+ solution abolishes nicotinic, but not muscarinic secretory responses, suggesting that Ca^{2+} influx through the muscarinic cation channel can directly contribute to catecholamine secretion.[141] However, this mechanism would not operate on its own, since the negative slope of the current-voltage relationship makes the muscarinic conductance very small at the resting membrane potential of –60 mV. On the other hand, the simultaneous activation of nicotinic receptors by the ACh released from preganglionic nerves and the ensued depolarization may shift the membrane potential to a level sufficient to bring into play the muscarinic conductance.

Similar to guinea pig chromaffin cells, the muscarinic response in chicken chromaffin cells also involves the slow activation of an inward current which reverses at near zero membrane potential. This finding is therefore consistent with the notion that muscarinic receptors depolarize the chicken cell by opening a nonselective cation channel.[142]

3.6. THE GABA-ERGIC RESPONSE

It is well known that chromaffin cells are capable of synthesizing, storing, and releasing GABA, and that there are GABA-containing nerve terminals within the splanchnic fibers innervating chromaffin cells.[143,144] Additionally, the existence of high-affinity binding sites for GABA agonists in membranes of chromaffin cells has been described.[145] Therefore it is not surprising that GABA modulates the electrical activity and the secretory function of chromaffin cells, and an inhibitory modulation of Ca^{2+} channel currents mediated by $GABA_B$ receptors has already been mentioned. Likewise, GABA, but not baclofen application, evokes transmembrane currents which are selective for Cl^- ions and can be blocked by bicuculline.[26] This indicates that GABA-activated channels are coupled to $GABA_A$ receptors. At GABA concentrations higher than 10 μM, the current typically desensitizes with a time course which is better described by the sum of two exponentials. As in the case for the nAChR, the recovery from the desensitized state of the $GABA_A$ receptor is approximated by a single exponential.[26]

In excised outside-out patches, GABA-activated channels usually display three different conductance levels, a behavior reminiscent of the multistate Cl^- channel activated by glycine in central neurons, though this amino acid does not open channels in chromaffin cells.[26] Pharmacologically, the $GABA_A$ receptor expressed in chromaffin cells appears very similar to the $GABA_A$ receptors located in the central nervous system. Hence, the convulsant agents picrotoxin and RU5135 inhibit the GABA-induced currents, whereas benzodiazepines, barbiturates, some general anesthetics like propanidid, as well as endogenous and synthetic steroids, enhance the amplitude of the responses elicited by GABA.[26,146-149]

GABA is supposed to play a dual role in catecholamine secretion via $GABA_A$ receptors. On a resting chromaffin cell, GABA application increases the rate of firing of action potentials recorded in cell-attached patches and also facilitates the spontaneous catecholamine release.[144,147] These observations indicate that GABA depolarizes the cell, which reflects that $E_{Cl}-$ is positive to the cell resting potential. However, it has also been found that GABA reduces the amplitude of the depolarization induced by ACh and suppresses the accompanying action potential discharge.[147] This effect is not due to a direct interaction of GABA with the nAChR because, under voltage clamp conditions, inward currents elicited by ACh are unaffected by GABA. On the contrary, it probably results from a functional antagonism, explained in terms of GABA shunting the cell input resistance while simultaneously driving the membrane potential towards E_{Cl^-}. More interestingly, this biophysical mechanism may underlie the inhibitory effects of GABA on neurally evoked release of catecholamines from adrenal glands *in situ*.[144]

Some steroids, endogenous to the adrenal cortex, are known to modulate $GABA_A$ receptor activity. The pregnane steroid 5α-pregnan-3α-ol-20-one enhances the size of responses mediated by $GABA_A$ receptor stimulation and, as several other allosteric modulators of this receptor-channel complex, is also able to directly activate the channel.[146,148-151] Since this substance is secreted by the adrenal glands, the possibility exists that the GABA-ergic regulation of catecholamine release is itself under the modulation of the adrenal cortex.

3.7. OTHER ION CHANNELS

3.7.1. Volume-Sensitive Cl⁻ Channel

Bovine chromaffin cells possess a Cl⁻ conductance which becomes activated during increases in cellular volume.[152] Both inflating the cells with pressure applied through the patch pipette and cell swelling during intracellular perfusion with hypertonic solutions cause the slow development of an inwardly directed transient current at a holding potential of –60 mV. The current gradually increases to reach a peak at 60–90 s after applying the pulse of pressure, and subsequently decays to almost its initial value within 5–10 min. The time course of the current goes in parallel with corresponding changes in the membrane conductance, which do not depend on membrane potential. The reversal potential of the current was close to E_{Cl^-} under different extracellular Cl⁻ concentrations; moreover, the inhibition of the volume-sensitive current by established blockers of Cl⁻ channels, like the stilbene derivatives DIDS and SITS, lends additional support to the notion that the underlying channels are primarily permeable to Cl⁻ ions. The single-channel conductance estimated from the ratio of the current variance to the current mean value is about 2 pS.[152]

Intracellular dialysis of GTP notably prolongs the duration of the volume-sensitive current, while the nonhydrolyzable analog of guanosine diphosphate, GDPβS, decreases the maximal rate of increase of the current. Indeed, this current is very similar if not identical to that generated following perfusion of chromaffin cells with GTPγS-containing intracellular solutions.[153] The slow time course and decline characteristic for both Cl⁻ currents favor the idea that G proteins involved in current generation do not act directly on Cl⁻-selective channels, but instead through some second messengers probably produced by G protein-activated enzymes. In this respect, activation of phospholipase A_2 and subsequent formation of lipoxygenase products of arachidonic acid metabolism, were proposed to play a role in activating the Cl⁻ conductance.[154]

All the properties of the volume-sensitive current in chromaffin cells appear similar to those of Cl⁻ currents involved in volume regulation in epithelial or red blood cells. In contrast to those cells, chromaffin cells are not supposed to be osmotically active, nor subjected to anisotonic media under physiological conditions. Therefore, the discovery of volume-sensitive ion channels in chromaffin cells is a somewhat unexpected finding, suggestive of a potential implication of these channels in the cellular surface and, possibly, volume changes occurring during the exocytotic release of catecholamines.

3.7.2. Voltage-Independent Ca²⁺ Channels

Using electrophysiological methods, Mochizuki-Oda and colleagues[155] have provided evidence indicating that PGE_2 activates voltage-insensitive Ca²⁺ channels to produce a sustained elevation of $[Ca^{2+}]_i$ in bovine chromaffin cells. Puff application of PGE_2 results in a prolonged depolarization in about 50% of the cells examined, and a corresponding long-lasting inward current in whole-cell voltage clamp recordings. This current persists in Na⁺-free medium, but is abolished by removal of Ca^{2+}. Single-channel recordings in the cell-attached configuration revealed inward currents reverting at about +40 mV, whose frequency was not affected by membrane potential changes. In inside-out membrane patches, IP_3 added to the cytoplasmic side activated the Ca^{2+} channel currents, but PGE_2 itself was ineffective. These results indicate that PGE_2 activates a Ca^{2+}-selective, voltage-insensitive channel via IP_3 generation.

This channel bears similitude with that activated by PTX treatment of bovine chromaffin cells.[156] Such a channel, termed G-type Ca^{2+} channel, is observed as brief current deflections in cell-attached patches, the frequency of openings being rather insensitive to membrane potential. The channel appears permeable to Ca^{2+} and Ba^{2+} and exhibits a slope conductance of 5 pS with 25 mM Ca^{2+} in the pipette. Since G-type channels can only be detected after PTX treatment, it is assumed that they are tonically inhibited by a G protein under control conditions. The activation of the G-type channel and the ensuing Ca^{2+} entry have been implicated in the potentiating effects of PTX on secretion induced by several secretagogues, as well as in the catecholamine secretion stimulated by PTX alone.[156]

3.7.3. Nonselective Channels

Fenwick et al.[44] described the presence in bovine chromaffin cells of inward currents with elementary amplitude of 0.5–2.0 pA at –60 mV, both with external Na^+ or Ba^{2+}. The current events, observed either in outside-out or whole-cell recordings, had a low probability of occurrence (0.01) which was significantly increased by raising the Ca^{2+} concentration in the pipette from 10^{-8} to 10^{-6} M. These Ca^{2+}-dependent cation channels are not blocked by TTX, and their activity does not depend on membrane potential. The functional significance of this channel is not known.

Mochizuki-Oda et al.[157] have also reported that bath application of arachidonic acid, but not of other fatty acids such as oleic acid or linoleic acid, induces a long-lasting inward current in chromaffin cells clamped at –50 mV. Arachidonic acid activates currents in inside-out patches even in the presence of inhibitors of cycloxygenase and lipoxygenase, suggesting a direct action of this substance on the channels. The channels are permeable to Na^+, Ca^{2+}, and Ba^{2+} and do not modify their opening probability to changes in membrane potential.

4. STIMULUS-SECRETION COUPLING AND THE ELECTRICAL BEHAVIOR OF CHROMAFFIN CELLS

Ca^{2+} ions serve as coupling agents between the stimulus and the exocytotic release of neurotransmitters and hormones. The efficacy of different substances in stimulating secretion from chromaffin cells depends on their ability to elevate $[Ca^{2+}]_i$ at the subplasmalemmal exocytotic sites. Studies in permeabilized bovine chromaffin cells have shown that secretion is triggered by low $[Ca^{2+}]_i$, around 1 μM, and saturates at Ca^{2+} concentrations well below 10 μM.[158-160] However, the low temporal resolution of these measurements limits the study of the kinetics of the secetory response. This aspect can be of relevance, since the rate of secretion rather than the total amount, as obtained in the permeabilization experiments, is a better indicator of the actual affinity of the secretory apparatus for Ca^{2+}. Using whole-cell patch clamp to measure membrane capacitance and to intracellularly deliver both Ca^{2+} and Ca^{2+}-sensitive indicator dyes has permitted the simultaneous measurement of the kinetics of secretion and $[Ca^{2+}]_i$ in single cells.[45] The results of such experiments indicate that the rate of secretion does depend upon $[Ca^{2+}]_i$, and accelerates more than three orders of magnitude as $[Ca^{2+}]_i$ at the release sites is elevated. Ca^{2+}-dependent secretion appears to require $[Ca^{2+}]_i$ levels above 0.2 μM, saturating at concentrations somewhere above 10 μM. Depolarization-induced rises in $[Ca^{2+}]_i$ also increase the rate of secretion. However, comparison of the rates achieved during depolarization (up to 500 vesicles/s) to those measured during Ca^{2+} dialysis (25 vesicles/s) as a function of spatially averaged measurements of $[Ca^{2+}]_i$ rendered much higher apparent Ca^{2+} sensitivities for the secretory process triggered by Ca^{2+} entry through voltage-dependent Ca^{2+} channels.[45] This result can be best explained by considering both the formation of pronounced spatial gradients of $[Ca^{2+}]_i$ during the depolarization,[161-163] giving rise to high Ca^{2+} levels at the exocytotic sites beneath the plasma membrane, and the existence of a small pool of release-ready vesicles that can be exocytosed at high rates.[161,164] Upon sustained stimulation, as is the case of intracellular dialysis of Ca^{2+} or prolonged depolarizations, depletion of the fast releasable pool of granules will occur, and the supply of new vesicles into such a dynamic pool becomes the rate-limiting step of the secretory process.[45,164] A consequence of this interpretation is that it should be possible to unravel different kinetic components in the capacitance increase corresponding to the process of initial depletion of the release-ready pool and the subsequent vesicle mobilization, respectively.

Recently, Neher and Zucker,[165] employing the technique of flash photolysis of caged Ca^{2+} using DM-nitrophen, have demonstrated the existence of multiple kinetic components in the secretory response. Step-like elevations of $[Ca^{2+}]_i$ above 10 μM triggered a fast component of capacitance rise with half-times of 1–2 s, followed by a slow exocytotic response. The fast rise may represent the release of granules already docked at the plasma membrane, whereas the slower one the transport of vesicles recruited from deep within the cell and their subsequent fusion. Notably, both secretion and movement of vesicles between different cellular compartments appear to be $[Ca^{2+}]_i$ sensitive. This Ca^{2+} dependency allows the size of the release-ready pool of vesicles to be regulated in a way that prolonged, but moderated increases of $[Ca^{2+}]_i$ leads to an increase in the size of the pool, and hence to an augmented secretory response to subsequent stimuli. However, strong stimuli can deplete this quickly releasable pool, leading to a depressed catecholamine response to further stimulation (Figure 13).[166]

It is now pertinent to review the mechanisms of $[Ca^{2+}]_i$ increase and their relation to the electrical activity of the cells. In chromaffin cells, receptor activation results in an elevation of $[Ca^{2+}]_i$ either by Ca^{2+} entry or by mobilization from internal stores. Ca^{2+} entry into the cells may be mediated by membrane depolarization and opening of voltage-gated Ca^{2+} channels, or it may take place via receptor-operated channels. Receptor-operated channels can be activated by the agonist directly, as is the case for the nAChR-channel complex, or indirectly by a second messenger as occurs with the PGE_2-generated, IP_3-activated Ca^{2+} channel. Ca^{2+} release from internal stores is capable of promoting catecholamine release from chromaffin cells,[45,167] and will also influence the electrical behavior of the cells by activating Ca^{2+}-dependent K^+ channels, particularly those of small conductance.[29,40]

ACh, the physiological stimulant of chromaffin cells, acting on nicotinic receptors produces a depolarizing current, which is carried in part by Ca^{2+}. This depolarization will then activate voltage-dependent Ca^{2+} channels, resulting in more Ca^{2+} influx. Moreover, ACh also binds to muscarinic receptors, thus activating phospholipase C and inducing IP_3 formation (and Ca^{2+} mobilization from cellular stores) as well as protein kinase C activation. In addition, muscarinic stimulation has been shown to activate, by a variety of voltage-sensitive and -resistant mechanisms, the entry of Ca^{2+} into the cells. Therefore, ACh stimulation encompasses most of the possibilities whereby any agonist can induce $[Ca^{2+}]_i$ elevation in a chromaffin cell. In the following sections I will outline the contribution of the different Ca^{2+} entry pathways activated by ACh to the secretion of catecholamines.

4.1. CA^{2+} ENTRY THROUGH VOLTAGE-DEPENDENT CA^{2+} CHANNELS

Activation of voltage-dependent Ca^{2+} channels is the most efficient way to stimulate Ca^{2+} influx and trigger secretion in chromaffin cells. Experimentally, these channels can be activated either electrically in a voltage-clamped cell, or by excess external K^+ in intact cells. Additionally, ACh-induced depolarization alone or with the assistance of voltage-gated Na^+ channels can also recruit HVA Ca^{2+} channels. The Na^+ component of the action potentials induced by ACh and, therefore, the TTX sensitivity of the stimulated catecholamine release,

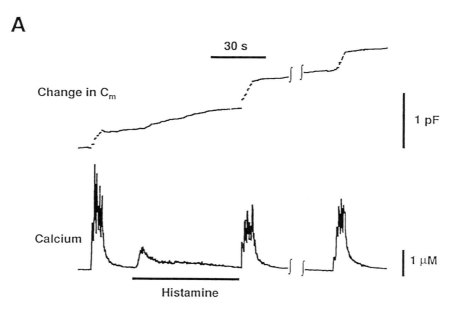

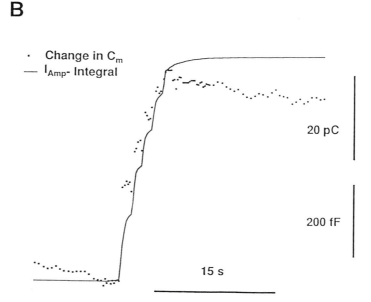

FIGURE 13.   Augmentation (*A*) and depression (*B*) of secretory responses. (*A*) Cytosolic calcium and change in membrane capacitance (C_m) for a time segment in which three trains of five depolarizing pulses (200 ms duration, 1 Hz) each were applied to a bovine chromaffin cell. Between the first and the second pulse trains, histamine (20 μM) was applied to the cell for 50 to 60 s as indicated by the bar. The interval between the third and fourth trains (not fully shown) had the same length (approximately 90 s) as the previous one. Notice that histamine causes the release of calcium followed by a long-lasting plateau of increased $[Ca^{2+}]_i$. The total C_m response to depolarizing pulses after the histamine application was larger than before, although the calcium signal was smaller and appreciable secretion occurred during histamine application. (*B*) The integral of an amperometric current (solid trace) compared with the simultaneously recorded capacitance increase (dotted trace) during a train of five depolarizing pulses. For both types of secretion assay, depression of the response was observed, being larger for the first or second pulse and smaller for later pulses.(From Rüden, L. V. and Neher, E., *Science*, 262, 1061, 1993. Copyright 1993 by the AAAS. With permission.)

will depend primarily on the nAChR density of the cells and on the concentration of the agonist used. Only for low concentrations of ACh or low densities of nAChR, like those existing in freshly isolated chromaffin cells, the Na^+ inward current seems to play a role in depolarizing the membrane to the threshold for activation of Ca^{2+} channels.[15] However, when large doses of ACh are used, direct nicotinic-mediated depolarization is sufficiently great to activate the Ca^{2+} channels directly, and also to induce the inactivation of voltage-gated Na^+ channels.[27]

Since, based on pharmacological grounds, there are at least three different types of HVA Ca^{2+} channels in chromaffin cells, the question arises as to what is their relative contribution to catecholamine secretion. Several reports have stressed the DHP sensitivity of the secretory response to high K^+ depolarization of bovine chromaffin cells.[52,74,168-170,172] On the other hand, CgTx-GVIA has little or no effect,[168-172] whereas FTX effectively blocks the K^+-evoked catecholamine release in bovine chromaffin cells.[169] The relative contribution of these three pharmacologically distinguishable components of the Ca^{2+} current to the chromaffin cell function has been addressed in a recent study in which whole-cell Ca^{2+} currents and membrane capacitance were measured.[70] It was concluded that although all three types of Ca^{2+} channels trigger secretion individually, the DHP-sensitive ("facilitation") Ca^{2+} channels produce much greater secretion for a given size of the Ca^{2+} current, indicating that they are coupled more efficiently to exocytosis. Recent availability of ω-conotoxin MVIIC has led to the recognition of another pharmacologically distinct component of the Ca^{2+} current in bovine chromaffin cells: the so-called Q-type channel. This toxin, together with blocking a fraction of the Ca^{2+} currents and of the high K^+-evoked rise in $[Ca^{2+}]_i$, also effectively inhibits catecholamine release induced by chemical depolarization.[73] Thus, it appears that both L-type and Q-type Ca^{2+} channels control catecholamine secretion from bovine chromaffin cells.

Evidence that neurosecretion is controlled by a given subtype of Ca^{2+} channel is also found in cat chromaffin cells.[173] Kinetic and pharmacological analyses of whole-cell Ca^{2+} currents have revealed the presence, in these cells, of mostly L-type and N-type Ca^{2+} channels.[75] The increase of $[Ca^{2+}]_i$ induced by 10-s pulses of 70 mM K^+ applied to fura-2-loaded cells was reduced by a DHP antagonist by 44%; CgTx-GVIA reduced it by 42%. When secretion was electrochemically monitored from intact adrenal glands or isolated chromaffin cells, the DHP blocked by over 95% the secretory response to the above mentioned stimulus, while CgTx-GVIA reduced secretion only by about 20%. It seems that although Ca^{2+} entry through both Ca^{2+} channel subtypes leads to similar increments on the averaged $[Ca^{2+}]_i$, Ca^{2+} entering through L-type Ca^{2+} channels is more relevant to induce secretion. Nevertheless, it should be noted that strong K^+ depolarizations may not reflect the conditions existing *in vivo*, where small synaptic potentials or the strong, but extremely short depolarizations of action potentials dominate the electrical activity of the cell upon ACh stimulation.[23,42] Certainly, experiments reproducing such conditions should be performed to fully test the hypothesis of a differential regulation of the secretory function by the different Ca^{2+} channel types.

At present, there is not a clear explanation for the functional specialization of L-type and Q-type Ca^{2+} channels in controlling catecholamine secretion. Similarly, it is not obvious which kind of signal is generated by Ca^{2+} penetrating the cell through the other channel types. By using flame-etched carbon fiber microelectrodes with 2-μm-diameter tips for the electro-chemical detection of catecholamine release from single bovine chromaffin cells, Schroeder et al.[174] have demonstrated the existence of localized hot spots where exocytosis occurs, alternating with other silent spots on the plasmalemma where secretion does not take place. Pulsed-laser Ca^{2+} imaging in chromaffin cells has also revealed focal domains of submembrane $[Ca^{2+}]_i$ elevation upon depolarization, suggestive of Ca^{2+} channel clustering in localized areas of the plasma membrane. Although in chromaffin cells there is no clear morphological evidence of active zones in the cell membrane,[3] and the existence of a long secretory delay between the depolarizing stimulus and the electrochemically detected secretory events does

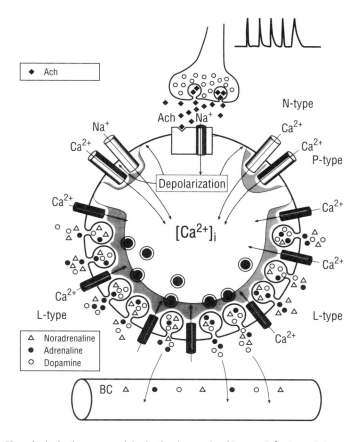

FIGURE 14. Hypothetical scheme to explain the dominant role of L-type Ca²⁺ channels in controlling depolarization-evoked secretion from chromaffin cells. L-type Ca²⁺ channels forming clusters in the membrane would allow Ca²⁺ ions passing across them to build up overlapping regions of elevated [Ca²⁺]ᵢ (darker shadowed areas) which trigger secretion more efficiently than those generated by N- and P-type Ca²⁺ channels. It is also speculated that L-type Ca²⁺ channels would be preferentially segregated to membrane areas of the secretory pole of the chromaffin cell facing the blood capillaries.

not suggest a strict colocalization of any type of Ca²⁺ channel with the secretory vesicles,[19] Ca²⁺ ions flowing through clustered Ca²⁺ channels may summate to build up regions of localized higher [Ca²⁺]ᵢ elevations, which would more efficiently trigger secretion. Thus, it seems reasonable to speculate that L-type and possibly Q-type Ca²⁺ channels would be preferentially segregated to specialized membrane areas, where Ca²⁺ channels cluster to regulate the release of catecholamines (Figure 14).

4.2. Ca²⁺ ENTRY THROUGH nAChR CHANNELS

Early electrophysiological studies showed that ACh causes a slight depolarization even in the absence of extracellular Na⁺, suggesting that Ca²⁺ ions can carry an inward depolarizing current through nAChR channels.[12,14,116] Ca²⁺ flux through these non-selective cation channels has been corroborated by applying reversal potential measurements of the nicotinic current[120,126] and with quantitative fluorescence microscopy using fura-2.[127] This last approach has led to the conclusion that under certain experimental conditions Ca²⁺ influx through the nicotinic channel can cause substantial rises in [Ca²⁺]ᵢ. For instance, for a nicotinic current of 1 nA, which is easily achieved with 1-s external application of ACh onto a typical chromaffin cell held at a −70 mV, the increase in [Ca²⁺]ᵢ due to Ca²⁺ influx through the nAChR can be about 2 μM.[127] This calculation assumes experimental conditions where a large number of nicotinic channels are open, and the membrane potential is maintained at hyperpolarized levels. Under these premises, secretion elicited by Ca^{2+} entering voltage-clamped cells through the nAChR has been observed.[176] When extrapolating these data to the intact gland, one should realize that the relative contribution of this Ca^{2+} entry pathway vs. voltage-gated Ca^{2+} channels activated by nicotinic depolarization, will depend on the density and degree of activation of nicotinic channels. In the intact gland, the entry of Ca^{2+} through nAChR channels is limited to small subsynaptic membrane areas where the receptors are clustered. Likewise, the degree of presynaptic stimulation will determine the number of nicotinic channels open and, consequently, the total amount of Ca^{2+} ions passing across them. By using a voltage-sensitive dye in a slice preparation of the rat adrenal medulla that was stimulated transynaptically, Iijima et al.[42] found that the duration of the synaptic potentials was about 60 ms, which is much longer than the synaptic potentials of a few ms in mammalian skeletal muscle. Although each presynaptic spike appeared to generate a spike in the postsynaptic chromaffin cell, the refractory period for the action potential discharge in chromaffin cells was longer compared to that of presynaptic splanchnic nerve. On the contrary, desensitization of the nicotinic postsynaptic potentials was not observed even with a stimulus interval of 10 ms, while the majority of cells failed to fire action potentials. These results, which have been confirmed in the bissected adrenal gland from the guinea pig by recording membrane potentials with conventional intracellular microelectrodes,[23] suggest that, under physiological conditions, and particularly at high levels of stimulation, Ca^{2+} influx through the nicotinic receptor can contribute significantly to secretion. This idea is supported by experiments conducted on perfused cat adrenal glands, where nicotinic agonists were still capable of inducing secretion in the presence of millimolar concentrations of Co^{2+}, which fully suppressed high K^+-evoked catecholamine release in this same preparation. [177]

4.3. Ca^{2+} ENTRY THROUGH RECEPTOR-OPERATED Ca^{2+} CHANNELS

Mochizuki-Oda et al.[155] have demonstrated that IP_3 directly opens a channel that conducts Ca^{2+} into bovine chromaffin cells. Although this study explicitly refers to the mechanism by which PGE_2 promotes external Ca^{2+}-dependent increase in $[Ca^{2+}]_i$, those results can be tentatively extrapolated to many other receptor agonists known to stimulate phospholipase C and the subsequent generation of IP_3 in chromaffin cells. Obviously, the potency of the different agonists in generating IP_3, and the time course of its generation, will presumably determine the extent and duration of this Ca^{2+} entry mechanism.

Nonselective cationic channels activated by muscarine, and possibly by other agonists including arachidonic acid, are also means of introducing Ca^{2+} into the cells following plasma membrane-receptor activation. The physiological importance of these Ca^{2+} entry pathways is not fully understood. Hormone-induced sustained activation of Ca^{2+} entry is expected to produce a plateau of increased $[Ca^{2+}]_i$, which has commonly been assumed to subserve the refilling of intracellular Ca^{2+} stores previously emptied by IP_3-mediated Ca^{2+} mobilization. However, several IP_3-generating hormones, like histamine or angiotensin II, are capable of triggering secretion that depends upon external Ca^{2+}.[178,179] Therefore, it seems reasonable to suggest that receptor-operated cation channels could allow Ca^{2+} entry responsible for both the agonist alone-induced secretion and the usually observed potentiating effects of this substance on the catecholamine release stimulated by more potent secretagogues.[180,181] Additionally, activation of these channels would act as a depolarizing drive to further recruit voltage-dependent Ca^{2+} entry and hence to facilitate the secretory response. Last, but not least, by causing a sustained $[Ca^{2+}]_i$ increase, receptor-operated Ca^{2+} influx would also serve for the Ca^{2+}-dependent replenishment, and eventually overfilling, of the release-ready pool of secretory granules. This action would be manifested as a conserved or enhanced secretion of catecholamines during repetitive stimulation of chromaffin cells.[164,166]

ACKNOWLEDGMENTS

I am much indebted to Professor Erwin Neher, without whose encouragement and support this work would not have been possible. I am particularly grateful to Drs. Erwin Neher, Anant Parekh, Robert Chow, Kevin Gillis, Jürgen Klingauf, and Stanley Misler for much help in the preparation of the manuscript and for helpful discussions. I also thank Drs. Cristina Artalejo and Antonio García for providing some unpublished material, and Drs. Geoffrey Fox, Zhuan Zhou, and Stanley Misler for supplying some of the illustrations.

REFERENCES

1. Hillarp, N.-Å. and Hökfelt, B., Evidence of adrenaline and noradrenaline in separate adrenal medullary cells, *Acta Physiol. Scand.*, 30, 55, 1953.
2. Coupland, R. E., Determining sizes and distributions of sizes of spherical bodies such as chromaffin granules in tissue sections, *Nature*, 217, 384, 1968.
3. Carmichael, S. W., Morphology and innervation of the adrenal medulla, in *Stimulus-Secretion Coupling*, Vol. 1, Rosenheck, K. and Lelkes, P., Eds., CRC Press, Boca Raton, Fl, 1986, 2.
4. Grynszpan-Wynograd, O., Ultrastructure of the chromaffin cell, in *Handbook of Physiology*, Sect. 7, Vol. 6, *Adrenal Gland*, Blaschko, H., Sayers, G., and Smith, A. D., Eds., Waverly Press, Baltimore, 1975, 295.
5. Schmidt, W., Patzak, A., Lingg, G., Winkler, H., and Plattner, H., Membrane events in adrenal chromaffin cells during exocytosis: a freeze-etching analysis after rapid cryofixation, *Eur. J. Cell Biol.*, 32, 31, 1983.
6. Brooks, J. C. and Carmichael, S. W., Ultrastructural demonstration of exocytosis in intact and permeabilized cultured bovine chromaffin cells, *Am. J. Anat.*, 178, 85, 1987.
7. Winkler, H., The composition of adrenal chromaffin granule: an assessment of controversial results, *Neuroscience*, 1, 65, 1976.
8. Annaert, W. G., Backer, A. C., Jacob, W. A., and De Potter, W. P., Catecholamines are present in a synaptic-like microvesicle-enriched fraction from bovine adrenal medulla, *J. Neurochem.*, 60, 1746, 1993.
9. Thomas-Reetz, A. C. and De Camilli, P., A role for synaptic vesicles in non-neuronal cells: clues from pancreatic β cells and from chromaffin cells, *FASEB J.*, 8, 209, 1994.
10. Douglas, W. W. and Rubin, R. P., The role of calcium in the secretory response of the adrenal medulla to acetylcholine, *J. Physiol.*, 159, 40, 1961.
11. Douglas, W. W., Kanno, T., and Sampson, S. R., Effects of acetylcholine and other medullary secretagogues and antagonists on the membrane potential of adrenal chromaffin cells: an analysis employing techniques of tissue culture, *J. Physiol.*, 188, 107, 1967.
12. Douglas, W. W., Kanno, T., and Sampson, S. R., Influence of the ionic environment on the membrane potential of adrenal chromaffin cells and on the depolarizing effect of acetylcholine, *J. Physiol.*, 191, 107, 1967.
13. Biales, B., Dichter, M. A., and Tischler, A., Electrical excitability of cultured adrenal chromaffin cells, *J. Physiol.*, 262, 743, 1976.
14. Brandt, B. L., Hagiwara, S., Kidokoro, Y., and Miyazaki, S., Action potentials in the rat chromaffin cells and effects of acetylcholine, *J. Physiol.*, 263, 417, 1976.
15. Kidokoro, Y. and Ritchie, A. K., Chromaffin cell action potentials and their possible role in adrenaline secretion from the rat adrenal medulla, *J. Physiol.*, 307, 199, 1980.
16. Hamill, O. P., Marty, A., Neher, E., Sakmann, B., and Sigworth, F., Improved patch-clamp techniques for high-resolution current recording from cells, and cell-free membrane patches, *Pflügers Arch.*, 391, 85, 1981.
17. Neher, E. and Marty, A., Discrete changes of cell membrane capacitance observed under conditions of enhanced secretion in bovine adrenal chromaffin cells, *Proc. Natl. Acad. Sci. U.S.A.*, 79, 6712, 1982.
18. Wightman, R. M., Jankowski, J. A., Kennedy, R. T., Kawagoe, K. T., Schroeder, T. J., Leszczyszyn, D. J., Near, J. A., Diliberto, E. J., and Viveros, O. H., Temporally resolved catecholamine spikes correspond to single vesicle release from individual chromaffin cells, *Proc. Natl. Acad. Sci. USA*, 88, 10754, 1991.
19. Chow, R. H., Rüden, L. v., and Neher, E., Delay in vesicle fusion revealed by electrochemical monitoring of single secretory events in adrenal chromaffin cells, *Nature*, 356, 60, 1992.
20. Fenwick, E. M., Marty, A., and Neher, E., A patch-clamp study of bovine chromaffin cells and of their sensitivity to acetylcholine, *J. Physiol.*, 331, 577, 1982.
21. Ishikawa, K. and Kanno, T., Influences of extracellular calcium and potassium concentrations on adrenaline release and membrane potential in the perfused adrenal medulla of the rat, *Jpn. J. Physiol.*, 28, 275, 1978.

22. Nassar-Gentina, V., Pollard, H. B., and Rojas, E., Electrical activity of intact mouse adrenal gland, *Am. J. Physiol.*, 254, C675, 1988.

23. Holman, M. E., Coleman, H. A., Tonta, M. A., and Parkington, H. C., Synaptic transmission from splanchnic nerves to the adrenal medulla of guinea-pigs, *J. Physiol.*, 478, 115, 1994.

24. Grynszpan-Wynograd, O. and Nicolas, G., Intercellular junctions in the adrenal medulla: a comparative freeze-fracture study, *Tissue Cell*, 12, 661, 1980.

25. Edwards, F. A., Konnerth, A., Sakmann, B., and Takahashi, T., A thin slice preparation for patch clamp recordings from neurones of the mammalian central nervous system, *Pflügers Arch.*, 414, 600, 1989.

26. Bormann, J. and Clapham, D., γ-Aminobutyric acid receptor channels in adrenal chromaffin cells: a patch clamp study, *Proc. Natl. Acad. Sci. U.S.A.*, 82, 2168, 1985.

27. Akaike, A., Mine, Y., Sasa, M., and Takaori, S., Voltage and current clamp studies of muscarinic and nicotinic excitation of the rat adrenal chromaffin cells, *J. Pharmacol. Exp. Ther.*, 255, 333, 1990.

28. González-García, C., Ceña, V., Keiser, H., and Rojas, E., Catecholamine secretion induced by tetraethylammonium from cultured bovine adrenal chromaffin cells, *Biochim. Biophys. Acta*, 1177, 99, 1993.

29. Artalejo, A. R., García, A. G., and Neher, E., Small-conductance Ca^{2+}-activated K^+ channels in bovine chromaffin cells, *Pflügers Arch.*, 423, 97, 1993.

30. Kubo, Y. and Kidokoro, Y., Potassium currents induced by muscarinic receptor activation in the rat adrenal chromaffin cell, *Biomed. Res.*, 10, 71, 1989.

31. Neely, A. and Lingle, C. J., Effects of muscarine on single rat adrenal chromaffin cells, *J. Physiol.*, 453, 133, 1992.

32. Rüden, L. v., García, A. G., and López, M. G., The mechanism of Ba^{2+}-induced exocytosis from single chromaffin cells, *FEBS lett.*, 336, 48, 1993.

33. Marty, A. and Neher, E., Potassium channels in cultured bovine adrenal chromaffin cells, *J. Physiol.*, 367, 117, 1985.

34. Sala, S. and Soria, B., Inactivation of delayed potassium current in cultured bovine chromaffin cells, *Eur. J. Neurosci.*, 3, 462, 1991.

35. Inoue, M. and Imanaga, I., G protein-mediated inhibition of inwardly rectifying K^+ channels in guinea pig chromaffin cells, Am. *J. Physiol.*, 265, C946, 1993.

36. Akaike, A., Mine, Y., Sasa, M., and Takaori, S., A patch clamp study of muscarinic excitation of the rat adrenal chromaffin cells, *J. Pharmacol. Exp. Ther.*, 255, 340, 1990.

37. Adams, P. R., Brown, D. A., and Constanti, A., M-Currents and other potassium currents in bullfrog sympathetic neurones, *J. Physiol.*, 330, 537, 1982.

38. Villarroel, A., Marrion, N. V., López, H., and Adams, P. R., Bradykinin inhibits a potassium M-like current in rat pheochromocytoma PC12 cells, *FEBS Lett.*, 255, 42, 1989.

39. Horn, R. and Marty, A., Muscarinic activation of ionic currents measured by a novel whole-cell recording method, *J. Gen. Physiol.*, 92, 145, 1988.

40. Neely, A. and Lingle, C. J., Two components of calcium-activated potassium current in rat adrenal chromaffin cells, *J. Physiol.*, 453, 97, 1992.

41. Dreyer, G. P., On secretory nerves to the suprarenal capsules, *Am. J. Physiol.*, 2, 203, 1899.

42. Iijima, T., Matsumoto, G., and Kidokoro, Y., Synaptic activation of rat adrenal medulla examined with a large photodiode array in combination with a voltage-sensitive dye, *Neuroscience*, 51, 211, 1992.

43. Hodgkin, A. L. and Huxley, A. F., A quantitative description of membrane current and its application to conduction and excitation in nerve, *J. Physiol.*, 177, 500, 1952.

44. Fenwick, E. M., Marty, A., and Neher, E., Sodium and calcium channels in bovine chromaffin cells, *J. Physiol.*, 331, 599, 1982.

45. Augustine, G. J. and Neher, E., Calcium requirements for secretion in bovine chromaffin cells, *J. Physiol.*, 450, 247, 1992.

46. Brehm, P. and Eckert, R., Calcium entry leads to inactivation of calcium channels in *Paramecium*, *Science*, 202, 1203, 1978.

47. Tillotson, D., Inactivation of Ca conductance dependent on entry of Ca ions in molluscan neurons, *Proc. Natl. Acad. Sci. U.S.A.*, 76, 1497, 1979.

48. Carbone, E. and Lux, H. D., A low voltage-activated, fully inactivating Ca channel in vertebrate sensory neurones, *Nature*, 310, 501, 1984.

49. Nowycky, M. C., Fox, A. P., and Tsien, R. W., Three types of calcium channel with different calcium agonist sensitivity, *Nature*, 316, 440, 1985.

50. Llinás, R., Sugimori, M., and Cherksey, B., Voltage-dependent calcium conductances in mammalian neurons: the P channel, *Ann. N.Y. Acad. Sci.*, 560, 103, 1989.

51. Diverse-Pierluissi, M., Dunlap, K., and Westhead, E., Multiples actions of extracellular ATP on calcium currents in cultured bovine chromaffin cells, *Proc. Natl. Acad. Sci. U.S.A.*, 88, 1261, 1991.

52. Ceña, V., Nicolas, G. P., Sánchez-García, P., Kirpekar, S. M., and García, A. G., Pharmacological dissection of receptor-associated and voltage-sensitive ionic channels involved in catecholamine release, *Neuroscience*, 10, 1455, 1983.

53. García, A. G., Sala, F., Reig, J. A., Viniegra, S., Frías, J., Fonteríz, R., and Gandía, L., Dihydropyridine Bay-K-8644 activates chromaffin cell calcium channels, *Nature*, 308, 69, 1984.

54. Hoshi, T. and Smith, S. J., Large depolarizations induce long openings of voltage-dependent calcium channels in adrenal chromaffin cells, *J. Neurosci.*, 7, 571, 1987.

55. Ceña, V., Stutzin, A., and Rojas, E., Effects of calcium and Bay K 8644 on calcium currents in adrenal medullary chromaffin cells, *J. Membr. Biol.*, 112, 255, 1989.

56. Artalejo, C. R., Ariano, M. A., Perlman, R. L., and Fox, A. P., Activation of facilitation calcium channels in chromaffin cells by D_1 dopamine receptors through a cAMP/protein kinase A-dependent mechanism, *Nature*, 348, 239, 1990.

57. Artalejo, C. R., Dahmer, M. K., Perlman, R. L., and Fox, A. P., Two types of Ca^{2+} currents are found in bovine chromaffin cells: facilitation is due to the recruitment of one type, *J. Physiol.*, 432, 681, 1991.

58. Artalejo, C. R., Mogul, D. J., Perlman, R. L., and Fox, A. P., Three types of bovine chromaffin cell Ca^{2+} channels: facilitaton increases the opening probability of a 27 pS channel, *J. Physiol.*, 444, 213, 1991.

59. Artalejo, C. R., Perlman, R. L., and Fox, A.P., ω-Conotoxin GVIA blocks a Ca^{2+} current in bovine chromaffin cells that is not of the "classic" N type, *Neuron*, 8, 85, 1992.

60. Bossu, J. L., De Waard, M., and Feltz, A., Inactivation characteristics reveal two calcium currents in adult bovine chromaffin cells, *J. Physiol.*, 437, 603, 1991.

61. Bossu, J. L., De Waard, M., and Feltz, A., Two types of calcium channels are expressed in adult bovine chromaffin cells, *J. Physiol.*, 437, 621, 1991.

62. Kleppisch, T., Ahnert-Hilger, G., Gollasch, M., Spicher, K., Hescheler, J., Schultz, G., and Rosenthal, W., Inhibition of voltage-dependent Ca^{2+} channels via α_2-adrenergic and opioid receptors in cultured bovine adrenal chromaffin cells, *Pflügers Arch.*, 421, 131, 1992.

63. Gandía, L., García, A. G., and Morad, M., ATP modulation of calcium channels in chromaffin cells, *J. Physiol.*, 470, 55, 1993.

64. Carbone, E., Sher, E., and Clementi, F., Ca currents in human neuroblastoma IMR32 cells: kinetics, permeability and pharmacology, *Pflügers Arch.*, 416, 170, 1990.

65. Pollo, A., Lovallo, M., Biancardi, E., Sher, E., Socci, C., and Carbone, E., Sensitivity to dihydropyridines, ω-conotoxin and noradrenaline reveals multiple high-voltage-activated Ca^{2+} channels in rat insulinoma and human pancreatic β-cells, *Pflügers Arch.*, 423, 462, 1993.

66. Ballesta, J. J., Palmero, M., Hidalgo, M. J., Gutiérrez, L. M., Reig, J. A., Viniegra, S., and García, A. G., Separate binding and functional sites for ω-conotoxin and nitrendipine suggest two types of calcium channels in bovine chromaffin cells, *J. Neurochem.*, 53, 1050, 1989.

67. Hans, M., Illes, P., and Takeda, K., The blocking effects of ω-conotoxin on Ca current in bovine chromaffin cells, *Neurosci. Lett.*, 114, 63, 1990.

68. Gandía, L., Albillos, A., and García, A. G., Bovine chromaffin cells possess FTX-sensitive calcium channels, *Biochem. Biophys. Res. Commun.*, 194, 671, 1993.

69. Albillos, A., García, A. G., and Gandía, L., ω-Agatoxin-IVA-sensitive calcium channels in bovine chromaffin cells, *FEBS Lett.*, 336, 259, 1993.

70. Artalejo, C. R., Adams, M. E., and Fox, A. P., Three types of Ca^{2+} channel trigger secretion with different efficacies in chromaffin cells, *Nature*, 367, 72, 1994.

71. Olivera, B. M., Miljanich, G., Ramachandran, J., and Adams, M. E., Calcium channel diversity and neurotransmitter release: the ω-conotoxins and ω-agatoxins, *Annu. Rev. Biochem.*, 63, 823, 1994.

72. Wheeler, D. B., Randall, A., and Tsien, R. W., Roles of N-type and Q-type Ca^{2+} channels in supporting hippocampal synaptic transmission, *Science*, 264, 107, 1994.

73. López, M. G., Villarroya, M., Lara, B., Martínez-Sierra, R., Albillos, A., García, A. G., and Gandía, L., Q- and L-type Ca^{2+} channels dominate the control of secretion in bovine chromaffin cells, *FEBS Lett.*, 349, 331, 1994.

74. Gandía, L., Michelena, P., De Pascual, R., López, M. G., and García, A. G., Different sensitivities to dihydropyridines of catecholamine release from cat and ox adrenals, *NeuroReport*, 1, 119, 1990.

75. Albillos, A., Artalejo, A. R., López, M. G., Gandía, L., García, A. G., and Carbone, E., Calcium channel subtypes in cat chromaffin cells, *J. Physiol.*, 477, 197, 1994.

76. Hoshi, T., Rothlein, J., and Smith, S. J., Facilitation of Ca^{2+} channel currents in bovine adrenal chromaffin cells, *Proc. Natl. Acad. Sci. U.S.A.*, 81, 5871, 1984.

77. Kleppisch, T., Pedersen, K., Strübing, C., Bosse-Doenecke, E., Flockerzi, V., Hofmann, F., and Hescheler, J., Double-pulse facilitation of smooth muscle α_1-subunit Ca^{2+} channels expressed in CHO cells, *EMBO J.*, 13, 2502, 1994.

78. Artalejo, C. R., Rossie, S., Perlman, R. L., and Fox, A. P., Voltage-dependent phosphorylation may recruit Ca^{2+} current facilitation in chromaffin cells, *Nature*, 358, 63, 1992.

79. Sculptoreanu, A., Rotman, E., Takahashi, M., Scheuer, T., and Catterall, W., Voltage-dependent potentiation of the activity of cardiac L-type calcium channel α_1 subunits due to phosphorylation by cAMP-dependent protein kinase, *Proc. Natl. Acad. Sci. U.S.A.*, 90, 10131, 1993.

80. Doupnik, C. A. and Pun, R. Y. K., G-Protein activation mediates prepulse facilitation of Ca^{2+} channel currents in bovine chromaffin cells, *J. Membr. Biol.*, 140, 47, 1994.

81. Yue, D. T., Herzig, S., and Marban, E., Beta-adrenergic stimulation of calcium channels occurs by potentiation of high-activity gating modes, *Proc. Natl. Acad. Sci. U.S.A.*, 87, 753, 1990.

82. Boksa, P., Dopamine release from bovine adrenal medullary cells in culture, *J. Auton. Nerv. Sys.*, 30, 63, 1990.

83. Chern, Y. J., Kim, K. T., Slakey, L. L., and Westhead, E., Adenosine receptors activate adenylate cyclase and enhance secretion from bovine adrenal chromaffin cells in the presence of forskolin, *J. Neurochem.*, 50, 1484, 1988.

84. Wakade, T. D., Blank, M. A., Malhotra, R. K., Pourcho, R., and Wakade, A. R., The peptide VIP is a neurotransmitter in rat adrenal medulla: physiological role in controlling catecholamine secretion, *J. Physiol.*, 444, 349, 1991.

85. Doupnik, C. A. and Pun, R. Y. K., Cyclic AMP-dependent phosphorylation modifies the gating properties of L-type Ca^{2+} channels in bovine chromaffin cells, *Pflügers Arch.*, 420, 61, 1992.

86. Schultz, G., Rosenthal, W., Trautwein, W., and Hescheler, J., Role of G proteins in calcium channel modulation, *Annu. Rev. Physiol.*, 52, 275, 1990.

87. Kleuss, C., Hescheler, J., Ewel, C., Rosenthal, W., Schultz, G., and Wittig, B., Assignment of G-protein subtypes to specific receptors inducing inhibition of calcium currents, *Nature*, 353, 43, 1991.

88. Callewaert, G., Johnson, R. G., and Morad, M., Regulation of the secretory response in bovine chromaffin cells, *Am. J. Physiol.*, 260, C851, 1991.

89. Artalejo, A. R., García, A. G., Montiel, C., and Sánchez-García, P., A dopaminergic receptor modulates catecolamine release from the cat adrenal gland, *J. Physiol.*, 362, 359, 1985.

90. Bigornia, L., Alen, C. N., Jan, C. R., Lyon, R. A., Titeler, M., and Schneider, A. S., D_2 dopamine receptors modulate calcium channel currents and catecholamine secretion in bovine adrenal chromaffin cells, *J. Pharmacol. Exp. Ther.*, 252, 586, 1990.

91. Sontag, J. M., Sanderson, P., Klepper, M., Aunis, D., Takeda, K., and Bader, M. F., Modulation of secretion by dopamine involves decreases in calcium and nicotinic currents in bovine chromaffin cells, *J. Physiol.*, 427, 495, 1990.

92. Twitchell, W. A. and Rane, S. G., Opioid peptide modulation of Ca^{2+}-dependent K^+ and voltage-activated Ca^{2+} currents in bovine adrenal chromaffin cells, *Neuron*, 10, 701, 1993.

93. Doroshenko, P. and Neher, E., Pertussis-toxin-sensitive inhibition by (-)baclofen of Ca signals in bovine chromaffin cells, *Pflügers Arch.*, 419, 444, 1991.

94. Kim, K. T. and Westhead, E., Cellular responses to Ca^{2+} from extracellular and intracellular sources are different as shown by simultaneous measurements of cytosolic Ca^{2+} and secretion from bovine chromaffin cells, *Proc. Natl. Acad. Sci. U.S.A.*, 86, 9881, 1989.

95. Nakazawa, K., Fujimori, K., Takanaka, A., and Inoue, K., An ATP-activated conductance in pheochromocytoma cells and its suppression by extracellular calcium, *J. Physiol.*, 428, 257, 1990.

96. Pintor, J., Torres, M., Castro, E., and Miras-Portugal, M. T., Characterization of diadenosine tetraphosphate (Ap_4A) binding sites in cultured chromaffin cells: evidence for a P_{2y} site, *Br. J. Pharmacol.*, 103, 1980, 1991.

97. Bean, B. P., Neurotransmitter inhibition of neuronal calcium currents by changes in channel voltage dependence, *Nature*, 340, 153, 1989.

98. Pollo, A., Lovallo, M., Sher, E., and Carbone, E., Voltage-dependent noradrenergic modulation of ω-conotoxin-sensitive Ca^{2+} channels in human neuroblastoma IMR32 cells, *Pflügers Arch.*, 422, 75, 1992.

99. Ikeda, S. R., Double-pulse calcium channel current facilitation in adult rat sympathetic neurones, *J. Physiol.*, 439, 181, 1991.

100. Klepper, M., Hans, M., and Takeda, K., Nicotinic cholinergic modulation of voltage-dependent calcium current in bovine adrenal chromaffin cells, *J. Physiol.*, 428, 545, 1990.

101. Johnson, R. G., Proton pumps and chemosmotic coupling as a generalized mechanism for neurotransmitter and hormone transport, *Ann. N.Y. Acad. Sci.*, 493, 162, 1987.

102. Kim, Y. I. and Neher, E., IgG from patients with Lambert-Eaton syndrome blocks voltage-dependent calcium channels, *Science*, 239, 405, 1988.

103. Pancrazio, J. J., Park, W. K., and Linch, C., Effects of enflurane on the voltage-gated membrane currents of bovine chromaffin cells, *Neurosci. Lett.*, 146, 147, 1992.

104. Pancrazio, J. J., Park, W. K., and Linch, C., Inhalational anesthetic actions on voltage-gated ion currents of bovine chromaffin cells, *Mol. Pharmacol.*, 43, 783, 1993.

105. Gomis, A., Gutiérrez, L. M., Sala, F., Viniegra, S., and Reig, J. A., Ruthenium red inhibits selectively chromaffin cell calcium channels, *Biochem. Pharmacol.*, 47, 225, 1994.

106. Garçez-Do-Carmo, L., Albillos, A., Artalejo, A. R., De la Fuente, M. T., López, M. G., Gandía, L., Michelena, P., and García, A. G., R56865 inhibits catecholamine release from bovine chromaffin cells by blocking calcium channels, *Br. J. Pharmacol.*, 110, 1149, 1993.

107. Cannon, S. D., Wilson, S. P., and Walsh, K. B., A G protein-activated K^+ current in bovine adrenal chromaffin cells: possible regulatory role in exocytosis, *Mol. Pharmacol.*, 45, 109, 1994.

108. Marty, A., Ca-dependent K channels with large unitary conductance in chromaffin cell membranes, *Nature*, 291, 497, 1981.

109. Solaro, C. and Lingle, C. J., Trypsin-sensitive, rapid inactivation of a calcium-activated potassium channel, *Science*, 257, 1694, 1992.

110. Glavinović, M. I. and Trifaró, J. M., Quinine blockade of currents through Ca^{2+}-activated K^+ channels in bovine chromaffin cells, *J. Physiol.*, 399, 139, 1988.

111. Yellen, G., Ionic permeation and blockade in Ca^{2+}-activated K^+ channels of bovine chromaffin cells, *J. Gen. Physiol.*, 84, 157, 1984.

112. Marty, A., Blocking of large unitary calcium-dependent potassium currents by internal sodium ions, *Pflügers Arch.*, 396, 179, 1983.

113. Zagotta, W. N., Hoshi, T., and Aldrich, R. W., Restoration of inactivation in mutants of Shaker potassium channels by a peptide derived from ShB, *Science*, 250, 568, 1990.

114. Kumakura, K., Karoum, F., Guidotti, A., and Costa, E., Modulation of nicotinic receptors by opiate receptor agonists in cultured adrenal chromaffin cells, *Nature*, 283, 489, 1980.

115. Uceda, G., Artalejo, A. R., López, M. G., Abad, F., Neher, E., and García, A. G., Ca^{2+}-activated K^+ channels modulate muscarinic secretion in cat chromaffin cells, *J. Physiol.*, 454, 213, 1992.

116. Kidokoro, Y., Miyazaki, S., and Ozawa, S., Acetylcholine-induced membrane depolarization and potential fluctuations in the rat adrenal chromaffin cell, *J. Physiol.*, 324, 203, 1982.

117. Clapham, D. E. and Neher, E., Substance P reduces acetylcholine-induced currents in isolated bovine chromaffin cells, *J. Physiol.*, 347, 255, 1984.

118. Cull-Candy, S. G., Mattie, A., and Powis, D., Acetylcholine receptor channels and their block by clonidine in cultured bovine chromaffin cells, *J. Physiol.*, 402, 255, 1988.

119. Maconochie, D. J. and Knight, D. E., A study of the bovine adrenal chromaffin nicotinic receptor using patch clamp and concentration-jump techniques, *J. Physiol.*, 454, 129, 1992.

120. Nooney, J. M., Peters, J. A., and Lambert, J. J., A patch clamp study of the nicotinic acetylcholine receptor of bovine adrenomedullary chromaffin cells in culture, *J. Physiol.*, 455, 503, 1992.

121. Kidokoro, Y., Mechanism of acetylcholine-induced membrane potential fluctuations in rat adrenal chromaffin cells, in *The Physiology of Excitable Cells*, Grinnell, A. D. and Moody, W. J., Eds., Alan R. Liss, New York, 1983, 127.

122. Inoue, M. and Kuriyama, H., Properties of the nicotinic-receptor-activated current in adrenal chromaffin cells of the guinea-pig, *Pflügers Arch.*, 419, 13, 1991.

123. Clapham, D. E. and Neher, E., Trifluoperazine reduces inward ionic currents and secretion by separate mechanisms in bovine chromaffin cells, *J. Physiol.*, 353, 541, 1984.

124. Sakmann, B., Patlak, J., and Neher, E., Single acetylcholine-activated channels show burst kinetics in the presence of desensitizing concentrations of agonist, *Nature*, 286, 71, 1980.

125. Hirano, T., Kidokoro, Y., and Ohmori, H., Acetylcholine dose-response relation and the effect of cesium ions in the rat adrenal chromaffin cell under voltage-clamp, *Pflügers Arch.*, 408, 401, 1987.

126. Vernino, S., Amador, M., Leuetje, C. W., Patrick, J., and Dani, J. A., Calcium modulation and high calcium permeability of neuronal nicotinic acetylcholine receptors, *Neuron*, 8, 127, 1992.

127. Zhou, Z. and Neher, E., Calcium permeability of nicotinic receptor channels in bovine chromaffin cells, *Pflügers Arch.*, 425, 511, 1993.

128. Nooney, J. M., Lambert, J. J., and Chiappinelli, V. A., The interaction of κ-bungarotoxin with the nicotinic receptor of bovine chromaffin cells, *Brain Res.*, 573, 77, 1992.

129. López, M. G., Fonteríz, R. I., Gandía, L., De la Fuente, M. T., Villarroya, M., García-Sancho, J., and García, A. G., The nicotinic acetylcholine receptor of the bovine chromaffin cell, a new target for dihydropyridines, *Eur. J. Pharmacol.*, 247, 199, 1993.

130. Inoue, M. and Kuriyama, H., Somatostatin inhibits the nicotinic receptor-activated inward currents in guinea pig chromaffin cells, *Biochem. Biophys. Res. Commun.*, 174, 750, 1991.

131. Bormann, J., Flügge, G., and Fuchs, E., Effect of atrial natriuretic factor (ANF) on nicotinic acetylcholine receptor channels in bovine chromaffin cells, *Pflügers Arch.*, 414, 11, 1989.

132. Adams, P. R., Acetylcholine receptor kinetics, *J. Membr. Biol.*, 58, 161, 1981.

133. Charlesworth, P., Jacobson, I., Pocock, G., and Richards, C. D., The mechanism by which procaine inhibits catecholamine secretion from bovine chromaffin cells, *Br. J. Pharmacol.*, 106, 802, 1992.

134. Nörenberg, W., Illes, P., and Takeda, K., Neuropeptide Y inhibits nicotinic cholinergic currents but not voltage-dependent calcium currents in bovine chromaffin cells, *Pflügers Arch.*, 418, 346, 1991.

135. Inoue, M. and Kuriyama, H., Glucocorticoids inhibit acetylcholine-induced current in chromaffin cells, *Am. J. Physiol.*, 257, C906, 1989.

136. Jacobson, I., Pocock, G., and Richards, C. D., Effects of pentobarbitone on the properties of nicotinic channels of chromaffin cells, *Eur. J. Pharmacol.*, 202, 331, 1991.

137. Boehm, S. and Huck, S., Methoxyverapamil reduction of nicotine-induced catecholamine release involves inhibition of nicotinic acetylcholine receptor currents, *Eur. J. Neurosci.*, 5, 1280, 1993.

138. Akaike, A., Sasa, M., Tamura, Y., Ujihara, H., and Takaori, S., Effects of protein kinase C on the muscarinic excitation of rat adrenal chromaffin cells, *Jpn. J. Pharmacol.*, 61, 145, 1993.

139. Inoue, M. and Kuriyama, H., Muscarinic receptor is coupled with a cation channel through a GTP-binding protein in guinea-pig chromaffin cells, *J. Physiol.*, 436, 511, 1991.

140. Inoue, M. and Imanaga, I., Phosphorylation-dependent regulation of nonselective cation channels in guinea pig chromaffin cells, *Am. J. Physiol.*, 265, C343, 1993.

141. Nakazato, Y., Ohga, A., Oleshansky, M., Tomita, U., and Yamada, Y., Voltage-independent catecholamine release mediated by the activation of muscarinic receptors in guinea pig-adrenal glands, *Br. J. Pharmacol.*, 93, 101, 1988.

142. Knight, D. E. and Machonochie, D. J., Muscarine induces an inward current in chicken chromaffin cells, *J. Physiol.*, 394, 147P, 1987.

143. Kataoka, Y., Gutman, Y., Guidotti, A., Panula, P., Wroblesky, J., Cosenza-Murphy, D., Wu, J. Y., and Costa, E., Intrinsic GABA-ergic system of adrenal chromaffin cells, *Proc. Natl. Acad. Sci. U.S.A.*, 81, 3218, 1984.

144. Kataoka, Y., Fujimoto, M., Alho, H., Guidotti, A., Geffard, M., Kelly, G. D., and Hanbauer, I., Intrinsic gamma aminobutyric acid receptors modulate the release of catecholamines from canine adrenal glands in situ, *J. Pharmacol. Exp. Ther.*, 239, 584, 1986.

145. Castro, E., Oset-Gasque, M. J., Cañadas, S., Giménez, G., and González, M. P., GABA$_A$ and GABA$_B$ sites in bovine adrenal medulla membranes, *J. Neurosci. Res.*, 20, 241, 1988.

146. Peters, J. A., Kirkness, E. F., Callachan, H., Lambert, J. J., and Turner, A. J., Modulation of the GABA$_A$ receptor by depressant barbiturates and pregnane steroids, *Br. J. Pharmacol.*, 94, 1257, 1988.

147. Peters, J. A., Lambert, J. J., and Cottrell, G. A., An electrophysiological investigation of the characteristics and function of GABA$_A$ receptors on bovine adrenomedullary chromaffin cells, *Pflügers Arch.*, 415, 95, 1989.

148. Cottrell, G. A., Lambert, J. J., and Peters, J. A., Modulation of GABA$_A$ receptor activity by alphaxalone, *Br. J. Pharmacol.*, 90, 491, 1987.

149. Lambert, J. J. and Peters, J. A., Steroidal modulation of the GABA$_A$/benzodiazepine receptor complex: an electrophysiological investigation, in *The Allosteric Modulation of Amino Acid Receptors and Its Therapeutic Implications*, Costa, E., and Barnard, E., Eds., Raven Press, New York, 1989, 139.

150. Hales, T. G. and Lambert, J. J., The actions of propofol on inhibitory aminoacid receptors of bovine adrenomedullary chromaffin cells and rodent central neurones, *Br. J. Pharmacol.*, 104, 619, 1991.

151. Hales, T. G. and Lambert, J. J., Modulation of GABA$_A$ and glycine receptors by chlormethiazole, *Eur. J. Pharmacol.*, 210, 239, 1992.

152. Doroshenko, P. and Neher, E., Volume-sensitive chloride conductance in bovine chromaffin cells, *J. Physiol.*, 449, 197, 1992.

153. Doroshenko P., Penner, R., and Neher, E., Novel chloride conductance in the membrane of bovine chromaffin cells activated by intracellular GTPγS, *J. Physiol.*, 436, 711, 1991.

154. Doroshenko, P., Second messengers mediating activation of chloride current by intracellular GTPγS in bovine chromaffin cells, *J. Physiol.*, 436, 725, 1991.

155. Mochizuki-Oda, N., Mori, M., Negishi, M., and Ito, S., Prostaglandin E$_2$ activates Ca^{2+} channels in bovine adrenal chromaffin cells, *J. Neurochem.*, 56, 541, 1991.

156. Ceña, V., Brocklehurst, K. W., Pollard, H. B., and Rojas, E., Pertussis toxin stimulation of catecholamine release from adrenal medullary chromaffin cells: mechanism may be by direct activation of L-type and G-type calcium channels, *J. Membr. Biol.*, 122, 23, 1991.

157. Mochizuki-Oda, N., Negishi, M., and Ito, S., Arachidonic acid activates cation channels in bovine chromaffin cells, *J. Neurochem.*, 61, 1882, 1993.

158. Knight, D. E. and Baker, P. F., Calcium-dependence of catecholamine release from bovine adrenal medullary cells after exposure to intense electrical fields, *J. Membr. Biol.*, 68, 107, 1982.

159. Dunn, L. A. and Holz, R. W., Catecholamine secretion from digitonin-treated adrenal medullary chromaffin cells, *J. Biol. Chem.*, 258, 4989, 1983.

160. Wilson, S. P. and Kirshner, N., Calcium-evoked secretion from digitonin-permeabilized adrenal medullary chromaffin cells, *J. Biol. Chem.*, 258, 4994, 1983.

161. Neher, E. and Augustine, G. J., Calcium gradients and buffers in bovine chromaffin cells, *J. Physiol.*, 450, 273, 1992.

162. O'Sullivan, A. J., Cheek, T. R., Moreton, R. B., Berridge, M. J., and Burgoyne, R. D., Localization and heterogeneity of agonist-induced changes in cytosolic calcium concentration in single bovine adrenal chromaffin cells, *EMBO J.*, 8, 401, 1989.

163. Cheek, T. R., O'Sullivan, A. J., Moreton, R. B., Berridge, M. J., and Burgoyne, R. D., Spatial localization of the stimulus-induced rise in cytosolic Ca^{2+} in bovine adrenal chromaffin cells: distinct nicotinic and muscarinic patterns, *FEBS Lett.*, 247, 429, 1989.

164. Heinemann, C., Rüden, L. v., Chow, R. H., and Neher, E., A two-step model of secretion control in neuroendocrine cells, *Pflügers Arch.*, 424, 105, 1993.

165. Neher, E. and Zucker, R. S., Multiple calcium-dependent processes related to secretion in bovine chromaffin cells, *Neuron*, 10, 21, 1993.

166. Rüden, L. v. and Neher, E., A Ca-dependent early step in the release of catecholamines from adrenal chromaffin cells, *Science*, 262, 1061, 1993.

167. Harish, O. E., Kao, L. S., Raffaniello, R., Wakade, A. R., and Schneider, A. S., Calcium dependence of muscarinic receptor-mediated catecholamine secretion from the perfused rat adrenal medulla, *J. Neurochem.*, 48, 1730, 1987.

168. Rosario, L., Soria, B., Feuerstein, G., and Pollard, H. B., Voltage-sensitive calcium flux into bovine chromaffin cells occurs through dihydropyridine- and ω-conotoxin-insensitive pathways, *Neuroscience*, 29, 735, 1989.

169. Duarte, C. B., Rosario, L. M., Sena, C. M., and Carvalho, A. P., A toxin fraction (FTX) from the funnel-web spider poison inhibits dihydropyridine-insensitive Ca^{2+} channels coupled to catecholamine release in bovine adrenal chromaffin cells, *J. Neurochem.*, 60, 908, 1993.

170. Jiménez, R., López, M. G., Sancho, C., Maroto, R., and García, A. G., A component of the catecholamine secretory response in the bovine adrenal gland is resistant to dihydropyridines and ω-conotoxin, *Biochem. Biophys. Res. Commun.*, 191, 1278, 1993.

171. Jan, C. R., Titeler, M., and Schneider, A. S., Identification of ω-conotoxin binding sites on adrenal medullary membranes: possibility of multiple calcium channels in chromaffin cells, *J. Neurochem.*, 54, 355, 1990.

172. Owen, P. J., Marriot, D. B., and Boarder, M. R., Evidence for a dihydropyridine-sensitive and conotoxin-insensitive release of noradrenaline and uptake of calcium in adrenal chromaffin cells, *Br. J. Pharmacol.*, 97, 133, 1989.

173. López, M. G., Albillos, A., De la Fuente, M. T., Borges, R., Gandía, L., Carbone, E., García, A. G., and Artalejo, A. R., Localized L-type calcium channels control exocytosis in cat chromaffin cells, *Pflügers Arch.*, 427, 348, 1994.

174. Schroeder, T. J., Jankowsky, J. A., Senyshyn, J., Holz, R., and Wightman, R. M., Zones of exocytotic release of bovine adrenal medullary cells in culture, *J. Biol. Chem.*, 269, 17215, 1994.

175. Monck, J. R., Robinson, I. M., Escobar, A. L., Vergara, J. L., and Fernández, J. M., Pulsed laser imaging of rapid Ca^{2+} signaling in excitable cells, *Biophys. J.*, 66, A351, 1994.

176. Mollard, P. E., Seward, E. P., and Nowicky, M. C., Activation of nicotinic receptors triggers exocytosis from bovine chromaffin cells in the absence of membrane depolarization. *Proc. Natl. Acad. Sci. U.S.A.* 92, 3065, 1995.

177. Montiel, C., Artalejo, A. R., Sánchez-García, P., and García, A. G., Two components in the adrenal nicotinic secretory response revealed by cobalt ramps, *Eur. J. Pharmacol.*, 230, 77, 1993.

178. Noble, E. P., Bommer, M., Liebisch, D., and Herz, A., H_1-histaminergic activation of catecholamine release by chromaffin cells, *Biochem. Pharmacol.*, 37, 221, 1988.

179. Stauderman, K. A., Murawsky, M. M., and Pruss, R. M., Agonist-dependent patterns of cytosolic Ca^{2+} changes in single bovine adrenal chromaffin cells: relationship to catecholamine release, *Cell Regul.*, 1, 683, 1990.

180. Forsberg, E. J., Rojas, E., and Pollard, H. B., Muscarinic receptor enhancement of nicotine-induced catecholamine secretion may be mediated by phosphoinositide metabolism in bovine adrenal chromaffin cells, *J. Biol. Chem.*, 261, 4915, 1986.

181. Yokohama, H., Tanaka, T., Ito, S., Negishi, M., Hayashi, H., and Hayaishi, O., Prostaglandin E receptor enhancement of catecholamine release may be mediated by phosphoinositide metabolism in bovine adrenal chromaffin cells, *J. Biol. Chem.*, 263, 1119, 1988.

Chapter 13

ION CHANNELS AND RENIN SECRETION FROM JUXTAGLOMERULAR CELLS*

Hartmut Osswald and Ulrich Quast

CONTENTS

1. SUMMARY

Renin is one element of the renin-angiotensin-aldosterone system (RAAS) which has a dominant role in the regulation of blood pressure and electrolyte balance of the organism. Renin is stored in and released from renin-containing cells in the kidney. The process of renin secretion is under control by signals from the macula densa cells, arterial wall tension, the endothelium of the vas afferens, nerve endings, and circulating hormones. At the level of the renin-secreting cells, these signals converge to modulate potassium, chloride, and, possibly,

---

* Dedicated to Prof. Dr. O. Heidenreich on the occasion of his 70th birthday.

0-8493-2477-7/95/$0.00+$.50
© 1995 by CRC Press Inc.

calcium-channels together with cytosolic factors like Ca^{2+}, cAMP, and cGMP, which are all involved in the cellular control of renin secretion.

2. INTRODUCTION

The aspartyl proteinase renin is part of the RAAS which has a central place in the homeostatic regulation of the electrolyte balance of the body. Arterial blood pressure regulation is also dependent on the activity of the RAAS. Renin liberates the decapeptide, angiotensin I (Ang I), from the renin substrate (angiotensinogen). The angiotensin converting enzyme, an ectoenzyme located at the membrane of endothelial cells, generates from Ang I the octapeptide, angiotensin II (Ang II), which is the vasoactive element in this cascade. Besides other actions, Ang II stimulates the release of the mineralocorticoid aldosterone from the zona glomerulosa of the adrenal cortex. Aldosterone, in turn, stimulates sodium and water reabsorption in the kidney, thereby reducing salt and water loss of the body. Although this short description contains a number of simplifications, it is obvious that the RAAS functions to control the sodium homeostasis of the organism. Fluid balance and arterial pressure control are more indirectly related to the RAAS. For more detailed information on various aspects of the RAAS the reader is referred to some excellent reviews.[1-5]

The renin-secreting cells are located in the wall of the afferent arterioles close to the entrance into the glomerulus of the kidney.[5-7] Almost all renin found in the systemic circulation and in extrarenal tissue is of kidney origin.[8] The renin-secreting cells are surrounded by several different cell types, including the cells of the extraglomerular mesangium, the macula densa cells at the end of the thick ascending limb of Henle's loop, sympathetic nerve endings, and the myocytes and endothelial cells of the vas afferens.[6,7] These different cells are components of the juxtaglomerular apparatus as shown schematically in Figure 1. Each of these cells can produce signals which are transmitted to the renin-secreting cells. In addition, circulating hormones like catecholamines, Ang II, vasopressin, atrial natriuretic factors, arterial blood pressure, and changes in the plasma and the local electrolyte concentration can modulate the secretion of renin.

The complex anatomy of the juxtaglomerular apparatus (JGA) makes it difficult to investigate the process of renin secretion by electrophysiological techniques. It is thus not surprising that only few studies are available that have characterized the electrical properties of the renin-secreting cells. Therefore we will first review the results of physiological studies on the control of renin secretion. These results are then used for the interpretation of the sparce results obtained by measurement of ion channels and intracellular Ca^{2+} concentration in single renin-containing cells.

3. PHYSIOLOGICAL REGULATION OF RENIN SECRETION

3.1 SYMPATHETIC NERVOUS SYSTEM

The innervation of the kidney with catecholamine-containing nerve endings is the morphological substrate for the effect that changes in efferent sympathetic tone can modulate renin release from the kidney. Renin-secreting cells have quantitatively more contacts with the sympathetic nerve endings than cells from vascular smooth muscle or from extraglomerular mesangium.[9] An increase in norepinephrine release induced by stimulation of efferent renal nerve bundles leads to an elevation of renin secretion.[10] Even at subthreshold levels, renal nerve stimulation potentiates the renin release induced by other interventions like the administration of furosemide or the reduction in renal perfusion pressure.[11] In addition, acute renal denervation reduces basal and nonneurally stimulated renin secretion.[12] The efferent sympathetic nerve activity reaching the kidney can be stimulated from brain regions associated with the cardiovascular control in the central nervous system (CNS). The receptors mediating sympathetic stimulation of renin release are of the β_1-adrenergic type.[5] The signal transduction

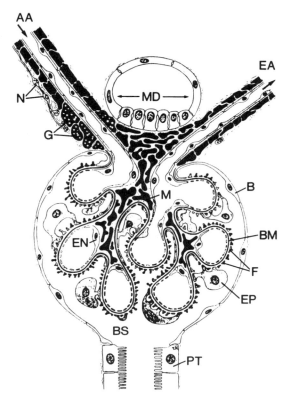

FIGURE 1. Schematic drawing of a cross section through a glomerulus. The macula densa (MD) segment of the tubular epithelium at the end of the thick ascending limb of Henle's loop is located at the vascular pole with the afferent arteriole (AA) entering and the efferent arteriole (EA) leaving the glomerulus. The renin-secreting granular cells (G) in the wall of the afferent arteriole are in close proximity to sympathetic nerve endings (N), and to the extraglomerular mesangium which is connected with the mesangium (M) of the glomerular interstitium and to MD cells. EN, fenestrated endothelial cells; BM, glomerular basement membrane; EP, epithelial podocytes with foot processes (F); B, BS, Bowman's capsule and space, respectively; PT, cell from the proximal tubule at its beginning. (From Koushanpour, E. and Kriz, W., *Renal Physiology: Principals, Structure, and Function,* 2nd ed., Springer-Verlag, New York, 1986. With permission.)

pathway most likely involves stimulation of adenylyl cyclase, leading to increased cAMP levels in the renin-secreting cells.[4,5] α-adrenergic receptor activation, however, leads to an inhibition of renin release probably via increased formation of inositol trisphosphate (IP_3), which mobilizes Ca^{2+} from intracellular stores of the renin-secreting cell.[4]

In conclusion, the sympathetic nervous system is an important factor in the extrarenal control of renin secretion; for integration of the sympathetic control into the sum of other important regulatory factors see also Section 3.4.4 below.

3.2 MYOGENIC ("BARORECEPTOR") CONTROL

The concept of a pressure-related signal (baroreceptor) that affects renin secretion has been reviewed by Blaine.[13] Since changes in renal perfusion pressure lead to a simultaneous alteration of the glomerular filtration rate, tubular NaCl load, and thus NaCl concentration at the macula densa, it was desirable to functionally isolate the renin-secreting cells from other stimuli besides renal perfusion pressure. For this purpose, Blaine[13] developed the model of a denervated, nonfiltering kidney and demonstrated an inverse relationship between renal perfusion pressure and renin release *in vivo*. It was assumed that the wall tension in the afferent arterioles is the sensor site for this pressure dependence of the renin secretion rate.[14] Also, the results of *in vitro* experiments in isolated juxtaglomerular cells which respond to mechanical distortion or stretch with an inhibition of renin release supported the idea of a myogenic pressure stretch-mediated control of renin secretion.[14,15] The signal transduction pathway of this myogenic control of renin secretion probably involves the activation of stretch channels; these channels could either directly admit Ca^{2+} to the cell and/or induce membrane depolarization with subsequent opening of voltage-dependent Ca^{2+} channels (VOCC).[4,15,16] The existence of voltage-gated Ca^{2+} channels in renin-secreting cells is, however, not proven (see Section 4.1. below).

More recently it was shown that myogenic and macula densa signals interact synergistically to keep glomerular capillary pressure and NaCl delivery to the distal tubule constant.[17-20] The interaction was shown by frequency analysis of the myogenic and macula densa response characteristics by external forcing.[17-19] In a different type of experiment, Schnermann and Briggs[20] demonstrated at the single nephron level that glomerular capillary pressure and filtration rate are perfectly maintained in response to changes in perfusion pressure when the macula densa signal was set maximal. At zero flow at the macula densa, however, pressure and filtration rate were poorly autoregulated.[20] These experiments strongly support the conclusion that a myogenic element exists which contributes together with other factors to the control of glomerular filtration and renin secretion (see also Figure 2).

3.3. MACULA DENSA CONTROL

Based on the close anatomical proximity of the macula densa cells to the renin-secreting cells (see Figure 1), Goormaghtigh[21] first proposed an influence of the tubular fluid composition on renin release. The sodium chloride (NaCl) concentration in the tubular fluid passing the macula densa segment at the end of the thick ascending limb of Henle's loop is about 30 mM under normal conditions, thus clearly below the plasma NaCl concentration.[3] The response of the JGA to an increase in the concentration of NaCl at the macula densa is twofold: constriction of the afferent arteriole and inhibition of renin secretion.[3] Inhibition of renin release by increasing the luminal NaCl concentration at the macula densa segment was demonstrated with the micropuncture technique *in vivo*[22] and *in vitro* in the isolated perfused nephron segment with the intact JGA attached.[23,24]

The mechanism(s) by which the macula densa cells transform the signal (i.e., the tubular NaCl concentration) into one or more mediators modulating renin release and afferent vascular tone is incompletely understood. As one possible mechanism it was proposed that adenosine released from macula densa cells and the tubular cells in close proximity to the JGA could be a mediator for both vasoconstriction and inhibition of renin secretion.[25] In experiments using isolated afferent arterioles with the macula densa attached, Itoh et al.[26] found evidence to support a role of adenosine in controlling renin secretion via a macula densa-dependent mechanism. Subsequent studies in the isolated perfused JGA demonstrated that the specific antagonist at adenosine A_1 receptors, 1,3-dipropyl-8-cyclopentylxanthine, blocks the macula densa-dependent inhibition of renin release.[27] Isolated juxtaglomerular renin-containing cells in culture also respond to adenosine with a reduced secretion rate.[28] Evidence for the presence of adenosine A_1 receptors was found in binding studies in dog glomerular membranes[29] and for A_1 receptor mRNA in the rat JGA using antisense mRNA.[30]

Although adenosine is considered an "attractive" candidate for mediating renin secretion inhibition in response to an increased NaCl concentration at the macula densa segment,[3] other factors, like prostaglandins and nitric oxide (NO), are also involved in this effect (see below).

3.4. LOCAL FACTORS AND CIRCULATING HORMONES
3.4.1. Syncytial Organization

The different cell types in the JGA have to integrate several completely different stimuli, i.e., mechanical, chemical, and electrical signals. The propagation of these locally generated signals across the extraglomerular mesangium from the macula densa cells to the vascular pole of the glomerulus requires special features, e.g., cell-to-cell contacts. In fact, extensive gap

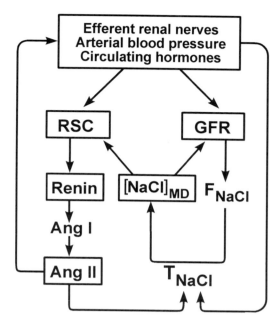

FIGURE 2. Scheme of the different loops that control renin secretion in the kidney. The sodium chloride concentration, [NaCl], in the tubular fluid passing the macula densa segment, MD (see also Figure 1), is the signal that affects the renin-secreting cells (RSC) and, via glomerular filtration rate (GFR), the tubular NaCl load, F_{NaCl}. The released renin leads to angiotensin II (Ang II) formation which acts as a humoral substance together with neurotransmitter release from renal nerve endings and with aldosterone to stimulate tubular NaCl reabsorption, T_{NaCl}. Arterial blood pressure can also contribute to the control of renin secretion via a myogenic mechanism associated with the so-called "baroreceptor" or stretch receptor. See text for further explanation.

junctional coupling has been described between the cells in the JGA, including renin-containing cells, mesangium cells inside and outside of the glomerulus, and vascular smooth muscle, cells of the afferent and efferent arterioles.[31-33] Hence, the mesangium, the smooth muscle and the granular renin-containing cells together form a functional unit allowing the simultaneous regulation of both glomerular filtration and renin secretion (see also Churchill).[4] However, gap junctions between the extraglomerular mesangium and the macula densa cells have not been found.

Also in culture, mesangial cells form a syncytium capable of propagating intracellular calcium transients from a single depolarized cell to its coupled neighbors.[34] Thus, the syncytial organization of the JGA cells derived from smooth muscle cells is likely to participate via mechanical and electrical signals in the control of renin secretion.

3.4.2. Chloride, Prostaglandins, and NO

Using the isolated perfused macula densa segment with its glomerulus attached, Lorenz et al.[23,24] demonstrated that the macula densa-mediated control of renin secretion is dependent on chloride transport. This clear *in vitro* demonstration corroborates previous evidence for a dominant role of chloride ions in the macula densa control of renin release.[35-38] Most recent investigations show that cultured mesangial cells respond to relatively small, physiologically relevant (≥ 13 mM) reductions of the ambient chloride concentration by a transient increase in the cytosolic calcium concentration and by generation of NO[39] and prostaglandins like PGE_2.[40, 41]

Under physiological conditions the chloride concentration in the extraglomerular mesangium is likely to fall below plasma levels in response to a reduction of the chloride concentration in the tubular fluid at the macula densa segment (see Section 3.3.). The increased production of NO and prostaglandins described above could help to explain the observed afferent arteriolar vasodilatation and stimulation of renin release after reduction of early distal tubular chloride load. The rapidly diffusing NO will induce relaxation of the vas afferens via an increase of cGMP in the smooth muscle cells of the arteriole and, in parallel, prostaglandins like PGE_2 via an increase in cAMP.[42] The dilatation of the vas afferens and the increased cAMP levels in the renin-secreting cells will, in turn, stimulate renin secretion in agreement with the observation that a reduction of NaCl concentration at the macula densa segment is accompanied by an increased rate of renin secretion.[3]

The role of NO in controlling renin secretion is, however, only incompletely defined.[5] A number of authors reported a stimulation of renin secretion with increased levels of NO,[43-45] others observed an inhibition after interventions which increased NO or cGMP.[28,46-49] The macula densa, too, releases NO in response to changes in tubular NaCl concentrations;[50-52] again, the role of this effect in the control of renin secretion has to be clarified.

3.4.3. Adenosine

Osswald et al.[53-55] observed that endogenous and exogenous adenosine inhibited renin release and reduced the glomerular filtration rate by afferent arteriolar vasoconstriction. This finding led to the concept that adenosine as a metabolic mediator may couple energy metabolism (ATP hydrolysis for tubular Na^+ transport) with the control of glomerular filtation rate and renin secretion.[25,56] The inhibitory effect of adenosine on renin release via A_1 receptors was confirmed by several investigators using isolated preparations.[26-28,57,58] A synergistic action of adenosine and Ang II on the resistance of afferent arterioles *in vivo* and *in vitro* was described recently by Weihprecht et al.[59]

These authors were able to show that adenosine A_1 receptor antagonists nearly completely blocked the vascular response to Ang II and that Ang II receptor antagonists attenuated markedly the vasoconstriction induced by adenosine A_1 agonists. Whether or not this synergism in the regulation of vascular resistance also extends to the regulation of renin secretion remains to be established. It is also important to note that adenosine inhibits norepinephrine release from sympathetic nerves in the kidney by a presynaptic action.[60]

3.4.4. Circulating hormones

A large number of hormones affect renin release; a general discusssion of these effects is beyond the scope of this article and the interested reader is referred to comprehensive reviews elsewhere.[5,8,61] In the present context it is important, however, to mention the catecholamines (epinephrine, dopamine, norepinephrine) which stimulate renin secretion via activation of β-adrenergic receptors. By far the most important circulating hormone inhibiting renin relase is Ang II, which acts directly via Ang II receptors in the plasma membrane of renin-secreting cells and indirectly by changes in glomerular filtration and tubular Na^+ reabsorption, and, possibly, via its effect on mesangial cells.

The scheme in Figure 2 summarizes the elements involved in the control of renin secretion. The anatomical proximity and the functional coupling of the cells in the JGA form the basis for a simultaneous regulation of renin secretion and glomerular filtration rate. In fact, both systemic and intrarenal factors have been found to influence the renin secretion and glomerular filtration rate (Figure 2). The immense amount of filtered fluid leaving the circulation across the glomerular capillary wall requires a very efficient autoregulation of the glomerular filtration rate. Correspondingly, the importance of the electrolyte and water balance requires an efficient control of sodium metabolism. The development of techniques that allow to perfuse isolated preparations of the glomerulus, the macula densa segment, or both together[23,62] was important for the understanding of the interdependence of the different loops that control glomerular filtration dynamics and renin secretion rates (Figure 2).

4. ELECTROPHYSIOLOGICAL STUDIES OF THE RENIN-SECRETING CELLS

4.1. BASIC ELECTROPHYSIOLOGICAL PROPERTIES OF RENIN-SECRETING CELLS

Despite reports on primary cultures of renin-secreting cells,[46,63,64] there are, to the best of our knowledge, no electrophysiological studies in isolated cells, perhaps due to the problem of contamination of the primary culture with other cells. All published studies have been conducted in the vas afferens from mouse kidney nephrons near the entry into the glomerulus where the density of renin-secreting cells in the tunica media is highest.[65,66] In some cases, glomeruli were prepared with the part of afferent arteriole still attached;[67-69] another group used isolated kidney halves from hydronephrotic mice.[65,66,70] Capacity measurements suggest a strong electrical coupling between the cells in the media of the afferent arteriole,[68] probably by gap junctions as shown by electron microscopy (see Section 2.4.1.). The electrical coupling of cells is reduced by $GTP_\gamma S$.[68]

There is agreement that the resting membrane potential of viable renin-secreting cells lies between −60 and −70 mV.[65-68,70] Increasing extracellular K^+ concentration depolarizes the cell in accordance with the Nernst equilibrium for K^+, showing the K^+ permeability of the membrane dominating.[67] Whole-cell clamp studies have shown that membrane potential is determined by the activity of inwardly rectifying K^+ channels[68] which stabilize the cell against hyperpolarizing and weak depolarizing stimuli.[71] In addition, renin-secreting cells are endowed with delayed-rectifier K^+ channels which are activated by strong depolarizations, thus repolarizing the cell. The resulting current-voltage relationship is shown in Figure 3. These cells also have Ca^{2+}-dependent Cl^- channels (Cl_{Ca}) which open massively as $[Ca^{2+}]_i$ reaches 200 nM; these channels show no voltage dependence (Figure 3).[68] The efflux of Cl^- (I_{Cl}) through these channels depolarizes the membrane.[68] There is indirect evidence[72,73] to suggest that the renin-secreting cells also have ATP-sensitive K^+ channels which are normally closed, but open when the quotient of ATP/ADP falls, thereby inducing cell hyperpolarization.[74]

Using a combination of whole-cell clamp and single-cell Ca^{2+} measurements in freshly prepared vas afferens from mouse kidney, Kurtz and colleagues[69] found no evidence for the existence of VOCC, although evidence for receptor-mediated Ca^{2+} entry was obtained.[68] In addition, neither in patch clamped nor in intact fura-2 acetoxymethylester-loaded cells did depolarization of the preparation by 80 mM KCl induce any increase in $[Ca^{2+}]_i$.[69] This lack of direct evidence for the existence of VOCC in renin-secreting cells is a most intriguing result for two reasons: first, there is massive (although indirect) evidence from functional studies in more complex preparations (tissue slices, perfused kidney) which strongly suggests the existence of such channels in renin-secreting cells.[4] Interestingly, Kurtz et al.[69] found that basal renin secretion from a suspension of glomeruli was not inhibited when $[K^+]_o$ was iso-osmotically increased to 56 mM, whereas in kidney slices, a (delayed) inhibition was observed. Second, the techniques employed[68,69] do not allow one to distinguish between renin-secreting cells and vascular myocytes and the authors most certainly hit both cell types.[68] It seems, however, very unlikely that freshly isolated smooth muscle cells would not possess VOCC. Interestingly, there is clear evidence for the existence of VOCC in cultured rat mesangial cells from fluorescence[75] and from electrophysiological[76] experiments.

4.2. EFFECTS OF ANGIOTENSIN II

4.2.1. Spontaneous Activity and Ang II-Induced Depolarizations

Membrane potential recordings from renin-secreting cells in halves from hydronephrotic kidneys showed spontaneous polymorphous depolarizations resembling excitatory junction potentials and which were attributed to spontaneous transmitter release from adrenergic terminals;[65,66,70] this spontaneous activity was essentially abolished by removal of extracellular

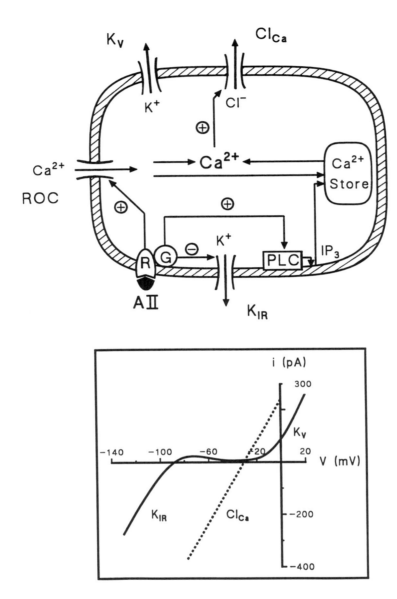

FIGURE 3. Properties of the renin-secreting cell as indicated by whole-cell clamp and microfluorimetric studies.[68,69] Upper panel: Whole-cell clamp studies have shown that renin-secreting cells are endowed with inwardly rectifying K⁺ channels (K$_{IR}$), delayed-rectifier K⁺ channels (K$_V$), and Ca²⁺-dependent chloride channels (Cl$_{Ca}$).[68] Angiotensin II (AII), by binding to its receptor, activates a G protein (G) which inhibits K$_{IR}$ and activates phospholipase C (PLC). PLC liberates inostol 1,4,5-trisphosphate (IP$_3$) from the cell membrane which releases Ca²⁺ from intracellular stores; increased [Ca²⁺]$_i$ activates the Cl$_{Ca}$ channels. The resulting Cl⁻ efflux and the inhibition of K$_{IR}$ lead to depolarization of the cell. Receptor-mediated Ca²⁺ entry is assumed here for simplicity to occur via receptor-operated ion channels (ROC). These studies[68,69] have not produced evidence for the existence of voltage-dependent Ca²⁺ channels or a Na⁺-Ca²⁺ exchange system (see text for details). Lower panel: Steady-state current-voltage relationships for the K⁺current (continuous curve) and the Ca²⁺-dependent Cl⁻ current (broken line) in renin-secreting cells. The K⁺ current was measured at 2.8 m*M* external K⁺; note the strong inward rectification by K$_{IR}$ and the outward rectification by K$_V$ resulting in the bump of the curve. The Ca²⁺-dependent Cl⁻ current passing through Cl$_{Ca}$ shows an Ohmian voltage dependence. (Redrawn after Kurtz[82]; Kurtz and Penner[68]).

Ca^{2+}.[66] Superfusion of epinephrine, norepinephrine, phenylephrine, arginine vasopressin, and Ang II first increased the spontaneous small depolarizations and then massively depolarized the renin-secreting cells by up to 40 mV, whereas cAMP-increasing agents like isoprenaline, histamine, and PGE_2, or the α_2 agonist clonidine were without effect.[70] However, another study has reported a hyperpolarization in response to epinephrine (100 μM).[67]

Analysis of the mechanism of the Ang II-induced depolarization revealed that the hormone blocked the inwardly rectifying K^+ channel which is responsible for the resting membrane potential of the renin-secreting cells.[68] This effect of Ang II was mimicked by GTPγS but not by cAMP, cGMP, IP_3, or a phorbol ester, suggesting that this inhibition was directly mediated by an activated G protein or an unusual second messenger.[68]

4.2.2. Ang II-Induced Oscillations

In voltage-clamped cells of the vas afferens tunica media, Ang II induced oscilliatory increases in $[Ca^{2+}]_i$ with a frequency that declined to zero upon increasing depolarization.[68] The major part of the Ca^{2+} required for these oscillations probably stems from intracellular stores from where it is mobilized via the IP_3 signaling chain; a smaller part may come from receptor-mediated Ca^{2+} entry, since removal of extracellular Ca^{2+} reduced the frequency of the $[Ca^{2+}]_i$ oscillations.[68] The oscillating $[Ca^{2+}]_i$ transients led, in turn, to oscillations of I_{Cl} passing through the Cl_{Ca} channels.[68]

Under physiological conditions Ang II (20 and 100 nM) depolarizes renin-secreting cells transiently to approximately −30 mV without noticeable oscillatory changes in membrane potential;[66,70] such oscillations would be expected, however, if oscillations in I_{Cl} would occur. The absence of oscillations in membrane potential may be explained by the fact that Ang II brings the membrane potential of the renin-secreting cells close to the chloride equilibrium potential;[77] hence, net Cl^- fluxes would be minimal. In addition, the $[Ca^{2+}]_i$ oscillations are abolished at these depolarized potentials.[68] Thus, the Ang II-induced oscillations in $[Ca^{2+}]_i$ and I_{Cl}, observed under voltage clamp conditions, may not be as relevant physiologically as originally thought.[68]

4.3. IMPLICATIONS FROM THE ELECTROPHYSIOLOGICAL AND MICROFLUORIMETRIC MEASUREMENTS

4.3.1. Ca^{2+} Homeostasis in Renin-Secreting Cells

The elements of Ca^{2+} handling in the renin-secreting cell, as far as they emerge from the voltage clamp and microfluorimetric studies discussed above,[68,69] are shown in Figure 3, upper panel. Little is known about the pathways of Ca^{2+} exchange of these cells with the extracellular space. The lack of direct electrophysiological evidence for VOCC[68,69] is a paradoxical result; however, there is evidence for receptor-mediated Ca^{2+} entry by pathways not yet defined, but designated as "receptor-operated channels" in Figure 3. There is a large body of indirect evidence for the existence of a Na^+-Ca^{2+} exchange system in these cells.[78] However, under KCl depolarization the exchanger should admit Ca^{2+} to the cell, but this has not been observed by Kurtz and Penner[68] in direct Ca^{2+} measurements. Interestingly, these measurements have shown that $[Ca^{2+}]_i$ in the renin-secreting cells increases with the extracellular Ca^{2+} concentration, $[Ca^{2+}]_o$.[68] If sufficiently sensitive, this coupling (the mechanism of which is still unknown) could modulate renin secretion in response to fluctuations of $[Ca^{2+}]_o$. Renin-secreting cells possess IP_3-sensitive Ca^{2+} stores.[68] The Ca^{2+} transients induced by Ang II were prevented by cAMP and by stimulation of protein kinase C with phorbol esters; cGMP, however, had no effect.[68]

4.3.2. I_{Cl} and the Ca^{2+} Paradox of Renin Secretion

It is well established that interventions that increase $[Ca^{2+}]_i$ in renin-secreting cells inhibit renin release, a phenomenon termed the "calcium paradox" (for reviews see Churchill,[4,78]

Hackenthal and Taugner).[79] A similar anomaly has been observed in the regulation of parathyroid hormone secretion,[80] a feature apparently linked to the existence of an unusual type of Ca^{2+}-dependent K^+ channels.[81] Based on their finding that in renin-secreting cells an increase in $[Ca^{2+}]_i$ induces a Cl^- efflux (which is balanced by a K^+ efflux for reasons of electroneutrality), Kurtz[82] has advanced the hypothesis that the ensuing loss of KCl and water leads to cell shrinkage. Shrinkage of these cells will inhibit swelling of the secretory granules[83] (but see Jensen and Skøtt)[84] and hence, renin secretion. This hypothesis could explain the inhibition of renin secretion by Ang II and other Ca^{2+}-mobilizing hormones like noradrenaline via the α_1 receptor.[82] Increased cAMP levels and a stimulation of protein kinase C both prevent the Ang II-induced Ca^{2+} transient[68] and subsequent cell shrinking, thereby des-inhibiting (stimulating) renin release.[82]

This hypothesis which inversely couples cell volume and renin secretion to $[Ca^{2+}]_i$ needs further experimental support. It rivals with a hypothesis derived by Taugner et al.[85] from the observation that in the tunica media of the vas afferens there is a continous spectrum of cells ranging from the fully granulated, epitheloid secretory type which completely lack myosin to pure smooth muscle cells with a well-developed contractile apparatus. In intermediate cells, renin-containing granula are at a certain distance from the sarcolemma and their mobility is impeded by a dense sublemmal myofilament network. Any increase in $[Ca^{2+}]_i$ tightens this network and prevents the granula from moving to the plasmalemma, thus reducing renin secretion from these cells.[85] This hypothesis, too, requires further experimental support.

4.3.3. Membrane Potential and Renin Secretion

Membrane potential is generally considered to be one of the prime determinants of the exocytotic activity of secretory cells.[86,87] In the renin-secreting cell, interventions which depolarize the cell (e.g., high extracellular K^+- or Ca^{2+}-mobilizing hormones) inhibit renin secretion. However, in the case of Ca^{2+}-mobilizing hormones it has been clearly shown that $[Ca^{2+}]_i$ in the renin-secreting cells is increased,[68] and, in the case of KCl, this seems likely[78] (but see Kurtz et al.).[69] Since an increase in $[Ca^{2+}]_i$ inhibits renin secretion, a direct effect of depolarization on the secretory process cannot be inferred from these experiments.

What is the evidence that hyperpolarization enhances renin secretion? The hyperpolarization of the cell in response to epinephrine observed by Fishman[67] (but not confirmed by others)[66,70] is expected to occur together with an increase in cAMP. Since cAMP is itself a strong stimulator of renin secretion,[78] a contribution of the epinephrine-induced hyperpolarization to the increase in renin secretion cannot be inferred. A stronger case can be made for the K_{ATP} channel openers, a group of compounds which hyperpolarize and relax smooth muscle by opening ATP-sensitive K^+ channels.[73,74,88] The prototype of these compounds, cromakalim, has been shown to increase renin secretion from renin-secreting cells in culture[72] and in the pithed rat preparation[73] where cardiovascular reflexes are abolished. Glibenclamide, the standard inhibitor of K_{ATP} channels,[74,88] reduced basal plasma renin activity in the pithed rat and blunted the response to cromakalim, providing additional evidence that renin-secreting cells possess K_{ATP} channels which modulate renin release.[73] In vascular smooth muscle, the K_{ATP} channel openers, via their hyperpolarizing action, reduce resting $[Ca^{2+}]_i$,[89,90] perhaps via the Na^+-Ca^{2+} exchange system and by increasing the fraction of cytosolic Ca^{2+} bound to the sarcolemma.[91] These compounds also inhibit the Ca^{2+} transient induced by Ca^{2+}-mobilizing hormones by reducing agonist-induced IP_3 generation, again by hyperpolarizing the membrane.[89,90] It is not known whether or not a similar coupling between membrane potential and $[Ca^{2+}]_i$ exists in renin-secreting cells which are known to be derived from vascular smooth muscle cells.[92] The hypothesis that hyperpolarization stimulates renin secretion certainly warrants further investigations.

5. CONCLUDING REMARKS

Over the last 30 years great progress has been made in elucidating the complex physiological pathways that regulate renin secretion; however, at the molecular level our understanding is still very incomplete. Despite recent progress, the electrophysiological characterization of the renin-secreting cell is still in its infancy. We know little about the ion channels in renin-secreting cells; in particular, no data at the single-channel level are available. The Ca^{2+}-dependent Cl^- current which has been proposed as a major regulator of cell volume and renin secretion is not characterized in detail and its physiological relevance has not yet been put to test. Both the influence of membrane potential on secretory activity and the hypothesis that the myogenic (baroreceptor) control of renin secretion is mediated by activation of stretch channels are not backed up by electrophysiological studies. Further open questions concern the consequences of the gap junctional coupling of renin-secreting cells to neighboring cells or the sensitivity to osmotic challenge. Although technical problems seemingly oppose the electrophysiological investigation of these cells, such studies are essential for a better understanding of the cellular control of renin secretion.

REFERENCES

1. Ballermann, B. J., Zeidel, M. L., Gunning, M. E., and Brenner, B. M., Vasoactive peptides and the kidney, in *The Kidney*, 4th ed., Brenner, B. M. and Rector, F. C., Eds., W. B. Saunders, Philadelphia, 1991, chap. 14.
2. Schnermann, J., Häberle, D. A., Davis, J. M., and Thurau, K., The tubuloglomerular feedback control of renal vascular resistance, in *Handbook of Physiology*, Section 8: Renal physiology, Windhager, E. E., Ed., Oxford University Press, New York, 1992, chap. 34.
3. Schnermann, J. and Briggs, J. P., Function of the juxtaglomerular apparatus. Control of glomerular hemodynamics and renin secretion, in *The Kidney: Physiology and Pathophysiology*, 2nd ed., Seldin, D. W. and Giebisch, G., Eds., Raven Press, New York, 1992, chap. 35.
4. Churchill, P. C., First and second messengers in renin secretion, in *Hypertension: Pathophysiology, Diagnosis and Management*, Laragh, J. H. and Brenner, B. M., Eds., Raven Press, New York, 1990, Chap. 77.
5. Hackenthal, E., Paul, M., Ganten, D., and Taugner, R., Morphology, physiology, and molecular biology of renin secretion, *Physiol. Rev.*, 70, 1067, 1990.
6. Koushanpour, E. and Kriz, W., *Renal Physiology: Principles, Structure and Function*, 2nd ed., Springer-Verlag, New York, 1986.
7. Taugner, R. and Hackenthal, E., *The Juxtaglomerular Apparatus. Structure and Function*, Springer-Verlag, Berlin, 1989.
8. Laragh, J. H. and Sealey, J. E., Renin-angiotensin-aldosterone system and the renal regulation of sodium, potassium and blood pressure homeostasis, in *Handbook of Physiology*, Section 8: Renal physiology, Windhager, E. E., Ed., Oxford University Press, New York, 1992, chap. 31. .
9. Barajas, L., The ultrastructure of the juxtaglomerular apparatus as disclosed by three-dimensional reconstructions from serial sections: the anatomical relationship between tubular and vascular components, *J. Ultrastruct. Res.*, 33, 116, 1970.
10. Kopp, W. G. and DiBona, G. F., Interaction between neural and non-neural mechanisms controlling renin release, *Am. J. Physiol.*, 246, F620, 1984.
11. Thames, M. D. and DiBona, G. F., Renal nerves modulate the secretion of renin mediated by non-neural mechanisms, *Circ. Res.*, 44, 645, 1979.
12. Bunag, R. D., Page, I. H., and Mc Cubbin, J. W., Neural stimulation of release of renin, *Circ. Res.*, 19, 851, 1966 .
13. Blaine, E. H., Development of the canine non-filtering kidney and its utility for studying the intrarenal mechanisms regulating renin secretion, *J. Hypertens.*, 2 (Suppl. 1), 13, 1984.
14. Fray, J. C. S., and Lush, D. J., Stretch receptor hypothesis for renin secretion: the role of calcium, *J. Hypertens.*, 2 (Suppl. 1), 19, 1984.
15. Fray, J. C. S., Lush, D. J., and Park, C. S., Interrelationship of blood flow, juxtaglomerular cells, and hypertension: role of physical equilibrium and Ca, *Am. J. Physiol.*, 251, R643, 1986.

16. Sachs, F., Baroreceptor mechanisms at the cellular level, *Fed. Am. Soc. Exp. Biol.*, 46, 12, 1987.
17. Chon, K. H., Chen, Y.-M., Marmarelis, V. Z., Marsh, D. J., and Holstein-Rathlon, N.-H., Detection of interactions between myogenic and TGF mechanisms using non-linear analysis, *Am. J. Physiol.*, 267, F160, 1994.
18. Holstein-Rathlon, N.-H. and Marsh, D. J., Tubuloglomerular feedback dynamics and renal blood flow autoregulation in rats, *Am. J. Physiol.*, 260, F53, 1991.
19. Yip, K. P., Holstein-Rathlon, N.-H., and Marsh, D. J., Mechanisms of temporal variations in single-nephron blood flow in rats, *Am. J. Physiol.*, 264, F427, 1993.
20. Schnermann, J. and Briggs, J. P., Interaction between loop of Henle flow and arterial pressure as determinants of glomerular pressure, *Am. J. Physiol.*, 256, F421, 1989.
21. Goormaghtigh, N., Une glande endocrine dans la paroi des arterioles rénales, *Bruxelles-Med.*, 19, 1541, 1939.
22. Leyssac, P. P., Changes in single nephron renin release are mediated by tubular fluid flow rate, *Kidney Int.*, 30, 332, 1986.
23. Lorenz, J. N., Weihprecht, H., Schnermann, J., Skøtt, O., and Briggs, J. P., Characterization of the macula densa stimulus for renin secretion, *Am. J. Physiol.*, 259, F186, 1990.
24. Lorenz, J. N., Weihprecht, H., Schnermann, J., Skøtt, O., and Briggs, J. P., Renin release from isolated juxtaglomerular apparatus depends on macula densa chloride transport, *Am. J. Physiol.*, 260, F486, 1991.
25. Osswald, H., Hermes, H. H., and Nabakowski, G., Role of adenosine in signal transmission of tubuloglomerular feedback, *Kidney Int.*, 22, S136, 1982.
26. Itoh, S., Carretero, O. A., and Murray, R. D., Possible role of adenosine in the macula densa mechanism of renin release in rabbits, *J. Clin. Invest.*, 76, 1412, 1985.
27. Weihprecht, H., Lorenz, J. N., Schnermann, J., Skøtt, O., and Briggs, J. P., Effect of adenosine$_1$-receptor blockade on renin release from rabbit isolated perfused juxtaglomerular apparatus, *J. Clin. Invest.*, 85, 1622, 1990.
28. Kurtz, A., Della Bruna, R., Pfeilschifter, J., and Bauer C., Role of cGMP as second messenger of adenosine in the inhibition of renin release, *Kidney Int.*, 33, 798, 1988.
29. Kreutz, R., Charakterisierung von Adenosinbindungsstellen in der Niere, Thesis, Rheinische Friedrich-Wilhelm-Universität, Bonn, 1986.
30. Weaver, D. R. and Reppert, S. M., Adenosine receptor gene expression in rat kidney, *Am. J. Physiol.*, 263, F991, 1992.
31. Forssmann, W. G. and Taugner, R., Studies on the juxtaglomerular apparatus. V. The juxtaglomerular apparatus in Tupaia with special reference to intercellular contacts, *Cell Tissue Res.*, 177, 291, 1977.
32. Taugner, R., Schiller, A., Kaissling, B., and Kriz, W., Gap junctional coupling between the juxtaglomerular apparatus and the glomerular tuft, *Cell Tissue Res.*, 186, 279, 1978.
33. Pricam, C., Humbert, F., Perrelet, A., and Orci, L., Gap junctions in mesangial and lacis cells, *J. Cell Biol.*, 63, 349, 1974.
34. Iijima, K., Moore, L. C., and Goligorsky, M. S., Syncytial organization of cultured rat mesangial cells, *Am. J. Physiol.*, 260, F848, 1991.
35. Schlatter, E., Salomonsson, M., Persson, A. E. G., and Greger, R., Macula densa cells sense luminal NaCl concentration via furosemide sensitive Na$^+$2Cl$^-$K$^+$ cotransport, *Pflügers Arch.*, 414, 286, 1989.
36. Gonzales, E., Salomonsson, M., Müller-Suur, C., and Persson, A. E. G., Measurements of macula densa cell volume changes in isolated and perfused rabbit cortical thick ascending limb. I. Isosmotic and anisosmotic cell volume changes, *Acta Physiol. Scand.*, 133, 149, 1988.
37. Abboud, H. E., Luke, R.-G., Galla, H. H., and Kotchen, T. A., Stimulation of renin by acute selective chloride depletion in the rat, *Circ. Res.*, 44, 815, 1979.
38. Kotchen, T. A., Krzyzaniak, K. E., Anderson, J. E., Ernst, C. B., Galla, J. H., and Luke, R.-G., Inhibition of renin secretion by HCl is related to chloride in both dog and rat, *Am. J. Physiol.*, 239, F44, 1980.
39. Tsukahara, H., Krivenko, Y., Moore, L. C., and Goligorsky, M. S., Decrease in ambient [Cl$^-$] stimulates nitric oxide release from cultured rat mesangial cells, *Am. J. Physiol.*, 267, F190, 1994.
40. Kurokawa, K. and Okuda, T., Calcium-activated chloride conductance of mesangial cells, *Kidney Int.* 38, 30, S48, 1990.
41. Okuda, T., Kojima, I., Ogata, E., and Kurokawa, K., Ambient Cl$^-$ ions modify rat mesangial cell contraction by modulating cell inositol trisphosphate and Ca^{2+} via enhanced prostaglandin E$_2$, *J. Clin. Invest.*, 84, 1866, 1989.
42. Walter, U., Nieberding, M., and Waldmann, R., Intracellular mechanism of action of vasodilators, *Eur. Heart J.*, 9 (Suppl. H), 1, 1988.
43. Scholz, H. and Kurtz, A., Involvement of endothelium-derived relaxing factor in the pressure control of renin secretion from isolated perfused kidney. *J. Clin. Invest.*, 91, 1088, 1993.
44. Johnson, R. A. and Freeman, A. H., Renin release in rats during blockade of nitric oxide synthesis, *Am. J. Physiol.*, 266, R1723, 1994.
45. Münter, K. and Hackenthal, E., The participation of the endothelium in the control of renin release *J. Hypertens.*, 9 (Suppl. 6), S236, 1991.

46. Kurtz, A., Della Bruna, R., Pfeilschifter, J., Taugner, R., and Bauer, C., Atrial natriuretic peptide inhibits renin release from juxtaglomerular cells by a cGMP-mediated process, *Proc. Natl. Acad. Sci. U.S.A.*, 83, 4769, 1986.

47. Henrich, W. L., McAllister, E. A., Smith, P. B., and Campbell, W. B., Guanosine 3′,5′-cyclic monophosphate as a mediator of inhibition of renin release, *Am. J. Physiol.*, 255, F474, 1988.

48. Campbell, W. B. and Henrich, W. L., Endothelial factors in the regulation of renin release. *Kidney Int.*, 38, 612, 1990.

49. Sigmon, D. H., Carretero, O. A., and Beierwaltes, W. H., Endothelium-derived relaxing factor regulates renin release in vivo, *Am. J. Physiol.*, 263, F256, 1992.

50. Itoh, S. and Ren, Y., Evidence for the role of nitric oxide in macula densa control of glomerular hemodynamics, *J. Clin. Invest.*, 92, 1093, 1993.

51. Mundel, P., Bachmann, S., Bader, M., Fischer, A., Kummer, W., Mayer, B., and Kriz, W., Expression of nitric oxide synthase in kidney macula densa cells, *Kidney Int.*, 42, 1017, 1992.

52. Wilcox, C. S., Welch, W. J., Murad, F., Gross, S. S., Taylor, G., Levi, R., and Schmidt, H. H. H. W., Nitric oxide synthese in macula densa regulates glomerular capillary pressure, *Proc. Natl. Acad. Sci. U.S.A.*, 89, 11993, 1992.

53. Osswald, H., Schmitz, H. J., and Kemper, R., Renal action of adenosine effect on renin secretion in the rat, *Naunyn-Schmiedeberg's Arch Pharmacol.*, 303, 95, 1978.

54. Osswald, H., Schmitz, H.-J., and Kemper, R., Tissue content of adenosine, inosine and hypoxanthine in the rat kidney after ischemia and postischemic recirculation, *Pflügers Arch.*, 371, 45, 1977.

55. Osswald, H., Spielman, W. S., and Knox, F. G., Mechanism of adenosine-mediated decrease in glomerular filtration rate, *Circ. Res.*, 43, 465, 1978.

56. Osswald, H., The role of adenosine in the regulation of glomerular filtration rate and renin secretion, *TIPS*, 5, 94, 1984.

57. Churchill, P. C. and Churchill, M. C., A1 and A2 adenosine receptor activation inhibits and stimulates renin secretion of rat renal cortical slices, *J. Pharmacol. Exp. Ther.*, 232, 589, 1985.

58. Lorenz, J. N., Weihprecht, H., He, X.-R., Skøtt, O., Briggs, J. P., and Schnermann, J., Effect of adenosine and angiotensin on macula densa-stimulated renin secretion, *Am. J. Physiol.*, 265, F187, 1993.

59. Weihprecht, H., Lorenz, J. N., Briggs, J. P., and Schnermann, J., Synergistic effects of angiotensin and adenosine in the renal microvasculature, *Am. J. Physiol.*, 266, F227, 1994.

60. Hedquist, P., Fredholm, B. B., and Ölundh, S., Antagonistic effects of theophylline and adenosine on adrenergic neuroeffector transmission in the rabbit kidney, *Circ. Res.*, 43, 592, 1978.

61. Keeton, T. K. and Campbell, W. B., The pharmacologic alteration of renin release, *Pharmacol. Rev.*, 32, 81, 1980.

62. Itoh, S., Carretero, O. A., and Murray, R. D., Renin release from isolated afferent arterioles, *Kidney Int.*, 27, 762, 1985.

63. Della Bruna, R., Pinet, F., Corvol, P., and Kurtz, A., Regulation of renin secretion and renin synthesis by second messengers in isolated mouse juxtaglomerular cells, *Cell. Physiol. Biochem.*, 1, 98, 1991.

64. Rightsel, W. A., Okamura, T., Inagami, T., Pitcock, J. A., Takii, Y., Brooks, B., Brown, P., and Muirhead, E. E., Juxtaglomerular cells grown as monolayer cell culture contain renin, angiotensin I-converting enzyme, and angiotensins I and II/III, *Circ. Res.*, 50, 822, 1982.

65. Bührle, C. P., Nobiling, R., Mannek, E., Schneider, D., Hackenthal, E., and Taugner, R., The afferent glomerular arteriole: immunocytochemical and electrophysiological investigations, *J. Cardiovasc. Pharmacol.*, 6, S383, 1984.

66. Bührle, C. P., Nobiling, R., and Taugner, R., Intracellular recordings from renin-positive cells of the afferent glomerular arteriole, *Am. J. Physiol.*, 249, F272, 1985.

67. Fishman, M. C., Membrane potential of juxtaglomerular cells, *Nature*, 260, 542, 1976.

68. Kurtz, A., and Penner, R., Angiotensin II induces oscillations of intracellular calcium and blocks anomalous inward rectifying potassium current in mouse renal juxtaglomerular cells, *Proc. Natl. Acad. Sci. U.S.A.*, 86, 3423, 1989.

69. Kurtz, A., Skøtt, O., Chegini, S., and Penner, R., Lack of direct evidence for a functional role of voltage-operated calcium channels in juxtaglomerular cells, *Pflügers Arch.*, 416, 282, 1990.

70. Bührle, C. P., Scholz, H., Hackenthal, E., Nobiling, R., and Taugner, R., Epithelioid cells: membrane potential changes induced by substances influencing renin secretion, *Mol. Cell. Endocrinol.*, 45, 37, 1986.

71. Hille, B., *Ionic Channels of Excitable Membranes*, 2nd ed., Sinauer Associates, Sunderland, MA, 1992, chap. 5.

72. Ferrier, C. P., Kurtz, A., Lehner, P., Shaw, S. G., Pusterla, C., Saxenhofer, H., and Weidmann, P., Stimulation of renin secretion by potassium-channel activation with cromakalim, *Eur. J. Clin. Pharmacol.*, 36, 443, 1989.

73. Richer, C., Pratz, J., Mulder, P., Mondot, S., Giudicelli, J. F., and Cavero, I., Cardiovascular and biological effects of K+ channel openers, a class of drugs with vasorelaxant and cardioprotective properties, *Life Sci.*, 47, 1693, 1990.

74. Edwards, G. and Weston, A. H., The pharmacology of ATP-sensitive potassium channels, *Annu. Rev. Pharmacol. Toxicol.*, 33, 597, 1993.
75. Takeda, K., Meyer-Lehnert, H., Kim, J. K., and Schrier, R. W., Effect of angiotensin II on Ca^{2+} kinetics and contraction in cultured rat glomerular mesangial cells, *Am. J. Physiol.*, 254, F254, 1988.
76. Nishio, M., Tsukahara, H., Hiroka, M., Sudo, M., Kigoshi, S., and Muramatsu, I., Calcium channel in cultured rat mesangial cells, *Mol. Pharmacol.*, 43, 96, 1992.
77. Aickin, C. C. and Brading, A. F., Advances in the understanding of transmembrane ionic gradients and permeabilities in smooth muscle obtained by using ion-selective micro-electrodes, *Exp. Basel*, 41, 879, 1985.
78. Churchill, P. C., Second messengers in renin secretion, *Am. J. Physiol.*, 249, F175, 1985.
79. Hackenthal, E. and Taugner, R., Hormonal signals and intracellular messengers for renin secretion, *Mol. Cell. Endocrinol.*, 47, 1, 1986.
80. Habener, J. F. and Potts, J. T., Jr., Relative effectiveness of magnesium and calcium on the secretion and biosynthesis of parathyroid hormone in vitro, *Endocrinology*, 98, 197, 1976.
81. Jia, M., Ehrenstein, G., and Iwasa, K., Unusual calcium-activated potassium channels of bovine parathyroid cells, *Proc. Natl. Acad. Sci. U.S.A.*, 85, 7236, 1988.
82. Kurtz, A., Do calcium-activated chloride channels control renin secretion? *News Physiol. Sci.*, 5, 43, 1990.
83. Skøtt, O., Do osmotic forces play a role in renin secretion? *Am. J. Physiol.*, 255, F1, 1988.
84. Jensen, B. L. and Skøtt, O., Osmotically sensitive renin release from permeabilized juxtaglomerular cells, *Am. J. Physiol.*, 265, F87, 1993.
85. Taugner, R., Nobiling, R., Metz, R., Taugner, F., Bührle, Ch., and Hackenthal, E., Hypothetical interpretation of the calcium paradox in renin secretion, *Cell Tissue Res.*, 252, 687, 1988.
86. Petersen, O. H. and Maruyama, Y., Calcium-activated potassium channels and their role in secretion, *Nature*, 307, 693, 1984.
87. Penner, R. and Neher, E., The role of calcium in stimulus-secretion coupling in excitable and non-excitable cells, *J. Exp. Biol.*, 139, 329, 1988.
88. Quast, U., Potassium channel openers: pharmacological and clinical aspects, *Fundam. Clin. Pharmacol.*, 6, 279, 1992.
89. Ito, S., Kajikuri, J., Itoh, T., and Kuriyama, H., Effects of lemakalim on changes in Ca^{2+} concentration and mechanical activity induced by noradrenaline in the rabbit mesenteric artery, *Br. J. Pharmacol.*, 104, 227, 1991.
90. Itoh, T., Seki, N., Suzuki, S., Ito, S., Kajikuri, J., and Kuriyama, H., Membrane hyperpolarization inhibits agonist-induced synthesis of inositol 1,4,5-trisphosphate in rabbit mesenteric artery, *J. Physiol.*, 451, 307, 1992.
91. Quast, U., Do the K^+ channel openers relax smooth muscle by opening K^+ channels? *Trends Pharmacol. Sci.*, 14, 332, 1993.
92. Cantin, M., Araujo-Nascimento, M. F., Benchimol, S., and Desormeaux, Y., Metaplasia of smooth muscle cells into juxtaglomerular cells in the juxtaglomerular apparatus, arteries, and arterioles of the ischemic (endocrine) kidney, *Am. J. Pathol.*, 87, 581, 1977.

Section VIII
Neuroendocrine Cells in Other Organs

Chapter 14

THE PHYSIOLOGY AND ION CHANNELS
OF THE MERKEL CELL

Norio Akaike and Yoshiro Yamashita

CONTENTS

1. SUMMARY

Merkel cells are specialized epidermal cells containing abundant dense-cored granules and protruding cylindrical cytoplasmic processes, and make synaptic contacts with mechanosensitive nerve endings. Merkel cells are widely believed to be mechanoreceptor cells, but there is yet no direct evidence to support this concept. Patch clamp recording studies elucidated that Merkel cells are excitable and have at least three distinct voltage-gated channels and one chemically gated channel. The possibility of a mechanically gated channel is discussed.

2. INTRODUCTION

The Tastzellen or tactile cells, first described by Merkel[1,2] in the skin of various vertebrates, are now called "Merkel cells". Later electron microscopic observations revealed that Merkel cells contain abundant dense-cored granules in the cytoplasm and make synaptic contacts with expanded nerve endings to form Merkel cell-axon complexes.[3-8] On the basis of these morphological findings and a locational correlation between a cluster of Merkel cells and a physiological mechanosensitive spot, Merkel cells are widely believed to be mechanoreceptors, acting via synaptic transmission on afferent nerve terminals. However, to date direct evidence is still lacking.

The Merkel cell has been structurally regarded as a member of the "diffuse neuroendocrine cell system" [9] or, in other terms, a paraneuron, i.e., a recepto-secretory cell.[10] Recent immunohistochemical studies have demonstrated the presence of serotonin and several peptides in the dense-cored granular regions of the Merkel cell (see later), but the roles of these substances also remain unclear.

Knowledge of the electrical properties of the Merkel cell is essential for the elucidation of Merkel cell function(s). However, conventional glass microelectrode recordings from Merkel cells are difficult because of their small cell size, invisibility in the living state, and relative inaccessibility. We have successfully dissociated Merkel cells from the rat footpad epidermis and applied patch clamp recording[11] to a single Merkel cell[12] identified by vital staining with quinacrine.[13,14] What is now known is that Merkel cells are electrically excitable and possess voltage- and chemically gated (or chemical receptor-operated) ionic channels. If the Merkel cell is a mechanoreceptor, mechanically gated channels can be expected to exist in the cell membrane, but there is still no direct evidence for such a channel.

In regard to ion channels in Merkel cells, except for our recent report[12] and unpublished data, there is little current information. There are several reviews concerning the structure and function of the Merkel cell-axon complex,[15-20] but all were published more than 5 years ago. Therefore, at the beginning we will review recent findings concerning the physiology of the Merkel cell and then introduce information about the ion channels present or presumed to exist on the Merkel cell membrane.

3. STRUCTURE AND PUTATIVE FUNCTION OF MERKEL CELL

The overall morphology of the Merkel cell was first elucidated with scanning electron microscopy (Figure 1, C, D, and E).[21] In addition to the morphological features mentioned above, Merkel cells are also characterized by straight cytoplasmic processes protruding from the cell surface.[5,6,20] The processes in rat Merkel cells range from 0.1 to 0.25 μm in diameter, and reach a maximum of 2.5 μm in length with a mode of about 1 μm.[21] Under light microscopic observation, it is impossible to identify Merkel cells by shape or size. The technique of labeling living Merkel cells with the fluorescent dye quinacrine made cell identification possible.[13,14]

In the hairy skin of mammals, Merkel cells are found in clusters in the basal epidermis under a dome-like elevation of the skin which is generally referred to as a *Haarscheibe* or touch dome (Figure 1, A and B).[6,22-24] Based on a regional association of the touch domes with the sensitive spots of slowly adapting type I cutaneous mechanoreceptor (SA I receptor) afferents, Merkel cell-axon complexes in touch domes were identified as SA I receptors.[6] This was also confirmed in different sites of mammalian skin such as the glabrous skin[25] and sinus hair follicle (vibrissae)[26] and in the warty skin of frog;[27] but the complexes in some amphibian skins (salamander[28] and *Xenopus*)[29] have been identified as rapidly adapting mechanoreceptors which respond to a suprathreshold sustained stimulus with a single spike.

There is no general agreement as to whether the proper mechano electric transducer site of this receptor resides in the Merkel cell or in the afferent nerve terminal. Gottschaldt and Vahle-Hinz[30] showed that the SA I receptor in cat vibrissae could respond with very precise phase locking to high-frequency vibratory stimuli, and then concluded that the transducer exists in the nerve ending because the measured receptor delay of 0.2 or even 0.3 m was far too short

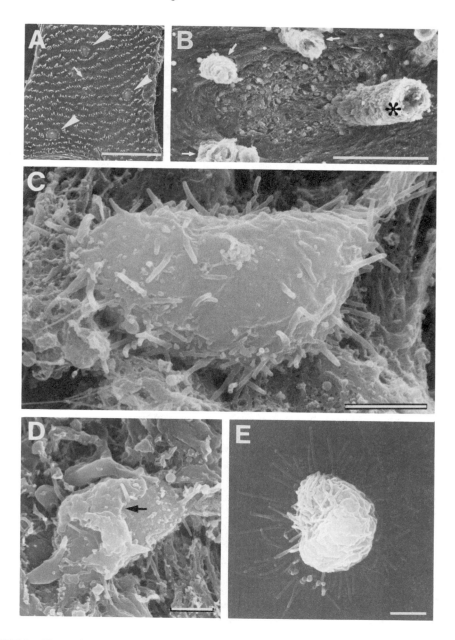

FIGURE 1. Observation of Merkel cells with scanning electron microscopy. (A) Basal view of the epidermis separated from the trunk skin of an 8-d-old rat. Three touch domes (arrowheads) are spaced 1–1.5 mm apart among rows of down hair follicles (arrow). (B) Basal view of rat touch dome from which a tylotrich hair follicle (∗) protrudes. Arrows: down hair follicles; 31-d-old rat. (C) A representative Merkel cell in the touch dome from a 35-d-old rat. Numerous cytoplasmic processes protrude straight from the cell surface. (D) Merkel cell with a nerve ending-like structure (arrow) in the touch dome from a 43-d-old rat. (E) Solitary Merkel cell dissociated from the footpad epidermis of a 12-d-old rat. Scale bars: 1 mm (A); 100 μm (B); 2 μm (C, D, E). (Figures A, B, C, E from Yamashita, Y., Toida, K., and Ogawa, H., *Neurosci. Lett.,* 159, 155, 1993. With permission; (D) from Yamashita, unpublished.)

for chemosynaptic transmission to occur. They proposed that the Merkel cell is a passive abutment for the mechanoreceptive nerve ending. Further, Diamond et al.[18] and Mearow and Diamond[29] introduced the technique of selectively destroying quinacrine-loaded Merkel cells by irradiation with exciting light, and then supported the "nerve ending" hypothesis because the neural response could still be elicited from the Merkel cell-axon complex following irradiation. However, Ikeda et al.[31] reexamined using a much higher irradiation level and better-defined mechanical stimuli, and showed the opposite result that the irradiation abolished the responses. Some investigators tested the effects of either hypoxic environment[32] or arterial infusion of Ca^{2+} antagonists such as verapamil,[33] Mn^{2+},[34] and aminoglycoside (neomycin)[35] on the response of the SA I receptor to mechanical stimulation: in the latter two cases a comparison was made with the response of SA type II receptors. The results suggested the interposition of chemical synaptic transmission in the transduction process of SA I receptors, and it was concluded that the transducer resides in the Merkel cell. However, to date there is no unambiguous evidence to prove this hypothesis.

In addition to the possible roles as mechanoreceptors or as passive abutments mentioned above, Merkel cells have also been proposed to function as the following: (1) targets for growing nerves,[18,36,37] (2) neuromodulators that affect the excitability of adjacent mechanoreceptive nerve endings by secreting modulating substances,[38] and (3) paracrine regulators of surrounding epidermal and adnexal structures.[10,17,39]

4. ION CHANNELS IN MERKEL CELL

4.1. GENERAL ELECTRICAL PROPERTIES OF MERKEL CELL

Merkel cells were dissociated enzymatically from the footpad epidermis of 10- to 20-d-old rats and were identified by quinacrine fluorescence. Exposure of the Merkel cell to the exciting light during cell identification was kept as short as possible, because an intense irradiation of light irreversibly reduces voltage-dependent currents[40] and causes cell death (our observation). Electrical recordings were performed in the whole-cell configuration using a conventional patch clamp technique. The mean resting membrane potential of quinacrine-fluorescent Merkel cells was –54.0 mV, and that of nonfluorescent, ordinary epidermal cells was –26.1 mV. No voltage-dependent channel was observed in ordinary epidermal cells.

Figure 2 shows the voltage responses of a Merkel cell to intracellular current injections for 200 ms at a resting membrane potential of –51 mV. The Merkel cells had no Na^+ spike in an external standard solution, but tetrodotoxin-resistant long-lasting action potentials were evoked by depolarization in an external solution containing Ba^{2+}, which effectively blocks the K^+ channel and passes through Ca^{2+} channels. This absence of voltage-dependent Na^+ channels in rat Merkel cells is similar to the type I cells of the rat carotid body,[41] though rabbit type I cells[42] possess this channel, suggesting species differences.

4.2. VOLTAGE-GATED CHANNELS
4.2.1. Delayed K^+ Current (I_{KD}) and Transient A Current (I_A)

In Merkel cells under voltage clamp, depolarizing step pulses from a holding potential (V_H) of –80 mV elicited predominantly outward K^+ currents composed of transient and sustained components: the former was selectively inhibited by 4-aminopyridine (4-AP), while the latter was inhibited by both tetraethylammonium (TEA) and quinacrine (Figure 3A). Quinacrine was more effective and selective than TEA in blocking the sustained K^+ current, but had no effect on the current at the low concentration ($10^{-7} M$ or $3 \times 10^{-7} M$) used for staining the Merkel cells. As shown in Figure 3B, the sustained outward K^+ current (I_{KD}) was activated at potentials more positive than –20 or –10 mV at a V_H of –50 mV, at which potential the transient outward K^+ channel was completely inactivated. The potential for half inactivation in the steady-state inactivation curve for I_{KD}, the amplitude of which was measured at the end (800 ms) of the test pulse, was –33 mV. On the other hand, the transient outward K^+ current (I_A) was activated at potentials more positive than –50 mV at a V_H of –80 mV (Figure 3C). The potential for half inactivation in the steady-state inactivation curve for I_A was –64 mV. The functional significance of the K^+ currents in Merkel cells is unknown. However, activation

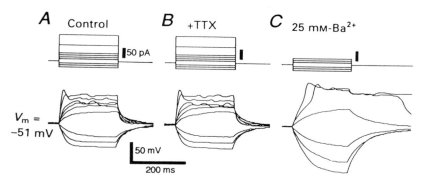

FIGURE 2. Membrane responses of a Merkel cell to intracellular current injections for 200 ms in each external solution. The resting membrane potential was –51 mV. (A) Abortive spikes evoked by passing constant-current steps through the patch pipette in a standard (control) external solution. (B) The addition of 0.1 μM tetrodotoxin (TTX) to the external solution led to no obvious change in the slow abortive potentials. (C) TTX-resistant long-lasting action potentials in a solution containing 25 mM Ba^{2+}. All recordings were obtained from the same cell. (From Yamashita, Y., Akaike, N., Wakamori, M., Ikeda, I., and Ogawa, H., *J. Physiol.*, 450, 143, 1992. With permission.)

of I_A lengthens the interspike interval and repolarizes the membrane following the action potential.[43] As in neurons, the activation of I_A and I_{KD} in Merkel cells would stabilize the membrane potential at a voltage close to the equilibrium potential for K$^+$.

4.2.2. High-Threshold Ca^{2+} Current

When the outward K$^+$ currents were blocked by adding both TEA and 4-AP, only a sustained inward Ca^{2+} current (I_{Ca}) was observed (Figure 4A). In the external solution containing 10 mM Ca^{2+}, I_{Ca} was evoked by potentials more positive than –20 mV at a V_H of –80 mV, and the maximum inward current appeared around +10 mV. Increases in the external Ca^{2+} concentration ($[Ca^{2+}]_o$) induced a hyperbolic increase in I_{Ca} and shifted the current-voltage relationship along the voltage axis in a more positive direction. Saturation of I_{Ca} occurred at about 25 mM $[Ca^{2+}]_o$. The selectivity of the Ca^{2+} channel for divalent cations was in the order Ba^{2+} > Sr^{2+} > Ca^{2+} (Figure 4B). In external solution containing 25 mM Ba^{2+}, the potential for half inactivation of the steady-state inactivation curve for I_{Ba} was –34 mV (Figure 4C). Organic and inorganic Ca^{2+} antagonists blocked I_{Ba} in a concentration-dependent manner, the potency of inhibition being in the order ω-conotoxin > flunarizine = nicardipine > diltiazem > verapamil, and La^{3+} > Cd^{2+} > Co^{2+} (Figure 4D).

The high-threshold (L-type) Ca^{2+} channel in Merkel cells may contribute to the release of the transmitter, as is the case for the same channel in sensory receptor cells such as hair cells[44,45] and taste cells.[46]

It is concluded that Merkel cells are excitable and have at least three distinct voltage-gated currents: a delayed K$^+$ current (I_{KD}), a transient A-current (I_A), and a high-threshold (L-type) Ca^{2+} current (I_{Ca}). The electrical and pharmacological properties of these voltage-gated channels are essentially similar to those of other excitable cells.

4.3. CHEMICALLY GATED CHANNELS

3.3.1. Immunoreactive Amine and Peptides in Merkel Cells

Immunohistochemical studies in Merkel cells have recently demonstrated immunoreactivity for serotonin,[47-49] met-enkephalin,[50,51] vasoactive intestinal polypeptide (VIP),[38,52,53] substance P,[53] and calcitonin gene-related peptide.[52,53] Merkel cells of some mammalian species (i.e., cat, dog, pig, and man) are exclusively immunoreactive to VIP, but not to met-enkephalin.[38] Cheng Chew and Leung[51] also demonstrated met-enkephalin-like immunoreactivity in the dense-cored granules in the Merkel cell of mouse, but failed to do so in those of hamster, guinea pig, rabbit, cat, and dog. In addition to such species differences, the peptide expression

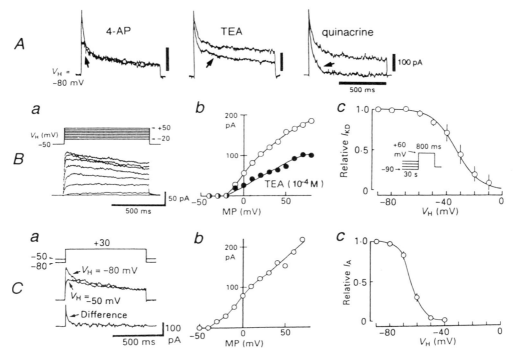

FIGURE 3. K⁺ currents in a Merkel cell. (A) Effect of K⁺ channel blockers on the outward K⁺ currents. The superimposed current traces show the responses to 800-ms depolarizing voltage steps to +50 mV from a V_H of −80 mV before (upper trace) and during (lower trace indicated by arrow) application of 4-AP (10^{-3} M), TEA (10^{-4} M), and quinacrine (10^{-5} M). (B) Delayed K⁺ current (I_{KD}). (Ba) Superimposed traces of I_{KD} evoked by a sequence of 800-ms depolarizing voltage pulses from a V_H of −50 mV. Voltage pulses were in 10-mV increments from −20 to +50 mV. (Bb) Current-voltage (I-V) relationships for the peak amplitude of I_{KD} before (○) and during (●) the application of 10^{-4} M TEA. (Bc) Steady-state inactivation (h_∞) curve of I_{KD} as a function of V_H. The inset shows the voltage protocol at different V_Hs (duration, 30 s). The amplitude of I_{KD} was measured at 800 ms. Each point is the average of four cells; bars indicate ± SEM when larger than the symbol size. The data points were fitted with a Boltzmann distribution equation with $V_{0.5} = -33.3$ mV and the slope factor $k = 8.7$ mV. (C) Transient outward K⁺ currents (I_A). (Ca) Uppermost, voltage protocol; middle, superimposed trace of the currents activated by step pulses to +30 mV from the V_H of −50 and −80 mV; bottom, the separated I_A, obtained by digital subtraction of the two current traces in the middle panel. (Cb) The peak I-V relationship of I_A. The threshold of activation was near −40 mV. (Cc) h_∞ curve of the peak I_A. Each point is the average of seven cells. The data points were also fitted by the Boltzmann distribution equation with $V_{0.5} = -63.8$ mV and $k = 4.5$ mV. (From Yamashita, Y., Akaike, N., Wakamori, M., Ikeda, I., and Ogawa, H., *J. Physiol.*, 450, 143, 1992. With permission.)

in the Merkel cells is also known to vary according to the developmental stage.[54] In the rat touch dome, on the other hand, serotonin was expressed in the adjoining nerve terminals as well as in Merkel cells.[49] It is interesting to note that Merkel-like basal cells in the taste bud of Necturus contain serotonin, which increases both receptor potentials and Ca²⁺ currents in adjoining taste receptor cells:[55] it has not yet been investigated how the serotonin acts on the postsynaptic afferent nerve ending. Merkel cells have been also suggested to contain purine such as ATP, due to an accumulation of quinacrine in this cell.[13,56]

4.3.2. Chemically Gated (or Chemical Receptor-Operated) Channels in the Merkel Cell

Presynaptic autoreceptors which regulate presynaptic release of neurotransmitters are known to exist in a variety of neurons.[57] Also, the presence of reciprocal synapse has been reported in the Merkel cell-axon complex, though such synapses are prominent in the complex of lower-vertebrate skin.[7,20] Further, sympathetic efferent activity has excitatory action on SA

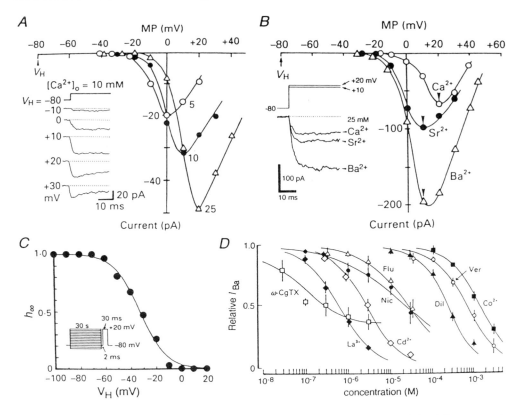

FIGURE 4. Ca^{2+} current (I_{Ca}) in Merkel cell. (A) I_{Ca}s were dependent upon the extracellular Ca^{2+} concentration ([Ca^{2+}]$_o$). The inset shows inward currents evoked at the indicated membrane potentials from a V_H of –80 mV. The [Ca^{2+}]$_o$ was 10 mM. The I-V relationships of the peak I_{Ca}s were obtained from the same cell in 5 (○), 10 (●), and 25 mM (△) [Ca^{2+}]$_o$. (B) Ionic selectivities of Ca^{2+} channel. I-V relationships of Ca^{2+} (○), Sr^{2+} (●), and Ba^{2+} (△) currents (I_{Ca}, I_{Sr} and I_{Ba}) were obtained from the same cell. The concentration of each extracellular divalent cation was 25 mM. The inset shows I_{Ca}, I_{Sr} and I_{Ba}, indicated by small arrowheads on the individual I-V relationships, which were evoked by step pulses from a V_H of –80 mV to +20, +10, and +10 mV, respectively. (C) Inactivation of I_{Ba} induced by a 30-s-long prepulse just before test depolarization to 20 mV from a V_H of –80 mV. Experimental protocol is shown in the inset. Data are representative of four cells. The data points were fitted by the Boltzmann distribution equation with $V_{0.5}$ = –32.3 mV and k = 10.0 mV. (D) Concentration-inhibition curves of Ca^{2+} antagonists on I_{Ba} evoked by voltage steps to +20 mV from a V_H of –80 mV. Application of each Ca^{2+} antagonist was started 2 min before step pulses. [Ba^{2+}]$_o$ was 25 mM. Data are mean values obtained from four to seven preparations. Vertical lines indicate ± SEM. ω-CgTX, Dil, Flu, Nic, and Ver are ω-conotoxin, diltiazem, flunarizine, nicardipine, and verapamil, respectively. (From Yamashita, Y., Akaike, N., Wakamori, M., Ikeda, I., and Ogawa, H., J. Physiol., 450, 143, 1992. With permission.)

I receptors in cat touch domes, and its action is blocked by α-adrenergic blockers, though catecholamine fluorescent fibers have not been found in the touch dome.[58] Therefore, to investigate what kind of chemically gated (or chemical receptor-operated) channel exists in the rat Merkel cell membrane, chemical solutions were rapidly applied to cells under voltage clamp condition by the "Y-tube method". By this technique, the solution surrounding a cell could be completely exchanged within 20 ms.[59]

4.3.2.1. Acetylcholine (ACh)-Induced Current

Application of ACh evoked an inward transient current accompanied by an increase in conductance (Figure 5A; our unpublished data). Nicotine at the same concentration produced a more rapidly desensitizing and larger inward current, but muscarine did not produce any current. The ACh-induced currents were inhibited by the nicotinic antagonist, D-tubocurarine.

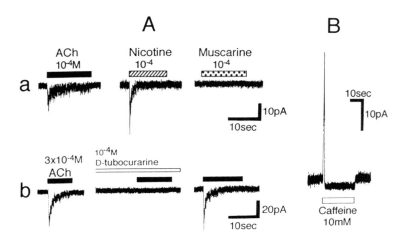

FIGURE 5. (A) Acetylcholine (ACh)-induced current. (Aa) The inward currents induced by ACh (10^{-4} *M*) and nicotine (10^{-4} *M*) at a V_H of –50 mV. Muscarine (10^{-4} *M*) induced no current at a V_H of –30 mV. The cells were exposed to each agonist for the periods indicated by a horizontal bar above the response. (Ab) Effect of the nicotinic antagonist, D-tubocurarine, on ACh-induced current. V_H was –50 mV. The D-tubocurarine (10^{-4} *M*) was applied 30 s before the simultaneous application of 3×10^{-4} *M* ACh and D-tubocurarine. (B) Caffeine-induced currents composed of an outward transient current followed by an inward sustained current. V_H was -20 mV. The concentration of caffeine was 10 m*M* (our unpublished data).

The result indicates that Merkel cells in rat footpad skin have nicotinic ACh receptors and transiently depolarize in response to ACh. It is interesting to note that acetylcholinesterase, an enzyme of ACh degradation, is present in the nerve terminals on the Merkel cells of rabbits.[60] Also, Smith and Creech[61] found that nicotine (0.1–1 mM) applied to the dome surface enhanced the afferent discharge by mechanical stimulation, and after 2–3 min blocked it. While it has been indicated that the excitatory effect of ACh upon mechano- and chemoreceptors may be due to either its depolarizing action on the sensory nerve ending or changes in the firing threshold of the first node of Ranvier,[62] ACh-induced responses similar to those of Merkel cells have been demonstrated in the paraneurons such as type I (glomus) cells of rat carotid body[63] and bovine chromaffin cells.[64] The functional role of the ACh response in the Merkel cell remains to be elucidated.

4.3.2.2. *Caffeine-Induced Currents*

Application of caffeine (10^{-2} *M*) produced an outward transient current followed by an inward sustained current with a decrease in membrane conductance (Figure 5B; our unpublished data). Since caffeine is a releasing agent of Ca^{2+} from intracellular Ca^{2+} pools such as the endoplasmic reticulum,[65] Ca^{2+}-activated channels are speculated to exist in the Merkel cell membrane as in neurons.[66] These currents in Merkel cells are not yet characterized, though it is well known that large-conductance Ca^{2+}-activated K^+ channels exist in the neuroendocrine cells, including chromaffin cells,[67] anterior pituitary cells,[68] and pancreatic β cells.[69]

Recently, neurotransmitters, neuromodulators, and second messengers have been demonstrated to regulate ion channel activities in a wide variety of cells.[43,70] To investigate the gating or modulating action of other chemical substances on ion channels in the Merkel cell is also a future subject.

5. PUTATIVE MECHANICALLY GATED CHANNEL AND FUTURE DIRECTIONS

To demonstrate that the Merkel cell is an actual mechanoreceptor cell it must be shown that there is a mechanically gated current in the Merkel cell membrane. In preliminary experiments

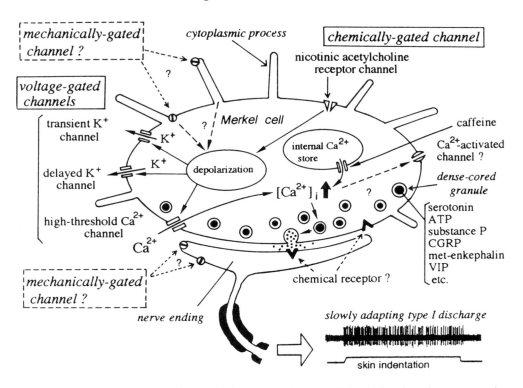

FIGURE 6. Schematic drawing of a Merkel cell-axon complex summarizing the ion channels, neurotransmitter, and neuromodulator candidates. The speculated locations of mechanically gated channels are indicated. See text for details.

we applied small pressure pulses of perfusion fluid to the isolated Merkel cell under whole-cell voltage clamp, but failed to elicit any mechanically gated currents, similar to the findings of Cooper and Nurse.[71] There are two possible reasons for a lack of such current: (i) mechanically gated channels are not present on the Merkel cell membrane or (ii) an isolated Merkel cell is an inappropriate preparation for showing the reception of mechanical stimulus, because its cylindrical cytoplasmic processes droop, as shown in Figure 1E. Since the cilia of the hair cell and microvilli of the taste receptor cell are transduction sites, it has been speculated that the cytoplasmic processes of Merkel cells may be also. We recently developed a new preparation of the Merkel cell with straight cylindrical processes as shown in Figure 1C and D. Such techniques as the slice patch recording[72] may now be available for the electrical recording from Merkel cells.

Figure 6 shows a schematic drawing of a Merkel cell-axon complex summarizing the ion channels, neurotransmitter, and neuromodulator candidates. A natural stimulus which causes the cell depolarization or the release of the content of the dense-cored granules is still unknown, though a mechanical one has been proposed. The activation of Ca^{2+} channels may contribute to the rise of $[Ca^{2+}]_i$ and then probably to the release of the content of the dense-cored granules. Also, the Ca^{2+} release from the intracellular store may play an important role in the cell function. The activation of K^+ channels may contribute to repolarize the cell to and beyond the steady-state potential. The physiological role of nicotinic ACh receptor channels on the Merkel cell membrane is unknown.

The electrophysiological studies of Merkel cells by means of patch clamp recording have just started. The expanded nerve endings attaching to the Merkel cell are occasionally observed in tissue preparations (Figure 1D). Such preparations may be more useful for the elucidation of the functional relationships between the Merkel cell and nerve endings.

ACKNOWLEDGMENTS

The authors thank B. Bell for helpful suggestions. This work was supported by a Grant-in-Aid for Scientific Research Nos. 04044029, 04304028, and 05271202 to N. Akaike from the Ministry of Education, Science and Culture, Japan, and by a grant to Y. Yamashita from the Okukubo Memorial Fund for Medical Research in Kumamoto University School of Medicine.

REFERENCES

1. Merkel, F., Tastzellen und Tastkörperchen bei den Hausthieren und beim Menschen, *Arch. Mikrosk. Anat.*, 11, 636, 1875.
2. Merkel, F., *Über die Endingungen der sensiblen Nerven in der Haut der Wirbelthiere*, H. Schmidt, Rostock, 1880.
3. Cauna, N., Functional significance of the submicroscopical, histochemical and microscopical organization of the cutaneous receptor organs, *Anat. Anz.*, 111 (Suppl. 2), 181, 1962.
4. Munger, B. L., The intraepidermal innervation of the snout skin of the opossum. A light and electron microscope study, with observations on the nature of Merkel's *Tastzellen, J. Cell Biol.*, 26, 79, 1965.
5. Andres, K. H., Über die Feinstruktur der Rezeptoren an Sinushaaren, *Z. Zellforsch., Mikrosk. Anat.*, 75, 339, 1966.
6. Iggo, A. and Muir, A. R., The structure and function of a slowly adapting touch corpuscle in hairy skin, *J. Physiol.*, 200, 763, 1969.
7. Mihara, M., Hashimoto, K., Ueda, K., and Kumakiri, M., The specialized junctions between Merkel cell and neurite: an electron microscopic study, *J. Invest. Dermatol.*, 73, 325, 1979.
8. Hartschuh, W. and Weihe, E., Fine structural analysis of the synaptic junction of Merkel cell-axon-complexes, *J. Invest. Dermatol.*, 75, 159, 1980.
9. Pearse, A. G. E., The diffuse neuroendocrine system: peptides, amines, placodes and the APUD theory, *Prog. Brain Res.*, 68, 25, 1986.
10. Fujita, T., Concept of paraneurons, *Arch. Histol. Jpn.*, 40 (Suppl.), 1, 1977.
11. Hamill, O. P., Marty, A., Neher, E., Sakmann, B., and Sigworth, F. J., Improved patch-clamp techniques for high-resolution current recording from cells and cell-free membrane patches, *Pflügers Arch.*, 391, 85, 1981.
12. Yamashita, Y., Akaike, N., Wakamori, M., Ikeda, I., and Ogawa, H., Voltage-dependent currents in isolated single Merkel cells of rats, *J. Physiol.*, 450, 143, 1992.
13. Crowe, R. and Whitear, M., Quinacrine fluorescence of Merkel cells in *Xenopus laevis, Cell Tissue Res.*, 190, 273, 1978.
14. Nurse, C. A., Mearow, K. M., Holmes, M., Visheau, B., and Diamond, J., Merkel cell distribution in the epidermis as determined by quinacrine fluorescence, *Cell Tissue Res.*, 228, 511, 1983.
15. Iggo, A. and Andres, K. H., Morphology of cutaneous receptors, *Annu. Rev. Neurosci.*, 5, 1, 1982.
16. Iggo, A. and Findlater, G. S., A review of Merkel cell mechanisms, in *Sensory Receptor Mechanisms*, Hamann, W. and Iggo, A., Eds., World Scientific, Singapore, 1984, 117.
17. Gould, V. E., Moll, R., Moll, I., Lee, I., and Franke, W. W., Neuroendocrine (Merkel) cells of the skin: Hyperplasias, dysplasias and neoplasms, *Lab. Invest.*, 52, 334, 1985.
18. Diamond, J., Mills, L. R., and Mearow, K. M., Evidence that the Merkel cell is not the transducer in the mechanosensory Merkel cell-neurite complex, *Prog. Brain Res.*, 74, 51, 1988.
19. Munger, B. L. and Ide, C., The structure and function of cutaneous sensory receptors, *Arch. Histol. Cytol.*, 51, 1, 1988.
20. Whitear, M., Merkel cells in lower vertebrates, *Arch. Histol. Cytol.*, 52 (Suppl.), 415, 1989.
21. Yamashita, Y., Toida, K., and Ogawa, H., Observation of Merkel cells with scanning electron microscopy, *Neurosci. Lett.*, 159, 155, 1993.
22. Pinkus, F., Über Hautsinnesorgane neben dem menschlichen Haar (Haarscheiben) und ihre vergleichend-anatomische Bedeutung, *Arch. Mikrosk. Anat. Entwicklungsmech.*, 65, 121, 1905.
23. Smith, K. R., Jr., The structure and function of the Haarscheibe, *J. Comp. Neurol.*, 131, 459, 1967.
24. English, K. B., Morphogenesis of Haarscheiben in rats, *J. Invest. Dermatol.*, 69, 58, 1977.
25. Munger, B. L. and Pubols, L. M., The sensorineural organization of the digital skin of the raccoon, *Brain Behav. Evol.*, 5, 367, 1972.
26. Gottschaldt, K.-M., Iggo, A., and Young, D. W., Functional characteristics of mechanoreceptors in sinus hair follicles of the cat, *J. Physiol.*, 235, 287, 1973.

27. Yamashita, Y. and Ogawa, H., Slowly adapting cutaneous mechanoreceptor afferent units associated with Merkel cells in frogs and effects of direct currents, *Somatosens. Mot. Res.,* 8, 87, 1991.

28. Parducz, A., Leslie, R. A., Cooper, E., Turner, C. J., and Diamond, J., The Merkel cells and the rapidly adapting mechanoreceptors of the salamander skin, *Neuroscience,* 2, 511, 1977.

29. Mearow, K. M. and Diamond, J., Merkel cells and the mechanosensitivity of normal and regenerating nerves in Xenopus skin, *Neuroscience,* 26, 695, 1988.

30. Gottschaldt, K.-M. and Vahle-Hinz, C., Merkel cell receptors: structure and transducer function, *Science,* 214, 183, 1981.

31. Ikeda, I., Yamashita, Y., Ono, T., and Ogawa, H., Selective phototoxic destruction of rat Merkel cells abolishes responses of slowly adapting type I mechanoreceptor units, *J. Physiol.,* 479, 247, 1994.

32. Findlater, G. S., Cooksey, E. J., Anand, A., Paintal, A. S., and Iggo, A., The effects of hypoxia on slowly adapting type I (s. a. I) cutaneous mechanoreceptors in the cat and rat, *Somatosens. Res.,* 5, 1, 1987.

33. Pacitti, E. G. and Findlater, G. S., Calcium channel blockers and Merkel cells, *Prog. Brain Res.,* 74, 37, 1988.

34. Yamashita, Y., Ogawa, H., and Taniguchi, K., Differential effects of manganese and magnesium on two types of slowly adapting cutaneous mechanoreceptor afferent units in frogs, *Pflügers Arch.,* 406, 218, 1986.

35. Baumann, K. I., Hamann, W., and Leung, M. S., Acute effects of neomycin on slowly adapting type I and type II cutaneous mechanoreceptors in the anaesthetized cat and rat, *J. Physiol.,* 425, 527, 1990.

36. Scott, S. A., Cooper, E., and Diamond, J., Merkel cells as targets of the mechanosensory nerves in salamander skin, *Proc. R. Soc. London,* B211, 455, 1981.

37. Vos, P., Stark, F., and Pittman, R. N., Merkel cells *in vitro:* production of nerve growth factor and selective interactions with sensory neurons, *Dev. Biol.,* 144, 281, 1991.

38. Hartschuh, W., Weihe, E., Yanaihara, N., and Reinecke, M., Immunohistochemical localization of vasoactive intestinal polypeptide (VIP) in Merkel cells of various mammals: evidence for a neuro-modulator function of the Merkel cell, *J. Invest. Dermatol.,* 81, 361, 1983.

39. Tachibana, T., Ishizeki, K., Sakakura, Y., and Nawa, T., Ultrastructural evidence for a possible secretory function of Merkel cells in the barbels of a teleost fish, *Cyprinus carpio, Cell Tissue Res.,* 235, 695, 1984.

40. DeCoursey, T. E., Jacobs, E. R., and Silver, M. R., Potassium currents in rat type II alveolar epithelial cells, *J. Physiol.,* 395, 487, 1988.

41. Fieber, L. A. and McCleskey, E. W., L-type calcium channels in type I cells of the rat carotid body, *J. Neurophysiol.,* 70, 1378, 1993.

42. Duchen, M. R., Caddy, K. W. T., Kirby, G. C., Patterson, D. L., Ponte, J., and Biscoe, T. J., Biophysical studies of the cellular elements of the rabbit carotid body, *Neuroscience,* 26, 291, 1988.

43. Rudy, B., Diversity and ubiquity of K channels, *Neuroscience,* 25, 729, 1988.

44. Ohmori, H., Studies of ionic currents in the isolated vestibular hair cell of the chick, *J. Physiol.,* 350, 561, 1984.

45. Hudspeth, A. J. and Lewis, L. S., Kinetic analysis of voltage- and ion-dependent conductances in saccular hair cells of the bullfrog *Rana catesbeiana, J. Physiol.,* 400, 237, 1988.

46. Kinnamon, S. C. and Roper, S. D., Membrane properties of isolated mudpuppy taste cells, *J. Gen. Physiol.,* 91, 351, 1988.

47. Zaccone, G., Neuron-specific enolase and serotonin in the Merkel cells of conger-eel (Conger conger) epidermis. An immunohistochemical study, *Histochemistry,* 85, 29, 1986.

48. Garcia-Caballero, T., Gallego, R., Rosón, E., Basanta, D., Morel, G., and Beiras, A., Localization of serotonin-like immunoreactivity in the Merkel cells of pig snout skin, *Anat. Rec.,* 225, 267, 1989.

49. English, K. B., Wang, Z-Z., Stayner, N., Stensaas, L. J., Martin, H., and Tuckett, R. P., Serotonin-like immunoreactivity in Merkel cells and their afferent neurons in touch domes from the hairy skin of rats, *Anat. Rec.,* 232, 112, 1992.

50. Hartschuh, W., Weihe, E., Büchler, M., Helmstaedter, V., Feurle, G. E., and Forssman, W. G., Met-enkephalin-like immunoreactivity in Merkel cells, *Cell Tissue Res.,* 201, 343, 1979.

51. Cheng Chew, S. B. and Leung, P. Y., Species variability in the expression of met- and leu-enkephalin-like immunoreactivity in mammalian Merkel cell dense-core granules. A light- and electron-microscopic immunohistochemical study, *Cell Tissue Res.,* 269, 347, 1992.

52. Alvarez, F. J., Cervantes, C., Villalba, R., Blasco, I., Martínez-Murillo, R., Polak, J. M., and Rodrigo, J., Immunocytochemical analysis of calcitonin gene-related peptide and vasoactive intestinal polypeptide in Merkel cells and cutaneous free nerve endings of cats, *Cell Tissue Res.,* 254, 429, 1988.

53. Gauweiler, B., Weihe, E., Hartschuh, W., and Yanaihara, N., Presence and coexistence of chromogranin A and multiple neuropeptides in Merkel cells of mammalian oral mucosa, *Neurosci. Lett.,* 89, 121, 1988.

54. Hartschuh, W. and Weihe, E., Multiple messenger candidates and marker substances in the mammalian Merkel cell-axon complex: a light and electron microscopic immunohistochemical study, *Prog. Brain Res.,* 74, 181, 1988.

55. Roper, S. D., Synaptic interactions in taste buds, in *Mechanisms of Taste Transduction,* Simon, S. A. and Roper, S. D., Eds., CRC Press, Boca Raton, FL, 1993, chap. 11.

56. Böck, P., Identification of paraneurons by labelling with quinacrine (Atebrin), *Arch. Histol. Jpn.*, 43, 35, 1980.

57. Starke, K., Göthert, M., and Kilbinger, H., Modulation of neurotransmitter release by presynaptic autoreceptors, *Physiol. Rev.*, 69, 864, 1989.

58. Roberts, W. J., Elardo, S. M., and King, K. A., Sympathetically induced changes in the responses of slowly adapting type I receptors in cat skin, *Somatosens. Res.*, 2, 223, 1985.

59. Murase, K., Randic, M., Shirasaki, T., Nakagawa, T., and Akaike, N., Serotonin suppresses N-methyl-D-aspartate responses in acutely isolated spinal dorsal horn neurons of the rat, *Brain Res.*, 525, 84, 1990.

60. Winkelmann, R. K., The Merkel cell system and a comparison between it and the neurosecretory or APUD cell system, *J. Invest. Dermatol*, 69, 41, 1977.

61. Smith, K. R., Jr. and Creech, B. J., Effects of pharmacological agents on the physiological responses of hair discs, *Exp. Neurol.*, 19, 477, 1967.

62. Akoev, G. N., Alekseev, N. P., and Krylov, B. V., *Mechanoreceptors*, Springer-Verlag, Berlin, 1988, chap. 5.

63. Wyatt, C. N. and Peers, C., Nicotinic acetylcholine receptors in isolated type I cells of the neonatal rat carotid body, *Neuroscience*, 54, 275, 1993.

64. Fenwick, E. M., Marty, A., and Neher, E., A patch-clamp study of bovine chromaffin cells and of their sensitivity to acetylcholine, *J. Physiol.*, 331, 577, 1982.

65. Berridge, M. J., Inositol trisphosphate and calcium signalling, *Nature*, 361, 315, 1993.

66. Akaike, N. and Sadoshima, J.-I., Caffeine affects four different ionic currents in the bull-frog sympathetic neurone, *J. Physiol.*, 412, 221, 1989.

67. Marty, A., Calcium-dependent channels with large unitary conductance in chromaffin cell membranes, *Nature*, 291, 497, 1981.

68. Wong, B. S., Lecar, H., and Adler, M., Single calcium-dependent potassium channels in clonal anterior pituitary cells, *Biophys. J.*, 39, 313, 1982.

69. Peterson, O. H. and Maruyama, Y., Calcium-activated potassium channels and their role in secretion, *Nature*, 307, 693, 1984.

70. Gandía, L., García, A. G., and Morad, M., ATP modulation of calcium channels in chromaffin cells, *J. Physiol.*, 470, 55, 1993.

71. Cooper, E. and Nurse, C. A., Studies on isolated Merkel cells using patch and whole cell recording, *Soc. Neurosci. Abstr.*, 12, 46, 1986.

72. Konnerth, A., Patch-clamping in slices of mammalian CNS, *Trends Neurosci.*, 13, 321, 1990.

Chapter 15

ELECTRICAL ACTIVITY AND HORMONE SECRETION IN SERTOLI CELLS

Patricia Grasso and Leo E. Reichert, Jr.

CONTENTS

1. ANATOMY AND GENERAL PHYSIOLOGY OF THE SERTOLI CELL

The Sertoli cell, the only somatic cell within the seminiferous epithelium of the testis, mediates, to a large extent, development and maintenance of spermatogenesis in mammals. The critical role of the Sertoli cell in this process has been illustrated by a number of different experimental strategies. Ultrastructural studies with the electron microscope have shown that contiguous Sertoli cell membranes form tight junctions which effectively divide the seminiferous tubule into a basal compartment, in which spermatogonia give rise to preleptotene primary spermatocytes, and an adluminal compartment which is occupied by germinal cells in more advanced stages of spermatogenesis.[1] Thus, the "blood-testis barrier" created by these tight junctions guarantees that substances carried by the blood gain ready access to the basal region of the tubule occupied by the spermatogonia, but that other substances transit the Sertoli cell cytoplasm before reaching spermatocytes, spermatids, and spermatozoa. In this way, Sertoli cells maintain a highly specialized microenvironment within the seminiferous tubule that supports meiosis and spermatid maturation.[2-4]

Influences of Sertoli cells on other cell types of the testis, i.e., germinal, peritubular myoid, and Leydig cells (reviewed in Reference 5), have been examined by a wide variety of organ and cell culture procedures, and have greatly advanced our understanding of the cellular and molecular aspects of testicular function. Although clonal Sertoli and Leydig cell lines have been developed,[6,7] most experiments dealing with cell-cell interactions in the testis have utilized primary cell cultures, and determination of the physiological relevance of such *in vitro* observations has been achieved by combination with *in vivo* analyses.[8,9]

The Sertoli cell is the primary target for follicle-stimulating hormone (FSH) in the male.[10] FSH binds to specific high-affinity ($K_d \simeq 10^{-10}$ to 10^{-12} M) G protein-coupled receptors on the Sertoli cell plasma membrane.[11] The hormone stimulates adenylyl cyclase and inhibits a calcium-dependent isoform of cyclic nucleotide-dependent phosphodiesterase.[12,13] These two effects bring about a net increase in adenosine 3',5'-monophosphate (cAMP) biosynthesis which in turn leads to protein kinase activation, protein phosphorylation, nuclear effects, and steroidogenesis.[14]

Although this pathway, i.e., cAMP as a second messenger, has long been thought to constitute the principle mechanism by which the FSH signal is transduced, recent reports suggest that there may be other second messengers involved in Sertoli cell responsiveness to FSH stimulation. This notion would not seem unreasonable given the number and variety of cellular responses FSH induces in the Sertoli cell, e.g., enzyme activation,[11-14] ion flux,[15-17] RNA and protein synthesis,[18-20] steroid secretion,[18,21,22] cell division,[23] and cell motility.[24,25] Modulation of intracellular free calcium by FSH,[26-28] as has recently been proposed for regulation of LH-stimulated steroidogenesis in rat Leydig[29] and granulosa[30] cells, and for ACTH-stimulated steroidogenesis in bovine adrenocortical cells,[31] may be another mechanism by which Sertoli cells respond to FSH stimulation.

The Sertoli cell is also a target for testosterone, and FSH and testosterone are essential for initiation and maintenance of spermatogenesis.[19,32,33] Adult Sertoli cells acquire the ability to synthesize small amounts of testosterone from pregnenolone or progesterone, but Leydig cells produce most of this androgen which, together with FSH, is necessary for spermatogenesis to be maintained.[34] Spermatogonia proliferate and advance to become preleptotene primary spermatocytes within the basal compartment of the seminiferous tubule, i.e., outside the blood-testis barrier, and a number of studies indicate that this process can continue without support of androgens or gonadotropins.[35,36] Development beyond this stage, which occurs within the adluminal compartment created by tight junctions of contiguous Sertoli cells, can be blocked by androgen deprivation.[36]

2. ELECTRICAL PROPERTIES OF THE SERTOLI CELL

Clearly, the electrophysiologic properties of the Sertoli cell have not been studied as intensively as those of other endocrine cells, as is evidenced by relevant chapters in this handbook. This is mainly because past interest has focused, for the most part, on the sequence of biochemical events initiated by the interaction of FSH with its Sertoli cell receptor, or on morphological changes induced by FSH stimulation. The electrophysiological consequences of FSH binding, however, were not examined until recently, when attempts were made to obtain complementary information on the mechanism of FSH stimulation by examining FSH-induced changes in Sertoli cell membrane potential.

Electrophysiological studies were first reported using whole seminiferous tubules of the rat,[37] and since then, intracellular recordings of membrane potential of rat Sertoli cells grown in monolayer culture have been made.[38-40] *In situ* and *in vitro* measurements were highly correlated, and membrane potential of unstimulated Sertoli cells was determined to be approximately −21.0 mV. FSH induced a dose-dependent hyperpolarization in cultured rat Sertoli cells (Figure 1) which was attributed to cAMP-dependent mechanisms, since similar changes were invoked by treatment with dibutyryl cyclic AMP.[39]

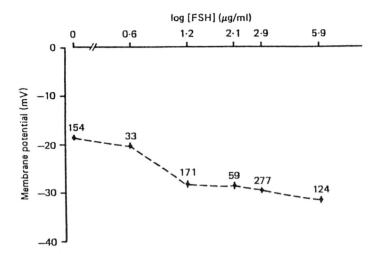

FIGURE 1. Dose-response relationship of FSH-induced hyperpolarization of Sertoli cells. Sertoli cells were isolated and incubated with varying FSH (NIAMDD oFSH S15) concentrations for 24 h before the experiment. Each point represents the mean potential ±SEM and the number of impaled cells (n). Data obtained from six different cultures. (From Joffre, M. and Roche, A., *J. Physiol.,* 400, 481–499, 1988. With permission.)

Additional studies demonstrated that FSH-stimulated hyperpolarization was reduced in low-sodium medium, by removal of external calcium, by increasing external potassium, and by treatment with ouabain, quinidine, or cobalt.[39] Taken together, these observations indicated that FSH-induced hyperpolarization was related to changes in membrane permeability to potassium, and probably the result of changes in Na^+/K^+-ATPase activity. Studies examining the effects of ouabain on basal and FSH-stimulated sodium-dependent calcium (as $^{45}Ca^{2+}$) uptake by Sertoli cell monolayers,[17,41] are in agreement with these observations.

In addition to their supportive and nutritive roles in spermatogenesis, Sertoli cells secrete a potassium-enriched fluid containing androgen-binding protein, a protein which increases the local concentration of testosterone necessary for sperm maturation, into the lumen of the seminiferous tubule.[42] Secretion of this fluid was found to be dependent on the formation of specialized junctional complexes between contiguous Sertoli cells.[43] Ultrastructural studies have shown that similar junctional complexes form in monolayer cultures of Sertoli cells.[44]

Variation in intracellular calcium concentration has been shown to regulate junctional permeability between neighboring cells in several tissue types; increasing intracellular calcium to relatively high levels, or decreasing extracellular calcium, reduces intercellular coupling (reviewed in Reference 45). When Sertoli cells were exposed to treatments known to decrease or suppress intercellular electrical coupling, e.g., treatments which elevated intracellular calcium (A23187, heptanol, carbonylcyanide *m*-chlorophenyl hydrozone) or reduced extracellular calcium (EGTA), rapid depolarization of FSH-stimulated cells to the unstimulated level was observed.[40] These results highlighted the importance of calcium in the regulation of electrical coupling between Sertoli cells and demonstrated that the calcium ion can modify FSH-stimulated hyperpolarization.

3. FUNCTIONAL STUDIES

The recently identified ability of FSH to bind calcium,[46] activate calcium channels,[47-49] and influence Na^+/Ca^{2+} exchange[17,41] in cultured rat Sertoli cells and FSH receptor-containing proteoliposomes suggests that the mechanism of FSH signal transduction in the testis is a process that may involve second messengers other than those associated with hormone-stimulated elevation of cAMP.[50] The involvement of FSH in regulation of calcium flux into Sertoli cells suggests cytosolic free calcium may be an important second messenger in FSH signal transduction. Evidence recently developed in this and other laboratories in support of this notion is reviewed below.

3.1. FSH-STIMULATED CALCIUM CHANNEL ACTIVITY

The ability of FSH to affect changes in intracellular free calcium in Sertoli cells is well documented,[50] although the mechanisms involved in this process have not been fully defined. We[51] and others[52] have shown that although the phosphatidyl inositol pathway, which is associated with mobilization of intracellular calcium,[53] is present in cultured Sertoli cells from immature (15- to 18-d-old) rats, it is unresponsive to FSH. Thus, the effect of FSH on cytosolic free calcium levels appears to be more directly related to internalization of extracellular calcium, rather than to mobilization from intracellular stores.

Successful incorporation into liposomes of hormone-responsive Triton® X-100 solubilized FSH receptors derived from bovine calf testes[54] provided a useful model with which to examine the effects of FSH on calcium flux. These studies indicated that hormone-specific and concentration-dependent uptake of calcium (as $^{45}Ca^{2+}$) in response to FSH stimulation could be inhibited by antagonists specific for either voltage-sensitive (methoxyverapamil and nifedipine) or voltage-independent (ruthenium red and gadolinium chloride) calcium channels.[47] Similar results were obtained when cultured Sertoli cells from 14- to 16-d-old rats were utilized as the experimental model (Figure 2). These observations, as well as data from other laboratories,[55] suggested that FSH-stimulated uptake of extracellular calcium is facilitated by channels gated by hormonally induced changes in membrane potential, as well as by channels which are insensitive to such changes.

A class of ion channels gated by specific signal-transducing membrane-associated G proteins has been identified.[56] These channels, when activated by nonhydrolyzable analogs of guanosine triphosphate (GTP), in the absence of receptor occupancy, mimic the response produced by a ligand-bound receptor. Several lines of evidence suggest that the calcium channel regulatory protein is G_s, the activator of adenylyl cyclase.[56]

The Triton® X-100-solubilized FSH receptor we incorporated into liposomes was determined to be coupled to a cholera toxin-sensitive G protein, presumably G_s.[57] If the calcium channels activated by FSH binding to its receptor were gated by G_s, it was expected that GTP, GTP$_\gamma$S or Gpp(NH)p would stimulate calcium (as $^{45}Ca^{2+}$) uptake by FSH receptor-containing proteoliposomes. Such stimulation of calcium uptake, however, was not observed,[47] nor was cholera toxin, which inhibits the inherent GTPase activity of G_s and constitutively activates G_s, able to induce uptake of extracellular calcium by monolayer cultures of rat Sertoli cells.[48] These observations suggested that receptor occupancy by FSH exerts a direct effect on calcium influx which is not mediated by activation of receptor-associated G_s protein in Sertoli cell monolayers.

In fura-2-AM fluorescence spectrophotometric studies with freshly isolated Sertoli cells from slightly older rats (17 to 20 d), however, treatment with forskolin, cholera toxin, or dibutyryl cAMP resulted in an increase in cytosolic free calcium similar in magnitude to that induced by FSH.[58] Since this study did not include data from parallel experiments with these agents in the presence of calcium channel antagonists or in calcium-free medium, it is difficult to unequivocally relate the observed rise in cytosolic free calcium in response to these treatments to an effect on G_s-gated plasma membrane calcium channels.

In Sertoli cell monolayers, dibutyryl cAMP, a membrane-permeable analog of cAMP, did not enhance $^{45}Ca^{2+}$ uptake over basal levels, although a concentration-related increase in conversion of androstenedione to estradiol, a Sertoli cell response known to be mediated by

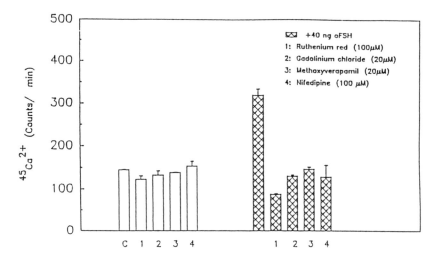

FIGURE 2. Effects of voltage-independent and voltage-activated calcium channel blocking agents on $^{45}Ca^{2+}$ uptake by cultured rat Sertoli cells. Sertoli cells (1.3×10^5 cells/well) were preincubated for 15 min in culture medium labeled with 0.4 μCi $^{45}CaCl_2$ and containing ruthenium red, (RR), $GdCl_3$, methoxyverapamil (D-600), or nifedipine before addition of oFSH. Cultures were subsequently incubated for 24 h. Each bar and vertical line represents the mean ±SD (n = 4) $^{45}Ca^{2+}$ uptake during the 24-h culture period. (C) Control cultures preincubated in $^{45}CaCl_2$-labeled culture medium alone. (From Grasso, P. and Reichert, L. E., Jr., *Endocrinology*, 125, 3029–3036, 1989. ©The Endocrine Society. With permission.)

activation of adenylyl cyclase, was evident.[48] These observations, together with the results of cholera toxin experiments, suggested that FSH stimulation of calcium channels in monolayer cultures of rat Sertoli cells does not involve gating by activated G_s protein either directly or indirectly through generation of second messengers.

In a recent study using fura-2-AM-loaded, freshly isolated rat Sertoli cells, however, treatment with MDL 12,330A, an inhibitor of adenylyl cylase, Rp-cAMP, a cAMP antagonist, or pertussis toxin, which uncouples G_i from its receptor, was observed to suppress FSH-stimulated increases in cytosolic free calcium.[28] These results suggested a relationship between adenylyl cyclase activation and FSH-induced elevation of intracellular calcium in this model system. The mechanism whereby pertussis toxin treatment antagonized G_s-coupled FSH receptors[57] rather than augmented FSH action via inhibition of G_i, as previously demonstrated for its effect on FSH-stimulated estradiol biosynthesis,[48] however, is unclear.

One explanation for FSH-stimulated calcium entry into Sertoli cells could be that the FSH receptor itself may function as a calcium channel. This suggestion, however, is not supported by recent studies.[59] Using whole-cell patch clamp techniques, neither FSH, forskolin, nor dibutyryl cAMP were able to induce inwardly directed calcium currents in human embryonic kidney 293F(wt1) cells (a clonal line of 293 cells which lack detectable inward calcium currents and have been stably transfected with the cDNA for rat FSH receptor).[59]

3.2. FSH EFFECTS ON SODIUM-DEPENDENT CALCIUM UPTAKE

When monolayer cultures of Sertoli cells from 14- to 16-d-old rats or FSH receptor-containing proteoliposomes were incubated in sodium-free (choline-substituted) buffer, calcium (as $^{45}Ca^{2+}$) uptake increased in a concentration-dependent manner as intracellular (or liposome-encapsulated) sodium levels were elevated.[41] FSH was found to reduce this effect in both model systems. Removal of the outwardly directed sodium gradient, achieved by increasing external sodium, decreased sodium-dependent $^{45}Ca^{2+}$ influx and eliminated the inhibitory effect of FSH (Figure 3).

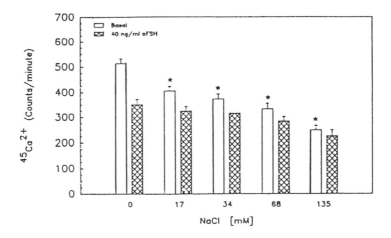

FIGURE 3. Effects of extracellular Na^+ on $^{45}Ca^{2+}$ uptake by cultured rat Sertoli cells. Sertoli cell monolayers were preincubated for 15 min in buffer containing 135 m*M* NaCl. The monolayers were then transferred to $^{45}CaCl_2$-labeled buffer containing ruthenium red and methoxyverapamil (RR/D600) (100 μ*M*/20 μ*M*) and 135 m*M* choline chloride (0 m*M* NaCl) or increasing concentrations of NaCl in the presence or absence of 40 ng/ml oFSH for 24 h. Each bar and vertical line represents mean ±SD (n = 4) $^{45}Ca^{2+}$ uptake during the 24-h culture period. *, $^{45}Ca^{2+}$ uptake significantly (*p* <0.05) lower than basal uptake in choline buffer. (From Grasso, P., Joseph, M. P., and Reichert, L. E., Jr., *Endocrinology,* 128, 158–164, 1991. ©The Endocrine Society. With permission.)

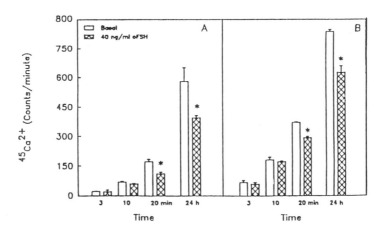

FIGURE 4. Temporal aspects of the augmenting effect of ouabain on basal and FSH-attenuated Na^+-dependent $^{45}Ca^{2+}$ uptake by cultured rat Sertoli cells. Sertoli cell monolayers were preincubated for 15 min in buffer containing 135 m*M* NaCl in the absence (panel A) or presence (panel B) of 1 m*M* ouabain and then transferred to $^{45}CaCl_2$-labeled buffer containing ruthenium red and methoxyverapamil (RR/D-600) (100 μ*M*/20 μ*M*) and 135 m*M* choline chloride in the absence or presence of 40 ng/ml oFSH for varying time periods. Each bar and vertical line represents mean ±SD (n = 4) $^{45}Ca^{2+}$ uptake. *, $^{45}Ca^{2+}$ uptake significantly (*p* <0.05) lower than basal level. (From Grasso, P., Joseph, M. P., and Reichert, L. E., Jr., *Endocrinology,* 128, 158–164, 1992. ©The Endocrine Society. With permission.)

Treatment of monolayer cultures of Sertoli cells with ouabain also resulted in significantly greater uptake of $^{45}Ca^{2+}$ when compared to untreated cells, presumably by preventing extrusion of intracellular sodium through Na^+/K^+-ATPase activity known to be present in Sertoli cell membranes[15] (Figure 4). The inhibitory effect of FSH on sodium-dependent $^{45}Ca^{2+}$ influx was also evident in the presence of ouabain. These results demonstrate that pharmacological manipulation of the electrochemical gradient for sodium caused increased uptake of calcium

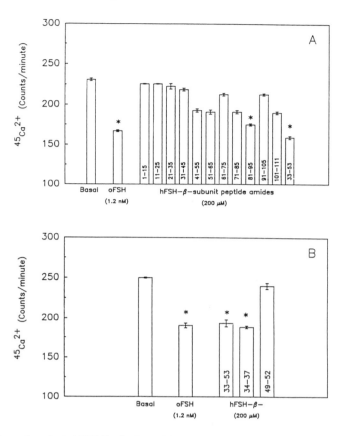

FIGURE 5. Effects of synthetic hFSH-β subunit peptide amides on Na$^+$-dependent ^{45}Ca^{2+} uptake by cultured rat Sertoli cells. Sertoli cell monolayers were preincubated for 15 min in buffer containing 135 mM NaCl. The monolayers were then transferred to buffer labeled with 0.4 µCi/ml ^{45}Ca^{2+} and containing 100 µM ruthenium red (RR), 20 µM methoxyverapamil (D-600), and 135 mM choline chloride in the absence or presence of 1.2 nM oFSH or 200 µM hFSH-β subunit peptide amides for 24 h. Each bar and vertical line represents mean ±SD (n = 4) ^{45}Ca^{2+} uptake by (A) overlapping peptide amides representing the entire primary structure of hFSH-β subunit, and (B) hFSH-β-(33-53), hFSH-β-(34-37), and hFSH-β-(49-52) peptides during the 24-h culture period. *, ^{45}Ca^{2+} uptake significantly (p <0.05) lower than basal uptake. (From Grasso, P., Joseph, M. P., and Reichert, L. E., Jr., *Mol. Cell. Endocrinol.* 96, 19–24, 1993. With permission.)

which was also sensitive to FSH inhibition, and provided a second line of evidence supporting FSH-mediated Na$^+$/Ca^{2+} exchange activity in the testis.

Using a synthetic peptide strategy, it was recently determined that the observed effect of FSH on sodium-dependent calcium uptake in monolayer cultures of Sertoli cells from immature rats involves a tetrapeptide contained within a receptor-binding domain of the β subunit.[17] When a series of overlapping peptide amides, representing the entire primary structure of the β subunit of hFSH, were screened for their effects on sodium-dependent calcium (as ^{45}Ca^{2+}) uptake, hFSH-β-(33-53), previously identified as a receptor-binding domain of hFSH-β subunit,[60] significantly reduced sodium-dependent calcium influx (Figure 5A). It was found that a region within this sequence, TRDL [hFSH-β-(34-37)], was equally as active as hFSH-β-(33-53), suggesting that the regulatory influence of hFSH-β-(33-53) on Na$^+$/Ca^{2+} exchange was due to this tetrapeptide (Figure 5B). As with FSH, the inhibitory effect of hFSH-β-(34-37) on sodium-dependent ^{45}Ca^{2+} influx could still be seen after Sertoli cell monolayers were treated with ouabain, or when extracellular sodium was replaced with equimolar concentrations of choline.

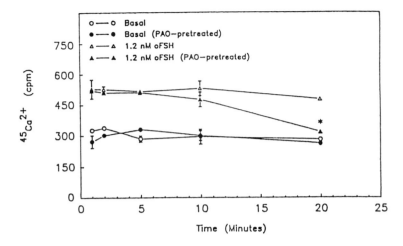

FIGURE 6.   Effects of phenylarsine oxide (PAO) on FSH-stimulated $^{45}Ca^{2+}$ uptake by cultured rat Sertoli cells. Sertoli cell monolayers (1.0×10^5 cells per well) were preincubated for 5 min with 80 μM PAO in DMEM/F-12 or with DMEM/F-12 alone, washed extensively, and incubated in DMEM/F-12 labeled with 0.4 μCi $^{45}Ca^{2+}$ in the absence or presence of oFSH (1.2 nM) for the indicated times. Each point represents mean $\pm$SD (n = 4) $^{45}Ca^{2+}$ uptake. *, $^{45}Ca^{2+}$ uptake significantly ($p <0.05$) different from FSH-stimulated uptake by untreated cells. (From Grasso, P., Santa-Coloma, T. A., and Reichert, L. E., Jr., *Endocrinology,* 131, 2622–2628, 1992. ©The Endocrine Society. With permission.)

The observed stimulatory influence of FSH on calcium channel activity and its inhibitory effect on sodium-dependent calcium uptake by cultured rat Sertoli cells and FSH receptor-containing proteoliposmes indicated that FSH-regulated influx of extracellular calcium involves more than one mechanism. Inhibition of the sodium-dependent component of calcium influx by FSH, in conjunction with its stimulatory effect on calcium channel activity, apparently helps to maintain the appropriate level of intracellular Ca^{2+} required for modulation of FSH receptor-mediated post-binding events in the Sertoli cell.

3.3.   VESICULAR UPTAKE OF EXTRACELLULAR CALCIUM BY FSH-STIMULATED SERTOLI CELLS

It has been shown that calcium is required for optimal binding of FSH to receptor,[61] and that stabilization of FSH-receptor complexes may involve calcium-dependent transglutaminase activation.[62,63] To determine the functional basis for these observations, eleven overlapping peptide amides representing the entire primary structure of hFSH-β subunit were screened for their ability to bind calcium (as $^{45}Ca^{2+}$).[42] These studies indicated that hFSH-β-(1-15), which contains an amino acid sequence similar to that found in the loop structures of the calcium-binding domains of calmodulin, bound significant amounts of $^{45}Ca^{2+}$ with a K_d of 1.2 ± 0.3 mM, an affinity similar to that reported for a peptide corresponding to calmodulin binding site III.[64]

The ability of hFSH-β-(1-15) to bind calcium correlated well with its ability to induce uptake of $^{45}Ca^{2+}$ by liposomes.[49] The physiological significance of these heretofore unrecognized properties of this peptide is unknown. We have suggested that, in addition to its effect on voltage-sensitive calcium channel activity, interaction of FSH with its receptor may induce a confirmational change in the hormone which results in the formation of calcium-conducting transmembrane channels.[49] Alternatively, fragments of FSH-β containing calcium-binding regions may be generated after FSH-receptor complex internalization and FSH degradation.[65] Since internalized vesicles contain higher calcium concentrations (similar to extracellular calcium) than cytosol, calcium-binding fragments of FSH resulting from proteolytic cleavage may be able to induce calcium entry into the cytosol via a process thermodynamically favored

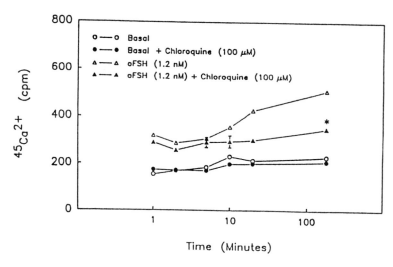

FIGURE 7. Effects of chloroquine on FSH-stimulated $^{45}Ca^{2+}$ uptake by cultured rat Sertoli cells. Sertoli cell monolayers (2.9×10^5 cells per well) were preincubated for 1 h in 100 μM chloroquine before addition of DMEM/F-12 labeled with 0.4 μCi $^{45}Ca^{2+}$ and containing oFSH (1.2 n*M*). The incubation was allowed to continue for the indicated times. Each point represents mean ±SD (n = 4) $^{45}Ca^{2+}$ uptake. *, $^{45}Ca^{2+}$ uptake significantly ($p < 0.05$) different from FSH-stimulated uptake by untreated cells. (From Grasso, P., Santa-Coloma, T. A., and Reichert, L. E., Jr., *Endocrinology*, 131, 2622–2628, 1992. ©The Endocrine Society. With permission.)

by these high levels of intravesicular calcium. Indeed, in a study designed to investigate this hypothesis,[66] we demonstrated that exposing Sertoli cell monolayers to agents which inhibited internalization (Figure 6) and endosomal/lysosomal degradation (Figure 7) of membrane-bound FSH blocked the sustained phase of FSH-stimulated calcium influx. Vesicular uptake of extracellular calcium, which accompanies receptor-mediated endocytosis of FSH-receptor complexes, and release of channel-forming peptides by proteolytic processing of FSH, suggests a novel mechanism by which FSH increases intracellular calcium in Sertoli cells.

3.4. TGF-β1 EFFECTS ON CALCIUM UPTAKE

There is a growing body of evidence suggesting that testicular function is regulated not only by the pituitary gonadotropins, but also by local modulators which include several regulatory peptides and growth factors (reviewed in Reference 67). Among these local regulators is transforming growth factor-β1 (TGF-β1). TGF-β1 is secreted in the testis[68] and has been shown to antagonize FSH-stimulated aromatase activity and lactate production in Sertoli cells.[69]

The mechanism of signal transduction by TGF-β1 in the testis, other than possible activation of a serine/threonine kinase,[70] is unknown. In an effort to determine whether or not Sertoli cell responsiveness to TGF-β1 might include changes in calcium metabolism, Sertoli cell monolayers were treated with TGF-β1 and its effects on calcium (as $^{45}Ca^{2+}$) influx were measured.[71] In this study, TGF-β1 stimulated uptake of extracellular calcium, but with a time course different from that observed in response to FSH.[48] In contrast to the rapid stimulation invoked by FSH (observed in less than 1 min), the calcium response to TGF-β1 required several hours and was abolished by treatment of the monolayers with actinomycin D. TGF-β1 also inhibited the initial surge of calcium influx induced by FSH binding. These results suggested that TGF-β1 may have short- as well as long-term effects on calcium metabolism in Sertoli cells, and that the increase in calcium influx observed in response to TGF-β1 stimulation requires gene expression. Furthermore, the ability of TGF-β1 to inhibit FSH-induced calcium influx suggests that it may exert important local regulatory influences on other, as yet undetermined, FSH effects in the Sertoli cell.

4. PHARMACOLOGY AND TOXICOLOGY

Several phthalate esters, including the widely used plasticizer di(2-ethylhexyl)phthalate (DEHP), are reproductive toxins in rats and mice.[72] Several studies have demonstrated that the Sertoli cell is the primary target for these compounds, which disrupt Sertoli cell-germ cell interactions essential for spermatogenesis.[73] Testicular lesions produced by active phthalates are characterized by structural changes in Sertoli cell plasma membranes, endoplasmic reticulum, and mitochondrial membranes, extensive vacuolation of Sertoli cell cytoplasm, and early release of spermatids and spermatocytes into the lumen of the seminiferous tubule.[74] The testicular toxicity of DEHP was mimicked *in vivo* by its active metabolite, MEHP [mono(2-ethylhexyl)phthalate].[75] *In vitro* studies utilizing Sertoli cell-germ cell cocultures showed that monoesters of active phthalates increased germ cell detachment, lactate secretion and intracellular lipid content and decreased ATP levels, mitochondrial dehydrogenase activity, and pyruvate secretion.[73]

Because testicular lesions induced by active phthalates were restricted to tubules at stages in the spermatogenic cycle most responsive to FSH,[76] it seemed reasonable to suspect that these agents might antagonize FSH action. In Sertoli cell monolayers, MEHP specifically inhibited FSH-induced cAMP biosynthesis, although it had no effect on cAMP accumulation in response to isoproterenol, forskolin, cholera toxin, or prostaglandin E.[77,78] These results suggested that the effects of phthalate monoesters on FSH-stimulated cAMP biosynthesis may be related to their testicular toxicity.

We have recently shown that the inhibitory effect of MEHP on FSH-induced cAMP accumulation in Sertoli cells is at the level of the cholera toxin-sensitive G protein[79] which couples the activated FSH receptor to adenylyl cyclase.[80] Exposure of Sertoli cell monolayers to MEHP resulted in a fourfold decrease in affinity of the receptor for FSH with no change in receptor concentration, and amplified the attenuating effect of GTP on FSH binding.[79] Thus, the ability of MEHP to reduce FSH-stimulated cAMP biosynthesis observed in earlier studies[77,78] may be related to decreased FSH binding, resulting from some as yet undetermined influence of MEHP on G_s.

5. RELEVANCE OF CALCIUM FLUX TO SERTOLI CELL FUNCTION

The recently identified ability of FSH to bind calcium[46] and to stimulate its entry by multiple transport mechanisms[41,47,66] provides convincing evidence that precise control of cytosolic free calcium is a critical requirement for Seroli cell function. Although the role of calcium in Sertoli cell responsiveness to FSH stimulation is just beginning to be eludicated, its importance in FSH binding[61-63] and steroidogenesis[48,81,82] is well documented.

In a recent study, a synthetic peptide approach was utilized to examine the role of Cys residues within hFSH-β subunit receptor-contact regions in both calcium flux and activation of adenylyl cyclase in monolayer cultures of rat Sertoli cells.[82] Analogs of previously identified receptor binding domains within the β subunit of hFSH[60,83] were synthesized, in which all Cys residues were replaced with Ser. The ability of the Ser analogs to stimulate uptake of extracellular calcium (as $^{45}Ca^{2+}$), cAMP biosynthesis, and androstenedione conversion to estradiol were then examined. The results indicated that while the native Cys-containing peptides were unable to initiate a calcium response, Ser-substituted peptides stimulated uptake of $^{45}Ca^{2+}$ significantly over basal levels. The concentration-dependent agonist effect previously demonstrated for the Cys-containing peptides,[60,83] however, was not observed for the Ser analogs. Although the Cys residues contained within receptor-binding domains of hFSH-β subunit were not required for binding of hormone to receptor,[84] the Ser analog study indicated that these residues have the ability to influence post-binding events associated with transduction of the FSH signal.

These observations suggested that elevation of intracellular free calcium may exert autocrine negative control of testicular steroidogenesis. The presence of Cys residues within receptor-contact regions of FSH-β subunit seems required for steroidogenesis to occur. Ser substitution provides additional negative charges and potential calcium chelating residues within the hydrophilic regions of these peptides, and may thereby increase calcium binding and transport abilities of the peptides. These substitutions, either alone or in association with as yet undetermined conformational or other changes, may convert the native Cys-containing peptides into calcium-mobilizing analogs.

6. SUMMARY

As evidenced from other contributions to this handbook, the electrophysiologic properties of the Sertoli cell have not been studied in as great detail as have a wide variety of other endocrine cell types. Calcium flux, however, can reasonably be concluded to strongly influence the electrophysiology of the Sertoli cell, so elucidation of factors influencing calcium flux should facilitate understanding of Sertoli cell function and its control.

A growing body of evidence indicates that precise control of cytosolic free calcium is critical to Sertoli cell function. The calcium requirement for maximal binding of FSH to receptor and to stabilize FSH-receptor complexes, the effects of calcium on FSH-stimulated hyperpolarization of Sertoli cells, and Sertoli cell junctional permeability, the ability of regions of FSH-β subunit to bind calcium and form transmembrane calcium-conducting pores, as well as the presence of multiple mechanisms for regulating calcium influx, e.g., FSH-sensitive calcium channel activity, sodium-dependent calcium entry, and vesicular uptake of extracellular calcium, clearly demonstrate the importance of calcium in FSH action, including the initiation and consequences of electrical activity in Sertoli cells (Figure 8).

The role of calcium in Sertoli cell function is under active investigation, and initial data indicate that it has a profound influence on FSH-stimulated steroidogenesis. Given the power of such experimental tools as solid-phase peptide synthesis and molecular cloning, these technologies, in combination with the more traditional methodologies used in functional studies, are likely to reveal other heretofore unknown influences that calcium may exert on Sertoli cell responsiveness to FSH stimulation.

ACKNOWLEDGMENT

Supported by NIH Grant HD-13938.

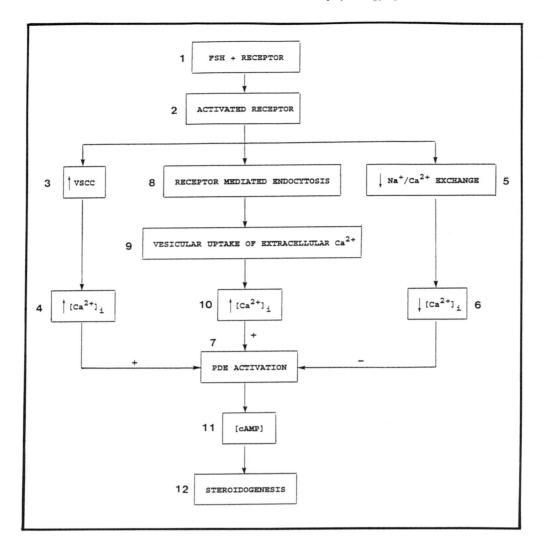

FIGURE 8. Functional relevance of FSH regulation of intracellular free calcium in Sertoli cells. Binding of FSH to its membrane receptor (1) results in receptor activation (2), hyperpolarization of the cell membrane, and activation of voltage-sensitive calcium channels (VSCC) (3). This sequence of events causes an immediate elevation in intracellular free calcium $[Ca^{2+}]_i$. (4). Sodium-calcium exchange is attenuated by FSH binding (5), and the sodium-dependent component of calcium entry is reduced (6). These two actions of FSH on calcium influx have opposite effects on calcium-dependent phosphodiesterase (PDE) (7), which hydrolyzes cAMP and is present in high concentrations in immature rat Sertoli cells. Receptor-mediated endocytosis of FSH-receptor complexes (8) facilitates vesicular uptake of extracellular calcium (9), and results in a secondary sustained elevation of intracellular free calcium (10). Phosphodiesterase activity modulates cAMP levels (11) and regulates steroidogenesis (12).

REFERENCES

1. Dym, M. and Fawcett, D. W., The blood-testis barrier in the rat and the physiological compartmentation of the seminiferous epithelium, *Biol. Reprod.*, 3, 308, 1970.
2. Aoki, A. and Fawcett, D. W., Impermeability of Sertoli cell junctions to prolonged exposure to peroxidase, *Andrologia*, 7, 63, 1975.
3. Setchell, B. P., The blood-testicular fluid barrier in sheep, *J. Physiol. (London)*, 189, 63, 1967.
4. Setchell, B. P.,The functional significance of the blood-testis barrier, *Andrology*, 1, 3, 1980.
5. Skinner, M. K., Cell-cell interactions in the testis, *Endocr. Rev.*, 12, 45, 1991.

6. Mather, J. P., Establishment and characterization of two distinct mouse testicular epithelial cell lines, *Biol. Reprod.*, 23, 243, 1980.

7. Ascoli, M., Characterization of several clonal lines of cultured Leydig tumor cells: gonadotropin receptors and steroidogenic responses, *Endocrinology*, 108, 88, 1981.

8. Conti, M., Toscano, M. V., Geremia, R., and Steganini, M., Follicle-stimulating hormone regulates in vivo testicular phosphodiesterase, *Mol. Cell. Endocrinol.*, 29, 79, 1983.

9. Means, A. R., Early sequence of biochemical events in the action of follicle-stimulating hormone on the testis, *Life Sci.*, 15, 371, 1974.

10. Steinberger, E., Hormonal control of mammalian spermatogenesis, *Physiol. Rev.*, 51, 1, 1971.

11. Dattatreyamurty, B., Zhang, S.-B., and Reichert, L. E., Jr., Purification of follitropin receptor from bovine calf testes, *J. Biol. Chem.*, 265, 5494, 1990.

12. Conti, M., Toscano, M. V., Petrelli, L., Geremia, R., and Stefanini, M., Regulation by follicle-stimulating hormone and dibutrylyl adenosine $3',5'$-monophosphate of a phosphodiesterase isoenzyme of the Sertoli cell, *Endocrinology*, 110, 1189, 1982.

13. Steinberger, A., Hintz, M., and Heindel, J. J., Changes in cyclic AMP responses to FSH in isolated rat Sertoli cells during sexual maturation, *Biol. Reprod.*, 19, 566, 1978.

14. Fekunding, J. L. and Means, A. R., Characterization of follicle-stimulating hormone activation of Sertoli cell cyclic AMP-dependent protein kinase, *Endocrinology*, 101, 1358, 1977.

15. Joffre, M. and Roche, A., Follicle-stimulating hormone induces hyperpolarization of immature rat Sertoli cells in monolayer culture, *J. Physiol. (London)*, 400, 481, 1988.

16. Grasso, P. and Reichert, L. E., Jr., Induction of calcium transport into cultured rat Sertoli cells and liposomes by follicle-stimulating hormone, *Rec. Prog. Horm. Res.*, 48, 517, 1993.

17. Grasso, P., Joseph, M. P., and Reichert, L. E., Jr., A tetrapeptide within a receptor-binding regions of human follicle-stimulating hormone beta-subunit, hFSH-ß-(34-37), regulates sodium-calcium exchange in Sertoli cells, *Mol. Cell. Endocrinol.*, 96, 19, 1993.

18. Foucault, P., Carreau, S., Kuczynski, W., Guillaumin, J. M., Bardos, P., and Drosdorosky, M. A., Human Sertoli cells in vitro lactate, estradiol-17ß and transferrin production, *J. Androl.*, 13, 361, 1992.

19. Louis, B. G. and Fritz, I. B., Follicle-stimulating hormone and testosterone independently increase the production of androgen-binding protein by Sertoli cells in culture, *Endocrinology*, 104, 454, 1979.

20. Lamb, D. J., Kessler, M. J., Shewach, D. S., and Steinberger, A., Characterization of Sertoli cell RNA synthetic activities in vitro at selected times during sexual maturation, *Biol. Reprod.*, 27, 374, 1982.

21. Dorrington, J. H., Fritz, I. B., and Armstrong, D. I., Control of testicular estrogen synthesis, *Biol. Reprod.*, 18, 55, 1978.

22. Reddy, P. R. K. and Vilee, L. E., Stimulation of ornithine decarboxylase activity by gonadotropic hormones and cyclic AMP in the testis of immature rats, *Biochem. Biophys. Res. Commun.*, 65, 1350, 1975.

23. Griswald, M. D., Solari, A., Tung, P. S., and Fritz, I. B., Stimulation by follicle-stimulating hormone of DNA synthesis and mitosis in cultured Sertoli cells prepared from testes of immature rats, *Mol. Cell. Endocrinol.*, 7, 151, 1977.

24. Marcum, J. M., Dedman, J. R., Binkley, B. R., and Means, A. R., Control of microtubule assembly-disassembly by Ca^{2+}-dependent regulator protein, *Proc. Natl. Acad. Sci. U.S.A.*, 75, 3771, 1978.

25. Solari, A. J. and Fritz, I. B., The ultrastructure of immature Sertoli cells. Maturation-like changes during culture and the maintenance of mitotic potentiality, *Biol. Reprod.*, 18, 329, 1978.

26. Grasso, P., Santa-Coloma, T. A., and Reichert, L. E., Jr., Synthetic peptides corresponding to human follicle stimulating hormone (hFSH)-β-(1-15) and hFSH-β-(51-65) induce uptake of $^{45}Ca^{++}$ by liposomes: evidence for calcium-conducting transmembrane channel formation, *Endocrinology*, 128, 2745, 1991.

27. Grasso, P., Santa-Coloma, T. A., and Reichert, L. E., Jr., Correlation of follicle-stimulating hormone (FSH) receptor complex internalization with the sustained phase of FSH-induced calcium uptake by cultured Sertoli cells, *Endocrinology*, 131, 2622, 1992.

28. Gorczynska, E., Spaliviero, J., and Handelsman, D. J., The relationship between $3',5'$-cyclic adenosine monophosphate and calcium in mediating follicle-stimulaing hormone signal transduction in Sertoli cells, *Endocrinology*, 134, 293, 1994.

29. Cooke, B. A., Is cyclic AMP an obligatory second messenger for luteinizing hormone? *Mol. Cell. Endocrinol.*, 69, C11, 1990.

30. Asem, E. K., Molnar, M., and Hertelendy, F., Luteinizing hormone-induced intracellular calcium mobilization in granulosa cells: comparison with forskolin and 8-bromo-adenosine $3'5'$-monophosphate, *Endocrinology*, 120, 853, 1987.

31. Li, Z-G., Park, D., and LaBella, F. S., Adrenocorticotropin$_{(1-10)}$ and $_{(11-24)}$ promote adrenal steroidogenesis by different mechanisms, *Endocrinology*, 125, 592, 1989.

32. Sandborn, B. M., Steinberger, A., Tcholakian, R. K., and Steinberger, E., Direct measurement of receptors in cultured Sertoli cells, *Steroids*, 29, 493, 1977.

33. Russell, L. D., Corbin, T. J., Borg, K. E., deFranca, L. R., Grasso, P., and Bartke, A., Recombinant human follicle stimulating hormone is capable of exerting a biological effect in the adult hypophysectomized rat by reducing the numbers of degenerating germ cells, *Endocrinology*, 133, 2062, 1993.

34. Hutson, J. C. and Stocco, D. M.,Specificity of hormone-induced responses of testicular cells in culture, *Biol. Reprod.*, 19, 768, 1978.

35. Parvinen, M., Wright, W. W., Phillips, D. M., Mather, J. P., Musto, N. A.and Bardin, C. W., Spermatogenesis in vitro: completion of early meiosis and early spermatogenesis, *Endocrinology*, 112, 1150, 1983.

36. Hage-van Noort, M., Puijk, W. C., Schaaper, W. M. M., Kuperus, D., Beckman, N. J. C. M., Plasman, H. H., Grootegoed, J. A., and Meloen, R. H.,Development of antagonists and agonists of follicle stimulating hormone, in *Spermatogenesis, Fertilization, Contraception: Molecular, Cellular and Endocrine Events in male Reproduction*, Nieschlag, E. and Habenicht, U.-F., Eds., Springer-Verlag, New York, 1992, 33.

37. Cuthbert, A. W. and Wong, P. Y. D., Intracellular potentials in cells of the seminiferous tubules of rats, *J. Physiol.*, 248, 173, 1975.

38. Roche, A. and Joffre, M., Effect of FSH on the membrane potential of cultured Sertoli cells from immature rat testis, *IRCS Med. Sci.*, 12, 570, 1984.

39. Joffre, M. and Roche, A., Follicle-stimulating hormone induces hyperpolarization of immature rat Sertoli cells in monolayer culture, *J. Physiol.*, 400, 481, 1988.

40. Roche, A. and Joffre, M., Effect of uncoupling treatments on FSH-induced hyperpolarization of immature rat Sertoli cells from Sertoli cell-enriched cultures, *J. Reprod. Fertil.*, 85, 343, 1989.

41. Grasso, P., Joseph, M. P., and Reichert, L. E., Jr., A new role for follicle-stimulating hormone in the regulation of calcium flux in Sertoli cells: inhibition of Na^+/Ca^{++} exchange, *Endocrinology*, 128, 158, 1991.

42. Waites, G. M. H. and Gladwell, K. T., Physiological significance of fluid secretion in the testis and blood testis barrier, *Physiol. Rev.*, 62, 624, 1982.

43. Vitale, R., Fawcett, D. W., and Dym, M., The normal development of the blood testis barrier and the effects of clomiphene and estrogen treatment, *Anat. Rec.*, 176, 333, 1973.

44. Hadley, M. A., Byers, S. W., Suarez-Quian, C. A., Kleinman, H. K., and Dym, M., Extracellular matrix regulates Sertoli cell differentiation, testicular cord formation, and germ cell development in vitro, *J. Cell. Biol.*, 101, 154, 1985.

45. Spray, D. C. and Bennett, M. V. L., Physiology and pharmacology of gap junctions, *Annu. Rev. Physiol.*, 47, 281, 1985.

46. Santa-Coloma, T. A., Grasso, P., and Reichert, L. E., Jr., Synthetic human follicle-stimulting hormone-β-(1-15) peptide amide binds Ca^{2+} and possesses sequence similarity to calcium binding sites of calmodulin, *Endocrinology*, 130, 1103, 1992.

Grasso, P. and Reichert, L. E., Jr., Follicle-stimulating hormone receptor-mediated uptake of $^{45}Ca^{2+}$ by oteoliposomes and cultured rat Sertoli cells: evidence for involvement of voltage-activated and voltage-pendent calcium channels, *Endocrinology*, 125, 3029, 1989.

, P. and Reichert, L. E., Jr., Follicle-stimulating hormone receptor-mediated uptake of $^{45}Ca^{2+}$ by rat Sertoli cells does not require activation of cholera toxin- or pertussis toxin-sensitive guanine binding proteins or adenylate cyclase, *Endocrinology*, 127, 949, 1990.

., Santa-Coloma, T. A., and Reichert, L. E., Jr., Synthetic peptides corresponding to hFSH-β-(1-15) -β-(51-65) induce uptake of $^{45}Ca^{2+}$ by liposomes: evidence for calcium-conducting transmembrane mation, *Endocrinology*, 128, 2745, 1991.

, Dedman, J. R., Tash, J. S., Tindall, D. J., van Sickle, M. and Welsh, M. J., Regulation of the ell by follicle-stimulating hormone, *Annu. Rev. Physiol.*, 42, 59, 1980.

d Reichert, L. E., Jr., Regulation of the phosphoinositide pathway in cultured Sertoli cells from ndocrinology, 123, 230, 1988.

o, S., and Conti, M., Follicle-stimulating hormone modulation of phosphoinositide turnover Sertoli cell in culture, *Endocrinology*, 123, 2032, 1988.

lular calcium homeostasis, *Annu. Rev. Biochem.*, 56, 395, 1987.

murty, B., and Reichert, L. E., Jr., Reconstitution of hormone-responsive detergent-ulating hormone receptors into liposomes, *Mol. Endocrinol.*, 2, 420, 1988.

P., and Stefanini, M., Voltage-gated calcium channels in rat Sertoli cells, *Biol.*

ner, L., Direct G protein gating of ion channels, *Am. J. Physiol.*, 254, H401, 1988.

. W., and Reichert, L.E., Jr., Physical and functional association of follitropin ensitive guanine nucleotide-binding protein, *J.Biol.Chem.*, 262, 11737, 1987.

n, D. J., The role of calcium in follicle- stimulating hormone signal transduc-m., 266, 23739, 1991.

, K. A., Collison, K. A., and Segaloff, D. L., Evidence that the FSH receptor docrinology, 131, 479, 1992.

60. Santa-Coloma, T. A., Dattatreyamurty, B., and Reichert, L. E., Jr., A synthetic peptide corresponding to human FSH β-subunit 33-53 binds to receptor, stimulates basal estradiol synthesis, and is a partial antagonist of FSH, *Biochemistry*, 29, 1194, 1990.

61. Andersen, T. T. and Reichert, L. E., Jr., Follitropin binding to receptors in testis, *J. Biol. Chem.*, 257, 11551, 1982.

62. Grasso, P., Dattatreyamurty, B., Dias, J. A., and Reichert, L. E., Jr., Transglutaminase activity in bovine calf testicular membranes: evidence for a possible role in the interaction of follicle- stimulating hormone with its receptor, *Endocrinology*, 121, 459, 1987.

63. Grasso, P. and Reichert, L. E., Jr., Stabilization of follicle-stimulating hormone-receptor complexes may involve calcium-dependent transglutaminase activation, *Mol. Cell. Endocrinol.*, 87, 49, 1992.

64. Reid,R. E., Synthetic fragments of calmodulin calcium-binding site. III. A test of the acid pair hypothesis, *J. Biol. Chem.*,265, 5971, 1990.

65. Fletcher, P. W. and Reichert, L. E., Jr., Cellular processing of follicle-stimulating hormone by Sertoli cells in serum-free culture, *Mol. Cell. Endocrinol.*, 34, 39, 1984.

66. Grasso, P., Santa-Coloma, T. A., and Reichert, L. E., Jr., Correlation of follicle-stimulating hormone (FSH)-receptor complexinternalization with the sustained phase of FSH-induced uptake by cultured rat Sertoli cells, *Endocrinology*, 131, 1622, 1992.

67. Bellve, A. R. and Zheng, W., Growth factors as autocrine and paracrine modulators of male gonadal function, *J. Reprod. Fertil.*, 85, 771, 1989.

68. Morera, A. M., Benahmed, M., Cochet, C., Chauvin, M. A., Chambaz, E., and Revol, A., A TGFβ-like peptide is a possible intratesticular modulator of steroidogenesis, *Ann. N.Y. Acad. Sci.*, 513, 494, 1987.

69. Morera, A. M., Esposito, G., Ghiglieri, C., Chauvin, M. A., Hartmann, D. J., and Benahmed, M., Transforming growth factor β1 inhibits gonadotropin action in cultured porcine Sertoli cells, *Endocrinology*, 130, 831, 1992.

70. Ohtsuki, M. and Massague, J., Evidence for the involvement of protein kinase activity in transforming growth factor-β signal transduction, *Mol. Cell. Biol.*, 12, 261, 1992.

71. Grasso, P., Reichert, L. E., Jr., Sporn, M. B., and Santa-Coloma, T. A., Transforming growth factor-β1 modulates calcium metabolism in Sertoli cells, *Endocrinology*, 132, 1745, 1993.

72. Gray, T. J. B. and Gangolli, S. D., Aspects of the testicular toxicity of phthalate esters, *Environ. Health Perspect.*, 65, 229, 1986.

73. Gray, T. B. J. and Beamond, J. A., Effect of some phthalate esters and other testicular toxins on primary cultures of testicular cells, *Food Chem. Toxicol.*, 22, 123, 1984.

74. Cater, B. R., Cook, M. W., Gangolli, S. D., and Grasso, P., Studies on dibutylphthalate-induced testicular atrophy in the rat: effect on zinc metabolism, *Toxicol. Appl. Pharmacol.*, 41, 609, 1977.

75. Sjoberg, P., Bondesson, U., Gray, T. J. B., and Ploen, L.,Effect of di(2-ethylhexyl)phthalate and five of its metabolites on rat testis in vivo and in vitro, *Acta Pharmacol. Toxicol.*, 58, 225, 1986.

76. Parvinen, M., Regulation of the seminiferous epithelium, *Endocrinol. Rev.*, 3, 404, 1982.

77. Lloyd, S. C. and Foster, P. M. D., Effect of mono(2-ethylhexyl)phthalate on follicle-stimulating hormone responsiveness of cultured rat Sertoli cells, *Toxicol. Appl. Pharmacol.*, 95, 484, 1988.

78. Heindel, J. J. and Chapin, R. E., Inhibition of FSH-stimulated cAMP accumulation by mono(2-ethylhexyl)phthalate in primary rat Sertoli cell cultures, *Toxicol. Appl. Pharmacol.*, 97, 377, 1989.

79. Grasso, P., Heindel, J. J., Powell, C. J., and Reichert, L. E., Jr., Effects of mono(2-ethylhexyl)phthalate, a testicular toxicant, on follicle-stimulating hormone binding to membranes from cultured rat Sertoli cells, *Biol. Reprod.*, 48, 454, 1993.

80. Zhang, S. B., Dattatreymurty, B., and Reichert, L. E., Jr., Regulation of follicle-stimulating hormone binding to receptors on bovine calf testis membranes by cholera toxin-sensitive guanine nucleotide binding protein, *Mol. Endocrinol.*, 88, 148, 1992.

81. Talbot, J. A., Lambert, A., Mitchell, R., Grabinski, M., Anderson, D. C., Tsatsoulis, A., Shalet, S. M., and Robertson, W. R., Follicle-stimulating hormone-dependent estrogen secretion by rat Sertoli cells in vitro: modulation by calcium, *Acta Endocrinol.*, 125, 280, 1991.

82. Grasso, P., Crabb, J. W., and Reichert, L. E., Jr., An explanation for the disparate effects of synthetic peptides corresponding to human follicle-stimulating hormone beta-subunit receptor binding regions (33-53) and (81-95) and their serine analogs on steroidogenesis in cultured rat Sertoli cells, *Biochem. Biophys. Res. Commun.*, 190, 65, 1993.

83. Santa-Coloma, T. A. and Reichert, L. E., Jr., Identification of a follicle-stimulating hormone receptor-binding region in FSH-β-(81-95) using synthetic peptides, *J. Biol. Chem.*, 265, 5037, 1990.

84. Santa-Coloma, T. A., Crabb, J. W., and Reichert, L. E., Jr., Serine analogues of hFSH-beta-(33-53) and hFSH-beta-(81-95) inhibit hFSH binding to receptor, *Biochem. Biophys. Res. Commun.*, 184, 1273, 1992.

Section IX
Carotid Body Type 1 Cells

Chapter 16

OXYGEN-SENSITIVE NEUROSECRETION OF CAROTID BODY TYPE I CELLS

Helmut Acker and Jürgen Hescheler

CONTENTS

1. SUMMARY

The carotid body belongs to the familiy of peripheral chemoreceptors which monitor PO_2 and PCO_2 of the arterial blood to regulate ventilation and blood circulation for avoiding hypoxic damage of the body. The oxygen-sensing process in the carotid body is composed of a PO_2-dependent neurotransmitter release from type I cells exciting synaptically connected nerve endings which transfer action potentials in the sinus and glossopharyngeal nerve to the brain stem. The oxygen-sensing process could be explained on the molecular level by hemoproteins cooperatively interacting with molecular oxygen and thereby influencing the open probability of potassium channels. This might lead to an enhanced calcium influx which is necessary for neurotransmitter release.

2. OXYGEN SENSING

To assert a constant oxygen supply to different organs and herewith a constant energy supply maintaining highly specialized organ functions, cells able to sense oxygen levels in the blood are situated at different locations in the body, stimulating various reflex pathways. This oxygen-sensing process comprises a hemoprotein which undergoes as a sensor conformational changes in dependence on oxygen and a signal cascade, which transfers the message stimulated by the sensor to ion channels or to specific gene regions (for review see Reference 1). For the last pathway, numerous examples are given in the literature, such as the $CoCl_2$-impedible induction of phosphoenolpyruvate carboxykinase by glucagon in hepatocytes being higher at 16% O2 than 8% O_2,[2] the regulation of the glutathione peroxidase content by oxygen tension at the transcriptional level with lower mRNA levels of this enzyme in cardiomyocytes at a PO_2 of 40 Torr,[3,4] the enhanced production of platelet-derived growth factor B, but not of platelet-derived growth factor A by human umbilical vein endothelial cells under hypoxia,[5] the enhanced gene expression for tyrosine hydroxylase in carotid body type I — and PC12 — cells[6] or the increased erythropoietin — as well as vascular endothelial growth factor — production in liver cells peaking at 1% O_2.[7] Well-known examples for the participation of specialized ion channels in the oxygen-sensing process are type I cells of the carotid body,[8] cells of neuroepithelial bodies of the lung[9] and smooth muscle cells of the lung vasculature.[10] These cells possess oxygen-sensitive potassium channels which decrease their open probability under hypoxia, leading to membrane potential depolarization and opening of voltage-sensitive calcium channels with a subsequent increase of the intracellular calcium level inducing neurotransmitter release or smooth muscle contraction. In all three cell types the participation of hemoproteins as oxygen sensors which produce oxygen radicals as second messengers to control the open probability of potassium channels is discussed (for review see Reference 1). It is the aim of this article to summarize the recent findings on the oxygen-sensing process in the carotid body type I cells to contribute to the relevance of oxygen-sensing mechanisms in other cell types.

3. CAROTID BODY PHYSIOLOGY AND ANATOMY

De Castro[11] was the first to propose the hypothesis that the carotid body is a chemoreceptor which perceives the concentration of oxygen and carbon dioxide within the blood. This theory was confirmed by Heymans et al.,[12] who obtained physiological evidence that the organ affects respiration and cardiovascular function via the sinus nerve (a branch of the glossopharyngeal nerve). Meanwhile, it is known that the carotid body detects changes in chemical composition, tonicity, and the temperature of its environment. Sensory discharges of the sinus nerve increases in frequency, when there is a fall in the environmental O_2 tension (PO_2) or pH, an increase in CO_2 tension (PCO_2), or when temperature or tonicity increase. Conversely, the discharge frequency decreases with increasing arterial PO_2, low PCO_2, alkalinity, a fall in environmental temperature, or when the medium is made hypo-osmotic. In addition, chemical substances such as NaCN or K^+ ions also stimulate these receptors. Different effects of chemoreceptor reflex pathways (for review see Reference 13) can be summarized as follows:

1. Hyperventilation, blood pressure increase, bradycardia
2. Peripheral vasoconstriction, coronary vasodilation, negative and positive ionotropic effect
3. Influence on oxygen supply of organs, such as muscle, kidney, liver, heart, brain
4. Increase of sodium and water excretion in the kidney
5. Vasopressin secretion
6. Elicitation of hypoxic defense mechanisms, such as anxiety and discomfort, blood pressure control during deep sleep, and arousal reactions.

The carotid bodies are small (about 1 mm³ in the cat) sensory organs bilaterally located next to the carotid sinuses at the junction of the external and common carotid arteries; their blood supply is provided by neighboring arterial vessels. The carotid body and sinus are innervated by fibers of the carotid sinus nerve, a branch of the glossopharyngeal (IXth cranial) nerve. The sensory cells of the IXth nerve are contained in the petrosal and superior ganglia. Most of the perikarya of the carotid nerve fibers are located in the petrosal ganglion; very few are in the superior ganglion.[14]

The carotid bodies are known as glomera because of their extensive capillary networks. They are organized in compact or disseminated lobules, islands, or "glomoids" around highly convoluted capillaries provided with a thin fenestrated endothelium and a thin basal lamina. The capillary network seems to be composed of high-flow and low-flow channels. The total blood flow of the carotid body measured at the venous outflow amounts to about 2000 ml/min/ 100 g. Calculations of the specific tissue blood flow, taking into account published values of the mean small vessel area of the carotid body, the mean local blood flow velocity,[15] and the weight of the perfused tissue, amount to a value of about 65 ml/min/100 g.[16] This means that only 3% of the total flow passes the specific carotid body tissue, whereas 97% is shunt flow. This unusual flow heterogeneity is likely to enable the carotid body to maintain an arterial PO_2 of about 100 Torr, a tissue PO_2 distribution with a mean value between 20 and 30 Torr, similar to other organs.[1] The carotid body tissue, therefore, seems to mirror the oxygen supply situation of other organs for conveying a suitable signal to the oxygen sensor inside the glomoids in case of a hypoxic challenge.

The glomoids (Figure 1) consist of many glomus (type I) cells covered by a few sustentacular (type II) cells. A few ganglion cells (sympathetic or parasympathetic) occur around the carotid body. Their numbers vary in different mammalian species. In the rat carotid body the proportion between glomus, sustentacular, and ganglion cells is 348:92:1.[17] Glomus cells are small (10 μm) ovoid structures with large nuclei that frequently possess cytoplasmic processes of different lengths (Figure 1). The cells present many mitochondria, a well-developed Golgi apparatus, a granular endoplasmic reticulum (usually dispersed, but sometimes aggregated as a Nissl body), microtubules, and usually a cilium; small (60-nm) clear-core vesicles can be found close to the plasma membrane. A salient feature of glomus cells is the presence of abundant and relatively large (70 to 200-nm) dense-core vesicles or granules similar to but smaller than those in the adrenal medulla. These dense-core vesicles contain calcium-binding sites that appear as electron-dense particles (20–30 nm) within the vesicles.[17]

Most type I cells are innervated by one, two, or three sensory nerve endings, usually derived from one parent axon. However, some type I cells may be innervated only by preganglionic sympathetic axons, and analyses of serial sections have shown that some glomus cells are not innervated at all. Sensory nerve endings range in shape from small boutons, which contact only a tiny portion of the type I cell, to large calyces, which cover as much as half of the surface of a glomus cell. All sensory nerve endings contain small clear-cored vesicles and large densed-cored vesicles. Hypoxia and hypercapnia reduce the number of vesicles in the sensory nerve endings. Most of the sensory nerve endings are postsynaptic to type I cells, as demonstrated by the presence of large dense-cored vesicles and small vesicles in type I cells near presynaptic dense projections. A small proportion of sensory nerve endings (cat 15%, rat 6%) exhibit morphological features of being presynaptic to type I cells. Some sensory nerve endings are interconnected with type I cells by reciprocal synapses. Such nerve terminals are presynaptic and postsynaptic to the same type I cell. Preganglionic sympathetic axons, which resemble the nerve terminals on chromaffin cells of the adrenal medulla, are only presynaptic to some type I cells. Nerve endings that are morphologically similar to baroreceptors of the carotis sinus, encircle some arteries and arterioles of the carotid body.[17] An extensive plexus of nerve fibers capable of synthesizing nitric oxide was demonstrated in the cat carotid body by immunocytochemical and biochemical studies of nitric oxide synthase.[18] Denervation experiments indicated that the axons originate from: (i) microganglial neurons located within the carotid body and along the glossopharyngeal and carotid sinus nerves, whose ramifications primarily innervate carotid body blood vessels; and (ii) sensory neurons in the petrosal ganglion, whose terminals end in association with lobules of type I cells. In the *in vitro* superfused cat carotid body, the nitric oxide synthase substrate, L-arginine, induced a dose-dependent inhibition of carotid sinus nerve discharge evoked by hypoxia. In contrast, the nitric oxide synthase inhibitor, L-NG-nitroarginine methylester, augmented the chemoreceptor response to hypoxia, and this effect was markedly enhanced when the preparation was both perfused and superfused *in vitro*. The nitric oxide donor, nitroglycerine, inhibited carotid sinus nerve discharge, and immunocytochemistry revealed that this drug stimulated the formation of cyclic 3′, 5′-guanosine monophosphate in both type I cells and blood vessels. These data indicate that nitric oxide is an inhibitory neuronal messenger in the carotid body, which affects the process of chemoreceptor transduction transmission via actions on both the receptor elements and their associated blood vessels.[18]

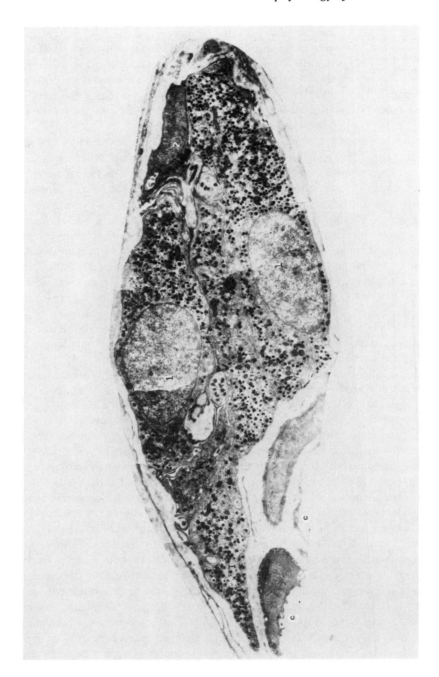

FIGURE 1.   Electron micrograph of a glomoid of a cat carotid body. Type I cells (1), type II cells (2) and vessels (c) are to be seen. Collagen and nerves are located in the spaces between the glomoid cells. 75.3% of this glomoid was occupied by type I cells, 10.7% by type II cells, 2.5% by nerves, and 11.5% by collagen. (Kindly provided by Dr. D. Schäfer MPI für Molekulare Physiologie Dortmund.)

Some glomus cells are linked to other glomus cells by chemical synapses, and an undetermined number of glomus cells are apparently electrically coupled to one another by gap junctions. Numerous putative transmitters have been detected in the carotid body like acetylcholine, dopamine, noradrenaline, adrenaline, 5-hydroxytryptamine, adenosine, and methionine enkephalin, substance P, vasoactive intestinal polypeptide, taurine, and glutamine. There is a general agreement (see for review Reference 19) that the sensory nerve endings contain substance P, the sympathetic nerve ending noradrenaline and acetylcholine, and the type I cells dopamine, noradrenaline, acetylcholine, substance P, as well as met-enkephalin and 5-hydroxy-tryptamine. Experiments by Eyzaguirre, Fidone, McQueen and co-workers (see for review References 13 and 19) have shown that the acetylcholine receptors are nicotinic in the cat and muscarinic in the rabbit carotid body, and predominantly located at type I cells. Inward currents, as measured by patch clamp technique, were evoked in rat type I cells clamped at –70 mV in response to carbachol and nicotine and dimethylphenylpiperazinium. Muscarine failed to produce any change in membrane current. Currents evoked by nicotine were reduced or abolished in the presence of mecamylamine and also by high concentrations of atropine (10 or 100 μM). Activation of these receptors could therefore lead to excitation by either of two possible mechanisms: depolarization of type I cells sufficient to open voltage-gated Ca^{2+} channels, or Ca^{2+} influx through the receptor pore itself. Either (or both) mechanisms could trigger catecholamine release from type I cells.[20] Two types of dopamine receptors have been described in the carotid body[19] with inhibitory or excitatory effect on the nervous discharge, mostly located in the nerve terminals. Most of the hormones of the carotid body therefore fulfill the criteria for the identification as synaptic transmitters, such as the presence of synthesizing enzymes, inactivating enzymes, metabolic precursors, and reduced efficacy after inhibition. An exogenous application of hormones mimics the effect of presynaptic stimulation, specific blockage by drugs, and collectability of transmitters. Following this direction, Hanbauer and Hellström[21] could show that short hypoxic stimulation leads to a decline of the dopamine content of the rat carotid body, whereas a longer hypoxic exposure leads to an increase of the dopamine content. Also, other hormones, such as polypeptides, are influenced in their carotid body tissue concentration by hypoxia and other stimuli. However, the sympathetic and sensory nerve endings of the carotid body have a profound effect on the tissue level of different hormones and interact with different stimuli. Table 1 summarizes these effects, mostly according to the work of Fidone and co-workers.[22]

4. ION CHANNELS OF CAROTID BODY TYPE I CELLS

Numerous patch clamp studies using the whole-cell variant or single membrane patches have been carried out to study the electrical properties of ion channels in adult and embryonic type I cells (see for review References 8 and 14). Sodium, calcium, chloride, and potassium currents were found.

The sodium current has a fast activation time course and an activation threshold at about –40 mV. At all voltages inactivation follows a single exponential time course with a time constant of 0.67 ms at 0 mV and a half steady-state inactivation at –50 mV.[23] The Na^+ current density in hypoxia-treated type I cells (6% O_2 over 2 weeks) increases significantly, reaching values up to six times that seen in normoxic (20%) controls. In addition, the whole-cell capacitance, an indicator of cell size, is also significantly larger (three to four times control) in type I cells exposed to chronic hypoxia. Both effects are mimicked qualitatively by chronic treatment of normoxic cultures with $N^6,O^{2'}$-dibutyryl adenosine $3',5'$-cyclic monophosphate, but not nerve growth factor, which is known to induce similar changes in the chromaffin cell line PC12. Thus, the physiological and morphological effects of chronic hypoxia on the carotid body *in vivo* may be due in part to a cAMP-mediated stimulation of Na^+ channel expression and hypertrophy in the chemosensory type I cells.[24]

The calcium current is almost totally abolished when most of the external calcium is replaced by magnesium. The activation threshold of this current is at –40 mV and at 0 mV it reaches a peak amplitude in 6–8 ms. The calcium current inactivates very slowly and only decreases to 27% of the maximal value at the end of 300-ms pulses to 40 mV. The calcium

TABLE 1
Transmitter and Corresponding Enzymes of Carotid Body under Different Stimulatory Conditions

	Natural Stimulus Hypoxia, Hypercapnia	CSN Denervation	Sympathectomy
Substance P (located in sensory nerve endings, type I cells)	Increased released	Increased level; no effect on hypoxic release	Increased level; increased hypoxic release
Met-enkephalin (located in type I cells)	Increased release	Blocks hypoxic release	Blocks hypoxic release
Acetylcholine (located in type I cells in sympathetic nerve endings)	Increase of synthesis	No effects on levels or synthesis	No effects on levels or synthesis
Catecholamines (located in type I cells in sympathetic nerve endings)	• 2-G phosphorylation increase • Free tyrosine level increase • TH mRNA increase (24–48 h) • DA synthesis increase • Increased DA and NE release	• Increased TH level • Increased hypoxic TH induction • Increased NE synthesis • Decreased DA release; no effect on DA level or synthesis	• Blocks hypoxic TH induction • Decreased DA release • Decreased NE level and synthesis • No effect on DA synthesis or level

current is about two times larger when barium ions are used as charge carriers. Deactivation kinetics of the calcium current follows a biphasic time course well fitted by the sum of two exponentials, suggesting that type I cells have predominantly fast deactivating calcium channels.[23] Open probability of rat type I cells is increased by the dihydropyridine calcium channel agonist BAY K 8644 and decreased by the antagonist nifedipine, hinting to L-type calcium channels.[25] In type I cells from adult rabbits it was found that calcium currents were reversibly inhibited by application of prostaglandin E_2. A good parallel exists between the dose-response curves for prostaglandin E_2 inhibition of calcium current in type I cells and high extracellular potassium- or hypoxia-evoked release of catecholamines from the whole carotid body. When patch clamp recordings were made with an internal electrode solution lacking GTP and containing 100 μM GDPβS, a GDP analog which inhibits G protein cycling, prostaglandin E_2 did not inhibit the calcium current in type I cells. Prostaglandin E_2 seems to inhibit the release of catecholamines induced by hypoxic and high extracellular potassium stimulation in type I cells by reducing the entry of calcium through voltage-dependent calcium channels mediated by a G protein-dependent mechanism.[26]

Chloride channels have a large conductance of 296 pS with random open and closed kinetics in inside-out patches.[27] The open-state probability (PO; mean = 0.61) is hardly affected by membrane potential (-50 to +50 mV) and cytoplasmic calcium (0–1 mM). Similarly, the channel does not appear to be regulated by cytoplasmic nucleotides (1 mM) or pH (6.5–8). Ion-substitution experiments yielded the following selectivity sequence: chloride > bicarbonate > sulfate > glutamate > sodium. Single-channel currents are reversibly reduced or blocked by anthracene-9-carboxylic acid (5–10 mM), but were unaffected by stilbene derivatives (0.5–1 mM), by furosemide (1 mM), and by 5-nitro-2-(3-phenyl-propyia-mino)benzoic acid (0.01 mM). Because type I cells have been shown to express carbonic anhydrase,[17] it is inferred that the chloride channels may play an important role in the physiology of glomus cells by aiding in the regulation of pH_i and the resting potential via bicarbonate and chloride permeability.[27]

Abolishing current flow through sodium — and calcium — channels, three classes of voltage-gated potassium channels can be distinguished by patch clamp technique in rabbit type I cells that differ in their single-channel conductance, dependence on internal calcium, and sensitivity to changes in PO_2. Calcium-activated potassium channels with a conductance of about 210 pS in symmetrical solutions are observed when internal calcium is below 0.1 μM. Small conductance channels with a value of about 16 pS are not affected by internal calcium, and they exhibit slow activation and inactivation time courses. In these two channel types open probability is unaffected when exposed to normoxic (PO_2 = 140 Torr) or hypoxic (PO_2 = 5–10 Torr) external solutions. A third channel type having an intermediate conductance of about 40 pS was the most frequently recorded. These channels are steeply voltage dependent and not affected by internal calcium in the rabbit carotid body type I cells. They inactivate almost completely in less than 500 ms, and their open probability reversibly decreases upon exposure to low PO_2. The effect of low PO_2 is voltage dependent, being more pronounced at moderately depolarized voltages. At 0 mV, for example, the open probability diminishes to about 40% of the control value. The time course of ensemble current averages of these channels is remarkably similar to that of the O_2-sensitive potassium current. In addition, ensemble average and macroscopic potassium currents are affected similarly by low PO_2. These observations strongly suggest that these potassium channels are the main contributors to the macroscopic potassium current of type I cells. The reversible inhibition of potassium channel activity by low PO_2 does not desensitize and is not related to the presence of F^-, ATP, and $GTP_\gamma S$ at the internal face of the membrane.[28] The maximal inhibition is achieved at a PO_2 of about 90 Torr.[29] Raising the intracellular level of cAMP in type I cells shifts the PO_2 sensitivity of the potassium channels to lower values.[30]

Electrophysiological data obtained from type I cells that have been isolated from embryos neonatal animals are different from those just described. Cells from rabbit embryos lack TTX-sensitive inward currents and show no spontaneous action potentials. However, calcium and potassium currents are mostly similar to those obtained in adult chemoreceptor cells, including the sensitivity of the potassium current to low PO_2. In addition, embryonic type I cells seem to exhibit an inward rectifying potassium current that appears to be active at negative potentials (in the range of the estimated membrane potential of –50 mV), and that is also reversibly inhibited by low PO_2.[31,32] This is shown in Figure 2. Reducing the PO_2 from 20.0 to 3.7 kPa (constant PCO_2 and pH) induced an initial flickering of the membrane potential followed by a delayed depolarization to 0 mV. The depolarization outlasted the hypoxic period, but recovered after normoxia (Figure 2A). A possible mechanism underlying the membrane depolarisation is the closing of a potassium channel. In the cell-attached patch configuration, a large conductance potassium channel was determined with a single-channel conductance of 137 pS under symmetrical potassium conditions, which responded to hypoxia with a drastic decrease of the open probability (see Figure 2B). A similar potassium channel with 250 pS was found in cells from rat embryos. At present it is not clear whether or not this channel corresponds to the intermediate conductance channel described by Ganfornina and López-Barneo[29] in adult rabbit glomus cells.

5. MOLECULAR MECHANISMS OF OXYGEN SENSING

There are basically three different views of the oxygen-sensing mechanism in the carotid body, which is hypothesized to be composed of the following steps: The membrane depolarization necessary for transmitter release under hypoxia could be accomplished by the outward rectifying potassium currents of type I cells which are reduced under hypoxia as described above. Membrane depolarization would open voltage-dependent calcium channels, increasing

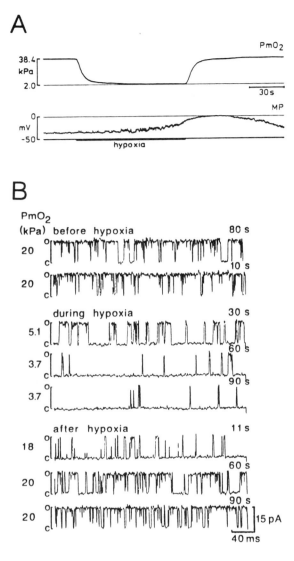

FIGURE 2. Patch clamp experiments were carried out on carotid body type I cells from rabbit embryos. The cells were superfused with a physiological salt solution. The pipette solution contained, among others, 130 mM potassium. (A): Fast hypoxic stimulation was produced by switching between two reservoirs equilibrated with different PO_2 concentrations. The actual PO_2 was continuously monitored (upper panel). The membrane potential was determined by the patch clamp technique in the current clamp mode (lower level). (B): In the cell-attached patch configuration using symetrical potassium conditions (130 mM on both sides), the activity of a single potassium channel was recorded under normoxia (upper two panels), under hypoxia (middle three panels), and after hypoxia (lower three panels); c and o correspond to the closed and opened states, respectively.

calcium influx as well as cytosolic calcium levels, facilitating transmitter release. It could be shown by Buckler and Vaughan-Jones[33] that graded reductions in PO_2 from 160 Torr to 38, 19, 8, 5, and 0 Torr induced a graded rise of intracellular calcium in type I cells. The rise of intracellular calcium in response to anoxia was 98% inhibited by removal of external calcium, indicating the probable involvement of calcium influx from the external medium in mediating the anoxic calcium response. The L-type calcium channel antagonist nicardipine (10 μM) inhibited the anoxic intracellular calcium response by 67%, and the nonselective calcium channel antagonist nickel (2 mM) inhibited the response by 77%. Under voltage recording conditions, anoxia induced a reversible membrane depolarization (or receptor potential) accompanied, in many cases, by trains of action potentials. These electrical events were coincident with a rapid rise of intracellular calcium. When cells were voltage clamped close to their resting potential (−40 to −60mV), the intracellular calcium response to anoxia was greatly reduced and its onset was much slower. Under voltage clamp conditions, anoxia also induced a small inward shift in holding current. It seems, therefore, that anoxia promotes a rise of intracellular calcium in type I cells, principally through voltage-gated Ca2+ entry, which occurs in response to the receptor potential and/or concomitant electrical activity.

1. However, the participation of intracellular calcium stores, especially by mitochondria, in regulating the cytosolic calcium-content has also been discussed.[34] In their classical paper, Mills and Jöbsis[35] analyzed photometrically a cytochrome aa_3 with a low as well as with a high O_2 affinity component in the cat carotid body, probably explaining a decrease of the mitochondrial membrane potential with a concomitant calcium release which is necessary for transmitter release starting at high PO_2. This is in accordance with tissue PO_2 measurements in the carotid body as published by Nair et al.[36] with values between 60 and 90 Torr. Duchen and Biscoe[37] supported the idea of a specialized cytochrome aa_3 by a model which located the O_2 sensor in the carotid body mitochondria, responding to oxygen changes due to a low-O_2 affinity far above the critical mitochondrial PO_2 with a mitochondrial membrane depolarization and a subsequent calcium release. Under most conditions the control of electron flux by oxygen concentration in both mitochondria and cells appears to be minimal. The apparent K_m of respiration for oxygen in cells and in isolated mitochondria is less than 0.7 Torr, although respiratory chain intermediates do show redox responses to changes in PO_2 of 70 Torr or even higher.[38] Chance[39] pointed out that explanations of cytochrome responses to high PO_2 (20–70 Torr) observed by spectroscopy are provided by the following factors: interference from hemoglobin deoxygenation, high O_2 gradients particularly in rapidly metabolizing tissues and aggregated cells, nonsteadystates of control chemicals and mitochondrial substrates, as well as the presence of another pigment beside cytochrome aa_3. These arguments question, therefore, low affinity PO_2 values of a specialized cytochrome aa_3 and mitochondrial calcium release at high PO_2 values for elucidating an oxygen sensor mechanism.

2. Recent studies as published by Lahiri et al.[40] see the regulation of ATP production under hypoxia as a key process for oxygen sensing. With a newly developed optical method they could measure under normoxic conditions of the inflowing medium (PO_2: 103.4 ± 4.1 Torr) intercapillary PO_2 values of about 52.5 ± 3.6 Torr in the carotid body tissue, confirming PO_2 measurements with microelectrodes in the carotid body tissue as published by Acker et al.[41] In the experiments as published by Rumsey et al.[42] chemosensory discharge rose slowly as intercapillary PO_2 steadily declined to values of 10 Torr. Between 10 and 3 Torr, chemosensory discharge increased strikingly, concomitant with an enhanced rate of oxygen disappearance. As PO_2 fell below 3 Torr, oxygen disappearance slowed and neural activity decayed. This low range of PO_2 would be expected to affect oxygen metabolism and thus the metabolic state in the cells of the carotid body. It is conceivable that the decrease in the tissue PO_2 of the carotid body produces a decline in the cytosolic phosphorylation potential and consequent adjustments in the redox state of the intramitochondrial pyridine nucleotides.[42] A fall in the cytosolic phosphorylation potential will stimulate in the presence of oxygen oxidative phosphorylation to meet the energy demand of the different ATP-driven reactions inside the cells necessary to initiate and maintain nervous chemoreceptor activity. This view is in line with the stimulatory effect of cyanide or other blocking agents of the respiratory chain on carotid body nervous discharge with a concomitant transmitter release and

an enhanced level of intracellular calcium in type I cells. These agents, however, do not mimic the effect of hypoxia, but seem to have their own mode of action, because cyanide increases the potassium current and decreases the calcium current of type I cells.[32,43] Furthermore, blocking the energy production of the respiratory chain by different drugs, Mulligan et al.[44] could impede the hypoxic responsiveness of the carotid body. Measuring an enhanced glucose uptake of the neurotransmitter containing type I cells of the carotid body under low PO_2, Obeso et al.[45] could confirm that the utilization of metabolic energy is an integral component of the chemoreceptor response to hypoxia. Also, Donnelly[46] underlined the importance of an unimpaired oxidative metabolism for chemoreception of the carotid body by showing that chemoreceptor discharge and catecholamine release of the rat carotid body increase concomitantly with falling PO_2, but that chemoreceptor discharge declines while catecholamine release further increases when PO_2 reaches very low levels. One should conclude from these experiments that energy production by the respiratory chain is necessary for a regular nervous response curve of carotid body cells with respect to hypoxia, which does not necessarily imply that the mitochondrial respiratory chain also functions as an oxygen sensor. It seems reasonable to assume that the chemoreceptor process needs a stable energetic base which is guaranteed by the mitochondrial ATP production.

3.   The molecular mechanism of the inhibitory effect of low PO_2 on potassium channel conductivity and thus on the chemoreceptor properties of type I cells is unknown, but the involvement of a heme-type PO_2 sensor protein has been suggested.[47] The idea of heme proteins acting as PO_2 sensors in the carotid body has been published by several groups (see Reference 1 for review). Cross et al.[48] carried out a detailed photometric analysis of the rat carotid body to gain more information about hemeprotein characteristics in this tissue. They detected, besides typical absorbtion peaks hinting to the different cytochromes of the respiratory chain, a measurable heme signal with absorbance maxima at 559, 518, and 425 nm, suggesting the presence of a b-type cytochrome. This was confirmed by pyridine hemochrome and CO spectra. This heme protein is capable of H_2O_2 formation and seems to possess, therefore, similarities with cytochrome b_{558} of the NAD(P)H oxidase in neutrophils.[49] Cytochrome b_{558} consisting of 91 kDa ($gp91_{phox}$) and 22 kDa ($p22_{phox}$) subunits resides in the plasma membrane and membrane of neutrophil-specific granules, where it serves as the terminal electron carrier of the oxidase. Two additional cofactors, $p47_{phox}$ and $p67_{phox}$, are cytosolic and may exist as preformed complexes which translocate to the plasma membrane upon phagocyte activation, becoming integral parts of the active oxidase.[50] $p22_{phox}$, $gp91_{phox}$, $p47_{phox}$, and $p67_{phox}$ could be identified immunohistochemically in type I cells of the human, rat, and guinea pig carotid body by Kummer et al.[51] highlighting the probability of the involvement of an NAD(P)H oxidase or a related isoform in the cellular oxygen sensing of the carotid body. Of special interest in this context are findings as published by López-López and González[52] showing that the hypoxia-induced decrease of the activity of potassium channels of type I cells can be inhibited by CO. This might be interpreted as CO is inducing an oxidation of a hemeprotein which interacts with potassium channels in the cell membrane of type I cells

Figure 3 tries to give a model for the involvement of H_2O_2-generating NAD(P)H oxidase in the cellular oxygen-sensing process influencing potassium channels or erythropoietin production. The NAD(P)H oxidase with cytochrome b_{558} is shown to produce oxygen radicals that are dismutated to hydrogen peroxide, acting as a second messenger by scavenging through catalase. This leads to the formation of the heme-containing catalase-compound I complex which activates the heme-containing guanylate cyclase by interaction of the two heme groups as described for smooth muscle cells of the lung vasculature.[53] The enhanced cGMP level,

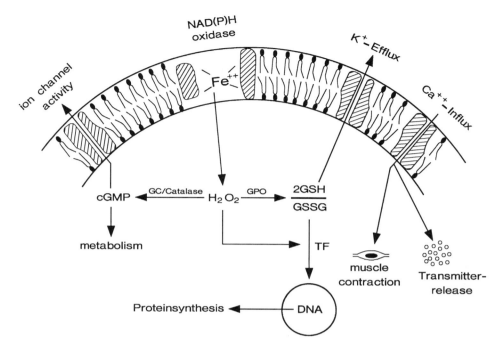

FIGURE 3. Scheme of an oxygen sensor (NAD(P)H oxidase) localized in the cell membrane generating H_2O_2 as a second messenger in dependence on PO_2. Scavenging H_2O_2 through glutathione peroxidase (GPO) the quotient reduced vs. oxidized glutathione (2GSH/GSSG) determines, on the one hand, the open probability of potassium channels and therefore the calcium influx triggering transmitter release or muscle contraction, on the other hand, the genetic expression and herewith the protein synthesis by interaction with transcription factors (TF). H_2O_2 might also influence the cGMP level with consequences for ion channel activity and metabolism by being scavenged through catalase and guanylate cyclase (GC).

which is furthermore controlled by the NO synthase in the carotid body,[18] might lead to an activation of the different cytosolic activator proteins of the NAD(P)H oxidase.[50] The availability of oxygen provided by the oxygen supply determines the amount of hydrogen peroxide formed by the oxidase with a declining formation under hypoxia. This mechanism would also lead to a PO_2-dependent cGMP level, which is known to vanish in carotid body type I cells under hypoxia,[54] influencing perhaps enzyme activities or different ion channel currents. The level of cAMP, which is known to increase in type I cells under hypoxia,[54] might enhance the PO_2 sensitivity of potassium channels by inhibiting the NAD(P)H oxidase.[50] Hydrogen peroxide can also be scavenged by glutathione peroxidase[55] changing the ratio 2GSH/GSSG. The higher level of GSH under hypoxia could lead to a closing of GSH-sensitive potassium channels,[56] which might be similar to the PO_2-sensitive potassium channels of type I cells, since the sulfydryl reagent, *p*-chloromercuribenzenesulfonic acid (PCMBS), selectively and irreversibly inhibits the calcium-activated potassium current of rat type I cells in a dose-dependent manner (0.01–1 m*M*). The same concentrations of PCMBS do not affect the calcium-independent potassium current, but cause a transient enhancement of the calcium current. The inhibition of the calcium-activated potassium current by PCMBS is similar to the previously reported effects of hypoxia, and suggests a central role for the channels underlying this current in the oxygen-sensing process of type I cells.[57] A change in the 2GSH/GSSG ratio with a concomitant variation of the chemoreceptor discharge by using H_2O_2 or organic hydroperoxides could be measured by Acker et al.[58] on the superfused rat carotid body *in vitro*.

This scheme might also explain the PO_2-dependent serotonin secretion of cells of the neuroepithelial bodies in the lung[9] and the hypoxia-induced vasoconstriction of the lung vasculature.[10] Even the hypoxia-stimulated erythropoietin production seems to be explainable by the participation of an H_2O_2 producing NAD(P)H oxidase[59] due to an enhanced activity of transcription factors, like AP1 binding to the erythropoietin gene.[60]

REFERENCES

1. Acker, H., Mechanisms and meaning of cellular oxygen sensing in the organism, *Respir. Physiol.,* 95, 1-10, 1994.
2. Kietzmann, Th., Schmidt, H., Probst, I., and Jungermann, K., Modulation of the glucagon-dependent activation of the phosphoenolpyruvate carboxykinase gene by oxygen in rat hepatocyte cultures *FEBS Lett.,* 311, 3, 251-255, 1992.
3. Cowan, D.B., Weisel, R.D., Williams, W.G., and Mickle, D.A.G., The regulation of glutathione peroxidase gene expression by oxygen tension in cultured human cardiomyocytes. *J. Mol. Cell. Cardiol.,* 24, 423-433, 1992.
4. Cowan, D.B., Weisel, R.D. Williams, W.G. and Mickle, D.A.G., Identification of oxygen responsive elements in the 5' flanking region of the human glutathion peroxidase gene, *J. Biol. Chem.,* 268, 26904-26910, 1993.
5. Roszinzki, S. and Jelkmann, W., Effect of PO_2 on prostaglandin E_2 production in renal cell cultures, *Respir. Physiol.,* 70, 131-141, 1987.
6. Czykzyk-Krzeska, M., Furnari, B.A. Lawson, E.E. and Millhorn, D.E., Hypoxia increases rate of transcription and stability of tyrosine hydroxylase mRNA in pheochromocytoma (PC12) cells, *J. Biol. Chem.,* 7, 760-764, 1994.
7. Goldberg, M.A. and Schneider, Th.J., Similarities between the oxygen sensing mechanisms regulating the expression of vascular endothelial growth factor and erythropoietin, *J. Biol. Chem.,* 269, 4355-4359, 1994.
8. López-Barneo, J., Benot, A.R., and Urena, J., Oxygen sensing and the electrophysiology of arterial chemoreceptor cells, *NIPS,* 8, 191-195, 1993.
9. Youngson, Ch., Nurse, C., Yeger, H., and Cutz, E. Oxygen sensing in airway chemoreceptors, *Nature,* 365, 153-155, 1993.
10. Yuan, X.J., Goldmann, W.F., Tod, M.L., Rubin, L.J., and Blaustein, M.P., Hypoxia reduces potassium currents in cultured rat pulmonary but not mesenteric arterial myocytes, *Am. J. Physiol.,* 264, 116-123, 1993.
11. De Castro, F., Sur la structure et l'innervation de la glande intercarotidienne (glomus caroticum) de l'homme et de mammifères et sur un noveau système d'innervation autonome du nerf glossopharyngien. Études anatomiques et experimentales. *Trab. Lab. Invest. Biol. Univ. Madr.,* 24, 365-432, 1926.
12. Heymans, C., Bouchaert, J.J., and Dautrebande, L., Sinus carotidien et reflexes respiratoires. II. Influences repiratoires reflexes de l'acidose, de l'alcalose, de l'anhydride carbonique, de l'ion hydrogéne et de l'anoxémie. Sinus carotidiens et échanges respiratoires dans les poumons et au delà poumons, *Arch. Int. Pharmacodyn. Ther.,* 39, 400-448, 1930.
13. Acker, H., The involvement of nerve terminals in the paraganglionic chemoreceptor system, in *The terminal Nerve, Structure, Function and Evolution,* (Demski, L.S. and Schwanzel-Fukuda, M.). *Ann. N.Y. Acad. Sci.,* 519, 369-384, 1987.
14. González, C., Almaraz, L., and Rigual, R., Oxygen and acid chemoreception in the carotid body chemoreceptors, *TINS,* 15, 4, 146-153, 1992.
15. Hilsmann, J., Degner, F., and Acker, H., Local flow velocities in the cat carotid body tissue, *Pflügers Arch.,* 410, 204-211, 1987.
16. Clarke, J.A., de Burgh Daly, M., Ead, H.W., and Krevolic, G., A morphological study of the size of the vascualr compartment of the carotid body in a non human primate (cercopithecus ethiopus) and a comparison with the cat and rat, *Acta Anat.,* 147, 240-247, 1993.
17. Eyzaguirre, C., Fitzgerald, R.S., Lahiri, S., and Zapata, P., Arterial chemoreceptors, in *Handbook of Physiology,* (Shepherd, J.T., Abboud, F.M. Geiger, St.R., Eds.,) The cardiovascular system, Vol. III, American Physiological Society, Williams & Wilkins, Baltimore, 1983, 557-621.
18. Wang, Z.Z., Stensaas, L.J. Bredt, D.S., Diinger, B. and Fidone, S.J., Localization and actions of nitric oxide in the cat carotid body, *Neuroscience,* 60, 275-286, 1994.
19. Eyzaguirre, C. and Zapata, P., Perspectives in carotid body research, *J. Appl. Physiol.,* 57, 931-957, 1984.
20. Wyatt, C.N. and Peers, C. Nicotinic acetylcholine receptors in isolated type I cells of the neonatal rat carotid body, *Neuroscience,* 54, 275-281, 1993.
21. Hanbauer, I. and Hellström, S., The regulation of dopamine and noradrenaline in the rat carotid body and its modification by denervation and by hypoxia, *J. Physiol.,* 282, 21-34, 1978.

22. Fidone, S.J., Stensaas, L.J. and Zapata, P., Sites of synthesis, storage, release and recognition of biogenic amines in carotid bodies, in Physiology of *Peripheral Chemoreceptors,* (Acker, H. and O´Regan, R.G., Eds.) Elsevier, Amsterdam, 1983, 21-44.

23. Urena, J., López-López, J., Gonzalez, C., and López-Barneo, J., Ionic currents in dispersed chemoreceptor cells of the mammalian carotid body, *J. Gen. Physiol.,* 93, 979-999, 1989.

24. Stea, A., Jackson, A., and Nurse, C.A., Hypoxia and N[6],O[2']-dibutyryladenosine 3′,5′- cyclic monophosphate, but not nerve growth factor, induce Na$^+$ channels and hypertrophy in chromaffin-like arterial chemoreceptors, *PNAS,* 89, 9469-9473, 1992.

25. Fieber, L.A. and McCleskey, E.W., L-Type calcium channels in type I cells of the rat carotid body, *J. Neurophysiol.,* 70, 4, 1378-1384, 1993.

26. Gómez-Nino, A., López-López, J.R., Almaraz, L. and González, C., Inhibition of [^{3}H] catecholamine release and Ca^{2+} currents by prostaglandin E$_2$ in rabbit carotid body chemoreceptor cells, *J. Physiol.,* 476, 269-277, 1994.

27. Stea, A. and Nurse, C.A., Chloride channels in cultured glomus cells of the rat carotid body, *Am. J. Physiol.,* 257, C174-C181, 1989.

28. Ganfornina, M.D. and López-Barneo, J., Potassium channel types in arterial chemoreceptor cells and their selective modulation by oxygen, *J. Gen. Physiol.,* 100, 401-426, 1992.

29. Ganfornina, M.D. and López-Barneo, J., Single K$^+$ channels in membrane patches of arterial chemoreceptor cells are modulated by O$_2$ tension, *PNAS,* 88, 2927-2930, 1991.

30. López-López, J.R., DeLuis, D.A., and González, C., Properties of a transient K$^+$ current in chemoreceptor cells of rabbit carotid body, *J. Physiol.,* 15-32, 1992.

31. Delpiano, M.A. and Hescheler, J., Evidence for a PO$_2$ sensitive K$^+$ channel in the type I cell of the rabbit carotid body. *FEBS Lett.,* 249, 2, 195-198, 1989.

32. Hescheler, J., Delpiano, M.A., Acker, H., and Pietruschka, F., Ionic currents on type-I cells of the rabbit carotid body measured by voltage clamp experiments and the effect of hypoxia, *Brain Res.,* 486, 79-88, 1989.

33. Buckler, K.J. and Vaughan-Jones, R.D., Effects of hypoxia on membrane potential and intracellular calcium in rat neonatal carotid body type I cells, *J. Physiol.,* 476, 3, 423-428, 1994.

34. Duchen, M.R. and Biscoe, T.J., Relative mitochondrial membrane potential and [Ca^{2+}] in type I cells isolated from the rabbit carotid body, *J. Physiol.,* 450, 33-61, 1992.

35. Mills, E. and Jöbsis, F.F., Mitochondrial respiratory chain of carotid body and chemoreceptor response to changes in oxygen tension, *Neurophysiol.,* J., 35, 405-428, 1972.

36. Nair, P.K., Buerk, D.G., and Whalen, W.J., Cat carotid body oxygen metabolism and chemoreception described by a two cytochrome model, *Am. J. Physiol.,* 19, H202-H207, 1986.

37. Duchen, M.R. and Biscoe, T.J., Mitochondrial function in type I cells isolated from rabbit arterial chemoreceptors, *J. Physiol.,* 450, 13-31, 1992.

38. Brand, M.D. and Murphy, M.P., Control of electron flux through the respiratory chain in mitochondria and cells, *Biol. Rev.,* 62, 141-193, 1987.

39. Chance, B., Early reduction of cytochrome c in hypoxia, *FEBS Lett.,* 226, 2, 343-346, 1988.

40. Lahiri, S., Rumsey, W.L., Wilson, D.F., and Iturriaga, R. Contribution of in vivo microvascular PO$_2$ in the cat carotid body chemotransduction, *J. Appl. Physiol.,* 75, 3, 1035-1043, 1993.

41. Acker, H., Lübbers, D.W. and Purves, M.J., Local oxygen tension field in the glomus caroticum of the cat and its change at changing arterial PO$_2$. *Pflügers Arch.,* 329, 136-155, 1971.

42. Rumsey, W.L., Iturriaga, R., Spergel, D., Lahiri, S., and Wilson, D.F., Optical measurements of the dependence of chemoreception on oxygen pressure in the cat carotid body, *Am.J. Physiol.,* 261(30), C614-C622, 1991.

43. Duchen, M.R., Caddy, K.W.T., Kirby, G.C., Patterson, D.L., Ponte, J., and Biscoe, T.J., Biophysical studies of the cellular elements of the rabbit carotid body, *Neuroscience,* 26, 1, 291-311, 1988.

44. Mulligan, E., Lahiri, S., and Storey, B.T. Carotid body O$_2$ chemoreception and mitochondrial oxidative phosphorylation, *J. Appl. Physiol.,* 51, 438-446, 1981.

45. Obeso A., Gonzalez, C., Rigual, R., Dinger, B., and Fidone, S., Effetc of low O$_2$ on glucose uptake in rabbit carotid body. *J. Appl. Physiol.,* 74, 2387-2393, 1993.

46. Donnelly D.F., Electrochemical detection of catecholamine release from rat carotid body in vitro. *J. Appl. Physiol.,* 74, 2330-2337, 1993.

47. Lloyd, B.B., Cunningham, D.J.C., and Goode, R.C., Depression of hypoxic hyperventilation in man by sudden inspiration of carbon monoxide, in *Arterial Chemoreceptors,* Torrance, R.W., Ed., Blackwell Scientific, Oxford, 1968, 145-148.

48. Cross, A.R., Henderson, L., Jones, O.T.G., Delpiano, M.A., Hentschel, J. and Acker, H., Involvement of an NAD(P)H oxidase as a PO2 sensor protein in the rat carotid body, *Biochem. J.,* 272, 743-747, 1990.

49. Jones, O.T.G., Cross, A.R., Hancock, J.T., Henderson, L.M., and O'Donnel, V.B., Inhibitors of NAD(P)H oxidase as guides to its mechanism, *Biochem. Soc. Trans.,* 19, 70-72, 1991.

50. Bokoch G.M., Biology of the Rap proteins, members of the ras superfamily of GTP-binding proteins, *Biochem. J.,* 289, 17-24, 1993.

51. Kummer, W., Habeck, J.O., Koesling, D., Quinn, M., Acker, H., Immunohistochemical analysis of components of the oxygen sensing cascade in the human carotid body, *Pflügers Arch.*, 422, R129, 1993.

52. López-López, J.R. and González, C., Time course of K$^+$ current inhibition by low oxygen in chemoreceptor cells of adult rabbit carotid body. Effects of carbon monoxide, *FEBS Lett.*, 299, 3, 251-254, 1992.

53. Cherry, P.D., Omar, H.A., Farrell, K.A., Stuart, J.S., and Wolin, M.S., Superoxide anion inhibits cGMP associated bovine pulmonary relaxation, *Am. J. Physiol.*, 259(28), H1056-H1062, 1990.

54. Wang, W.J., Stensaas, L.J., de Vente, J., Dinger, B., and Fidone, S.J., Immunocytochemical localization of cAMP and cGMP in cells of the rat carotid body following natural and pharmacological stimulation, *Histochemistry*, 96, 523-530, 1991.

55. Sies, H., Gerstenecker, Ch., Menzel, H., and Flohe, L., Oxidation in the NADP system and release of GSSG from hemoglobin free perfused rat liver during peroxidatic oxidation of glutathion by hydroperoxides, *FEBS Lett.*, 27, 171-175, 1972.

56. Ruppersberg, J.P., Stocker, M., Pongs, O.P., Heinemann, St., Frank, R., and Koenen, M., Regulation of fast inactivation of cloned mammalian I$_k$ (A) channels by cysteine oxidation, *Nature*, 352, 711-714, 1991.

57. Wyatt, C.N. and Peers, C., Modulation of ionic currents in isolated type I cells of the neonatal rat carotid body by p-chloromercuribenzenesulfonic, *Brain Res.*, 591, 341-344, 1992.

58. Acker H., Bölling, B., Delpiano, M.A., Dufau, E., Görlach, A., and Holtermann, G., The meaning of H$_2$O$_2$ generation in carotid body cells for PO$_2$ chemoreception, *J. Auton. Nerv. Syst.*, 41, 41-52, 1992.

59. Görlach, A., Jelkmann, W., Hancock, J.T., Jones, S.A., Jones, O.T.G., Acker, H., Photometric characteristics of heme proteins in erythropoietin producing hepatoma cells (HepG2), *Biochem. J.*, 290, 771-776, 1993.

60. Lee-Huang, S., Liin, J.J., Kung, H.-F., Huang, P.L., Lee, L., and Huang, P.L., The 3′ flanking region of the human erythropoietin-encoding gene contains nitrogen-regulatory/oxygen sensing consensus sequences and tissue transcriptional regulatory elements, *Gene*, 137, 203-210, 1993.

INDEX

A

A cells, 210–211, 212, 226–227, see also B cells; Pancreatic islet cells

Acetylcholine (ACh)
  adrenal chromaffin cell
    catecholamine release, 262
    muscarinic component, 283–285
    nicotinic component, 279–283
  carotid body, 350–352
  GH cells, 112
  magnocellular neurosecretory cells, 94
  Merkel cells, 323–324

Acetylcholinesterase (AChE), 5, 324

ACTH, see Adreno-corticotrophin hormone

Actinomycin D, 337

Action potential
  adrenal chromaffin cells, 262, 263, 264–265, 284
  atrial myocardium, 191, 193
  AtT-20/D16-16 cells, 127
  calcium flux, 25
  carotid body type I cells, 355
  corticotrophs, 126
  GH cell lines, 106, 108
  gonadotrophs, 123–124
  ion channels activated during
    calcium, 265, 267–275
    potassium, 275–279
    sodium, 265, 266
  lactotrophs, 117, 118
  magnocellular neurosecretory cells, 91
  melanotrophs, 128–129
  Merkel cells, 320
  pancreas
    A cells, 210–211
    B cells, 63–65, 212, 219, 225, 228
  pituitary cells, 103
  plasma membrane modulation, 27–29
  somatotrophs, 114–115
  thyroid C cells, 151

Adenine nucleotides, 274

Adenosine, 250, 304, 306, 350

Adenosine A_1 receptors, 304, 306, see also Renin

Adenosine diphosphate (ADP), 214–215, see also B cells

Adenosine triphosphate (ATP)
  adrenal chromaffin cells, 273
  B cells, 214–216, 231–232
  enterochromaffin cells, 250
  hypoxic conditions for oxygen sensing, 355–356

Adenosine triphosphate-sensitive channels, see also Calcium; Potassium
  B cells, 208, 212, 214–216
    cyclic variations in activity and bursting, 223–224

sulfonylurea role, 234, 235
  renin-secreting cells, 307, 308, 310

Adenylyl cyclase, 156, 251, 252, 271, 330

ADP, see Adenosine diphosphate

Adrenal chromaffin cells
  electrical properties
    acetylcholine response, 279–285
    action potential, 264–265, 266
    GABAergic response, 285–286
    ion channels activated during action potential, 265, 267–279
    membrane passive, 262–263
    other ion channels, 286–287
    resting membrane potential and current-voltage relation, 263–264
  overview, 260–262
  stimulus-secretion coupling/electrical behavior
    calcium entry through nAChR channels, 291–292
    calcium entry through receptor-operated calcium channels, 292
    calcium entry through voltage-dependent calcium channels, 288–291
    overview, 287–288
  summary, 260

Adrenal gland, 74–75, see also Gap junctions

Adrenaline, 194, 221, 260, 261, 350

Adreno-corticotrophin hormone (ACTH), 75, 126, 127

Adrenomedullary chromaffin cells, 260

Adrenoreceptors, 251

Aequorin, 195

Afterhyperpolarization (AHP), 91, 224, 279, see also Hyperpolarization

Aldolase C, 43

Aldosterone, 302

Alkaline phosphatase, 54

αT3-1 cells, 125

Amine Precursors Uptake and Decarboxylation (APUD) cells, 3–4, 6–7

Amines, 321–322, see also Merkel cells

Amino acids, 17, 93–94, see also Magnocellular neurosecretory cells

L-Amino acid carboxylase, 4–5

γ-Aminobutyric acid (GABA)
  A cells, 212, 227
  adrenal chromaffin cells, 273
  -benzodiazepine receptors in enterochromaffin cells, 250
  lactotrophs, 117
  magnocellular neurosecretory cells, 94
  melanotrophs, 128, 130, 131
  neuroendocrine cells, 17

Aminophosphonovaleric acid, 93

Amperometry, 226

Androgen-binding protein, 331

Anesthetics, 55